T0138032

Studies in Computational Intelligence

Volume 514

Series editor

Janusz Kacprzyk, Polish Academy of Sciences, Warsaw, Poland
e-mail: kacprzyk@ibspan.waw.pl

For further volumes:
http://www.springer.com/series/7092

About this Series

The series "Studies in Computational Intelligence" (SCI) publishes new developments and advances in the various areas of computational intelligence—quickly and with a high quality. The intent is to cover the theory, applications, and design methods of computational intelligence, as embedded in the fields of engineering, computer science, physics and life sciences, as well as the methodologies behind them. The series contains monographs, lecture notes and edited volumes in computational intelligence spanning the areas of neural networks, connectionist systems, genetic algorithms, evolutionary computation, artificial intelligence, cellular automata, self-organizing systems, soft computing, fuzzy systems, and hybrid intelligent systems. Of particular value to both the contributors and the readership are the short publication timeframe and the world-wide distribution, which enable both wide and rapid dissemination of research output.

Colette Faucher · Lakhmi C. Jain
Editors

Innovations in Intelligent Machines-4

Recent Advances in Knowledge Engineering

Editors
Colette Faucher
Ecole Polytechnique Universitaire de Marseille
Aix-Marseille University
Marseille
France

Lakhmi C. Jain
Faculty of Education, Science, Technology and Mathematics
University of Canberra
Canberra
Australia

ISSN 1860-949X ISSN 1860-9503 (electronic)
ISBN 978-3-319-34989-3 ISBN 978-3-319-01866-9 (eBook)
DOI 10.1007/978-3-319-01866-9
Springer Cham Heidelberg New York Dordrecht London

Printed on acid-free paper

Springer is part of Springer Science+Business Media (www.springer.com)

For Nick, my son, my inspiration
and the source of my fighting spirit

C. F.

Foreword

Quo vadis, Knowledge Engineering? Where are you going?

If you want interesting answers to this question, who better to ask than my respected friends Colette Faucher and Lakhmi C. Jain? Colette is a Professor of Computer Science and Artificial Intelligence at Polytech'Marseille, Marseille, France and Lakhmi is a Professor and the Founder of globally renewed Knowledge Engineering International, KES, www.kesinternatinal.org. Combined, their expertise in the field is truly second to none. We all should be grateful for their continuing efforts to share this outstanding expertise with us through, among others, the series of edited books published by Springer. This is their fourth book that they have edited together and is titled *Recent Advances in Knowledge Engineering*. Each part of this book (*Big Data and Ontologies*; *Knowledge-Based Systems*; *Applications*), and actually each chapter of all 13 that were collected, abounds with insights, and incremental journey toward the understanding of the cutting edge issues destined for Knowledge Engineering (KE). This book is also a unique acknowledgment of the scientific endeavor of researchers who were involved in writing separate book chapters. Exceptionally smart combination of these chapters (I would not expect anything less from Colette and Lakhmi) and their grouping into three logically intervened parts, provides an intellectual feast for the reader guiding him or her through probably most exciting areas in KE: data handling and its modeling as the "origin of things" (Part I), enhancement of systems through knowledge-based features "down the road" of smart systems development (Part II), and ultimate real-life applications which is the "destiny" of what we try to do in knowledge engineering (Part III).

The additional overall benefit of reading this book is that it provides guidelines to develop tools and approaches to smart engineering of knowledge, data, and information. Understandably, the guide does not presume to give ultimate answers. Rather, it poses ideas and case studies to explore, and the complexities and challenges of modern knowledge engineering issues. It also encourages its reader to become aware of the multifaceted interdisciplinary character of such issues. One of the unique, and for me personally, probably the most exciting aims of this book is that its reader will leave it with a heightened ability to think—in different ways—about developing, evaluating, and supporting smart knowledge engineering systems in real-life based environment. The Authors are to be congratulated on such a comprehensive and truly successful effort to achieve this aim.

Newcastle, Australia, 9 August 2013 E. Szczerbicki

Preface

Knowledge engineering has been recognized as one of the most important areas of research universally. This research volume is a continuation of our previous volumes on intelligent machines. We laid the foundation for intelligent machines in SCI Series Volume 70 by including the possible and successful applications of computational intelligence paradigms in machines for mimicking the human behavior.

In SCI Series Volume 376, a number of innovative applications of intelligent paradigms such as document processing, language translation, English academic writing, crawling system for web pages, web-page retrieval technique, aggregate k-Nearest Neighbor for answering queries, context-aware guide, recommendation system for museum, meta-learning environment, case-based reasoning approach for adaptive modeling in exploratory learning, discussion support system for understanding research papers, system for recommending e-Learning courses, community site for supporting multiple motor-skill development, community size estimation of internet forum, lightweight reprogramming for wireless sensor networks, adaptive traffic signal controller, and virtual disaster simulation system.

The SCI Series Volume 442 presents further advances in the theoretical and application-oriented intelligent systems research mainly in the field of neural computing.

The present volume is divided into three parts. Part I deals with *Big Data and Ontologies*. It includes examples related to the text mining, rule mining, and ontology. Part II is on *knowledge-Based Systems*. It includes context-centered systems, knowledge discovery, interoperability, consistency, and systems of systems. The final part is on *Applications*. The applications involve prediction, decision optimization, and assessment.

We believe that this research volume will prove useful to Engineers, Scientists, and Researchers as a reference to explore the field of knowledge engineering further.

It has been our pleasure to work with the Contributors and Reviewers. This book would not have existed without their very active contributions.

Our special thanks are due to Prof. Dr. Janusz Kacprzyk, the Series Editor, for his vision and direction during the development phase of this book.

We are grateful to Dr. Thomas Ditzinger, Holger Schaepe, and their team for their excellent cooperation right from the beginning to the completion of this book project.

Marseille, France Colette Faucher
Canberra, Australia Lakhmi C. Jain

Books on Innovations in Intelligent Machines

- Chahl, J.S., Jain, L.C., Mizutani, A. and Sato-Ilic, M., Innovations in Intelligent Machines 1, Springer-Verlag, Germany, 2007.
- Watanabe, T. and Jain, L.C., Innovations in Intelligent Machines 2: Intelligent Paradigms and Applications, Springer-Verlag, Germany, 2012.
- Jordanov, I. and Jain, L.C., Innovations in Intelligent Machines 3: Contemporary Achievements in Intelligent Systems, Springer-Verlag, Germany, 2012.
- Faucher, C. and Jain, L.C., Innovations in Intelligent Machines 4: Recent Advances in Knowledge Engineering, Springer-Verlag, Germany, 2013.

Contents

Part I
Big Data and Ontologies

Chapter 1
Large Scale Text Mining Approaches for Information Retrieval and Extraction

Patrice Bellot, Ludovic Bonnefoy, Vincent Bouvier, Frédéric Duvert and Young-Min Kim

Abstract The issues for Natural Language Processing and Information Retrieval have been studied for long time but the recent availability of very large resources (Web pages, digital documents…) and the development of statistical machine learning methods exploiting annotated texts (manual encoding by crowdsourcing is a new major way) have transformed these fields. This allows not limiting these approaches to highly specialized domains and reducing the cost of their implementation. For this chapter, our aim is to present some popular text-mining statistical approaches for information retrieval and information extraction and to discuss the practical limits of actual systems that introduce challenges for future.

P. Bellot (✉) · V. Bouvier · Y.-M. Kim
CNRS, Aix-Marseille Université, LSIS UMR 7296, Av. Esc. Normandie-Niemen, 13397, Marseille cedex 20, France
e-mail: patrice.bellot@lsis.org

V. Bouvier
e-mail: vincent.bouvier@lsis.org

Y.-M. Kim
e-mail: young-min.kim@eric.univ-lyon2.fr

L. Bonnefoy · V. Bouvier · F. Duvert
iSmart, 565 rue M. Berthelot, 13851, Aix-en-Provence cedex 3, France
e-mail: ludovic.bonnefoy@ismart.fr

F. Duvert
e-mail: frederic.duvert@ismart.fr

L. Bonnefoy
LIA, Université d'Avignon et des Pays de Vaucluse, Agroparc, 84911, Avignon cedex 9, France

C. Faucher and L. C. Jain (eds.), *Innovations in Intelligent Machines-4*,
Studies in Computational Intelligence 514, DOI: 10.1007/978-3-319-01866-9_1,

1.1 Introduction

Real text mining systems have been developed for finding precise and specific information on large collections of texts from keyword-based or natural language questions. The development of Web searching and question-answering systems that aim to retrieve precise answers corresponds to the combination of methods from computational linguistics, graph analysis, machine learning and, of course, information extraction and information retrieval. The most recent systems tend to combine statistical machine learning methods along with symbolic and rule-based approaches. The challenge is to get robust efficient systems that can self-adapt to users' needs and profiles, to dynamic collections composed of unstructured or weakly structured and interconnected documents.

Numerical methods based on a statistical study of corpus, have proven their ability to adapt quickly to different themes and languages and are very popular on Web search systems. This was done despite some approximation in the results that would be more accurate by using symbolic rule-based approaches as long as the rules would be fair, complete and unambiguous... while natural language is absolutely ambiguous. On the other hand, Natural Language Processing (NLP) tasks such named entity recognition (automatic extraction of the names of "things" in texts: person or company names, locations...) and part-of-speech tagging (automatic labeling of words according to their lexical classes: noun, verbs, adjectives...) can be realized both by means of symbolic approach that rely on hand-coded rules, grammar and gazetteers and by following supervised machine learning-based approaches that need large quantities of manually annotated corpus. For Named Entity Recognition, symbolic systems tend to produce better results on well-formed texts [55] whereas statistical approaches show best results on low-quality texts such as automatic transcripts of speech recording [5].

This chapter is composed of two main sections dedicated to text mining, information retrieval and information extraction. We begin by a general discussion about symbolic and numerical approaches for Natural Language Processing then we present the most popular tasks that have been evaluated through annual conferences for twenty years along with some resources one can employ to process texts. In the first section, we present some classical information retrieval models for document and Web searching and the principles of semantic information retrieval that can exploit specialized lexicon, thesaurus or ontologies. The second section introduces natural language processing for information extraction. We outline the most effective approaches for question-answering systems and semantic tagging of texts (named entities recognition, pattern extraction...): rule and lexicon based approaches and machine learning approaches (Hidden Markov Models and Conditional Random Fields). Then, we present some approaches that aim to find information about entities and to populate knowledge bases. In this section, we describe the approaches we proposed and experimented in the last few years. The last section is dedicated to some industrial applications we work on and that respectively relate to digital libraries, marketing and dialog systems.

1.2 Symbolic and Numerical Approaches for Natural Language Processing

Symbolic and numerical approaches for natural language processing (NLP), have long been opposed, both linked to scientific communities distinguished by differing goals between limited but accurate prototypes and rough but functional systems and by attitudes more or less pragmatic. This opposition joined in some way the one that opposed and opposes always, some linguists and philosophers on the nature of language and its acquisition that is, in simplified terms, the pre-existence of a system (cognitive one) generating rules of possible sentences in a language (or at least the degree of pre-existence). This debate on the nature and role of grammar has its origins in the 17th century between proponents of empiricism (the human being is born "empty" and is fully shaped by experience) and rationalism (the man can not be reduced to experience). In the 1950s, the behaviorists, empiricist, attempted to define the acquisition of language learning as a form of chain reaction from positive reinforcement or negative one. In contrast, Chomsky proposed the pre-existence of mechanisms in the human brain that are specific to language and that could distinguish humans from other species. That suggests language is *something* really organic [48] even in the very beginning of the life and that language learning does not only rely on association between stimuli and responses.

A direct consequence of the pre-existence of a 'minimalist program' (generative grammar) in language acquisition has been to define both a universal grammar expressing linguistic universals and particular grammars for the specificities of given languages [22]. In this view, called cognitive linguistics, language acquisition was seen as a process of rule induction: the child, provided the general structure of universal grammar that defines a certain class of grammar, needs to discover the special rules that generate the particular language to which (s)he is exposed [88]. Subsequently, specific grammars have been reduced to specific values of parameters of universal grammar, acquisition of appropriate language and in setting these parameters [23]. There may be no need to induce any rules [88], induction can be replaced by a process of identifying and selecting among all a priori possible linguistic productions [66].

The learning process can be viewed from two points of view, namely statistical learning or "analytical" (both unconscious and possibly combined in human mind). In the first case, the child has to observe which language productions lead to the goal (s)he fixed and, on a rolling basis, to accumulate a kind of accounting of what succeeds and what fails and allows him/her to achieve a selection of possibilities. Computational neural networks, involving different layers more or less explicit, linking lexicon, concepts and sounds, might simulate this kind of learning. In that sense, any communicational intention (any goal) and any linguistic production might be seen as specific paths within the neural network. The success or failure would result in strengthening or weakening connections and the issue of convergence during learning step arises as well as the reduction of combinatorial.

In the second case, learning is a progressive refinement of the value of the parameters of universal grammar allowing the production and the understanding, of the statements it can generate. Note that in the first case (statistical learning), a rule-based grammar can still be induced once the system is stabilized. It would allow producing explicit structural patterns, which would be necessary to improve consistency over time and mutual understanding between two dialoging people.

Finally, proposals and models arising from the work of N. Chomsky do not seem incompatible with a contextual behaviorist point of view [51]. One can postulate the pre-existence of a specific neural network (or of any other biological element) with an initial structure facilitating language acquisition, both in compliance with an universal grammar and with the need to interconnect brain language areas with all the cognitive areas: the network can still be seen as a whole and one can begin to model "human neuronal dynamics" in which reinforcement learning (selection process) is constrained by the genetic code while being sensitive to the experience.

In the following, we do not pretend to decide between one approach or another for modeling the way a human brain works but we wish to present approaches that allow a computer *to understand* a text at best for tasks related to information retrieval and extraction in very large collections. We believe that one of the main achievements of natural language processing in recent years is that it is necessary, at least in the present state of knowledge and computational capacity of the computers, not to be confined in a fully statistical approach nor in a purely symbolic one. Combining the best of both worlds has to be considered for the tasks we want to resolve by means of computers. In a few words, statistics and probability theory allow to process by analogy (pattern analysis and matching) on a large scale and to manipulate uncertainty[1] whereas symbolic approaches can facilitate inferences in domain specific texts (textual entailment), preprocessing texts (e.g. for lexical normalization and disambiguation) and filtering obvious abusive analogies but also offer a better human readability of processing rules.

1.3 Information Retrieval Models

An information retrieval model is intended to provide a formalization of the process of information search. He must perform many roles: the most important is to provide a theoretical framework for modeling the relevance measure involving a query from a user on one hand and a document or a passage of a document on the other hand. A classical information retrieval system provides indeed the user with a list of documents or a list of Web pages in response to a query (usually composed

[1] Alternatives to the use of probability and to Bayesian networks or other probabilistic graphic models for dealing with uncertainty have been proposed. Among them fuzzy logic and Dempster-Shafer theory.

of few keywords) and not with precise and unique *information*. Information retrieval is then a generic term that designates document retrieval, passage retrieval and Web page retrieval among others. In that sense, a document can broadly correspond to any text (article, press release, blog entry, book…) that can be more or less structured and more or less respectful of language and of good writing rules (spelling, syntax, discourse structure). In the case of multimodal information retrieval, the retrieval can be obtained by handling speech or handwriting recognition systems.

Depending on its complexity level, the information retrieval model can take into account different types of descriptors and textual clues, lexical markers, syntactic, semantic and structural (or even prosodic in the case of searching for documents audio). Each clue can be obtained through a surface analysis or depth linguistic analysis resolving dependencies and references among others. Most information retrieval models characterize a document according to the collection from which it comes, taking into account global indices such as average length of documents in the collection and the average frequency distribution of words. The combination of theses indices involves a large number of parameters to be set and adapted to the corpus or learned according to user feedback.

Because of the network structure of Web pages (hyperlinks), Web search engines also take into account non-linguistic cues for estimating relevance scores: number of pages that link to a given page, the probability to access this page by "random walk" etc. These hints are the basis of the famous PageRank algorithm popularized by the Google Web search engine.

For information need and a set of documents given, the whole question is to determine which model, what parameters but also which external resources (dictionaries of inflected forms, thesaurus, knowledge bases, ontologies, online encyclopedias…) will be most effective in helping the search.

1.3.1 Original Boolean Model

The Boolean model is the simplest model for information retrieval. It was the most popular model before the advent of vector and probabilistic models in the 1960s and 1970s. It is based on the theory of sets and Boolean algebra and considers a query as a logical combination of terms or keywords. The Boolean model assumes that the query terms are present or absent in a document and that their importance is equal. The formulation of the query has a very high impact and its words must be chosen very carefully, ideally from a controlled vocabulary. Either the query is a conjunction of terms allowing to retrieve documents with high precision (the number of non relevant documents retrieved is low) but with low recall (a large number of relevant documents using different words to those of the query are not retrieved), either the query is a disjunction and the precision tends to be low and the recall high.[2] In its classic form, the Boolean model does not order the documents retrieved because two documents having an equal number of common words with the query have the same score and cannot be differentiated. Only a non-binary weighting of the query words would differentiate and then order these documents (see extensions of the Boolean model in [91, 14] for a fuzzy approach generalizing the Boolean model). The number of retrieved documents from any query on the Web is so important that it is mandatory to propose methods to order documents. In the other hand, indexing large collections and the emergence of the Web has made more complex the use of controlled vocabularies despite folksonomies and cooperative normalization of hashtags in social networks [21, 109].

1.3.2 Vector-Space Model

The vector space model represents documents and queries by vectors of terms or keywords [92]. These keywords are the index terms that are extracted from the texts during indexing and that may be individual words, composed words or phrases. In a controlled vocabulary context, they represent a thematic or a domain. In full text and non-controlled approaches, every word is an indexing unit (preprocessing such lemmatization. For each document, a weight is assigned to each index entry it contains. Vectors represent every query and every document in a unique vector space and they can be compared with each other: comparing a query and documents allows to rank documents and comparing a document to other ones allows to suggest similar documents for reading and to cluster them. This comparison can be conducted by computing and by ordering the values of a similarity (or distance) such as cosine function or Euclidian distance. In the basic vector-space model, words are supposed to occur independently of each other while this is clearly not the case. Newer models propose to take into account this dependence, as we shall see later in this chapter.

The aim of weighting scheme is to identify the words that are the most discriminant in a document, in respect to a user's query and according to a collection or to a sub-collection. In other words, a word is important (its weight must be high) if its presence in a document is a strong indicator of the relevance of this document. Sparck-Jones [101] proposed to express term weighting by combining term frequency *tf* in a document and inverse document frequency *idf* in a collection. The idea is that more a word occurs in a document more important it is, and that this has to be weighted inversely proportional to the number of documents that contain the word in the collection (a very frequent word in the collection is less informative and less discriminant than a rare one: they tend to refer to low specific

[2] Precision is the fraction of retrieved items that are relevant or well classified while recall is the fraction of relevant items that are retrieved and provided as result. F-score is the harmonic mean of precision and recall.

concepts). Even if original *tf.idf* weighting is still used and may be a good starting point to build a search engine, more efficient weighting schemes were proposed. For example, other weightings help to differentiate words according to their average use in a collection.

In the vector-space model, the cosine similarity is defined as:ir average use in a collection.

In the vector-space model, the cosine similarity is defined as:

$$\mathrm{sim}(q,d) = \frac{\sum_{i=1}^{n} q_i \times d_i}{\sqrt{\sum_{i=1}^{n} q_i^2} \cdot \sqrt{\sum_{i=1}^{n} d_i^2}} \tag{1.1}$$

with q_i the weight of term/word i in a query q containing n different terms and d_i the weight of i in a document d.

According to the basic term-weighting scheme described above, three factors are involved: the frequency of a term in a document and in the query, the frequency of a term in the whole collection and the weight of a document (normalization by taking into account the length of the document or the sum of the weights of the words it contains). A popular weighting scheme is *Lnu.ltc* [97] that incorporates two parameters that have to be optimized for the considered collection: *pivot* and *slope*. They prevent cosine function from preferring shorter documents [99] and allow to take into account the fact that relevance and document length are not independent. In that way, the relevance score of a document d for a query q is:

$$\mathrm{sim}(q,d) = \sum_{i \in q \cap d} \frac{\frac{1+\log(\mathrm{freq}_{i,d})}{1+\log(\mathrm{avg}_{j\in d}\,\mathrm{freg}_{i,d})} \cdot \left(\frac{\mathrm{freg}_{i,q}}{\max_{j\in q} \mathrm{freq}_{i,q}} \cdot \log\left(\frac{N}{n_i}\right)\right)}{((1-slope).pivot + slope \times uniqueWords_d).\sqrt{\sum_{i\in q}\left(\frac{freq_{i,q}}{\max_{j\in q}\mathrm{freq}_{i,q}} \cdot \log\left(\frac{N}{n_i}\right)\right)^2}} \tag{1.2}$$

with $\mathrm{freq}_{i,d}$ the number of occurrences of i in a document d (resp. in the query q).

Some years later, Deerwester et al. [31] proposed Latent Semantic Indexing (LSI) in order to reduce the dimensionality of the vector space and the mismatch between query and document terms by means of singular-value decomposition (that is computationally costly). By employing an LSI indexing approach, documents can be retrieved and be conceptually similar even if they do not share words/terms with the query.

Much other works have been conducted to integrate the result of linguistic analysis (anaphora resolution [69, 44], disambiguation [103], lemmatization, semantic role labeling…). Unfortunately, their impact seems to be limited and only stemming[3] is an effective procedure [54] often used but only for certain languages.

---

[3] Stemming consists in reducing words according to their morphological variants and roots. See for example Snowball that makes light stemming available for many languages (http://snowball.tartarus.org). Lemmatization can be seen as an advanced stemming.

1.3.3 Probabilistic Models

Some information retrieval models exploit user feedback on the retrieved documents for improving searching by using probability theory [90]. It allows to represent the research process as a decision-making process: for the user, the cost associated with the retrieval (downloading and reading time) of a document must be as low as possible. In other words, the decision rule is equivalent to propose a document only if the ratio of the probability that it is relevant (interesting for the user) and that it is not, is greater than a given threshold. One then seeks to model the set of relevant documents, that is to say, to estimate the probability that a document is relevant to a user i.e. that the given query words appear or do not appear in these documents. For a document d and a query q and with ‘relevant’ the set of relevant documents, this probability is estimated as:

$$P(\text{relevant}|d) = \frac{P(\text{relevant}) \times P(d|\text{relevant})}{P(d)} \tag{1.3}$$

By incorporating the probability of non-relevance and from the Bayes rule:

$$S(d,q) = \frac{P(d|\text{relevant}) \times P(\text{relevant})}{P(d|\text{relevant}) \times P(\text{relevant})} \approx \frac{P(d|\text{relevant})}{P(d|\text{relevant})} \tag{1.4}$$

By assuming that words are independent from each other (bag of words models), one can set, for i a term in q:

$$P(d|\text{relevant}) = \prod_{i=1}^{n} P(d_i|\text{relevant}) \tag{1.5}$$

With $p_i = P(i \in d|\text{relevant})$ and $q_i = P(i \in d|\overline{\text{relevant}})$, one obtains:

$$s(d,q) \approx \sum_{i \in d \cap q} \log \frac{p_i(1-q_i)}{q_i(1-p_i)} \tag{1.6}$$

After integrating non-binary weights of terms in the query and in the documents, the relevance score is:

$$s(d,q) \approx \sum_{i \in d \cap q} w_{i,d} \cdot w_{i,q} \cdot \log \frac{p_i(1-q_i)}{q_i(1-p_i)} \tag{1.7}$$

The solution to this problem needs an unsupervised estimation or, on the contrary, can be realized by following an iterative process in which the user (or a pseudo-user by assuming that the first retrieved documents are relevant) selects relevant documents in a limited ranked list (e.g. the first ten documents found by a vector-space model). The ranking function BM25 [89] is defined in that sense and it is yet very popular:

$$s(d,q) \approx \sum_{i \in d \cap q} w_i \times \frac{(k_1+1)\cdot \mathrm{freq}_{i,d}}{K+\mathrm{freq}_{i,d}} \times \frac{(k_3+1)\cdot \mathrm{freq}_{i,q}}{k_3+\mathrm{freq}_{i,q}} \tag{1.8}$$

$$w_i = \log\left(\frac{N-n_i+0.5}{n_i+0.5}\right), \quad K = k_i\left((1-b)+b\cdot\frac{l_d}{\overline{l}}\right)$$

k_1, k_3 and b are parameters, l_d the length of d, $\overline{l}$ the average document length, n_i the number of documents containing i and N the total number of documents.

A lot of models, more or less personalized, can be estimated automatically based on user behaviors on the Web. These models take into account the number of Web pages to which the user accesses from the list of retrieved results, the time between the recovery of each page, the number and the nature of queries entered after the initial query and the explicit expression of the relevancy by means of an adapted user-interface (e.g. star ratings). This leads researchers to propose to learn ranking functions [42] based on large sets of training data and features (query logs).

One of the models that emerged in the late 1990s is the language model introduced by Ponte and Croft [77]. This probabilistic model is based on common linguistic models that attempt to capture the regularities of a language (probable sequence of words or bi-grams or tri-grams of words) by observing a training corpus. A language model then estimates the likelihood (probability) to have a given sequence of words in a given language (or for a given topic in a given language). Its use in information retrieval consists in considering that every document is represented by its language model and is the basis for generating it. The relevance of a document in respect to a query is seen as the probability that the query is generated from the same model language model than the one of the document. In such a model, dependency between terms can be taken into account [by using word n-gram models—a word n-gram is an ordered set of n contiguous words in a sentence[4]—and proximity operators for example [67] and often outperforms models that assume independency.

The models that have been presented are not able to take into account the semantic relationships between lexical items without the use of pre-processing and post-processing (synonyms and semantic proximity, hierarchical links), no more than the expression of negation in queries (expression in natural language of what is *not* searched) and in the documents. To partially answer these problems, the Web search engines offer users to use Boolean operators. Another approach consists in expanding queries by employing some external semantic resources [32] and methods of disambiguation (based on statistical co-occurrences of words).

In the following, information retrieval models are an essential piece of information extraction systems. They allow to filter the collections of documents

[4] Google Books Ngram (http://books.google.com/ngrams) and n-grams from the Corpus of Contemporary American English COCA (http://www.ngrams.info/) are two popular and freely downloadable word n-grams sets for English.

according to an information need and to restrict deeper and costly linguistic analysis to a sub-collection.

1.4 Information Extraction

Information extraction is about automatically extracting structured information from unstructured sources. It has its origin in the natural language processing community [65, 96]. Historically, information extraction was focused on the identification of named entities (limited to persons, companies or locations) from texts and then finding relationships between these entities. Two evaluation campaigns strongly encouraged and deeply influenced research on these particular topics: the Message Understanding Conference (MUC) in 1995 [47] and the Automatic Content Extraction (ACE) in 1999 [33]. Both asked participants to retrieve persons, locations and companies in texts but also temporal information and quantities and some relationships among them. Information extraction was then performed in collection of news articles exclusively. Open information extraction [37, 39] is about modeling how relationships are expressed in general between an object and its properties and so being able to extract automatically entities of non-predefined classes and new kind of relations.

Locating named entities in texts, a task called named entity recognition (NER) [72], seems to be an easy task. But, this is not always as easy as we can first think. Let consider the following sentence: "*Jackson did a great show yesterday in Neverland*". One can guess that "*Jackson*" refers to a person because a show is likely to be performed by a person (let us remark that Jackson could designates an animal, a software or the name of a band and not of a single human person). Anyway, without more contextual information—what is *Neverland*? A location, but of what kind: a city or a concert hall?—, we cannot guess what kind of person is "*Jackson*": is he a politician, an actor or a singer? In this example, one might assume that Jackson refers to "*Mickael Jackson*" knowing that he owned a theme park called "*Neverland*". This example shows that without contextual knowledge and by limiting analysis to a single sentence, associating a semantic type to an entity may be very difficult. Such knowledge is of course difficult to gather and formalize in order to be used by a named entity recognizer. Another clue for semantic labeling is linked to character properties (such as capitalized letters). Let consider: "*yesterday, I saw a Jaguar*". Are we talking about a car maker ("*a Jaguar*") or an animal (either way, almost every Jaguar car owns a jaguar—not a real jaguar even if it is a real thing—on the hood and its logo)? Delimiting boundaries is another challenge for named entity recognition. Is "*Jackson, Alabama*" correspond to one entity or to two separate entities: one people and one city, two people, two cities? Actually, there is no a single definitive answer. It completely depends on the context.

Relationship extraction is the second historical task in Information Extraction [16, 27, 40, 68]. First implementations focused on a set of predefined relation between entities of predefined types. One example is "x *is the CEO of* y" where *x* is a person and *y* an organization. As for named entity recognition, choices on expected form of relations have to be made and are dependent of the objectives.

With the democratization of Internet, a wide variety of new information extraction problems raised. Indeed, the massive use of Internet allows companies to gather or to observe big data generated by Internet users. Brands are for instance interested to collect opinion expressed on the Internet by their customers about products. Collecting and even indexing this information may be easy but using them for analytic purpose and decision-making may be difficult. Let consider the following problem: a company wants to monitor what their customers think about their new smartphone. They obviously cannot read each review or comment about on the Web and one needs to exploit information retrieval (filtering the Web) and information extraction (opinion mining). In order to accomplish this task, a system has to automatically detect what are the main features of the phone (e.g. size of the screen, autonomy, OS...), to locate segment of texts dealing with each feature then to extract from these segments the expressed opinion. Each of these three subtasks is an information extraction problem, more or less related to a list of couples of precise information requests (e.g. "*what is the size of the screen... what the Web users think about it*").

Over the years, the scope of information extraction became wider. More types of entities had to be recognized, more complex kind of relations has to be found (e.g. not only unary relations). Nowadays, most of the available information extraction systems are domain-specific because building and maintaining ordered set of types and relations for larger applications is a hard problem and because the availability of semantic resources.

Historical applications of information extraction have greatly benefit from existing works in Natural Language Processing. As one knows, information extraction systems can be separated in two coarse classes: rules based approaches and statistical approaches. In this section we will see pros and cons and talk about results obtained with manual, supervised and weakly supervised approaches.

1.4.1 Symbolic Rule-Based Approaches

Early works on information extraction relied exclusively on manually created set of rules as recalled in [26, 71]. A rule generally takes the form of a condition (or pattern) that has to be found in the text and something to do if so. For instance let consider the following rule which would process any sequence of capitalized words, preceded by a title, as the name of a person: "*Mr.* [Word starting with a capital letter]+ –> label the word sequence as a person name". To find what kind of relation have a person (PERS) with a given organization (ORG), one could use the next rule: "PERS is REL of ORG -> rel(pers, org)". The pattern used in this rule may be instantiated to find out the name of the CEO of an organization: "PERS *is CEO of* ORG -> ceo_of(pers, org)".

Now, suppose that we want to label as a person name all sequences of words beginning with a capitalized letter and following title information. One may build a pattern that contains all the titles (or build one rule by title). A generic rule could be represented by the following pseudo-regular expression: "[Title][.]? [Word starting with a capital letter]+ –> label the words as a person name". This single rule will be able to identify "*Terry Pratchett*" as a person whatever his title and in both phrases "*Mr. Terry Pratchett*" and "*Sir Terry Pratchett*". To improve robustness and generality, a lot of rule-based systems rely on the use of dictionaries, called gazetteers, which are lists of person names, organizations etc. Such resources may be hard to build and to maintain. In the same time, rules become more and more complex and maintaining them may be very costly. However, manually created symbolic rules have the advantage to be easy to interpret, much more than any numeric model.

One alternative to manually created rules consists in employing rule induction algorithms [24, 59]. Like every supervised machine learning algorithm, they require training data. For instance, for named entity recognition, these data will be sentences with named entities labeled with their type and their boundaries. These algorithms will select seed instances from the training data to create new rules. An instance is selected as a seed depending on whether it is already covered by an existing rule. There are two classes of rule induction approaches: bottom-up algorithms and top-down algorithms. In bottom-up algorithms, one (or sometimes more) uncovered instance is selected as a seed. First, a new rule is created to exactly match this instance. Then, this rule will generalize more and more in order to increase its recall. However, precision will be lowered and a tradeoff has to be made between precision and recall. Example 4.1.1 shows one example of the creation of one rule from a seed instance. At the contrary, top-down approaches start with a rule with a high coverage but a low precision and then specialized it to improve precision. Specialization is realized with the objective that the rules cover the seed instance. Process is stopped when a predefined threshold is reached.

Example Suppose we have the following sentence: "Dr. Isaac Asimov was a writer." The rule induction (generalization) might follow:

- "[Dr.] [Isaac] [Asimov] [was] [a] [writer] -> label "Isaac Asimov" as a person"
- "[Dr.] [Capitalized word] [Capitalized word] [was] [a] [writer] -> label the two capitalized words as a person"
- "[Dr.] [Capitalized word] [Capitalized word] [be] [a] [writer] -> label capitalized words as a person"
- "[Title] [Capitalized word] [Capitalized word] [be] [a] [writer] -> label capitalized words as a person"
- "[Title] [Capitalized word] [Capitalized word] [be] [a] [Profession] -> label capitalized words as a person".

Going from 2. to 3. requires to lemmatize words, from 3. to 4. requires to have a predefined list of titles and going from 4. to 5. requires a list of professions.

Going from 2. to 3. requires to lemmatize words, from 3. to 4. requires to have a predefined list of titles and going from 4. to 5. requires a list of professions.

Rule-based systems are often composed of a large set of rules that can be in conflict. One of the central components of a rule-induction deals with resolution of these conflicts. Many strategies may be used but most of them are either based on the a posteriori performance of each rule (recall, precision) or either by following priorities defined by a set of policies.

1.4.2 Statistical Approaches: HMM and CRF

The other point of view for information extraction is to treat the extraction problem as a statistical sequence-labeling task that basically involves a numerical learning procedure.

Let us take again the example "*Dr. Isaac Asimov was a writer*". With the statistical viewpoint, this sentence is a sequence instance, which consists of six ordered words $w_i : w_1 = Dr., w_2 = Isaac, w_3 = Asimov, w_4 = was, w_5 = a$, and $w_6 = writer$. Named entity recognition becomes a special case of sequence labeling, where the aim is to predict the most probable label for each word from a set of predefined class labels. For example, if we want to distinguish three types of named entities—person (PERS), organization (ORG), and location (LOC)—, one can develop a system that takes into account those three classes more an additional label for non-named entity words (NA). The previous example corresponds to the following sequence of labels: NA, PERS, PERS, NA, NA, and NA.

Hidden Markov Model (HMM) [81, 82] is one of the traditional techniques for sequence labeling [57] and Conditional Random Field (CRF) [58] overcomes a restriction of HMM by allowing more informative input features in modeling [98, 75]. Apart from these two direct sequence labeling methods, many other classification and clustering algorithms have been applied to the labeling task [19, 25, 52] but direct labeling approaches are preferred because of simplicity and the fact that they usually provide good result.

A statistical model is described by a set of mathematical equations including an objective probability function to be maximized in a learning algorithm. The objective function consists of random variables representing input sequence words and their labels, and model parameters indicating the distributional information about the model. In the learning algorithm, the model parameters and the joint or conditional distributions of variables are repeatedly updated. We briefly introduce two major sequence labeling models, HMM and CRF.

The objective function of HMM is the joint distribution of inputs and labels. It means that we want to find and optimize the set of parameters and distributions that maximizes the distribution of co-occurring words and labels fitting training data. An input sequence corresponding to a sentence is called a sequence of observations $w = \{w_1, w_2, \ldots, w_T\}$ and a label sequence $y = \{y_1, y_2, \ldots, y_T\}$ is called a sequence of states. In HMM, we simplify the relation among random variables such that each state y_t depends only on its previous state y_{t-1} and each observation x_t depends only on the corresponding state y_t. We can also say that observations in a HMM are generated by hidden states. The word 'hidden' does not mean that the label values are not given in the training set. Instead, it signifies the nature of state sequence that we model: we assume that the states are invisible because they are not observed directly.

All the relations are represented by means of a directed graphical model (Fig. 1.1a). The joint distribution of an observation sequence and a state sequence can be written as:

$$P(y, w) = \prod_{t=1}^{T} \mathrm{P}(y_t|y_{t-1}) \cdot \mathrm{P}(\mathrm{w}_t|\mathrm{y}_t) \tag{1.9}$$

By applying an appropriate algorithm, we can compute the most probable label assignment for a given observation sequence: $\boldsymbol{y}^* = \mathrm{argmax}_y\ \mathrm{P}(y|x)$.

Let us return to the example. To predict the labels of "*Dr. Isaac Asimov was a writer*", we first compute the probability $\mathrm{P}(y_1|$ '*Dr.*' for all candidate labels using the learned parameters and distributions. Then one chooses the most probable label for the observed word '*Dr.*'. With a recursive computing, we find the most probable labels, y_1^*, $y_2^* \ldots$, y_6^*.

Conditional Random Field (CRF) can be thought as a discriminative version of HMM. A HMM is a generative model because sequence observations are generated from probability distributions. Since a generative model maximizes the joint distribution, we need to make statistical modeling for all random variables. CRFs overcome this restriction by directly modeling the conditional distribution $\mathrm{P}(y|w)$ instead of modeling the joint distribution $\mathrm{P}(y, w)$ (see Fig. 1.1b). Therefore in a CRF, we do not need to make a supposition over the distribution of observations, we just need to model $\mathrm{P}(y|w)$. This is a great advantage in modeling because in many information extraction tasks input data has rich information that may be useful for labeling. Suppose that we apply a HMM for a NER system where we can only use the sentence words as input. If we want to encode the specific characteristics of each word such as capitalized character, we now need to consider the relations of these characteristics and words w for modeling $\mathrm{P}(w)$. But in a CRF, modeling $\mathrm{P}(w)$ is unnecessary, thereby we can add any features describing the input words without complicating the model. This kind of models is called discriminative model against generative model. The conditional distribution of a CRF is written as the following equation where f_k is a feature function related to a specific composition of w_t, y_{t-1}, and y_t values.

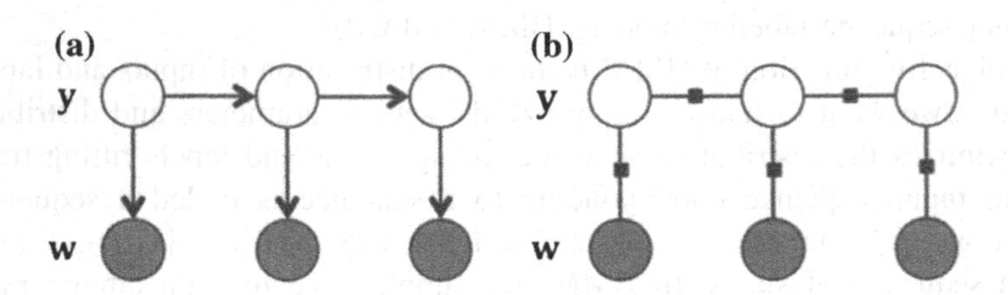

Fig. 1.1 Hidden-Markov model (HMM) **(a)** and conditional random fields (CRF) **(b)**

$$P(y|w) = \frac{1}{Z(w)} \prod_{t=1}^{T} \exp\left\{ \sum_{k=1}^{K} \theta_k f_k(y_t, y_{t-1}, w_t) \right\} \tag{1.10}$$

$$Z(w) = \sum_{y} \prod_{t=1}^{T} \exp\left\{ \sum_{k=1}^{K} \theta_k f_k(y_t, y_{t-1}, w_t) \right\} \tag{1.11}$$

With this model, we can enrich the input features of the word '*Isaac*' in the previous example. For instance, we can add a feature called 'firstcap', which indicates whether the first letter in the word is capitalized. Now instead of considering only word itself, we also take account this feature to calculate the most probable label of '*Isaac*'. An important thing is that candidate features are not limited to the current word, but previous or next words (and their characteristics) are also acceptable as features. For example, we can add the information that the current word '*Isaac*' follows an abbreviation, a capitalized word or anything else could be useful. Unlike rule-based approaches, this extra knowledge is not directly used to select a label but is just an additional clue. The co-occurrence of features and words affect the parameter estimation. For example, a capitalized word is more probable to be labeled as a named entity because named entities in training data are in general capitalized.

Core techniques in current named-entity recognition systems using statistical approaches converge into CRF thanks to its evident advantage described above. These NER systems especially focus on finding effective features to enhance system performance. Another important factor is training data as for other statistical learning problems. It is well known that we cannot construct a totally universal model for a learning task that works well for any new instances. It is unavoidable to restrict target-training data and we usually evaluate a constructed system on a closed dataset, which has same distributions with training set. Nevertheless, to design a system less dependent on change of training data is always an interesting subject. And that is why a well-designed system ponders on an efficient way of selecting and updating training data.

1.5 From Question-Answering to Knowledge Base Population

Since 1992, international evaluation campaigns are organized by NIST through the Text REtrieval Conferences TREC [108] in which specific retrieval tasks are defined. They allow evaluating several information retrieval approaches over the same large size collections of documents and queries by means of some standard criterion. For example, the 2012 Knowledge Base Acceleration Track of TREC[5]

[5] http://trec.nist.gov/tracks.html

aimed to develop systems helping human knowledge base curators to maintain databases. On the other side, the purpose of the Entity Track[6] was to perform entity-oriented searches on the Web [4] and the purpose of the Question-Answering track [105] was to answer to natural language questions by mining large document collections. For each task, the ability to recognize named entities in text is a fundamental problem.

1.5.1 Named-Entity Recognition

For named-entity recognition, a wide variety of clues are used like presence of words, part-of-speech, case etc. State-of-the-art approaches achieve really good results (more than 0.90 for F-measure) on coarse-grained classes (*person*, *organization* and *location*) in well-formatted documents like news articles.[7] However, most of them are not adapted for either more complex and/or fined-grained types of entities or for more challenging sources of Web documents.

Indeed, for named entity recognition on the Web, many new challenges raised, both in terms of scale and scope: systems have to be fast, in order to be able to deal with the huge quantity of texts and dynamic aspects. Manually created rules or building training data for fine grained types of named entities and for every Web source is impractical (What are the types one needs? For what purpose? How could we annotate Web pages efficiently for training?).

Downey et al. [34] deal with locating complex named entities like book or film titles, for which named entity recognition systems often fail. They consider named entities as species of multiword units, which can be detected by accumulating n-gram statistics over the Web. When their system tries to extend named entity bounds to adjacent n-grams, it is made according to Web statistics: do these n-grams co-occur frequently on the Web? With this approach, they achieve a F1 score 50 % higher than most supervised techniques trained on a limited corpus of Web pages.

Whitelaw et al. [110] present an approach generating training data automatically. They propose a multi-class online classification training method that learns to recognize broad categories such as place and person, but also more fine-grained categories such as soccer players, birds, and universities. The resulting system obtains precision and recall performance comparable to that obtained for more limited entity types in much more structured domains such as company recognition in newswire, even though Web documents often lack consistent capitalization and grammatical sentence construction.

---

[6] http://ilps.science.uva.nl/trec-entity/

[7] During CoNLL 2003 (Conference on Computational Natural Language Learning) a challenge that concerned language-independent named entity recognition was organized. Many other tasks related to Natural Language Processing have been organized in the context of CoNLL conferences: grammatical error correction, multilingual parsing, analysis of dependencies… (http://www.clips.ua.ac.be/conll/).

Pasca [74] chooses to focus on types of named entities for which users of information retrieval search engines may want to look for. The extraction is guided by a small set of seed named entities, without any need for handcrafted extraction patterns or domain-specific knowledge, allowing for the acquisition of named entities pertaining to various classes of interest for Web users.

These methods are focused on the precision of the extraction regardless to its recall (the recall is difficult to estimate because the lack of large pre-annotated date). However, Etzioni et al. [38] propose the Know-ItAll system that is able to extract thousands of named entities, without any hand-labeled training data and obtains a high precision and a good recall too (it is estimated by comparing extracted named entities to gazetteers). The Know-ItAll system is based on learning domain-specific extraction rules and is able to automatically identify sub-classes in order to boost recall. Moreover, this system is able to take advantage of lists of class instances by learning a wrapper for each list.

For most of the NER approaches, types are pre-defined and users cannot search for entity types that were not included in the learning phase (with statistical approaches, adding a new type in the hierarchy of types means that training data must be revisited). To answer this problem, we propose in [12] an unsupervised way to determine to what extent an entity belongs to any arbitrary given very fine-grained conceptual class (or type). This fully automatic method is mainly inspired by the idea of "distributional hypothesis" which state that words that occur in the same contexts tend to have similar meanings. From this, we propose to measure to what extent an entity belongs to a type according to how much the entity's context is similar to the one of the given type. Our idea is that we could do it by comparing the word distribution in Web pages related to an entity to the one in Web pages related to the type in the sense that every entity (instance) characterizes its own type (concept). Related Web pages are retrieved by mean of a Web search engine and by querying it twice: first with the entity name as query and the second with the type as query. Word distribution are then estimated from the top retrieved documents and compared to each other. The evaluation of this approach gives promising results and, as expected, shows that it works particularly well for highly specific types (e.g. scotch whisky distilleries) and less for broad ones like "person". It appears as a good and fast method when others fail (i.e. when the requested type was not learned before).

New challenges arise with social networks. Ritter et al. [87] show that most of state-of-the-art named entity recognizers failed for tweets that are short (144 characters at most) and employing specific language. They address this issue by re-building the NLP pipeline (part-of-speech tagging, chunking, named-entity recognition). They leverage the redundancy in tweets to achieve this task, using LabeledLDA [84] to exploit Freebase[8] structured data as a resource during learning.

[8] Freebase (https://developers.google.com/freebase/) contains in June 2013 more than 37 million entities, 1,998 types and 30,000 properties.

Their system doubled the F1 score compared to the well-known Stanford NER module.[9]

1.5.2 Retrieving Precise Answers in Documents

Question-answering corresponds to provide users with precise answers to their natural language questions by extracting answers from large document collections or the Web [60, 62, 64, 85, 106]. This task imply semantic analysis of questions and documents (Web pages, microblog messages, articles, books…) in order to determine *who* did *what*, *where* and *when*.

A classical architecture for question-answering systems involves many steps such as question and document analysis (Natural Language Processing: part-of-speech tagging, syntactic parsing, named-entity recognition, semantic role labeling [63, 73]…), document and passage retrieval (information retrieval approaches) and information extraction (logic or rule based, employing machine-learning or not) from selected passages. This corresponds to a sequential approach of question-answering that reduces the search field from the collection as a whole to a precise answer through a limited set of documents then a set of passages and a set of sentences extracted from these passages. This architecture has a double origin: the pre-existence of each module (evaluated and tuned through evaluation campaigns: TREC ad-hoc, MUC…) and the difficulty of deep analysis techniques to operate quickly on large corpora.

The analysis of natural language queries corresponds to the application of several more or less optional treatments: spelling, linguistic normalization or expansion, focus extraction [28, 41, 70, 80], identification of constraints (dates, places…), guessing the type of needed answers (factual—quantity, name of a place, name of an organization …—, definitional, yes/no…). For this last operation, most systems employ categorizers based on lexical patterns from simple heuristics on keywords but Qanda employs a base of several thousand nominal phrases for robustness [17]. The patterns can be handwritten or built by means of machine learning (Naïve Bayes, Decision Trees…) and can be lexical only or a combination of words, part-of-speech and semantic classes.

Passage retrieval can be performed by computing a lexical density score for each sentence in the set of the documents retrieved from a question by an IR system. Such a score [9] measures how much the words of the question are far away from the other ones. Then it allows to point at the centers of the document areas where the words of the topic are most present by taking into account the number of different lemmas $|w|$ in the topic, the number of topic lemmas $|w, d|$ occurring in the currently processed document d and a distance $\mu(o_w)$ that equals

[9] http://nlp.stanford.edu/software/CRF-NER.shtml

the average number of words from o_w to the other topic lemmas in d (in case of multiple occurrences of a lemma, only the nearest occurrence to o_w is considered).

Let s(o_w, d) be the density score of o_w in document d:

$$s(o_w, d) = \frac{\log(\mu(o_w) + (|w| - |w,d|.p)}{|w|} \tag{1.12}$$

where p is an empirically fixed penalty aimed to prefer or to not prefer few common words with the topic that are close to each other or many words that are distant to each other.

Secondly, a score is computed for each sentence S in a document d. The score of a sentence is the maximum density score of the topic lemmas it contains:

$$\mathrm{s}(S, d) = \max_{o_w \in S} \mathrm{s}(o_w, d) \tag{1.13}$$

Once passages are retrieved, they are mined to extract precise answers by matching question and answer syntactic trees, by exploiting semantic word labeling, surface analysis or logical inference. Redundancy in text collection often participates in ranking candidate answers [62].

The first TREC evaluation campaign[10] in question-answering took place in 1999 [107]. TREC-QA successive tracks evolved over the years to explore new issues. In TREC-8, the participants had to provide 50-bytes or 250-bytes document passages (from a corpus 500,000 documents—2 GB) containing answers to some factoid questions. For TREC-9, questions were derived from real user questions and a set of variants was proposed in order to study the impact of formulating questions in different ways. In TREC-2001, required passage size of answers was reduced to 50-bytes and list questions were introduced [105]. For this kind of questions, the answers should be mined in several documents and the questions specified the number of instances of items to be retrieved. Context questions were introduced for TREC-2001 as some sets of questions related to each other. In TREC-2002, systems were required to return exact answers rather than text passages. In TREC-2003, definition questions (e.g. "*Who is Colin Powell?*") appeared. In TREC-2004, more difficult questions were introduced including temporal constraints, more anaphors and references to previous questions. The collection was the AQUAINT Corpus of English News Text[11] consists of newswire text data in English, drawn from three sources: the Xinhua News Service, the New York Times News Service, and the Associated Press Worldstream News Service (375 million words, about 3 GB of data).

The question-answering TREC track last ran in 2007. However, in recent years, TREC evaluations have led to some new tracks that involved precise information retrieval and deep text mining. Among them, Entity Track for performing entity-related search on Web data, Recognizing Textual Entailment for determining

[10] http://trec.nist.gov/tracks.html

[11] LDC catalog number LDC2002T31 (http://www.ldc.upenn.edu).

whether a text entails the meaning of another one. In the same period, for question-answering, NIST encouraged the use of documents extracted from blogs instead of newspaper articles and sub-tasks dedicated to opinion mining and automatic summarization. This led to some new evaluation conferences such as Text Analysis Conference (TAC).[12] In 2008, the answers had to be either named-entities or complex explanatory answers. The tests were realized on the Blog06 corpus that is composed of 3.2 million texts extracted from more than 100,000 blogs.[13]

The INEX 2009-10 QA@INEX track we organized[14] aimed to estimate the performance of question-answering, passage retrieval and automatic summarization systems together on an encyclopedic resource such as Wikipedia [94]. The track considered two types of questions: factual questions which require a single precise answer to be found in the corpus if it exists and more complex questions whose answers require the aggregation of several passages (summarization of multiple documents). In order to consider more difficult texts than news articles, we have been organizing the Tweet Contextualization task of INEX[15] since 2011 [8]. Given a new tweet, participating systems must provide some context about the subject of a tweet, in order to help the reader to understand it. In this task, contextualizing tweets consists in answering questions of the form "what is this tweet about?" which can be answered by several sentences or by an aggregation of texts from different documents of the Wikipedia. The summaries are evaluated according to their informativeness (the way they overlap with relevant passages) and to their readability (linguistic quality). Informativeness is measured as a variant of absolute log-diff_ between term frequencies in the text reference and the proposed summary. Maximal informativeness scores obtained by participants from 19 different groups are between 10 and 14 %. This task corresponds to a real challenge [95].

1.5.3 Opinion Mining and Sentiment Analysis in Question Answering

The objective of the opinion question-answering task of the Text Analysis Conference TAC is to accurately answer questions on an opinion expressed in a document. For example: "*Who likes Trader Joe's?*" or "*Why do people like Trader Joe's?*" that are two questions about the chain of food stores "Trader Joe's" and calling either named entities or explanatory answers. In the latter case, the answers must contain the information nuggets differentiating essential

[12] http://www.nist.gov/tac/

[13] http://ir.dcs.gla.ac.uk/test_collections/blog06info.html (about 40 GB of data for feeds only)

[14] http://www.inex.otago.ac.nz/tracks/qa/qa.asp

[15] https://inex.mmci.uni-saarland.de/tracks/qa/

information and accurate information but not essential. Automatic software had to retrieve different aspects of opinion in respect of a particular polarity.

In 2008, 9 teams participated in the opinion question-answering task in TAC [29]. For the 90 sets of rigid type questions, the best system achieved an F-score of 0.156. In comparison, the scoring of the manual reference was 0.559. The scores of the other 8 teams in 2008 ranged between 0.131 and 0.011. The best scores were obtained by systems of the THU Tsinghua University (China) [61] and IIITHy-derabad (India) [104].

The first, THU Quanta was based on the question-answering system Quanta enriched with a vocabulary expressing feelings. The authors have tried several lexical databases such as Wordnet[16] but without much success because they did not establish the polarity of a word in context (e.g. the word big can be positive or negative). Depending on the type of question, the answer extraction was based either on analysis of the frequency of occurrences of words in the query, in selected sentences and in the title documents and on the number of opinion words or by using a probabilistic information retrieval model associated with a measure of density of the query words in the retrieved passages (traditional approach for question-answering—see above). The extraction of nuggets was achieved by combining pattern matching (for *why* questions for example) and external knowledge (such as list of actors and films extracted from the IMDB database).

The second system with the highest performance employed Lucene information retrieval engine[17] and machine learning approaches. The retrieved documents were classified into two categories according to their polarity and matched with the question polarity. To determine the polarity, the authors have used lists of positive and negative words and established classification rules. To calculate the polarity of the documents, they have chosen to create two Bayesian classifiers: one to recognize opinion sentences (regardless of the polarity of opinions) and second to differentiate positive and negative opinions.

In continuation of the work described above, NIST introduced in 2008 a task entitled Opinion Summarization in TAC campaigns. The aim was to produce texts of limited length to 7,000 or 14,000 characters summarizing the multiple opinions encountered in some blogs related to specific topics, themselves expressed as sets of questions in natural language. For example, related to Gregory Peck, the issues were:

What reasons did people give for liking Gregory Peck's movies?
What reasons did people give for not liking Gregory Peck's movies?

Here are some excerpts of text that could be extracted from blogs and that carry an opinion answering the questions:

– I've always been a big Peck fan since I saw him inTo Kill a Mockingbird. A Charmed Life is a fine biography of one my favorite actors.

[16] http://wordnet.princeton.edu

[17] http://lucene.apache.org

- Gregory Peck can be seen playing his first character without absolutely any redeeming quality in this thriller (Boys from Brazil).
- after half an hour of the slow-paced antics of Roman Holiday, I voted to stop watching it, so I oinked it from the DVD player, sealed up the disc in its return envelope

The methods used by the system IIITHyderabad are very similar to those described in the previous section for the opinion QA task. The main differences lie in the classifier that determines the polarity of sentences—Support Vector Machines (SVM) rather than Naïve Bayes classifier—and further exploitation of the SentiWordNet[18] lexical resource that assigns to each synset of Wordnet three sentiment scores (positive, negative or neutral) [3]. The descriptions of the methods used by each system and the results are available on the websites of TREC and TAC (see above).

Sentiment Analysis in Twitter became an important task and was the focus of the International Workshop on Semantic Evaluation that organized SemEval-2013 Exercices.[19] We proposed [50] to use many features and resources in order to improve a trained classifier of Twitter messages. These features extend the unigram model with the concepts extracted from DBpedia,[20] verb groups and similar adjectives extracted from WordNet, Senti-features extracted from SentiWordNet. We also employed a homemade dictionary of emotions, abbreviation and slang words frequent in tweets in order to normalize them. Adding these features has improved the f-measure accuracy 2 % from SVM with words only and 4 % from a Naïve Bayes classifier.

1.5.4 Automatic Knowledge Base Population

Knowledge bases (KB) are considered in a variety of domains as a massive resource of information. Two main issues arise. First, building such base is a lot of effort since it has to be populated enough to be really useful. In addition, to make a KB usable, the veracity of the information must be guarantee. So the first issue could be "*How to automatically build a reliable knowledge base?*". Second, maintaining a KB is also very challenging as information may change along the time. This gives the second problem: "*How a knowledge base can be kept up to date?*". The aim of some evaluation campaigns is to address these issues and participation systems use some well-known IR methods.

For example, the 2012 Knowledge Base Acceleration Track of TREC (Text Retrieval Conference organized by NIST) aims to develop systems helping human

[18] http://sentiwordnet.isti.cnr.it

[19] http://www.cs.york.ac.uk/semeval-2013/

[20] DBPedia is a large knowledge base (more than 3.77 million things are classified in an ontology) localized in 111 languages built by extracting structured information from Wikipedia (http://dbpedia.org)—June 2013

knowledge base curators to maintain databases. The Text Analysis Conference (TAC) has been initiated in 2008. In 2009 a track named Knowledge Base Population (KBP) has been launched. Its aim was to extract information from unstructured texts to create nodes in a given ontology, it was subdivided into three tasks:

- Entity Linking: the aim of this task is to find the KB node that is being referred by the entity in a query. The query is composed of a name-string (an entity) and a document id, which refers to a given document in a large text collection. Each name-string occurs in the associated document. The document can help to disambiguate the name-string because some entities may share a confusable name (e.g., George Washington could refer to the president, the university, or the jazzman). An entity may occur in multiple queries under different name-strings that must be detected as variants of the entity (e.g., Elvis, The King).
- Slot Filling: the task is more about information extraction where entities' attributes (called slots) must be found in the collection. A set of generic attributes is defined for each type of entity (i.e., person, organization, geo-political entity…). It is not expected though to correct or modify values from the reference KB, but only to add information. A slot can be either single-valued or list-valued (i.e., accept more than one value).
- Cold Start Knowledge Base Population: this task starts with a knowledge base schema that describes the facts and the relations that will compose the knowledge base and that is initially unpopulated. Participants have to build software that will create an entire knowledge base that will be then evaluated.

Run for the first time in 2012, the Knowledge Base Acceleration (KBA) track has been introduced as a KBP Entity Linking (EL) reverse process. EL is to find KB node that matches the tuple name-document. KBA starts from the KB node (called *topic*) and is to retrieve information and classify documents about that topic. When classified with a high degree of relevancy, information has to be extracted from the document with the aim of suggesting KB node edition. The 2012 collection is a stream of documents where every document has a time stamp which starts from October 2011 to May 2012. It is made up of three types of documents: news (from public news websites), socials (from blogs) and links (from bit.ly database). In 2012, KBA track started with only classifying documents into four classes that defined the degree of relevancy of the documents:

- Garbage: not relevant, e.g., spam;
- Neutral: not relevant, no information could be deduced about the entity, or only pertains to community of target such that no information could be learned about the entity;
- Relevant: relates indirectly; tangential with substantive implications;
- Central: relates directly to target such that you would cite it in the Wikipedia article.

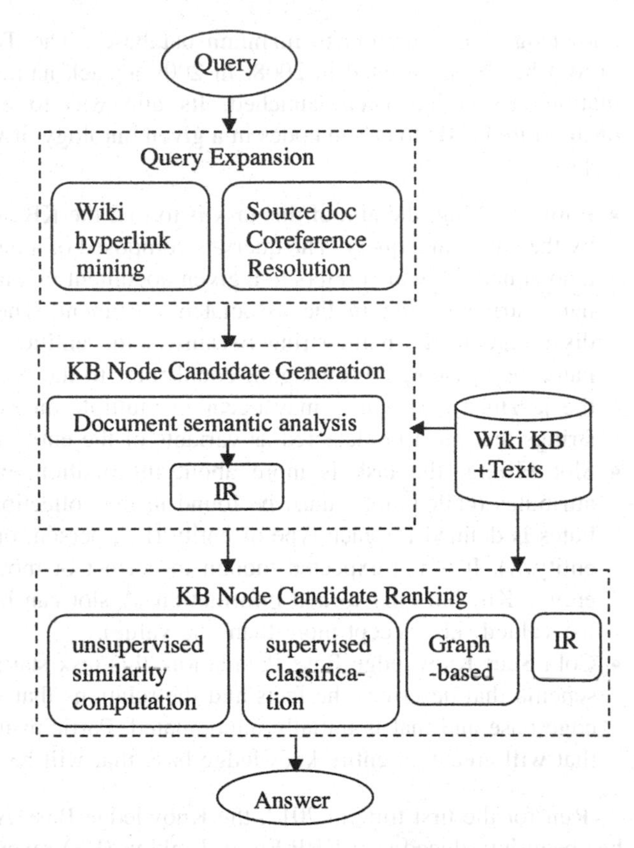

Fig. 1.2 Knowledge base population system architecture for entity linking [53]

The document collection for the KBP 2013 Entity Linking tasks are composed of English, Spanish, and Chinese documents including approximately half a million discussion forum posts, 1 million web texts and 2 million newswire articles.

1.5.4.1 Architectures and Approaches

There have been more than forty teams participating to KBP track over the last 4 years and there were eleven teams on the first KBA session in 2012. With a general improvement for KBP over the last years, the best system reaches the 85.78 % micro average accuracy. Most KBP Entity Linking architectures include 3 main steps: query expansion, KB Node candidate generation, KB Node candidate ranking (see Fig. 1.2). On KBA, the approaches are quite different, since the entity is already given but the document collection[21] is much larger (134 million news wires, 322 million feeds from blogs, 135 million texts that were shortened at

[21] http://trec.nist.gov/data/kba.html

bitly.com, 40 % in English, approximately 9 TB of 'raw' text). Teams must use entity KBNode information to retrieve centrally relevant documents.

Many methods involving supervised, semi-supervised or unsupervised systems have been developed and they all have pros and cons. When reading proceedings from TAC KBP,[22] it is interesting to notice that the best teams (in term of system performance) start with the same schema. Most approaches are recall oriented at first, and then the precision comes after going through different sets of filters that may provide eventually a result. In order to obtain as much answers as possible the name-string (considered as a query) must be analyzed to obtain information about the entities such as variants, aliases, and acronym complete form when the name-string is an acronym. All variants are often used as new queries to complete the original query (query expansion). Then, the resulting queries are processed by an Information Retrieval system and the output corresponds to a ranked list of KBNode candidates.

A KBA system starts with a name of an entity and a KBNode that is in our case a Wikipedia entry. Then, as the stream of document goes by, the system is to find whether a document in this stream concerns the entity. Even if the both evaluation campaigns KBA and KBP start with different inputs, some common points arise such as finding all possible aliases for an entity to sort of build an entity profile. This profile is then useful to assess whether a document is relevant. Then the document must be classified into one of the four classes.

1.5.4.2 Wikipedia as a Resource for Entity Disambiguation

One of the first things to do when dealing with an entity is to gather data about it. This helps to ensure that a document is really about that entity. It also helps to disambiguate or ensure there is no ambiguity with another entity. A knowledge base such as Wikipedia is really convenient for this task since both its structure and its content provide much information about entities. Every page on Wikipedia has a unique name that obviously corresponds to the main object of the page. For homonyms, it exists different alternative for naming pages such as:

- *Disambiguation pages* show all alternative uses of a word. There is no other content. The page name is always suffixed with_(disambiguation) in the English-speaking version of Wikipedia.
- *Redirect pages* are to redirect a user to another page where content relates the current page. It is often used for acronym such as UK page that is redirected to United Kingdom page. This particular feature also helps for handling acronyms.
- *Hat notes* are used when there is one really common usage of the word (for instance Rice) but other senses exist. The main pages are filled with common sense content and the hat notes contains an hyperlink that points to the disambiguation page or if only one other sense is known, it points directly to the other sense page.

[22] http://www.nist.gov/tac/publications/index.html

- Bold texts in the first paragraph may help for disambiguation since it often contains alias names or full names.

On KBP sides, participants often build their own knowledge repository with all those information gathered from the given set of entities. Moreover, the document attached to the query can also be used. On KBA side, the input already provides the Wikipedia page linked to an entity. However, KBA participants show that a name may not be enough to ensure a document is about a particular entity. They also use disambiguation pages, redirect pages hat notes and bold text. Some participants also use links from the other pages that points to the entity Wikipedia page to try to find even more aliases.

1.5.4.3 Automatic Query Expansion

The query expansion is to build a set of queries that describes different facets of the original query. It is usually used in recall-oriented system to gather as much documents as possible in order not to miss any relevant document. Xu and Croft [111] divided the automatic query expansion mechanism into two classes: global analysis of a whole document collection, and local analysis where only documents retrieved from the original query are used to expand it. The local mechanism is also known as pseudo relevance feedback (PRF). The last method is more efficient than the global one but it is not without any risk since a substantial amount of non-relevant document found from the original query may lead to a semantic drift.

Many KBP EL participants used the information gathered from the entity disambiguation process to generate queries and obtain a collection of candidates related to entities. Then, they proceed to a candidate disambiguation to select the correct KBNode. One of the best KBP EL system [20], build up from the original name-string a collection of queries based on:

- Whether the original name-string corresponds to a Wikipedia redirect page. If so, the title of the page pointed by the redirection is used as a query,
- Whether the title followed by_(disambiguation) exists in the Wikipedia collection. If so, every titles from disambiguation page are added to the set of query,
- The original name-string is considered as an acronym if it contains only capitalized letters. The full name is then searched in the attached document and is added to the set of query when found.

1.5.4.4 Results

In both KBP EL and KBA evaluation campaigns a confrontation between a query and the result is mandatory to assess whether a document deals with the entity in the original query. In KBP EL, the input is a query (name-string) and a document. Then, the knowledge base is queried using the different queries to generate a pool of candidates and to rank them eventually. Different ranking approaches have been used such as IR oriented approach where the attached document is used as a single query and the aim is to retrieve most relevant candidates. Another interesting unsupervised approach, measures the similarity between the attached document and the candidate. Some teams make this method weakly supervised with annotated data for tuning.

Using query expansion based on Wikipedia hyperlinks, title, disambiguation and the usage of a supervised classifier allows the Standford-UBC team to improve significantly the micro-averaged accuracy from 74.85 to 82.15 % where other unsupervised systems obtain up to 59.91 %. For unsupervised methods, efficient selection of features is a very important step. It is shown in Ji, et al. that unsupervised methods obtain significant improvement when using semantic relation features or context inference, even get better when using Semantic Relation Features, from 58.93 to 73.29 % micro-averaged accuracy for BuptPris system.

In KBA the inputs are an entity name and a knowledge base node (a Wikipedia page). For KBA 2012, we proposed an original approach [13] that obtained the 3rd score by introducing some new features and a new weakly supervised classification process. We subdivided the first problematic into a two-step classification process. For both classifiers we estimated a set of features:

- Document Centric Features: estimated on single documents such as Term Frequency (TF), TF in the title, whether a document has a title, TF on first 10 and 20 % of the document, word entropy…
- Entity Related Features: estimated with the help of the KBNode. This is the only needed input provided by a tier in our solution. A cosine similarity (or any other relevance score) is computed between the KBNode document and the candidate document.
- Time Features: their first purpose was to characterize burst effect and to estimate their impact in document relevancy. They are used to evaluate the quantity of relevant documents about an entity within a day or within a week.

KBA is a really new track, but a lot can be expected from it. In 2012, 11 teams participated and, for centrally relevant classification, best F-measure obtained was 0.359, 0.289 for the median and 0.220 for the average.

1.5.4.5 Automatic Extraction of Attributes Related to Entities

Poesio and Almuhareb [76] defined a named entity as a complex object characterized by its attributes (or features) and their value. If this assertion is widely accepted by the community and confirmed by many works, the definition of an attribute is still subject to discussion. Guarino [49] defined two types of attributes: relational attributes like qualities (color, position, role, spouse etc.) and non-relational ones like parts (wheels, engine). Many other definitions have been proposed but Poesio and Almuhareb [76] proposed the broadest one: in addition to

Guarino's definition, they added related objects (e.g. nest for a bird), related activities (reading or writing for a book), related agents (reader, writer). However, each existing work presents more or less its own definition of what an attribute is, depending mostly on the aimed task. For instance, for product attributes extraction, attributes are mainly qualities but for populating lexical bases like Wordnet, the extended definition of Poesio may be followed. Attributes and values have been showed to effectively describe and characterize entities [2].

Many works rely on rules to extract attributes. Berland and Charniak [10] introduced patterns for extracting meronyms (parts) like "NP*'s* NP" (phone's battery) or "NP *of* NPs" (battery of phones). Poesio and Almuhareb [76] proposed additional patterns and refined those proposed by Berland ("NP's NP {*is* | *are* | *was* | *were*}"). Sánchez [93] added patterns with verbs indicating inclusion like "NP {*come* | *comes* | *came*} *with* NP}" (phone come with battery). Zhao et al. [114] learned syntactic structures from examples and then compared them to syntactic trees in texts to find attributes.

Web documents are widely used to perform attribute extraction. For instance, [115] detected in web documents, regions of interests (minimum set of HTML nodes containing attributes) and then labeling elements of this region as attributes. They achieved good performances by using a "vision-tree" based on the HTML tag tree enhanced with information like color or font type.

Query logs have also been proved to be a really good source for attribute extraction. Pasca [74] started to exploit them with the intuition that if an attribute is relevant and central for an entity, people must have looked for it on the Web before. Attributes are then extracted from query logs with the help of patterns and then ranked according to their frequencies. They showed that using query logs produces high precision set of attributes, better than ones produced from Web pages by employing similar approaches. However, recall is lower. Finally, they showed that using both Web pages and query logs lead to the best results.

Merging different labels that refer to the same real-world attributes (e.g. size of the screen and length of the screen) is one the challenging problem in attribute extraction. Yao et al. [112] tackled it by looking at co-occurrences of the label (with the assumption that if two labels appear in the same documents they are probably not referring to the same attribute) and values of the attributes (two labels with the same values for the same entity are probably referring to the same attribute). Raju et al. [83] used Group Average Agglomerative Clustering, based on commons n-grams and Dice similarity to group labels together. One representative label is then selected.

When attributes are extracted, they have to be ranked according to their importance for the corresponding entity. Effective approaches exploit the number of times the attribute has been extracted, popularity of documents from where the extraction was made according to query logs [100] and measures of association between the attribute and the entity (like the PMI) computed by means of Web search engine hits [93].

Last step is value extraction. For most scenarios, value extraction is more or less straightforward: it may be done during record extraction and using patterns has been proved to be effective [93] (e.g. "NP*'s* NP *is*" - > phone's battery life is 24 h). For more difficult cases, Davidov and Rappoport [30] proposed an approach to check whether extracted values are likely to be correct: they compare the extracted value to maximum and minimum values found for similar entities and compare between entities. If the extracted value does not match these constraints then it is discarded. Their approaches rely on manually crafted patterns and conversion tables for units.

1.6 Industrial Applications

Of course, there are many industrial applications based on natural language processing and information retrieval. In this section, we present three applications we work on. The first is dedicated to digital libraries, the second to spoken language understanding for speech systems and the last to a marketing purpose.

1.6.1 Structuring Bibliographic References in Papers

In this section we present a real-world application of information extraction. BILBO is an automatic bibliographic reference parsing system that has been started by a research and development project supported by Google Digital Humanities Research Awards.[23] The system deals with in-text bibliographic references published on CLEO's OpenEdition[24] Web platforms for electronic articles, books, scholarly blogs and resources in the humanities and social sciences.

1.6.1.1 Reference Data

Most of earlier studies on bibliographic references parsing (annotation) are intended for the bibliography part at the end of scientific articles that has a simple structure and relatively regular format. Automated bibliographic reference annotation is comparably recent issue in the sequence labeling problems [75, 98]. It aims to separately recognize different reference fields that consist of author, title, date etc. This is involved in citation analysis, which has already started several decades ago [43], intended for finding patterns among references such as bibliographical coupling and also for computing impact of articles via citation frequency and patterns. A practical example is CiteSeer [45], a public search engine and digital library for scientific and academic papers that first realized automatic

[23] http://lab.hypotheses.org

[24] http://openedition.org

citation indexing. Recent works using citations concentrate in taking advantage of extracted citation information, for example on scientific paper summarization [78], text retrieval [86] and document clustering [1]. The success of these applications priory depends on the well-recognized reference fields.

We are interested in the annotation of much less structured bibliographic references than the usual data. OpenEdition on-line platform consists of more than 330 journals in social sciences. The mostly used language is French, which is the original language of the platform but it has been designed for the electronic publishing in the global scale and the quantity of papers in English or in Portuguese is growing fast. The rate of articles written in a different language than French is 10 %, of which more than half are in English. The articles have a diverse range of formats in their bibliographical references not only according to journal types, but also to authors in a same journal. We distinguish references into the following three levels according to the difficulty of manual annotation. We construct three training sets with the different levels, since the same difficulties hold true for automatic annotation (see Fig. 1.3):

- Level 1: references are at the end of the article in a heading 'Bibliography'. Manual annotation is comparably simple,
- Level 2: references are in the notes and they are less formulaic compared to that of corpus level 1. The extraction of bibliographical part is necessary,
- Level 3: references are in the body of articles. The identification and annotation are complex. Even finding a beginning and an end of references is difficult.

1.6.1.2 Presentation of BILBO, a Software for Information Extraction

BILBO is an automatic reference annotation tool based on Conditional Random Fields (CRFs). It provides both training and labeling services. For the former, manually annotated XML files are necessary as input data and BILBO automatically extracts labels for each reference field and uses them to learn a CRF model. Once BILBO finishes training a model, we can run it for labeling new data.

We try to encode as much information as possible during the manual annotation of OpenEdition training data using TEI Guidelines.[25] TEI is a consortium that develops, defines and maintains a markup language for describing structural and conceptual features of texts. By the way, rich information is not always useful for training a model and unnecessarily too detailed output labels, which are reference fields in our case, can complicate the learning process of a CRF model and increase the risk of over-fitting. Choosing appropriate output labels (see Table 1.1) is important before applying a CRF model. Meanwhile, the superiority of a CRF compared to other sequence models comes from its capacity to encode very diverse properties of input through feature functions. Feature selection is also an

[25] Text Encoding Initiative (http://www.tei-c.org/Guidelines/).

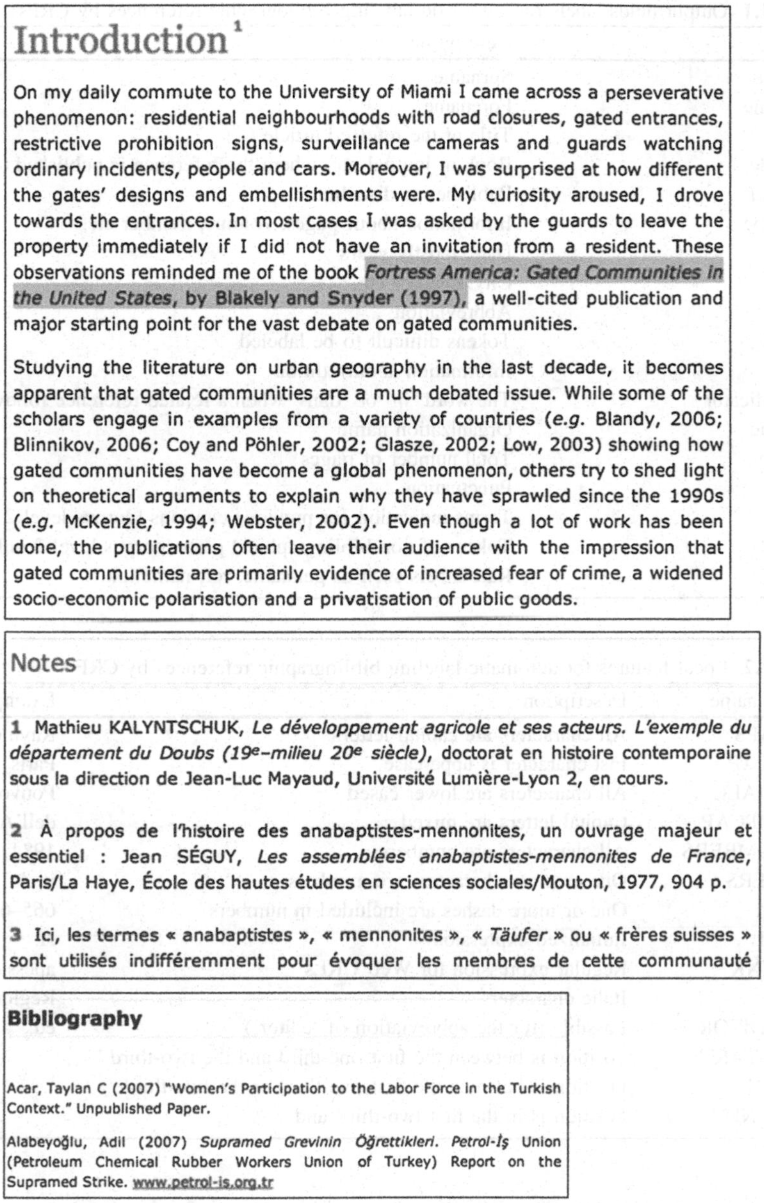

Introduction[1]

On my daily commute to the University of Miami I came across a perseverative phenomenon: residential neighbourhoods with road closures, gated entrances, restrictive prohibition signs, surveillance cameras and guards watching ordinary incidents, people and cars. Moreover, I was surprised at how different the gates' designs and embellishments were. My curiosity aroused, I drove towards the entrances. In most cases I was asked by the guards to leave the property immediately if I did not have an invitation from a resident. These observations reminded me of the book *Fortress America: Gated Communities in the United States*, by Blakely and Snyder (1997), a well-cited publication and major starting point for the vast debate on gated communities.

Studying the literature on urban geography in the last decade, it becomes apparent that gated communities are a much debated issue. While some of the scholars engage in examples from a variety of countries (*e.g.* Blandy, 2006; Blinnikov, 2006; Coy and Pöhler, 2002; Glasze, 2002; Low, 2003) showing how gated communities have become a global phenomenon, others try to shed light on theoretical arguments to explain why they have sprawled since the 1990s (*e.g.* McKenzie, 1994; Webster, 2002). Even though a lot of work has been done, the publications often leave their audience with the impression that gated communities are primarily evidence of increased fear of crime, a widened socio-economic polarisation and a privatisation of public goods.

Notes

1 Mathieu KALYNTSCHUK, *Le développement agricole et ses acteurs. L'exemple du département du Doubs (19e–milieu 20e siècle)*, doctorat en histoire contemporaine sous la direction de Jean-Luc Mayaud, Université Lumière-Lyon 2, en cours.

2 À propos de l'histoire des anabaptistes-mennonites, un ouvrage majeur et essentiel : Jean SÉGUY, *Les assemblées anabaptistes-mennonites de France*, Paris/La Haye, École des hautes études en sciences sociales/Mouton, 1977, 904 p.

3 Ici, les termes « anabaptistes », « mennonites », « *Täufer* » ou « frères suisses » sont utilisés indifféremment pour évoquer les membres de cette communauté

Bibliography

Acar, Taylan C (2007) "Women's Participation to the Labor Force in the Turkish Context." Unpublished Paper.

Alabeyoğlu, Adil (2007) *Supramed Grevinin Öğrettikleri*. *Petrol-İş* Union (Petroleum Chemical Rubber Workers Union of Turkey) Report on the Supramed Strike. www.petrol-is.org.tr

Fig. 1.3 Three levels of bibliographic references and three corpora: references can be in the body of papers, in the footnotes and in a specific 'Bibliography' section

essential process (see Table 1.2). To avoid confusing input tokens and features describing the characteristics of the input, we call the latter *local features*. Because of the heterogeneity of references in different levels (see Fig. 1.3), we learned a different model for each one.

Table 1.1 Output fields labels for automatic labeling bibliographic references by CRFs

Labels	Description
Surname	Surname
Forename	Forename
Title	Title of the referred article
Booktitle	Book or journal etc. where the reference is published
Publisher	Publisher or distributor
Biblscope	Information about pages, volume, number etc.
Date	Date, mostly years
Place	City, country etc.
Abbr	Abbreviation
Nolabel	Tokens difficult to be labeled
Edition	Information about edition
Bookindicator	The word 'in' or 'dans' when a related reference follows
Orgname	Organization name
Extent	Total number of pages
Punc.	Punctuation
w	Terms indication for previous citations (corpus level 2 only)
Nonbibl	Tokens of non-bibliographical parts (corpus level 2 only)
Others	Rare labels such as genname, ref, namelink…

Table 1.2 Local features for automatic labeling bibliographic references by CRFs

Feature name	Description	Example
ALLCAPS	All characters are capital letters	Raymond
FIRSTCAP	Fist character is uppercase	Paris
ALLSMALL	All characters are lower cased	Pouvoirs
NONIMPCAP	Capital letters are mixed	dell'Ateneo
ALLNUMBERS	All characters are numbers	1984
NUMBERS	One or more characters are numbers	in-4
DASH	One or more dashes are included in numbers	665–680
INITIAL	Initialized expression	H.
WEBLINK	Regular expression for Web URLs	apcss.org
ITALIC	Italic characters	Regional
POSSEDITOR	Possible (for the abbreviation of 'editor')	ed.
BIBL_START	Position is between the first one-third and the two-third	
BIBL_IN	Position is between the first one-third or the two-third	
BIBL_END	Position is in the first two-third and	

We have tested more than 40 different combinations of tokenization method, output labels and local features. 70 % of the manually annotated references are used as training data and the remaining 30 % are used for test. We obtained about 90 % of overall accuracy on test data for the level 1 dataset. Compared to the scientific reference data used in the work of Peng and McCallum [75], our level 1 dataset is much more diverse in terms of reference styles. We obtained a successful result in annotation accuracy, especially on surname, forename and title fields (92,

90, and 86 % of precision respectively). They are somewhat less than the previous work of Peng (95 % overall accuracy) but considering the difficulty of our corpus, the current result is very encouraging. The state of the art methods always learn and evaluate their systems with a well-structured data having rather simple and homogeneous reference style sheets (that is a reason why the parsing of new reference different from a regular format does not work well).

We tested labeling performance on the level 2 dataset as well. We obtained 84 % of overall accuracy using the same strategy and the same setting as the previous evaluation. The precision on the three most important fields was around 82 %. We have compared BILBO to different other online systems and shown that BILBO outperforms them especially on the level 2 dataset [56]. Using the same strategy and model as that applied for OpenEdition data, BILBO obtained 96.5 % of precision on the CORA dataset. That underlines the robustness of our approach as well as its capacity to adapt to multiple languages.

1.6.2 The LUNA Project for Spoken Language Understanding

The LUNA[26] (for *spoken Language UNderstanding in multilinguAl communication systems*) project has for main goal to develop and deploy component for robust phones services, in natural spoken language. Its purpose is to bring user with a complete interactive service and interfaces and a full understanding systems, not only with keywords like most automatic operating systems phone services, for example like in [15]. This project contains many interesting components relative to machine learning such as, spoken language understanding, speech to text, dialogue interpretation. Here, we will discuss only about the semantic composition component.

This component is an important process that makes emerge sense from the transcribed speech signal that contains recognition errors often because recording noise, poor pronunciation… The interpretation (semantic analysis) of the transcribed speech is also difficult because it is derived from a spoken dialogue, subject to the disfluency of speech, self-correction… The produced text is often ungrammatical because the spoken discourse itself is ungrammatical. Indeed the application of classical automatic grammatical analysis methods does not produce good results and does not help to improve the outcome of speech transcription [6]. In particular, hope to use deep syntactic analysis must be abandoned in favor of superficial analysis.

A primary objective is to provide a semantic representation of the meaning of the spoken messages. Ontologies are considered [36, 79] to conceptualize the

[26] This project was supported by the 6th Framework Research Programme of the European Union (EU), Project LUNA, IST contract no 33549 (www.ist-luna.eu).

knowledge related to the situation (context) of the recording. One can expresses the semantic components by means of first order logic with predicates. In the work described here, we represent the semantic elements by frames (FrameNet[27] is a lexical database that is both human and machine-readable, based on more than 170,000 manually annotated sentences that are examples of how words are used in actual texts). The frames are hierarchical structures and fragments of knowledge that can be inserted in the transcription or employed to infer other fragments of knowledge.

It is then proposed a system for speech understanding from logical rules with the support of an ontology in order to create links from semantic components [36]. We conducted a study on the discovery of syntactic/semantic relationships. A proposal was done for a compositional semantics experience to enrich the basic semantic components. A detection system for lambda-expression hypothesis to find the relationship through discourse is used at the end of the process [35]. Then, a semantic knowledge is retrieved from a transcribed spoken text and transmitted to a dialogue process that remains to be analyzed [46]. The system can finally transform a human request into a computer understandable request with predicate logic derivations for instance.

The process for retrieving users' intentions, wishes and requests is driven by defining an appropriate ontology [18]. It is important for machine learning (learning correct word associations and linking them to the ontology) to have a good training corpus. For this, training data has to be manually annotated. Figure 1.4 shows the result of the formalization of a spoken sentence after an automatic speech recognition process.

To achieve the best semantic representation (frame detection and frame composition), it is important to make correct requests to the dialogue manager and to other information retrieval systems involved. Some results of the correctness of this process can be found in [5] and are briefly exposed here. About 3,000 dialogue turns have been processed in two ways: a reference dialogue was built with a manual transcription and an automatic transcription with a state-of-the-art speech recognition system (ASR) that obtained about 27.4 % word error rate on these difficult data. Results presented in Table 1.3. Concern frame detection and frame composition F-measure when semantic analysis is applied on the reference corpus or on the automatically transcribed corpus. As expected on the latter, the quality of detection decreases, but with 27.4 % word error rate, one can see that frame detection is not so bad meaning that one can infer some missing information from badly formed data transcription.

In a more general point of view training data is valuable information for a tasks involving machine learning and text mining. Here we have data that link speech signals and transcriptions to 'knowledge chunks' that can be used for other semantic annotation tasks.

[27] https://framenet.icsi.berkeley.edu/fndrupal/

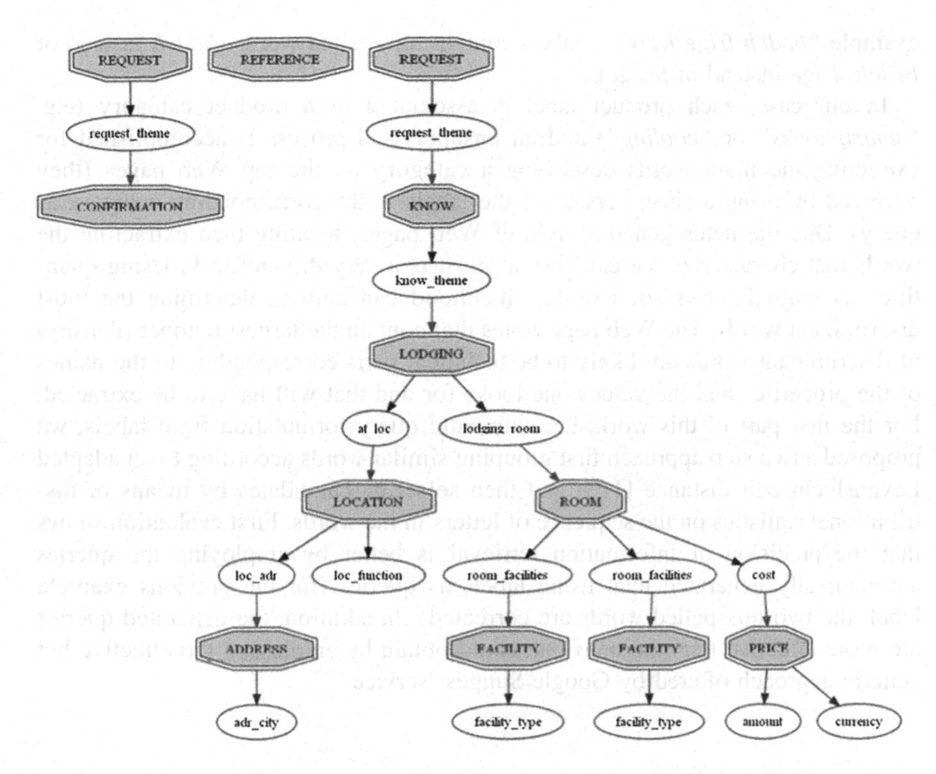

Fig. 1.4 The semantic representation of the sentence, included in the French MEDIA Corpus [11] "*Vous pouvez me confirmer tous les éléments, ils ont bien une baignoire un téléphone et c'est à quatre-vingt-dix euros à côté de Paris*" (*Can you confirm all these, they have a bath, a phone and it's 90 euros near Paris*)

Table 1.3 Quality of automatic frame detection and frame composition on manual and automatic transcription of speech

Level	F-measure on manual transcription (%)	F-measure on automatically transcribed text (%)
Frame detection	85.34	77.01
Frame composition	74.40	66.94

1.6.3 Marketing Product Characterization

The last industrial application we want to write about is under development by the young company KWare. It can be seen as an information retrieval and extraction process allowing to collect information on the Web that characterizes any consumer product. It differs from online well-known shopping services in that retrieval and extraction have to be realized in an unsupervised way from very short textual *labels* employed by distributors as only input. As one can see in the example "*bodch frige KDV3*", labels contain misspelled words (*bodch* instead of *Bosch*, *frige* instead of *fridge*).

In our case, each product label is associated to a product category (e.g. "*smartphones*" or "*cooling*") and an unsupervised process is accomplished for extracting the main words describing a category on the top Web pages (they retrieved by using a clean version of the labels in the corresponding category as query). Due the heterogeneous style of Web pages, locating then extracting the words that characterize the category at most is a very difficult task. Using quantities as mutual information or log-likelihood can help to determine the most discriminant words. The Web page zones that contain the largest number (density) of discriminant words are likely to be technical parts corresponding to the names of the properties and the values one looks for and that will have to be extracted. For the first part of this work, i.e. automatic query formulation from labels, we proposed a two step approach first grouping similar words according to an adapted Levenshtein edit distance [113] and then selecting candidates by means of distributional statistics on the sequence of letters in the words. First evaluation shows that the precision of information retrieval is better by employing the queries automatically generated than using labels as queries (for the previous example label, the two misspelled words are corrected). In addition, the generated queries are more effective than the ones one could obtain by employing the effective but generic approach offered by Google Suggest service.

1.7 Conclusion

We have presented some popular numerical models for information retrieval or extraction, several challenges organized in conferences yearly and three different applications related to information extraction. As discussed before, one difficulty is having enough data and sufficient information for employing machine-learning approaches. Despite the recent advances and good practical results, improvements remain to be achieved. Some approaches have been proposed so far for adapting information retrieval models to data (e.g. book retrieval vs. article retrieval) but have shown only small improvements, not always significant. For book retrieval, we know that some extra-linguistic features are specific to the books: the organization of the text and the logical structure of chapters and sections, the physical structure and paging, the types (novels, short stories, scientific books…), their length and the temporal evolution of themes covered, the presence of recurring characters and dialogue (possibly interviews), how the books relate to each other are all aspects that are studied for critical analysis and that are not taken into account by information retrieval systems. We believe that this is all part of which current models of information processing should be studied.

The robustness of processing has also to be improved for being able to consider documents with low-level language quality or language specific syntax and lexicon. This may include blog and micro-blog entries containing opinion expressed by users, audio transcriptions and OCR-processed (historical documents, for example). Improving robustness of information retrieval systems by providing efficient processing of noisy data is likely to be better than trying to correct or normalize the texts.

On another level, usual information retrieval models rank documents according to how much information they convey in respect to a query a user has typed, while taking into account, in the best case, the quantity of new information obtained. This is a purely informational definition of relevance based on the hypothesis that the greater the amount of new information, the more the document is likely to interest the user. This is true to a certain extent, but ignores the fact that the needs are different depending on the level of expertise of the user; a novice will certainly be more interested in an overview document than in a comprehensive study where the vocabulary and structure are more complex. This is even truer for people with reading difficulties such as dyslexics. Then, we will have to define new measures taking into account this aspect while providing the opportunity to present the most relevant AND the most *simple* documents first (the most *readable*). Belkin [7] cites Karen Sparck-Jones during the 1988 ACM SIGIR conference [102]:

> As it is, it is impossible not to feel that continuing research on probabilistic weighting in the style in which it has been conducted, however good in itself in aims and conduct, is just bombinating in the void…
> The current interest […] is in integrated, personalisable information management systems.

The issue of customization and the inclusion of the user in finding information, naturally refers to the well broadest bases of language processing at the intersection of linguistics and computer science, both joined by psychology to the study of individual behavior, neuroscience to study brain and physiological roots of language but also in sociology and semiotics for global analysis of needs, attitudes and meanings. A such cross-discipline approach is a major challenge for years to come if we want to go beyond, to quote K. Sparck-Jones, the only hope (and it still does just a hope without being convinced of the significance of earnings) of picking up a few points of precision for searching.

The latest advances in technology and medical imaging provide plausible models of our cognitive functions that can inspire us to simulate the human in such areas as language and thought. We think we have to pay attention simultaneously to work from neurosciences and linguistics, although it must be aware that the transfer from one discipline to another can be realized in the long term and that it is discussed for a long time. The way to combine these different pieces of information is a major issue, not to illuminate the human capacity (it is certainly not the purpose of computer science), but to develop software able to meet the many challenges of the Information Society.

References

1. Aljaber, B., Stokes, N., Bailey, J., Pei, J.: Document clustering of scientific texts using citation contexts. Inf. Retrieval **13**, 101–131 (2009). (Kluwer Academic Pub.)
2. Almuhareb, A., Poesio, M.: Attribute-based and value-based clustering: an evaluation. In: Proceedings of EMNLP, pp. 158–165 (2004)
3. Baccianella, S., Esuli, A., Sebastiani, F.: Sentiwordnet 3.0: an enhanced lexical resource for sentiment analysis and opinion mining. In: Proceedings of the 7th Conference on International Language Resources and Evaluation (LREC'10), Valletta, Malta (May, 2010)
4. Balog, K., Serdyukov, P., Vries, A.P.D.: Overview of the TREC 2010 entity track. DTIC document, (2010)
5. Béchet, F., Charton, E.: Unsupervised knowledge acquisition for extracting named entities from speech. IEEE International Conference on Acoustics Speech and Signal Processing (ICASSP 2010), pp. 5338–5341 (2010)
6. Béchet, F., Raymond, C., Duvert, F., de Mori, R.: Frame based interpretation of conversational speech. Spoken Language Technology Workshop (SLT), 2010 IEEE, pp. 401–406 (2010)
7. Belkin, N.J.: Some (what) grand challenges for information retrieval. SIGIR Forum **42**, 47–54 (2008)
8. Bellot, P., Chappell, T., Doucet, A., Geva, S., Gurajada, S., Kamps, J., Kazai, G., Koolen, M., Landoni, M., Marx, M., Mishra, A., Moriceau, V., Mothe, J., Preminger, M., Ramírez, G., Sanderson, M., Sanjuan, E., Scholer, F., Schuh, A., Tannier, X., Theobald, M., Trappett, M., Trotman, A., Wang, Q.: Report on INEX 2012. SIGIR Forum **46**, 50–59 (2012)
9. Bellot, P., Crestan, E., El-bèze, M., Gillard, L., de Loupy, C.: Coupling named entity recognition, vector-space model and knowledge bases for TREC-11 question-answering track. In: Proceedings of the Twelfth Text Retrieval Conference (TREC 2003), NIST Special publication, pp. 500–251 (2003)
10. Berland, M., Charniak, E.: Finding parts in very large corpora. In: Proceedings of the 37th Annual Meeting of the Association for Computational Linguistics on Computational Linguistics. Association for Computational Linguistics, pp. 57–64 (1999)
11. Bonneau-maynard, H., Rosset, S., Ayache, C., Kuhn, A., Mostefa, D.: Semantic annotation of the French media dialog corpus. In: Proceedings of Ninth European Conference on Speech Communication and Technology, Lisboa, Portugal (2005)
12. Bonnefoy, L., Bellot, P., Benoit, M.: The Web as a source of evidence for filtering candidate answers to natural language questions. In: IEEE/WIC/ACM International Conference on Web Intelligence and Intelligent Agent Technology (WI-IAT), pp. 63–66 (2011)
13. Bonnefoy, L., Bouvier, V., Bellot, P.: LSIS/LIA at TREC 2012 knowledge base acceleration. In: Proceedings of the Twenty-First Text REtrieval Conference (TREC 2012), pp. 500–298. NIST Special Publication SP, Gaithersburg, USA (2013)
14. Bordogna, G., Pasi, G.: A fuzzy linguistic approach generalizing Boolean information retrieval: a model and its evaluation. JASIS **44**, 70–82 (1993)
15. Brocki, Ł., Koržinek, D., Marasek, K.: Telephony based voice portal for a University. Appl. Syst. Homel. Secur. (2008)
16. Bunescu, R., Mooney, R.: Subsequence kernels for relation extraction. Adv. Neural Inf. Process. Syst. **18**, 171 (2006)
17. Burger, J.D.: Mitre's quanda at trec-12. In: Proceedings of the Twenty-First Text REtrieval Conference (TREC 2012), pp. 500–298. NIST Special Publication SP, Gaithersburg, USA (2003)
18. Camelin, N., Bechet, F., Damnati, G., de Mori, R.: Detection and interpretation of opinion expressions in spoken surveys. IEEE Trans. Audio Speech Lang. Process. **18**, 369–381 (2010)

19. Carreras, X., Marquez, L., Padró, L.: Named entity extraction using AdaBoost. In: Proceedings of the 6th Conference on Natural Language Learning-Volume 20, pp. 1–4. Association for Computational Linguistics (2002)
20. Cassidy, T., Zheng, C., Artiles, J., Ji, H., Deng, H., Ratinov, L.-A., Zheng, J., Han, J., Roth, D.: CUNY-UIUC-SRI TAC-KBP2011 entity linking system description. In: Proceedings of Text Analysis Conference (TAC2011), (2010)
21. Chang, H.C.: A new perspective on twitter hashtag use: diffusion of innovation theory. Proc. Am. Soc. Inform. Sci. Technol. **47**, 1–4 (2010)
22. Chomsky, N.: Current issues in linguistic theory. In: Fodor, J., Katz, B. (eds.) The Structure of Language. Prentice Hall, New York (1964)
23. Chomsky, N.: Lectures in Government and Binding. Foris Publications, Dordrecht (1981)
24. Ciravegna, D.: Adaptive information extraction from text by rule induction and generalisation. In: Proceedings 17th International Joint Conference on Artificial Intelligence (IJCAI 2001), Seattle (2001)
25. Collins, M., Singer, Y. Unsupervised models for named entity classification. In: Proceedings of the Joint SIGDAT Conference on Empirical Methods in Natural Language Processing and Very Large Corpora, pp. 189–196 (1999)
26. Cowie, J., Lehnert, W.: Information extraction. Commun. ACM **39**, 80–91 (1996)
27. Culotta, A., Sorensen, J.: Dependency tree kernels for relation extraction. In: Proceedings of the 42nd Annual Meeting on Association for Computational Linguistics, p. 423. Association for Computational Linguistics (2004)
28. Cutler, A., Fodor, J.A.: Semantic focus and sentence comprehension. Cognition **7**, 49–59 (1979)
29. Dang, H.T., Owczarzak, K.: Overview of the TAC 2008 opinion question answering and summarization tasks. In: Proceedings of the First Text Analysis Conference, (2008)
30. Davidov, D., Rappoport, A.: Extraction and approximation of numerical attributes from the Web. In: Proceedings of the 48th Annual Meeting of the Association for Computational Linguistics, pp. 1308–1317. Association for Computational Linguistics (2010)
31. Deerwester, S.C., Dumais, S., Landauer, T.K., Furnas, G.W., Harshman, R.A.: Indexing by latent semantic analysis. J. Am. Soc. Inf. Sci. **41**, 391–407 (1990)
32. Deveaud, R., Avignon, F., Sanjuan, E., Bellot, P.: LIA at TREC 2011 Web track: experiments on the combination of online resources. In: Proceedings of the Twentieth Text REtrieval Conference (TREC 2011), pp. 500–596. NIST Special Publication SP, Gaithersburg, USA (2011)
33. Doddington, G., Mitchell, A., Przybocki, M., Ramshaw, L., Strassel, S., Weischedel, R.: The automatic content extraction (ACE) program-tasks, data, and evaluation. In: Proceedings of LREC, pp. 837–840. Citeseer (2004)
34. Downey, D., Broadhead, M., Etzioni, O.: Locating complex named entities in web text. In: Proceedings of the 20th International Joint Conference on Artificial Intelligence, pp. 2733–2739 (2007)
35. Duvert, F., de Mori, R.: Conditional models for detecting lambda-functions in a spoken language understanding system. In: Eleventh Annual Conference of the International Speech Communication Association, (2010)
36. Duvert, F., Meurs, M.-J., Servan, C., Béchet, F., Lefevre, F., de Mori, R.: Semantic composition process in a speech understanding system. In: Proceedings of IEEE International Conference on Acoustics, Speech and Signal Processing, ICASSP 2008, pp. 5029–5032 (2008)
37. Etzioni, O., Banko, M., Soderland, S., Weld, D.S.: Open information extraction from the web. Commun. ACM **51**, 68–74 (2008)
38. Etzioni, O., Cafarella, M., Downey, D., Popescu, A.-M., Shaked, T., Soderland, S., Weld, D.S., Yates, A.: Unsupervised named-entity extraction from the web: an experimental study. Artif. Intell. **165**, 91–134 (2005)

39. Etzioni, O., Fader, A., Christensen, J., Soderland, S., Mausam, M.: Open information extraction: the second generation. In: Proceedings of the Twenty-Second International Joint Conference on Artificial Intelligence-Volume, vol. 1, pp. 3–10. AAAI Press (2011)
40. Fader, A., Soderland, S, Etzioni, O.: Identifying relations for open information extraction. In: Proceedings of the Conference on Empirical Methods in Natural Language Processing, pp. 1535–1545. Association for Computational Linguistics (2011)
41. Ferret, O., Grau, B., Hurault-plantet, M., Illouz, G., Monceaux, L., Robba, I., Vilnat, A.: Finding an answer based on the recognition of the question focus. In: Proceedings of the Tenth Text REtrieval Conference (TREC 2001), 2002 Gaithersburg, Maryland, USA (2002)
42. Fuhr, N., Buckley, C.: A probabilistic learning approach for document indexing. ACM Trans. Inf. Syst. (TOIS) **9**, 223–248 (1991)
43. Garfield, E.: Citation analysis as a tool in journal evaluation. Science **178**, 471–479 (1972)
44. Ge, N., Hale, J., Charniak, E.: A statistical approach to anaphora resolution. In: Proceedings of the Sixth Workshop on Very Large Corpora, pp. 161–170 (1998)
45. Giles, C.L., Bollacker, K., Lawrence, S.: CiteSeer: an automatic citation indexing system. In: Proceedings of the Third ACM Conference on Digital Libraries, pp. 89–98. ACM, Pittsburgh, Pennsylvania, USA (1998)
46. Griol, D., Riccardi, G., Sanchis, E.: A statistical dialog manager for the LUNA project. In: Proceedings of interspeech/ICSLP, pp. 272–275 (2009)
47. Grishman, R., Sundheim, B.: Message understanding conference-6: a brief history. In: Proceedings of COLING, pp. 466–471 (1996)
48. Grodzinsky, Y.: La syntaxe générative dans le cerveau. In: Bricmont, J., Franck, J. (eds.) Chomsky (Les Cahiers de l'Herne). Editions de l'Herne, Paris (2007)
49. Guarino, N.: Concepts, attributes and arbitrary relations: some linguistic and ontological criteria for structuring knowledge bases. Data Knowl. Eng. **8**, 249–261 (1992)
50. Hamdan, H., Béchet, F., Bellot, P.: Experiments with DBpedia, WordNet and SentiWordNet as re-sources for sentiment analysis in micro-blogging. In: International Workshop on Semantic Evaluation SemEval-2013 (NAACL Workshop), Atlanta, Georgia, USA (2013)
51. Harth, E.: The Creative Loop: How the Brain Makes a Mind. Addison-Wesley, New-York (1993)
52. Isozaki, H., Kazawa, H.: Efficient support vector classifiers for named entity recognition. In: Proceedings of the 19th International Conference on Computational Linguistics, pp. 1–7. Association for Computational Linguistics (2002)
53. Ji, H., Grishman, R.: Knowledge base population: Successful approaches and challenges. In: Proceedings of the 49th Annual Meeting of the Association for Computational Linguistics: Human Language Technologies, pp. 1148–1158 (2011)
54. Kantrowitz, M., Mohit, B., Mittal, V.: Stemming and its effects on TFIDF ranking (poster session). In: Proceedings of the 23rd Annual International ACM SIGIR Conference on Research and Development in Information Retrieval, pp. 357–359. ACM Press (2000)
55. Kim, J.-H., Woodland, P.: A rule-based named entity recognition system for speech input. In: Proceedings of the 6th International Conference on Spoken Language Processing, (2000)
56. Kim, Y.-M., Bellot, P., Tavernier, J., Faath, E., Dacos, M.: Evaluation of BILBO reference parsing in digital humanities via a comparison of different tools. In: Proceedings of the 2012 ACM Symposium on Document Engineering, pp. 209–212. ACM Press, Paris, France (2012)
57. Krogh, A. Hidden Markov models for labeled sequences. In: Proceedings of the IEEE 12th IAPR International. Conference on Pattern Recognition, Vol. 2-Conference B: Computer Vision and Image Processing, pp. 140–144 (1994)
58. Lafferty, J., Mccallum, A., Pereira, F.C.: Conditional random fields: Probabilistic models for segmenting and labeling sequence data. In: Proceedings of the 18th International Conference on Machine Learning 2001 (ICML 2001), pp. 282–289 (2001)
59. Langley, P., Simon, H.A.: Applications of machine learning and rule induction. Commun. ACM **38**, 54–64 (1995)

60. Lehnert, W.: The Process of Question Answering: A Computer Simulation of Cognition. Lawrence Erlbaum Associates, Hillsdale (1978)
61. Li, F., Zheng, Z., Yang, T., Bu, F., Ge, R., Zhu, X., Zhang, X., Huang, M.: Thu quanta at TAC 2008 qa and rte track. In: Proceedings of Human Language Technologies Conference/ Conference on Empirical Methods in Natural Language Processing (HLT/EMNLP), Vancouver, BC, Canada (2008)
62. Lin, J.: An exploration of the principles underlying redundancy-based factoid question answering. ACM Trans. Inf. Syst. **25**, 4–53 (2007)
63. Màrquez, L., Carreras, X., Litkowski, K.C., Stevenson, S.: Semantic role labeling: an introduction to the special issue. Comput. Linguis. **34**, 145–159 (2008)
64. Maybury, M.T.: New Directions in Question Answering. The MIT Press, Menlo Park (2004)
65. McCallum, A.: Information extraction: distilling structured data from unstructured text. Queue **3**, 48–57 (2005)
66. Mehler, J., Dupoux, E.: Naître Humain. Odile Jacob, Paris (1992)
67. Metzler, D., Croft, W.B.: A Markov random field model for term dependencies. In: Proceedings of the 28th Annual International ACM SIGIR Conference on Research and Development in Information Retrieval, pp. 472–479. ACM Press (2005)
68. Mintz, M., Bills, S., Snow, R., Jurafsky, D.: Distant supervision for relation extraction without labeled data. In: Proceedings of the Joint Conference of the 47th Annual Meeting of the ACL and the 4th International Joint Conference on Natural Language Processing of the AFNLP, pp. 1003–1011, Association for Computational Linguistics (2009)
69. Mitkov, R.: Anaphora Resolution. Pearson Education ESL, Boston (2002)
70. Moldovan, D., Harabagiu, S., Pasca, M., Mihalcea, R., Girju, R., Goodrum, R., Rus, V.: The structure and performance of an open-domain question answering system. In: Proceedings of the 38th Annual Meeting on Association for Computational Linguistics, pp. 563–570. Association for Computational Linguistics (2000)
71. Muslea, I.: Extraction patterns for information extraction tasks: a survey. The AAAI-99 workshop on machine learning for information extraction, 1999
72. Nadeau, D., Sekine, S.: A survey of named entity recognition and classification. Lingvisticae Investigationes **30**, 3–26 (2007)
73. Palmer, M., Gildea, D., Xue, N.: Semantic Role Labeling. Morgan & Claypool, Waterloo (2010)
74. PASCA, M.: Weakly-supervised discovery of named entities using web search queries. In: Proceedings of the Sixteenth ACM Conference on Information and Knowledge Management, ACM press, Lisbon, Portugal (2007)
75. Peng, F., McCallum, A.: Information extraction from research papers using conditional random fields. Inf. Process. Manage. **42**, 963–979 (2006)
76. Poesio, M., Almuhareb, A.: Extracting concept descriptions from the Web: the importance of attributes and values. In: Proceedings of the Conference on Ontology Learning and Population: Bridging the Gap between Text and Knowledge, pp. 29–44. Citeseer (2008)
77. Ponte, J.M., Croft, W.B. A language modeling approach to information retrieval. In: Proceedings of the 21st Annual International ACM SIGIR Conference on Research and Development in Information Retrieval, pp. 275–281. ACM Press, Melbourne, Australia (1998)
78. Qazvinian, V., Radev, D.R.: Scientific paper summarization using citation summary networks. In: Proceedings of the 22nd International Conference on Computational Linguistics, vol. 1, pp. 689–696. Association for Computational Linguistics (2008)
79. Quarteroni, S., Riccardi, G., Dinarelli, M.: What's in an ontology for spoken language understanding. In: Proceedings of Interspeech, pp. 1023–1026 (2009)
80. Quintard, L., Galibert, O., Adda, G., Grau, B., Laurent, D., Moriceau, V., Rosset, S., Tannier, X., Vilnat, A.: Question answering on web data: the qa evaluation in quæro. In: Proceedings of the Seventh Conference on International Language Resources and Evaluation (LREC'10), Valletta, Malta (2010)

81. Rabiner, L., Juang, B.: An introduction to hidden Markov models. IEEE ASSP Mag. **3**, 4–16 (1986)
82. Rabiner, L.R.: A tutorial on hidden Markov models and selected applications in speech recognition. Proc. IEEE **77**, 257–286 (1989)
83. Raju, S., Pingali, P., Varma, V.: An Unsupervised Approach to Product Attribute Extraction. Springer, Berlin Heidelberg (2009). (Advances in Information Retrieval)
84. Ramage, D., Hall, D., Nallapati, R., Manning, C.D.: Labeled LDA: a supervised topic model for credit attribution in multi-labeled corpora. In: Proceedings of the 2009 Conference on Empirical Methods in Natural Language Processing, vol. 1, pp. 248–256. Association for Computational Linguistics (2009)
85. Ramakrishnan, G., Chakrabarti, S., Paranjpe, D., Bhattacharya, P.: Is question answering an acquired skill? In: Proceedings of the 13th International Conference on World Wide Web, ACM Press, New York, NY, USA (2004)
86. Ritchie, A., Robertson, S., Teufel, S.: Comparing citation contexts for information retrieval. In: Proceedings of the 17th ACM Conference on Information and Knowledge Management, pp. 213–222. ACM Press (2008)
87. Ritter, A., Clark, S., Etzioni, O.: Named entity recognition in tweets: an experimental study. In: Proceedings of the Conference on Empirical Methods in Natural Language Processing (EMNLP), pp. 1524–1534. Association for Computational Linguistics (2011)
88. Rizzi, L.: L'acquisition de la langue et la faculté de langage. In: Bricmont, J., Franck, J. (eds.) Chomsky (Les Cahiers de l'Herne). Editions de l'Herne, Paris (2007)
89. Robertson, S., Zaragoza, H., Taylor, M.: Simple BM25 extension to multiple weighted fields. In: Proceedings of the Thirteenth ACM International Conference on INFORMATION and Knowledge Management %@ 1-58113-874-1, pp. 42-49. ACM Press, Washington, DC, USA (2004)
90. Robertson, S.E.: The probability ranking principle in IR. J. Doc. **33**, 294–304 (1977)
91. Salton, G., Fox, E., Wu, H.: Extended Boolean information retrieval. Commun. ACM **31**, 1002–1036 (1983)
92. Salton, G., Wong, A., Yang, C.-S.: A vector space model for automatic indexing. Commun. ACM **18**, 613–620 (1975)
93. Sánchez, D.: A methodology to learn ontological attributes from the Web. Data Knowl. Eng. **69**, 573–597 (2010)
94. Sanjuan, E., Bellot, P., Moriceau, V., Tannier, X.: Overview of the INEX 2010 question answering track (QA@INEX). In: Proceedings of the 9th International Conference on Initiative for the Evaluation of XML Retrieval: Comparative Evaluation of Focused Retrieval, Springer, Vught, The Netherland (2011)
95. Sanjuan, E., Moriceau, V., Tannier, X., Bellot, P., Mothe, J.: Overview of the INEX 2012 tweet contextualization track. Initiative for XML Retrieval INEX 2012, p. 148. Roma, Italia (2012)
96. Sarawagi, S.: Information extraction. Foundations and trends in databases **1**, 261–377 (2008)
97. Savoy, J., Le Calvé, A., Vrajitoru, D.: Report on the TREC-5 experiment: data fusion and collection fusion. In: Proceedings of the Fifth Text REtrieval Conference (TREC-5), pp. 500–538, 489–502. NIST Special Publication (1997)
98. Seymore, K., Mccallum, A., Rosenfeld, R.: Learning hidden Markov model structure for information extraction. AAAI-99 Workshop on Machine Learning for Information Extraction, pp. 37–42 (1999)
99. Singhal, A., Buckley, C., Mitra, M.: Pivoted document length normalization. In: Proceedings of the 19th Annual International ACM SIGIR Conference on Research and Development in Information Retrieval, pp. 21–29. ACM Press (1996)
100. Solomon, M., Yu, C., Gravano, L.: Popularity-guided top-k extraction of entity attributes. In: Proceedings of the 13th International Workshop on the Web and Databases (WebDB), p. 9. ACM Press, Indianapolis, IN, USA (2010)

101. Sparck-Jones, K.: A statistical interpretation of term specificity and its application in retrieval. J. Doc. **28**, 11–21 (1972)
102. Sparck-jones, K.: A look back and a look forward. In: Proceedings of the 11th Annual International ACM SIGIR Conference on Research and Development in Information Retrieval, pp. 13–29. ACM Press, Grenoble, France
103. Stokoe, C., Oakes, M.P., Tait, J.: Word sense disambiguation in information retrieval revisited. In: Proceedings of the 26th Annual International ACM SIGIR Conference on Research and Development in Information Retrieval, pp. 159–166. ACM Press (2003)
104. Varma, V., Pingali, P., Katragadda, S., Krishna, R., Ganesh, S., Sarvabhotla, K.H.G., Gopisetty, H., Reddy, K., Bharadwaj, R.: IIIT hyderabad at TAC 2009. In: Proceedings of Test Analysis Conference 2008 (TAC 2008), NIST, Gaithersburg, USA (2008)
105. Voorhees, E.M.: Overview of the TREC 2001 question answering track. In: Proceedings of the Tenth Text Retrieval Conference (TREC 2001), pp. 500–551, 42–50. NIST Special Publication (2001)
106. Voorhees, E.M.: Question answering in TREC. In: Voorhees, E.M., Harman, D.K. (eds.) TREC—Experiment and Evaluation in Information Retrieval. The MIT Press, Cambridge (2005)
107. Voorhees, E.M., Harman, D.K.: Overview of the eighth text retrieval conference (TREC-8). In: Proceedings of the Eighth Text REtrieval Conference (TREC 8), pp. 500–546, 1–24. NIST Special Publication (1999)
108. Voorhees, E.M., Harman, D.K.: TREC—Experiment and Evaluation in Information Retrieval. The MIT Press, Cambridge (2005)
109. Weerkamp, W., Carter, S., Tsagkias, M.: How people use twitter in different languages. ACM Web Science 2011, 2011, p. 2. Koblenz, Germany (2011)
110. Whitelaw, C., Kehlenbeck, A., Petrovic, N., Ungar, L.: Web-scale named entity recognition. In: Proceedings of the 17th ACM Conference on Information and Knowledge Management (CIKM 2008), pp. 123–132. ACM Press, Napa Valley, California, USA (2008)
111. Xu, J., Croft, W.B.: Query expansion using local and global document analysis. In: ACM-SIGIR Conference on Research and Development in Information Retrieval, pp. 4–11. ACM Press, Zurich, Suisse (1996)
112. Yao, C., Yu, Y., Shou, S., Li, X.: Towards a global schema for web entities. In: Proceedings of the 17th international Conference on World Wide Web, pp. 999–1008. ACM Press (2008)
113. Yujian, L., Bo, L.: A normalized Levenshtein distance metric. IEEE Trans. Pattern Anal. Mach. Intell. **29**, 1091–1095 (2007)
114. Zhao, Y., Qin, B., Hu, S., Liu, T.: Generalizing syntactic structures for product attribute candidate extraction. In: Proceedings of Human Language Technologies: The 2010 Annual Conference of the North American Chapter of the Association for Computational Linguistics, pp. 377–380. Association for Computational Linguistics (2010)
115. Zhu, J., Nie, Z., Wen, J.-R., Zhang, B., Ma, W.-Y.: Simultaneous record detection and attribute labeling in web data extraction. In: Proceedings of the 12th ACM SIGKDD international Conference on Knowledge Discovery and Data Mining, pp. 494–503. ACM Press (2006)

Chapter 2
Interestingness Measures for Multi-Level Association Rules

Gavin Shaw, Yue Xu and Shlomo Geva

Abstract Association rule mining is one technique that is widely used when querying databases, especially those that are transactional, in order to obtain useful associations or correlations among sets of items. Much work has been done focusing on efficiency, effectiveness and redundancy. There has also been a focusing on the quality of rules from single level datasets with many interestingness measures proposed. However, with multi-level datasets now being common there is a lack of interestingness measures developed for multi-level and cross-level rules. Single level measures do not take into account the hierarchy found in a multi-level dataset. This leaves the Support-Confidence approach, which does not consider the hierarchy anyway and has other drawbacks, as one of the few measures available. In this chapter we propose two approaches which measure multi-level association rules to help evaluate their interestingness by considering the database's underlying taxonomy. These measures of diversity and peculiarity can be used to help identify those rules from multi-level datasets that are potentially useful.

G. Shaw · Y. Xu (✉) · S. Geva
School of Electronic Engineering and Computer Science,
Queensland University of Technology, Queensland, Australia
e-mail: yue.xu@qut.edu.au

G. Shaw
e-mail: gavin.shaw@qut.edu.au

S. Geva
e-mail: s.geva@qut.edu.au

C. Faucher and L. C. Jain (eds.), *Innovations in Intelligent Machines-4*,
Studies in Computational Intelligence 514, DOI: 10.1007/978-3-319-01866-9_2,

2.1 Introduction

Association rule mining was first introduced in [1] and since then has become both an important and widespread tool in use. It allows associations between a set of items in large datasets to be discovered and often a huge amount of associations are found. Thus in order for a user to be able to handle the discovered rules it is necessary to be able to screen/measure the rules so that only those that are interesting are presented to the user. This is the role interestingness measures play. In an effort to help discover the interesting rules, work has focused on measuring rules in various ways from both objective and subjective points of view [2, 3]. The most common measure is the support confidence approach [1, 4, 5], but there are numerous other measures [2, 4, 5] to name a few. All of these measures were proposed for association rules derived from single level or flat datasets, which were most commonly transactional datasets and traditionally were the most common.

Today multi-level datasets are more common in many domains. With this increase in usage there is a big demand for techniques to discover multi-level and cross-level association rules and also techniques to measure interestingness of rules derived from multi-level datasets. Some approaches for multi-level and cross-level frequent itemset discovery (the first step in rule mining) have been proposed [6–8]. However, multi-level datasets are often a source of numerous rules and in fact the rules can be so numerous it can be much more difficult to determine which ones are interesting [1, 4]. Moreover, the existing interestingness measures for single level association rules cannot accurately measure the interestingness of multi-level rules since they do not take into consideration the concept of the hierarchical taxonomic structure that exists in multi-level datasets.

In this chapter, as our contribution, we propose measures particularly for assessing the interestingness of multi-level association rules by examining the diversity and distance among rules. Our proposed measures are objective based and use the taxonomy that underlies multi-level datasets to determine the relationships between the items within a rule in order to measure the diversity between those items. The distance between two rules is also determined by considering the taxonomy and relationships between items across the two rules. These measures can be determined during rule discovery phase for use during post-processing to help users determine the interesting rules. To the authors' best knowledge, this chapter is the first attempt to investigate the interestingness measures focused on multi-level datasets. We believe that this chapter is also the first to propose an interestingness measure for association rules that takes into consideration the taxonomic structure and information available. Our experiments show that the proposed measures can identify association rules that are interesting (based on the proposed measures) that support and confidence do not discover showing the potential of being able to make new knowledge available to a user. We can also show a use for the proposed diversity measure in improving recommendation system performance for users in terms of making better recommendations.

The chapter is organised as follows. Section 2.2 discusses related works on interestingness measures. Section 2.3 briefly looks at the area of multi-level association rule mining and related areas when it comes to handling rule sets. Our proposed interestingness measures, along with assumptions and background are presented in Sect. 2.4. Experiments and their results are presented in Sect. 2.5. Finally, Sect. 2.6 concludes the chapter.

2.2 Related Work

In this section, we give an indepth review of interestingness measures in the area of multi-level association rules mining. This sets the groundwork for the work we present here and shows where our work fits in.

For as long as association rule mining has been around, there has been a need to determine which rules are interesting. Originally an interesting association rule was determined through the concepts of support and confidence [1]. The Support-Confidence approach is appealing due to the antimonotonicity property of the support. However, the support component will ignore itemsets with a low support, even if those itemsets may generate rules with a high confidence level [5]. Also, the Support-Confidence approach does not necessarily ensure that the rules are truly interesting, especially when the confidence is equal to the marginal frequency of the consequent [5]. Furthermore, examples in [9] show that high confidence does not always equate to or equal high correlation or even positive correlation of the items in the rule. Work in [10, 11] also criticised the usage of the Support-Confidence approach. Based on this, other measures for determining the interestingness of an association rule are needed and have since been proposed [2, 4, 5].

Broadly speaking, all existing interestingness measures fall into one or more of the three following categories; objective based measures (based on the raw data), subjective based (based on the raw data and the user) and semantic based measures (considers the semantics and explanations of the patterns) [2]. Work in [2] listed nine criteria that can be used to determine if an association rule is interesting. For objective based measures the following five criteria are applicable; conciseness, coverage, reliability, peculiarity and diversity. For subjective based measures, the following two criteria are applicable: novelty and surprisingness. Finally, for semantic based measures the following two criteria are applicable: utility and actionability or applicability. Currently, there is no widely agreed upon definition of what interestingness is in the context of patterns and association rules [2]. Several surveys and studies have been undertaken focusing on measures[2, 3, 5, 12]. The survey in [3] evaluated the strengths and weaknesses of measures from the point of view of the level or extent of user interaction. The study in [12] attempted to classify measures into formal and experimental classes, along with evaluation properties. However, surveys like those in [2, 3, 5, 12] result in different outcomes over how useful or suitable etc., an interestingness measure is.

Furthermore, all of the previous works mentioned have focused purely on measuring for association rules derived from single level (flat) datasets. While they can be used to measure association rules that come from a multi-level dataset (rules from different concept levels or rules that contain items from different concept levels), they will treat the rules and/or items within as though they came from a single level dataset. Thus, the taxonomy structure behind a multi-level dataset is lost or ignored. Our research has found that little work has been done when it comes to interestingness measures for association rules derived from multi-level datasets, especially, when it comes to measures that use the dataset's taxonomical structure.

In this chapter, we will propose two interestingness measures for association rules from multi-level datasets. The first is a diversity-based measure and the second is a peculiarity (or distance) based measure. These measures will be objective, relying on only the data. The concepts of diversity and peculiarity are discussed in further detail in the following two sections.

2.2.1 Diversity-Based Measures

Diversity is one of the nine criteria listed in [2] that can be used to measure the interestingness of an association rule. Here it was argued that a pattern would be considered to be diverse if the elements making up that pattern differed significantly from one another. It is further claimed that a summary could be considered diverse if the probability distribution of that summary is far from the uniform distribution [2]. Also, it was claimed in [2] that diversity for summaries, while being difficult to define, was accepted to be determined by the following two factors: proportional distribution of classes in the population and the number of classes. The survey then went on to list 19 different measures for determining the diversity of summaries [2]. In concluding, the survey in [2] noted that at the time they were unaware of any existing work on applying or using diversity measures to determine the interestingness of association rules.

More recent work has been undertaken proposing diversity based interestingness measures for association rules similar to those proposed for summaries. Work in [13] proposed using variance and Claude Shannon Entropy (due to widespread use) to determine the diversity of a set of association rules (not the diversity of an individual rule). Here, in this chapter, we aim to show a diversity measure for determining the diversity of individual rules, as opposed to rule sets. Similarly, work presented in [14] considered using diversity to measure association rules in the application of spatial gene expression. Like the work in [13], it was proposed that variance and entropy could be used to measure the diversity of a set of association rules.

Despite the approaches to diversity presented in [13, 14], we believe that there is still a need for a different diversity measure. As stated in both [2] and [13] and here previously, diversity measures the difference between items. Here, in this chapter, we propose a measure to do that using the taxonomic information available in a multi-level dataset. By utilizing the taxonomic information we can determine the relationship between items and the strength of that relationship. Through this, we can then determine how much two or more items differ by and hence (as per [2] and [13] definition) measure their diversity. This approach will allow for the measuring of the diversity of individual rules, based on the difference between the items that make up that rule.

2.2.2 Peculiarity-Based Measures

Peculiarity is another one of the nine criteria listed in [2] that can be used to objectively measure the interestingness of association rules. Peculiarity was said to represent how far away a rule is/was from other rules based on a distance measure [2]. Hence, a rule with high peculiarity is far away from the rest of the rule set and most likely was generated from outlying data points. Usually rules with high peculiarity are small in number and may be unknown to a user. It is this property that makes them potentially interesting to a user.

One proposed approach to measuring the peculiarity of a rule was presented in [4] and was called *neighbourhood-based unexpectedness*. The basic premise was that if a rule had a different consequent to the neighbouring rules near it, then it was peculiar and thus interesting. The approach presented in [4] measured the peculiarity distance by determining the symmetric difference between rules. Using this, a neighbourhood around a given rule can be formed and then peculiar rules can be identified through a measure known as unexpected confidence (where the confidence of one rule is far from the average confidence of the rules in the neighbourhood/cluster) or through the sparsity of the neighbourhood (where the number of rules in a neighbourhood is significantly less than the number of potential rules for that neighbourhood) [4]. This approach works on the assumption that there is no relationship between items, which is not always true in a multi-level dataset, as, through the taxonomic structure, items can be related to each other and are not completely unique.

Early work in [15] looked at identifying outlying data in a database through a distance based approach. It is argued that these points are often treated as noise when in fact they may be useful and/or valuable to another user, or certain situations or applications may prefer these types of peculiar rules. The authors go on to propose several algorithms for locating and identifying outliers in a dataset. The work in [15] attempts to show why and how outlying data can be important and meaningful semantically. It also shows that detecting outliers is an important task.

Further work in [16] proposed that 'peculiarity rule' are discovered from peculiar data and data is considered to be peculiar if it represents a case that is described by a small number of objects (or transactions) in a dataset and these objects/transactions are very different from the other objects/transactions. The work in [16] also argues that 'peculiarity rule' are often hidden in a large amount of scientific, statistical and transaction based datasets. The approach proposed first finds the peculiar data in the dataset and then derives rules from it (with a check for relevance). Initial experiments showed the potential for the discovery of unexpected, but interesting rules. Thus the use of a peculiarity measure can allow a user to discover rules that may normally be missed. Work in [17] continued this into the area of multi-database mining: finding peculiar data from across multiple datasets/databases and the finding relevant 'peculiarity rule'. The author in [17] also stated that 'peculiarity rule' has a low support value. This would also show why they are often missed.

Related works in [18, 19] looked at discovering anomalous rules (ones hidden by dominant rules) and outlier detection algorithms. While neither works focused specifically on peculiarity they both show the need for approaches that find association rules that the classical approaches do not. The work in [19], focusing on outlier detection: through which peculiar rules may be discovered as they can come from outlying data, looks at several algorithms that are designed to detect outlying data quickly and efficiently. It was concluded that each method had its own strengths and weaknesses for discovering outliers from different types of datasets.

Recent work in [20] proposed a distance-based Local Peculiarity Factor (D-record LPF) to discover peculiar data and rules. The D-record LPF is a summation of the distances from a point to its nearest neighbours. The work proves mathematically that the proposed approach can accurately characterize the probability density function in a continuous m-dimensional distribution [20]. This helps to prove some of the existing peculiarity detection techniques that are distance based and the usefulness of peculiarity as a measure/approach to discovering interesting association rules. The work in [20] focuses on record based LPF which is used to represent multi-dimensional data. For single-dimensional data, which is the type of data we focus on, attribute-based LPF and PF (peculiarity factor) is used.

The works presented here show that there is a need and a use for a peculiarity based interestingness measure for association rule mining. By having one, a set of potentially interesting peculiar rules can be found which may otherwise have been missed. However, the work presented has focused on single level datasets, with the inclusion of multiple dimensions or even multiple databases, and not multi-level datasets. To address this, we will propose a peculiarity interestingness measure that considered the dataset's taxonomic structure when determining the distance between rules.

2.3 Introduction to Association Rule Mining

In this section we provide a brief introduction to multi-level association rules, the issue of presentation of discovered association rules and redundancy in discovered association rule sets.

2.3.1 Multi-Level Association Rules

Association rule mining was first presented by Agrawal et al. in [1]. Association rules are generally in the form $X \Rightarrow Y$, where both X and Y are items or itemsets that are wholly contained within the dataset and $X \cap Y = \phi$. In this form, X is the antecedent and Y is the consequent and thus the previous rule implies that whenever X is present, Y will also be present or that X implies Y.

Traditionally, association rule mining has been performed at a single concept or abstract level, usually either the lowest level (primitive level) or at the highest concept level [6, 21]. It is widely accepted that single level rule mining has two major problems: first it is difficult to find strong associations at low levels due to sparseness and second, mining high concept levels may result in rules containing general common knowledge [6, 21, 22]. One of the major arguments for the use of multi-level or cross-level rule mining is the potential for undiscovered knowledge to be discovered [6, 21, 23].

Multi-level and cross-level rule mining require a dataset with a taxonomy; either built beforehand or dynamically as the rule mining process executes [6, 8, 21]. A taxonomy means that the items within the dataset can be structured and related to each other. This allows for the rules to make sense and contain useful information/knowledge.

Multi-level datasets (or those with a taxonomy) are becoming more common and hence multi-level and cross-level rule mining will become increasingly more important. Because of this, measures to determine whether a rule from such a dataset is interesting will also become more important. The taxonomy underlying the datasets contains new information that could be potentially useful when determining the interestingness level of such a rule. Using this taxonomy information is the focus of this chapter.

2.3.2 Presentation of Association Rules

One of the biggest problems with association rule mining is the number of potentially interesting rules being discovered. Having such a large number makes it difficult for a user to inspect them and discover those that are truly interesting or important. This results in a gap between having information and being able to understand and utilise that information.

One view taken to deal with the number of rules is, instead of determining if there are redundant rules or using other interestingness measures, to find important rules, but rather organise the presentation of the rule set better. Work in [24] argues that the number of rules is not a problem, but rather their organisation. The approach in [24] proposes a hierarchical type organisation through the use of General rule, Summary and Exceptions (GSE), where rules are grouped under a more general rule, summary or exception.

There are two major issues with the approach in [24]; firstly it can be difficult to determine what general rules should be used to cover the rule set and determining the exceptions to these general rules. The second drawback is that this approach does nothing about redundant rules or the determination of which rules are interesting from the point of view of an interestingness measure.

2.3.3 Redundancy in Association Rules

As mentioned in the previous Sect. 2.3.2, association rule mining often discovers a large number of rules and usually it is more than the user can handle. Also, not all of the rules discovered are actually unique and thus redundancy exists. It was stated in [25] that the extent of the redundancy in rule mining is greater than expected. A redundant rule is one that gives no new information to a user as another rule has already conveyed the same information. It is also said the problem of redundant rules is greater when the dataset is dense or correlated [26].

Removing redundant rules makes it easier for a user to handle the resulting rule set and use it effectively. It was argued in [26] that redundant rules can also be misleading and hence removing them is good practice. This is usually done by generating a condensed rule set. Because of the importance of removing redundant rules, much work has been done. Most have focused on generating a condensed rule set by maximizing the information in each rule. Work includes [26, 27–29, 25] for flat or single level datasets. Recently, work in [30–32] extended redundant rule removal to multi-level datasets by using the hierarchical relationships between the items. Despite this, redundant rule removal, even in multi-level datasets, does not address the problem of ensuring the user receives interesting rules.

2.4 Concepts and Calculations of the Proposed Interestingness Measures

In this section we present the key parts of the theory and background and formula behind our proposed measures. We also present the assumptions we have made for our measures.

2.4.1 Assumptions and Definitions

Here we outline the assumptions we have made. Figure 2.1 depicts an example of the general structure of a multi-level dataset. As shown, there is a tree-like hierarchical structure to the concepts or items involved in the dataset. Thus items at the bottom are descendant from higher level items. An item at a higher level can

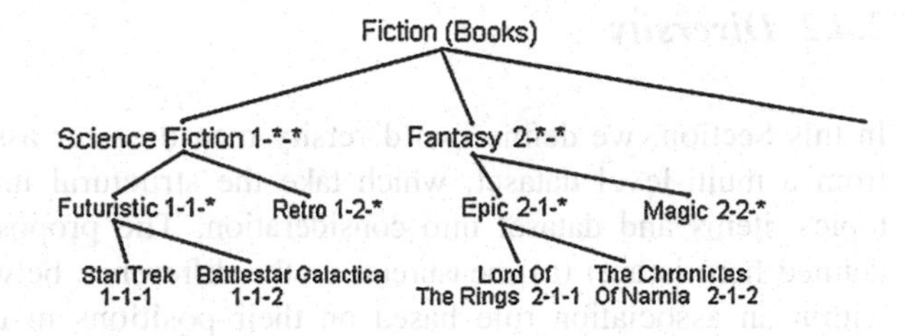

Fig. 2.1 Example of a multi-level dataset

contain multiple lower level items. With this hierarchy, we have made the three following assumptions.

1. That each step in the hierarchy tree is of equal length/weight. Thus the step from 1-*-* to 1-1-* is of equal distance to the step from 2-*-* to 2-1-* or 1-1-* to 1-1-1.
2. That the order of sibling items is not important and the order could be changed (along with any subtree) without any effect.
3. That each concept has one and only one ancestor concept (except for the root which does not have any ancestor).

The interestingness measures defined here have a single limitation on their use. Namely, they can only be used on association rules derived from a multi-level dataset for which a taxonomic structure exists. These measures will not work on association rules derived from a single level dataset as a taxonomic structure with 'is-a' relations is required in order to calculate diversity and peculiarity distance. Also, we define several terms that are used in the proposed interestingness measures here.

- C is a set of all concepts in the multi-level dataset.
- $ca\ (n_i, n_j)$ is the closest common ancestor to both n_i and n_j in the hierarchy, where $n_i, n_j \in C$ and neither of n_i, n_j is the root in the hierarchy. If n_i or n_j is the ancestor of the other, then the common ancestor will be n_i if n_i is an ancestor of n_j or n_j if n_j is an ancestor of n_i.
- *TreeHeight* is the maximum number of concepts on a path in the multi-level data*set* (not counting the root) from the root to a concept located at the lowest concept level in the dataset.
- *Hierarchy level of an item* is the depth of the item in the hierarchical tree, i.e., the hierarchy level of the root is 0 and the hierarchy level of an item is larger than the level of its parent by 1.
- $NLD(n_i, n_j)$ denotes the *Number of levels Difference* between two concepts n_i and n_j which is defined as the number of hierarchy concept levels between n_i and n_j as shown below:

$$NLD(n_i, n_j) = |hierarchy\ level\ of\ n_i - hierarchy\ level\ of\ n_j| \tag{2.1}$$

2.4.2 Diversity

In this Section, we define two diversity measures for association rules extracted from a multi-level dataset, which take the structural information of the nodes, topics, items and dataset into consideration. The proposed diversity measure(s) defined here is(are) (a) measure(s) of the difference between the items or topics within an association rule based on their positions in the multi-level dataset's hierarchy. Association rules are used to derive associated concepts or items with some known concepts or items (antecedent). In some cases, users would like to discover items which are quite different from the known items (e.g. users want something surprising or unexpected). In this case, this measure would allow us to choose rules which can help user to derive these associated items. Also, a diverse rule may be interesting because the uniform distribution does not hold due to the items within being significantly different from each other.

We propose that the diversity of an association rule can be measured using two different approaches: overall diversity and antecedent-consequent diversity. These approaches are presented in detail in the following sections. Before defining the diversity measures, we first define two distances, Hierarchical Relationship Distance (HRD) between items or topics and Concept Level Distance (LD) between items or topics, which are measured in order to determine the diversity of a rule.

2.4.2.1 Hierarchical Relationship Distance

The Hierarchical Relationship Distance (HRD) attempts to measure the strength of the relationship of two nodes n_i & n_j based on their ancestry. This distance is mainly measured based on the Horizontal Distance between two nodes which is defined as the distance between the two nodes via their common ancestor in the multi-level dataset's hierarchy. This measure is akin to family type relationships. The more distance (specifically horizontal distance) between two nodes, the further away their common ancestor is and the weaker their relationship. This would further imply that the further away their common ancestor is, the lower the similarity between the topics of the two nodes, and hence, more diverse. For example, sibling nodes have little diversity because they have the same parent node and the horizontal distance is minimal. Thus the similarity between their topics would be strong and the two would usually be considered to complement each other. However, as the distance to a common ancestor for two nodes increases, the horizontal separation between them also increases. Thus, the degree of similarity decreases and their diversity increases. The further away a common ancestor is, the less two nodes have in common. This is akin to family relationships where often distantly related members have little attachment to each other.

Maximum HRD is achieved when two nodes share no common ancestor (hence have nothing in common) and are both located at the lowest concept level in the hierarchy. This requirement for being at the lowest concept level helps ensure that the two nodes have the minimal degree of similarity (thus maximising the diversity) as at the lowest concept level the topics for the nodes are the most focused and specific.

HRD focuses on measuring the horizontal (or width) distance between two items. Usually the greater the horizontal distance, the greater the distance to a common ancestor and therefore the more diverse the two items are. Due to the second assumption (see Sect. 2.4.1), we cannot measure the horizontal distance without also utilising the vertical (height) distance. This is because we cannot directly measure how far apart two nodes are horizontally by just 'counting' the number of nodes that are between them. For example, sibling nodes (all from the same parent node) should have the same horizontal distance regardless of their order (with or without rearrangement), but if we 'count' across horizontally, then different pairs of sibling nodes will score different distances (which would change if the order of the siblings changed) with the maximum distance being achieved by the pairing of the first and last siblings.

Thus to determine the Hierarchical Relationship Distance (HRD) of two concepts $n_i, n_j \in C$, the following is proposed:

$$HRD(n_i, n_j) = \frac{NLD(n_i, ca(n_i, n_j)) + NLD(n_j, ca(n_i, n_j))}{2 \times TreeHeight} \tag{2.2}$$

The Hierarchical Relationship Distance between two items is defined as the ratio between the average number of levels between the two items and their common ancestor and the height of the tree. Thus if two items share a direct parent, the HRD value of the two items becomes the lowest value which is *1/TreeHeight*, while if the two items have no common ancestor or their common ancestor is the root, the HRD values of the two items can score high. Maximum HRD value, which is 1, is achieved when the two items have no common ancestor (or the common ancestor is the root) and both items are at the lowest concept level possible in the hierarchy. If the two items are the same, then HRD is also *1/TreeHeight*.

2.4.2.2 Concept Level Distance

The second aspect to be considered is the Concept Level Distance (LD) of two items and is based on the hierarchical levels of the two items. The idea is that the more levels between the two items, the more diverse they will be. Thus, two items on the same hierarchy level are considered to be not very diverse, but two items on different levels are more diverse as they have different degrees of specificity or abstractness. The LD is akin to measuring the generational difference between two members of a family. In this measure, no consideration is given to whether the two nodes are related by a common ancestor, that one is an ancestor of the other or that they do not share a common ancestor. LD differs from HRD in that HRD measures the distance two nodes are from a common ancestor item (or root) to measure the strength of their relationship, whereas LD measures the difference between the two items' specificity. LD focuses on measuring the distance between two items in terms of their height (vertical) difference only, whereas HRD considers the width (horizontal) distance when determining diversity.

Thus, we propose to use the ratio between the level difference (NLD) of two items and the height of the tree (e.g. the maximum level difference) to measure the Level Distance of the two items as defined as follows, where $n_i, n_j \in C$ are two concepts in the dataset:

$$LD(n_i, n_j) = \frac{NLD(n_i, n_j)}{(TreeHeight \; - \; 1)} \tag{2.3}$$

This means that two items on the same concept level will have a LD of 0, while an item at the highest concept level and another at the lowest concept level will have an LD of 1, as they are as far apart as possible in the given hierarchy.

In the following two sections, we will introduce our proposed diversity measures which are designed based on the proposed HRD and LD distances.

2.4.3 Antecedent-Consequent Diversity

The first proposed approach to measuring the diversity is known as the Antecedent-Consequent Diversity measure. This approach measures the diversity between the set of items/topics in the antecedent and the set of items/topics in the consequent of an association rule. Those rules which have a high difference between their antecedent and consequent will have a high antecedent-consequent diversity.

Association rules are used to derive associated concepts or items with some known concepts or items (antecedent). In some applications, it will be desirable to discover conclusions which are different than the expected (and thus may be a surprise to the user), or are seemingly unrelated to the initial conditions that lead to that conclusion(s) (such as all the initial conditions are close/similar, but the conclusion is significantly different). In this case, this measure would allow us to choose rules which can help user to derive these associated items.

The following formula sets out the proposed antecedent-consequent diversity measure. Let R be a rule R: $a_1, a_2, \ldots, a_n \rightarrow c_1, c_2, \ldots, c_m$, with n items/topics in the antecedent and m items/topics in the consequent, where $a_1, a_2, \ldots, a_n$, $c_1, c_2, \ldots, c_m \in C$, and D_{ACR} denote the antecedent to consequent diversity of R, the diversity of R can be determined as follows.

$$D_{ACR}(R) = \frac{\alpha \sum_{i=1}^{n} \sum_{j=1}^{m} HRD(a_i, c_j)}{nm} + \frac{\beta \sum_{i=1}^{n} \sum_{j=i}^{m} LD(a_i, c_j)}{nm} \tag{2.4}$$

where α and β are weighting factors such that $\alpha + \beta = 1$.

The first component in Eq. 2.4 is the average diversity between the items in the antecedent and the items in the consequent in terms of the HRD distance, and the second component is the average diversity between the antecedent and the consequent in terms of the LD distance. This measure allows us to determine the average diversity between the antecedent and consequent within a rule and thus we get an overall internal measure of the differences between the rule's antecedent and its consequent.

2.4.4 Overall Diversity

The second proposed approach to measuring the diversity is known as the Overall Diversity measure. This approach measures the overall diversity of the items and/or topics in an association rule by combining the antecedent and consequent into a single set of items and/or topics. If the contents of this set are different, the association rule will have a high overall diversity, regardless of which part of the rule the diverse items and/or topics came from.

The overall diversity differs from the previous antecedent-consequent diversity in that it considers all pairs of items in an association rule, including those pairs within the antecedent and/or consequent. This allows rules which have diverse items within the antecedent and/or consequent to be discovered. The antecedent-consequent diversity only discovers those rules whose consequent differs greatly from the antecedent. Overall diversity on the other hand will allow rules with diverse antecedents and/or consequents to be found. This allows those rules which have vastly different initial conditions, as well as those with vastly differing conclusions to be discovered.

The following formula sets out the proposed overall diversity measure. Let R be a rule with k items/topics, $n_1, n_2, \ldots\ldots, n_k$, and D_{OR} denotes the overall diversity of R, the diversity of R can be determined as follows.

$$D_{OR}(n_1, n_2, \ldots, n_k) = \frac{\alpha \sum_{i=1}^{k-1} \sum_{j=i+1}^{k} HRD(n_i, n_j)}{k(k-1)/2} + \frac{\beta \sum_{i=1}^{k-1} \sum_{j=i+1}^{k} LD(n_i, n_j)}{k(k-1)/2} \tag{2.5}$$

where α and β are weighting factors such that $\alpha + \beta = 1$. The values of α and β need to be determined experimentally and for our experiments are both set at 0.5.

The overall diversity considers the *HRD* and *LD* scores for every pair of topics in the rule. This measure allows us to determine the average diversity of the topics within the rule and thus we get an overall internal measure of these differences.

2.4.5 Peculiarity

Peculiarity is an objective measure that determines how far away one association rule is from others. The further away the rule is, the more peculiar. It is usually done through the use of a distance measure to determine how far apart rules are from each other. Peculiar rules are usually few in number (often generated from outlying data) and significantly different from the rest of the rule set. It is also possible that these peculiar rules can be interesting as they may derive some unknown or surprising consequents. One proposal for measuring peculiarity is the neighbourhood-based unexpectedness measure first proposed in [4]. Here we argue that a rule's interestingness can be influenced by the rules that surround it in its neighbourhood.

The measure is based on the idea of determining and measuring the symmetric difference between two rules, which forms the basis of the distance between them. From this, it was proposed in [4] that unexpected confidence (where the confidence of a rule R is far from the average confidence of the rules in *R's* neighbourhood) and sparsity (where the number of mined rules in a neighbourhood is far less than that of all the potential rules for that neighbourhood) could be determined, measured and used as interestingness measures [2, 4].

This measure [4] for determining the symmetric difference was developed for single level datasets where each item was equally weighted. Thus, the measure is actually a count of the number of items that are not common between the two rules. In a multi-level dataset, each item cannot be regarded as being equal due to the hierarchy. Thus the measure proposed in [4] needs to be enhanced to be useful with these datasets. Here we will present an enhancement as part of our work.

We believe it is possible to take the distance measure presented in [4] and enhance it for multi-level datasets. The original measure is a syntax-based distance metric is the following form:

$$P(R_1, R_2) = \delta_1 \times |(X_1 \cup Y_1)\Theta(X_2 \cup Y_2)| + \delta_2 \times |X_1 \Theta X_2| + \delta_3 \times |Y_1 \Theta Y_2| \quad (2.6)$$

where R_1 and R_2 are two rules, $R_1 : X_1 \rightarrow Y_1$ and $R_2 : X_2 \rightarrow Y_2$, X_1, Y_1, X_2, Y_2 are sets of items, the Θ operator denotes the symmetric difference between two item sets and thus $X \Theta Y$is equivalent to $X - Y \cup Y - X$. δ_1, δ_2 and δ_3 are the weighting factors to be applied to different parts of the rule. Equation 6 measures the peculiarity of two rules by a weighted sum of the cardinalities of the symmetric difference between the two rule's antecedents, consequents and the rules themselves.

We propose an enhancement to the measure given in Eq. (2.6) to allow it to handle multi-level datasets. Under the existing measure, every item is unique and therefore none shares any kind of 'syntax' similarity. However, we argue that the items 1-*-*-*, 1-1-*-*, 1-1-1-* and 1-1-1-1 (based on Fig. 2.2) all have a relationship with each other. Thus they are not completely different and should have a 'syntax' similarity due to their relation through the dataset's hierarchy.

The greater the $P(R_1, R_2)$ value is, the greater the difference (thus lower similarity) and so the greater the distance between those two rules. Therefore, the further apart the relation is between two items, the greater the difference and distance. Thus if we have,

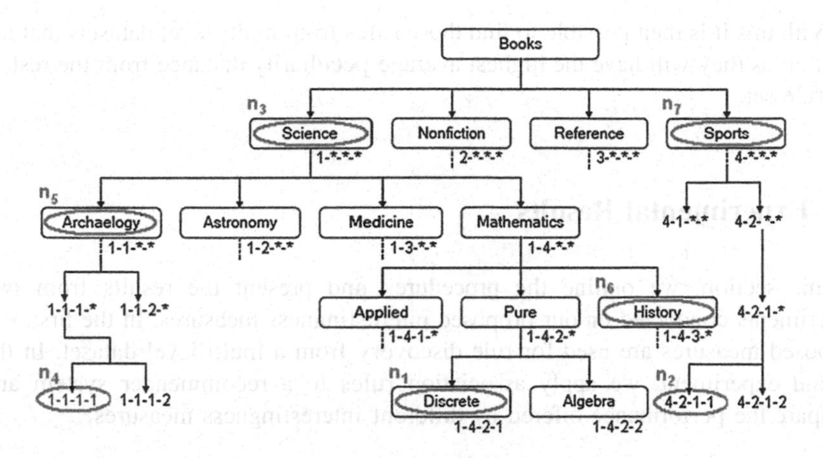

Fig. 2.2 Amazon taxonomy snippet showing highlighted topics for HRD and LD

$$
\begin{aligned}
R_1 &: \; 1-1-1-* \;\rightarrow\; 1-*-*-* \\
R_2 &: \; 1-1-*-* \rightarrow 1-*-*-* \\
R_3 &: \; 1-1-1-1 \;\rightarrow\; 1-*-*-*
\end{aligned}
$$

We believe that the following should hold; $P(R_1, R_3) < P(R_2, R_3)$ as 1-1-*-* and 1-1-1-1 are further apart from each other than 1-1-1-* and 1-1-1-1.

In order to achieve this we modify Eq. (2.6) by calculating the overall diversity of the symmetric difference between two rules instead of the cardinality of the symmetric difference as calculated in Eq. (2.6). The cardinality of the symmetric difference measures the difference between two rules in terms of the number of different items in the rules. The diversity of the symmetric difference takes into consideration the hierarchical difference of the items in the symmetric difference to measure the difference of the two rules.

By replacing the cardinality of the symmetric difference in Eq. 2.6 with the overall diversity of the symmetric difference, we propose the following equation to calculate the peculiarity distance between two rules $R_1 : X_1 \rightarrow Y_1$ and $R_2 : X_2 \rightarrow Y_2$, X_1, Y_1, X_2, Y_2 are sets of items:

$$
PM(R_1, R_2) = \delta_1 \times D_{OR}((X_1 \cup Y_1)\Theta(X_2 \cup Y_2)) + \delta_2 \times D_{OR}(X_1 \Theta X_2) + \delta_3 \times D_{OR}(Y_1 \Theta Y_2) \tag{2.7}
$$

To determine the average peculiarity distance between a rule and the entire ruleset, let RS be the ruleset of $\{R_1, R_2, \ldots, R_n\}$, then the peculiarity of a rule R_i with respect to the ruleset RS can be determined by the following:

$$
PE(R_i) = \frac{\sum_{R_j \in RS \text{ and } j \neq i} PM(R_i, R_j)}{|RS| - 1} \tag{2.8}
$$

With this it is then possible to find those rules from multi-level datasets that are peculiar as they will have the highest average peculiarity distance from the rest of the rule set.

2.5 Experimental Results

In this section, we outline the procedures and present the results from two experiments conducted on our proposed interestingness measures. In the first, our proposed measures are used for rule discovery from a multi-level dataset. In the second experiment, we apply association rules to a recommender system and compare the performance offered by different interestingness measures.

2.5.1 Interestingness Measure Comparison

For the first experiment, we use the existing measures of support and confidence, along with our proposed measures and discover association rules from a multi-level dataset. We look at the distribution curves of the measures and the trends of the measures when compared against another. This allows us to see how interesting rules can be identified and also discover if there is any relation between support, confidence and our proposed measures.

2.5.1.1 Concept Level Distance

The dataset used for our experiments is a real world dataset, the BookCrossing dataset (obtained from http://www.informatik.uni-freiburg.de/cziegler/BX/) [33]. From this dataset, we built a multi-level transactional dataset that contains 92,005 user records and 960 leaf items, with 3 concept/hierarchy levels.

To discover the frequent itemsets, we use the MLT2 L1 algorithm proposed in [6, 7] with each concept level having its own minimum support. From these frequent itemsets, we then derive the frequent closed itemsets and generators using the CLOSE+ algorithm proposed in [26]. From this, we then derive the non-redundant association rules using the MinMax (MM) rule mining algorithm [26].

For the experiment, we simply use the previously mentioned rule mining algorithm to extract the rules from the multi-level dataset. We assign a reducing minimum support threshold to each level. The minimum supports are set to 10 % for the first hierarchy level, 7.5 % for the second and 5 % for the third level (the lowest). During the rule extraction process, we determine the diversity and peculiarity distance of the rules that meet the confidence threshold. With two measures known for each rule, we are also able to determine the minimum, maximum and average diversity and peculiarity distance for the rule set.

2.5.1.2 Statistical Analysis

Firstly, we compare the distribution curves of the proposed measures (diversity and distance) against the distribution curves of support and confidence for the rule set. The distribution curves are shown in Fig. 2.3. The value of each measure ranges from 0 to 1. The values of the distance measure are based on the minimum distance (in this case 33,903.7) being equal to 0 and the maximum distance (in this case being 53,862.5) being equal to 1. The range between these two has been uniformly divided into 20 bins.

As Fig. 2.3 shows, the support curve shows that the majority of the association rules only have a support of between 0.05 and 0.1. Thus, for this dataset, distinguishing interesting rules based on their support would be difficult as the vast majority have very similar support values. This would mean the more interesting or important rules would be lost. The confidence curve shows that the rules are spread out from 0.5 (which is the minimum confidence threshold) up to close to 1. The distribution of rules in this area is fairly consistent and even, ranging from as low as 2,181 rules for 0.95–1, to as high as 4,430 rules for 0.85–0.9. Using confidence to determine the interesting rules is more practical than support, but still leaves over 2,000 rules in the top bin. The overall diversity curve shows that the majority of the rules (23,665) here have an average overall diversity value of between 0.3 and 0.4. The curve however, also shows that there are some rules which have an overall diversity value below the majority, in the range of 0.15–0.25 and some that are above the majority, in the range of 0.45 up to 0.7. The rules

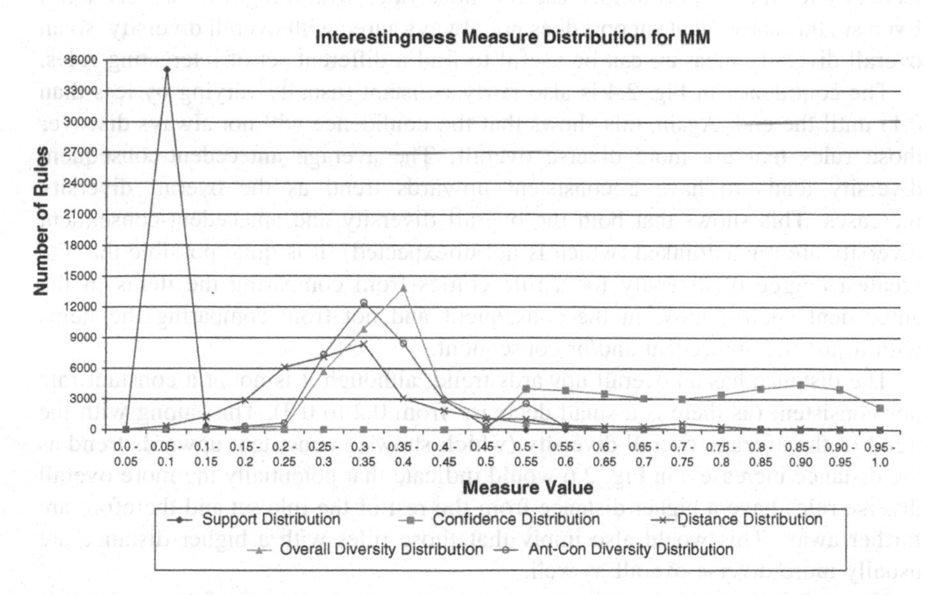

Fig. 2.3 Distribution curves for support, confidence, overall diversity, antecedent-consequent diversity and peculiarity distance for the Book-Crossing dataset using MinMax

located above the majority are different from the rules that make up the majority and could be of interest as these rules have a high overall diversity.

The antecedent-consequent diversity curve is similar to that of the overall diversity. It has a similar spread of rules, but the antecedent-consequent diversity curve peaks earlier at 0.3–0.35 (whereas the overall diversity curve peaks at 0.35–0.4), with 12,408 rules. The curve then drops down to a low number of rules at 0.45–0.5, before peaking again at 0.5–0.55, with 2,564 rules. The shape of this curve with that of the overall diversity seems to show that the two diversity approaches are related. Using the antecedent-consequent diversity allows rules with differing antecedents and consequents to be discovered when support and confidence will not identify them.

Lastly, the distance curve shows the largest spread of rules across a curve. There are rules which have a low distance from the rule set (0–0.1 which corresponds to a distance of 33,903.7–35,899.56) up to higher distances (such as 0.7 and above which corresponds to a distance of 47,874.88–53,862.52). The distance curve peaks at 0.3–0.35 (which is a distance of between 39,891.35 and 40,889.29). Using the distance curve to find interesting rules allows those that are close to the rule set (small distance away) or those that are much further away (greater distance) to be discovered.

Next, we look at the trends of the various measures when compared against the proposed diversity and distance measures.

Figure 2.4 shows the trend of the average support, average confidence, average antecedent-consequent diversity and average distance values against that of overall diversity. As can be seen, the average support remains fairly constant. There is a tendency for the support to increase for those rules with a high overall diversity. Even so, this shows that support does not always agree with overall diversity, so an overall diversity measure can be useful to find a different set of interesting rules.

The confidence in Fig. 2.4 is also fairly constant (usually varying by less than 0.1) until the end. Again, this shows that the confidence will not always discover those rules that are more diverse overall. The average antecedent-consequent diversity tends to have a consistent upwards trend as the overall diversity increases. This shows that both the overall diversity and antecedent-consequent diversity are related/linked (which is not unexpected). It is quite possible that the greatest degree of diversity for a rule comes from comparing the items in the antecedent against those in the consequent and not from comparing the items within just the antecedent and/or consequent.

The distance has an overall upwards trend, although it is not at a constant rate nor consistent (as there is a small decrease from 0.2 to 0.3). This, along with the trend of the average overall diversity (which shows a consistent upwards trend as the distance increases) in Fig. 2.6 would indicate that potentially the more overall diverse rules have a higher distance from the rest of the rule set and therefore are further away. This would also imply that those rules with a higher distance are usually more diverse overall as well.

Figure 2.5 shows the trends of average support, average confidence, average overall diversity and average distance against that of antecedent-consequent

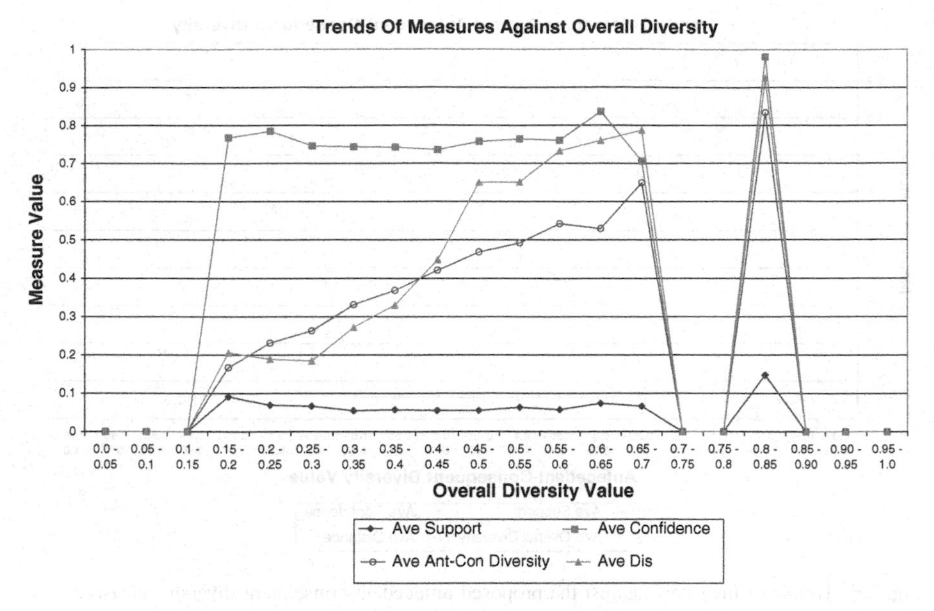

Fig. 2.4 Trends of measures against the proposed overall diversity measure

diversity. As in Fig. 2.4, the support remains fairly constant regardless of the antecedent-consequent diversity value. The confidence tends to decrease as the antecedent-consequent diversity increases, so this gives an indication that the more diverse rules (in terms of antecedent-consequent diversity) will not always be picked up by confidence. The overall diversity tends to increase as antecedent-consequent diversity increases (similar to Fig. 2.4). So again the biggest diversity in a rule is often in the difference between the antecedent and consequent. The distance also tends to increase (gradually at first for lower antecedent-consequent diversity values). There is a big jump in the distance trend when the antecedent-consequent diversity increases from 0.65 to 0.75. The highest distance values are achieved when the antecedent-consequent diversity reaches its highest values (this is also shown in Fig. 2.6).

Figure 2.6 shows the trend of the average support, average confidence, average overall diversity and average antecedent-consequent diversity values against that of distance. As shown, the support remains very constant regardless of the distance. This shows that support cannot be used to discover rules that have a low or high peculiarity distance.

Like the average overall diversity, the average antecedent-consequent diversity also trends upwards as the distance increases. The rate of rise is similar to that of the overall diversity initially at lower distance values, but becomes much steeper at high distance values. This shows that it seems the most distant rules also have the highest diversity between their antecedent and consequent as Fig. 2.6 shows the

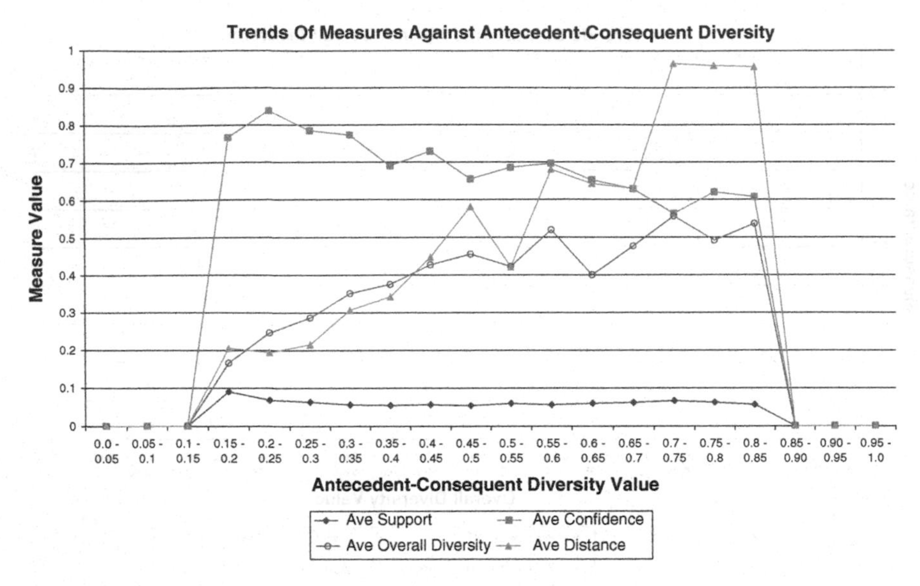

Fig. 2.5 Trends of measures against the proposed antecedent-consequent diversity measure

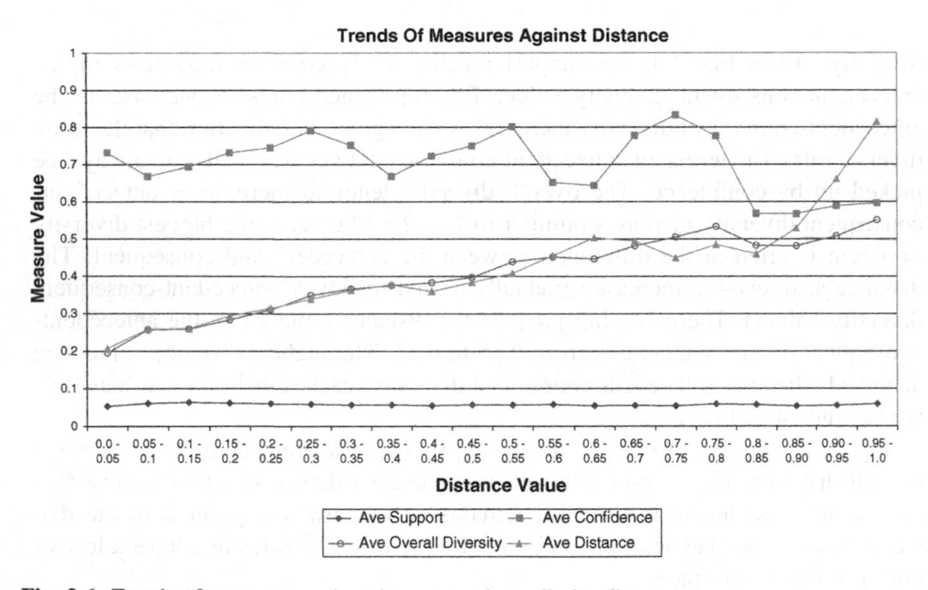

Fig. 2.6 Trends of measures against the proposed peculiarity distance measure

average antecedent-consequent diversity to be over 0.8 for the rules with the highest distance from the rest of the rule set.

The confidence trend in Fig. 2.6 also shows that confidence will not always discover those rules far away from the rule set (0.8/49,870.76 and above), as at these distances, the confidence values are at their lowest points. For rules with a low distance value, confidence may also not be the best measure as at these values (0/33,903.7–0.1/35,899.58) confidence values are not at their highest. The highest confidence value(s) occur when the distance is 47,874.86–48,872.82 (0.7–0.75) and thus does not occur for those rules with the lowest distance or the highest distance.

If we look closer at the discovered rules we find the following examples that show how diversity and peculiarity distance can be useful in identifying potentially interesting rules that would not normally be identified as such. (Note that the hyphen breaks the concept levels, while a comma indicates a new item).

2.5.2 Recommendation Making

For our second experiment, we refer to the work presented in [34]. Here it was proposed to use association rules from a multi-level dataset to improve recommendations. We will use this proposed approach, along with the experimental setup described as the basis for the experiment presented in this section. Due to space, we will omit most details of the work in [34]. As the work in [34] proposed, it is beneficial to rank relevant association rules from strongest to weakest (or best to worst) and then use the Top-n to expand short user profiles in a recommender system. The most common method for measuring association rules is confidence. However, it is possible that confidence may not always be the best measure to use to rank association rules from best to worst (or strongest to weakest). We believe it is necessary to test whether other measures may give a better performance improvement under the cold-start problem. Thus we will compare confidence against our proposed antecedent-consequent diversity and overall diversity measures.

The work proposed in [34] focused on improving recommendation quality for users with short profiles by expanding them. However, the recommendations made to other users by the recommender system will also be affected. We believe these users may also benefit from having a different measure used to rank the rules for short user profile expansion. One particular type of users we consider in this experiment is those with a diverse profile. A diverse user is one who has an interest in a variety of topics which appear to have little in common with each other. It would not be unexpected for these users to benefit more from a measure that somehow considers the diversity of the rules used during profile expansion. However, at the same time, we would still want to improve the quality of the recommendations that the cold-start users receive.

This experiment attempts to show a practical application where a diversity measure of the nature proposed in this chapter could be of use and give a better outcome than the traditional confidence measure which is widely used. The purpose of this experiment is not to prove the approach of using association rules to improve recommendation making (this is done in [34]), but to show that there is a use in certain situations (and applications) where our proposed diversity measures can be of use. (last sentence to be rephrased without using "use" twice)

2.5.2.1 Dataset and Setup

Again we use the BookCrossing dataset (like in Sect. 2.5.1) and follow the same setup as in [34]. The user profiles (containing ratings of items) are divided into training and test sets. The recommender system (we use the Taxonomy Product Recommender; TPR [35]) will use the training set to make recommendations and the recommendations will be evaluated against the test set.

Here we will use the MLT2 L1 and MinMax algorithms [6, 7, 26] to extract rules from the BookCrossing dataset (built in the same way as that in [34]) which will then be used to expand short user profiles. For our experiments here we have two rule sets: R1 with minimum support thresholds of 10, 7.5 and 5 %, containing 37,827 rules and R2 with minimum support thresholds of 10, 8 and 6 %, containing 9,520 rules. We compare the level of improvement from using confidence, antecedent-consequent diversity and overall diversity over no expansion at all (baseline). This will allow us to see whether our diversity measures could outperform confidence in this application, potentially showing an application where our diversity measures work best.

The performance of the recommender system will be measured using precision, recall and F1-measure as follows. For a user u_i and an item t, let T_i represent the set of items in the test dataset *P(test)* that have been rated by user u_i, where $rating(u_i, t)$ denotes the rating that the user gave to the item and $avg(u_i)$ denotes the average of the user's ratings. From this, we can define the set of items that are preferred by the user to be: $\hat{T}_i = \{t | t \in T_i, rating(u_i, t) > avg(u_i)\}$

The precision, recall and F1-measure can then be calculated using the following equations:

$$Precision = \frac{|\hat{T}_i \cap P_i|}{|P_i|}, \; Recall = \frac{|\hat{T}_i \cap P_i|}{|T_i|}$$

$$F1 - measure = \frac{2 \times Precision \times Recall}{Precision + Recall}$$

2.5.2.2 Recommender System Performance Analysis

Following the experiments in [34], we build a set of short user profiles (5 or less ratings) from the training set. These 15,912 user profiles we attempt to expand using the association rules extracted from the BookCrossing dataset a verb is missing ? After expansion we then make up to 10 recommendations for each user and evaluate performance. This is then compared against the same set of user profiles, but without any expansion.

To determine which rules to use in expanding a user's profile, we rank all suitable matching rules from the highest to lowest based on the score of their interestingness measure (confidence, overall diversity and antecedent-consequent diversity). The top-n are then used to expand the profile.

Figures 2.7 and 2.8 show the results of using the three interestingness measures to expand the short user profiles. The baseline represents the score (F1-measure) obtained by the recommender system if nothing is done to try and improve performance: e.g. no user profile expansion through association rules.

As the (suppress the) Figs. 2.7 and 2.8 show the use of anyone of these interestingness measures results in an improvement to the quality of the recommendations made to the set of short user profiles. As shown here and in [34], using confidence to rank the rules provides up to a 31.5 % increase (based on F1-measure) when using R1 and 20.1 % when using R2. We can also see that our proposed diversity measures achieve improvements of up to 26.1 % using R1 and 22.7 % using R2 for antecedent-consequent diversity and 14.2 % using R1 and 13 % using R2 for overall diversity. We can see that in Fig. 2.7 antecedent-consequent diversity outperforms confidence early when only the Top 1 to Top 3 rules are used. When the Top 5 rules are used antecedent-consequent diversity only trails confidence by 4.1 % (when the two are compared). For Fig. 2.8 we see the reverse is true; when the Top 1 to Top 3 rules are used confidence gives the best performance improvement, but when the Top 4 or Top 5 rules are used, antecedent-consequent gives the best improvement (by 5.4 % when the two are compared).

Typically, when a user finds something that interests them, they look for similar items. Usually these items are closely related (other books in a series, more books by the same author, books with a similar theme, etc.). Thus, often, there is not

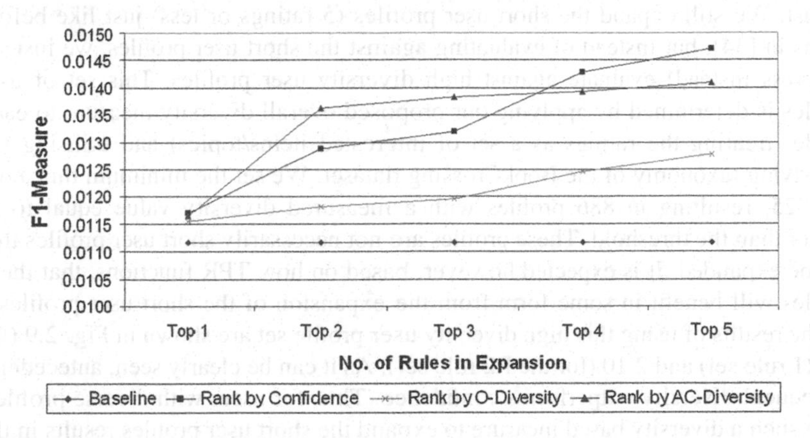

Fig. 2.7 Performance evaluation of TPR recommender system utilising different interestingness measures using full short user profile set (R1 rule set)

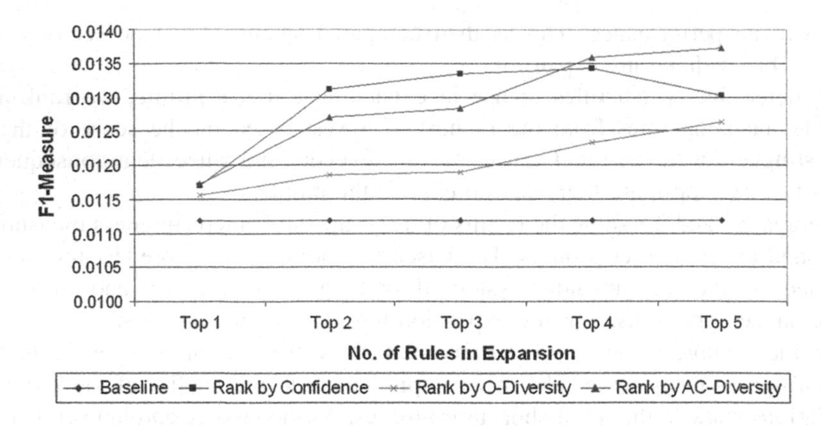

Fig. 2.8 Performance evaluation of TPR recommender system utilising different interestingness measures using full short user profile set (R2 rule set)

much diversity. This results in patterns where items come from the same sub-tree in the taxonomy to be the most common. Hence, the strongest rules in terms of confidence will have a tendency for their items to come from a small part of the taxonomy. Due to this, when confidence is used, short profiles are expanded to include very similar items (in terms of taxonomy). This results in the recommender system making stronger recommendations to users for similar items, matching with the common behavior of liking very similar items.

Based on these results (and the results of Figs. 2.9 and 2.10: which will be discussed later) it is indicated that the determination of which interestingness measure should be used to improve the performance of a recommender system is determined by several factors, including:

Following on from the first set of results in improving recommendations, we build a second set of user profiles to evaluate our proposed diversity measures against. We still expand the short user profiles (5 ratings or less) just like before and as in [34], but instead of evaluating against the short user profiles, we instead (suppress instead) evaluate against high diversity user profiles. This set of user profiles is determined by applying our proposed overall diversity measure to each profile (treating the ratings as a set of interested items/topics) and utilising the underlying taxonomy of the BookCrossing dataset. We set the minimum threshold to 0.325, resulting in 886 profiles with a measured diversity value equal to or greater than the threshold. These profiles are not necessarily short user profiles that will be expanded. It is expected however, based on how TPR functions, that these profiles will benefit in some form from the expansion of the short user profiles.

The results of using this high diversity user profile set are shown in Fig. 2.9 (for the R1 rule set) and 2.10 (for the R2 rule set). As it can be clearly seen, antecedent-consequent diversity outperforms confidence. Thus for users with diverse profiles, using such a diversity based measure to expand the short user profiles results in the best improvement for them. In Fig. 2.9, antecedent-consequent diversity results is

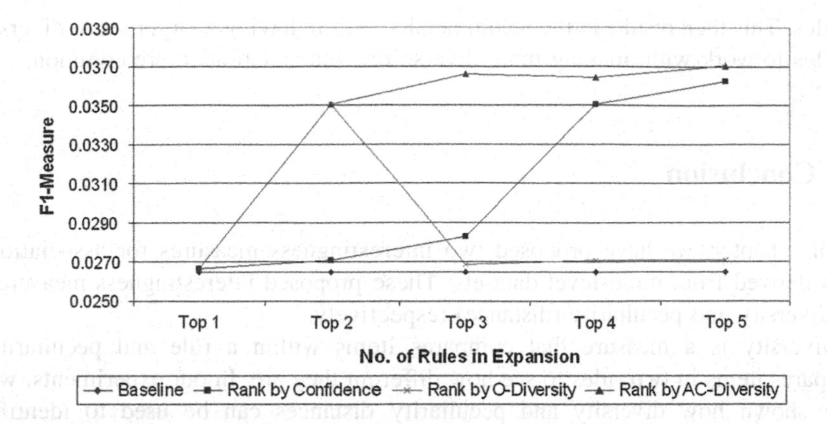

Fig. 2.9 Performance evaluation of TPR recommender system utilising different interestingness measures using high diversity user profile set (R1 rule set)

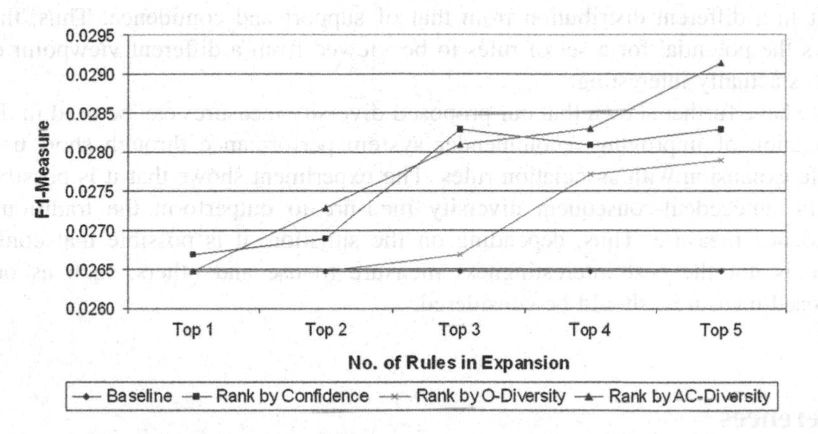

Fig. 2.10 Performance evaluation of TPR recommender system utilising different interestingness measures using high diversity user profile set (R2 rule set)

a noticeable improvement (32.3 %) when using just the Top 2 rules, while at that stage confidence only has resulted in an improvement of 2.3 %. Thus, here, if we wish to maximise the improvement and experience for users with diverse interests, our proposed antecedent-consequent diversity measure would be the best to use.

As outlined, antecedent-consequent diversity rates rules higher if their antecedents and consequents have little in common (weak relationship). Users whose profile is in the high diversity set have shown an interest in books from a wide variety of interests, not a handful of specific sub-topic areas (try to suppress one "interest"). The antecedent-consequent diversity measure matches better with (suppress with) this behaviour than confidence measures. By using antecedent-consequent diversity to expand user profiles, more diversity is added into those profiles. This then results in the recommender system having a larger set of diverse profiles to work with, making more diverse recommendations more common.

2.6 Conclusion

In this chapter, we have proposed two interestingness measures for association rules derived from multi-level datasets. These proposed interestingness measures are diversity and peculiarity (distance) respectively.

Diversity is a measure that compares items within a rule and peculiarity compares items in two rules to see how different they are. In our experiments, we have shown how diversity and peculiarity distances can be used to identify potentially interesting rules that normally would not be considered as interesting using the traditional support and confidence approaches. Through looking at the distribution of rules in our experiment, it can be seen that our proposed measures result in a different distribution from that of support and confidence. Thus, this shows the potential for a set of rules to be viewed from a different viewpoint of what is actually interesting.

We have further shown that our proposed diversity measures can be used in the application of improving recommender system performance through short user profile expansion with association rules. The experiment shows that it is possible for the antecedent-consequent diversity measure to outperform the traditional confidence measure. Thus, depending on the situation, it is possible that confidence is not the best interestingness measure to use and others, such as our proposed measures, should be considered.

References

1. Agrawal, R., Imielinski, T., Swami, A.: Mining association rules between sets of items in large databases. ACM SIGMOD International Conference on Management of Data (SIGMOD'93), pp. 207–216. Washington D.C., USA, May 1993
2. Geng, L., Hamilton, H.J.: Interestingness measures for data mining: a survey. ACM Comput. Surv. (CSUR) **38**(3), 9 (2006)
3. McGarry, K.: A survey of interestingness measures for knowledge discovery. Knowl. Eng. Rev. **20**(1), 39–61 (2005)
4. G. Dong and J. Li, Interestingness of Discovered Association Rules in terms of Neighbourhood-Based Unexpectedness, Second Pacific-Asia Conference on Knowledge Discovery and Data Mining (PAKDD'98), pp. 72-86, Apr 1998, Melbourne, Australia
5. Lallich, S., Teytaud, O., Prudhomme, E.: Association rule interestingness: measure and statistical validation. Qual. Measur. Data Min. **43**, 251–276 (2006)
6. Han, J., Fu, Y.: Discovery of multiple-level association rules from large databases. 21st International Conference on Very Large Databases (VLDB'95), pp. 420–431, Zurich, Switzerland, Sep 1995

7. Han, J., Fu, Y.: Mining multiple-level association rules in large databases. IEEE Trans. Knowl. Data Eng. **11**(5), 798–805 (1999)
8. Thakur, R.S., Jain, R.C., Pardasani, K.R.: Mining level-crossing association rules from large databases. J. Comput. Sci. **2**(1), 76–81 (2006)
9. Liu, J., Fan, X., Qu, Z.: A new interestingness measure of association rules. 2nd International Conference on Genetic and Evolutionary Computing, pp. 393–397, Sep 2008
10. Kotsiantis, S., Kanellopoulos, D.: Association rule mining: a recent overview. GESTS Int. Trans. Comput. Sci. Eng. **32**(1), 71–82 (2006)
11. Silberschatz, A., Tuzhilin, A.: What makes patterns interesting in knowledge discovery systems. IEEE Trans. Knowl. Data Eng. **8**(6), 970–974 (1996)
12. Lenca, P., Vaillant, B., Meyer, B., Lallich, S.: Association rule interestingness: experimental and theoretical studies. Stud. Comput. Intell. **43**, 51–76 (2007)
13. Huebner, R.A.: Diversity-based interestingness measures for association rule mining. 16th Annual Meeting of American Society of Business and Behavioral Sciences, vol. 16, no. 1, pp. 19–22. Las Vegas, USA, Feb 2009
14. Anandhavalli, M., Ghose, M.K., Gauthaman, K.: Interestingness measure for mining spatial gene expression data using association rule. J. Comput. **2**(1), 110–114 (2010)
15. Knorr, E.M., Ng, R.T., Tucakov, V.: Distance-based outliers: algorithms and applications. Int. J. Very Larg. Databases **8**(3–4), 237–253 (2000)
16. Zhong, N., Ohshima, M., Ohsuga, S.: Peculiarity oriented mining and its application for knowledge discovery in amino-acid data. 5th Pacific Asia Conference on Knowledge Discovery and Data Mining (PAKDD'01), pp. 260–269, (2001)
17. Zhong, N., Yao, Y.Y., Ohshima, M.: Peculiarity oriented multidatabase mining. IEEE Transactions on Knowledge and Data Engineering, vol. 15, no. 4, pp. 952–960, Jul/Aug 2003
18. Balderas, M.-A., Berzal, F., Cubero, J.-C., Eisman, E., Marin, N.: Discovering hidden association rules. International Workshop on Data Mining Methods for Anomaly Detection at the 11th ACM SIGKDD International Conference on Knowledge Discovery and Data Mining (KDD'05). Chicago, Illinois, USA, 21 Aug 2005
19. Otey, M.E., Parthasarathy, S., Ghoting, A.: An empirical comparison of outlier detection algorithms. International Workshop on Data Mining Methods for Anomaly Detection at the 11th ACM SIGKDD International Conference on Knowledge Discovery and Data Mining (KDD'05), Chicago, Illinois, USA, 21 Aug 2005
20. Yang, J., Zhong, N., Yao, Y., Wang, J.: Peculiarity analysis for classifications. 9th IEEE International Conference on Data Mining (ICDM'09), pp. 607–616, Miami, Florida, USA, 6–9 Dec 2009
21. Han, J.: Mining knowledge at multiple concept levels. 4th International Conference on Information and Knowledge Management, pp. 19–24. Baltimore, Maryland, USA (1995)
22. Han, J., Kamber, M.: Data mining: concepts and techniques, 2nd edn. Morgan Kaufmann Publishers, San Francisco (2006)
23. Ong, K.-L., Ng, W.-K., Lim, E.-P.: Mining multi-level rules with recurrent items using FP'-Tree. 3rd International Conference on Information, Communications and Signal Processing, Singapore, 2001
24. Liu, B., Hu, M., Hsu, W.: Multi-level organization and summarization of the discovered rules. Conference on Knowledge Discovery in Data (SIGKDD'00), pp. 208–217. Boston, Massachusetts, USA, 2000
25. Zaki, M.J.: Mining Non-Redundant Association Rules. Data Min. Knowl. Disc. **9**(3), 223–248 (2004)
26. Pasquier, N., Taouil, R., Bastide, Y., Stumme, G.: Generating a condensed representation for association rules. J. Intell. Inf. Syst. **24**(1), 29–60 (2005)
27. Xu, Y., Li, Y.: Generating concise association rules. 16th ACM Conference on Conference on Information and Knowledge Management (CIKM'07), pp. 781–790. Lisbon, Portugal, 6–8 Nov 2007
28. Xu, Y., Li, Y.: Mining non-redundant association rules based on concise bases. Int. J. Pattern Recognit Artif Intell. **21**(5), 659–675 (2007)

29. Xu, Y., Li, Y., Shaw, G.: Concise representations for approximate association rules. IEEE International Conference on Systems, Man & Cybernetics (SMC08), Singapore, 12–15 Oct 2008
30. Shaw, G., Xu, Y., Geva, S.: Eliminating redundant association rules in multi-level datasets. 4th International Conference on Data Mining (DMIN'08), Las Vegas, USA, 14–17 Jul 2008
31. Shaw, G., Xu, Y., Geva, S.: Deriving non-redundant approximate association rules from hierarchical datasets. ACM 17th Conference on Information and Knowledge Management (CIKM'08), pp. 1451–1452. Napa Valley, USA, 26–30 Oct 2008
32. Shaw, G., Xu, Y., Geva, S.: Extracting non-redundant approximate rules from multi-level datasets. 20th IEEE International Conference on Tools with Artificial Intelligence (ICTAI'08), pp. 333–340, Dayton, USA, 3–5 Nov 2008
33. Ziegler, C.-N., McNee, S.M., Konstan, J.A., Lausen, G.: Improving recommendation lists through topic diversification. 14th International Conference on World Wide Web WWW'05), pp. 22–32. Chiba, Japan, May 2005
34. Shaw, G., Xu, Y., Geva, S.: Using association rules to solve the cold-start problem in recommender systems. 14th Pacific-Asia Conference on Knowledge Discovery and Data Mining (PAKDD'10), pp., 21–24. Hyderabad, India, Jun 2010
35. Ziegler, C.-N., Lausen, G., Schmidt-Thieme, L.: Taxonomy-driven computation of product recommendations. International Conference on Information and Knowledge Management (CIKM04), pp. 406–415. Washington D.C., USA, Nov 2004

Chapter 3
Building a Schistosomiasis Process Ontology for an Epidemiological Monitoring System

Gaoussou Camara, Sylvie Despres, Rim Djedidi and Moussa Lo

Abstract This chapter describes the design of an ontology that aims to support a monitoring system for schistosomiasis. On one hand, a *domain ontology* for the schistosomiasis is built to support communication and collaborative work between domain experts along the monitoring steps. The domain ontology also supports data and application integration and allows some reasoning capabilities. On the other hand, a *process ontology* of the schistosomiasis spreading is built for explanations and decision-making. Furthermore, the possibilities of using this process ontology for prediction are also worth considering. Here, we have focused on the design of the process ontology of infectious disease spreading and its extension to schistosomiasis. We aim to provide a formal theory in the health domain to conceptualize the processes of the infectious disease spreading and to present reasoning capabilities on the disease occurrences within a population. We emphasize on the basic entities and their relations within the complex process of infectious disease spreading. A multi-level analysis of the global dynamics is provided, taking into account biomedical, clinical and epidemiological dependences. We then propose a formalization of the infectious disease spreading process. Finally, we extend the process ontology for schistosomiasis spreading in Senegal.

3.1 Introduction

The context of this chapter is the design of an epidemiological monitoring system whose main goals is to prevent and control an epidemiological phenomenon. A monitoring system aims to detect the event occurrences constituting risk factors for

G. Camara (✉) · S. Despres · R. Djedidi
LIM&BIO, Université Paris, 13, 74 rue Marcel Cachin 93017 Bobigny, France
e-mail: gaoussoucamara@gmail.com

G. Camara · M. Lo
LANI, Université Gaston Berger, B.P.234Saint-Louis, Sénégal

C. Faucher and L. C. Jain (eds.), *Innovations in Intelligent Machines-4*,
Studies in Computational Intelligence 514, DOI: 10.1007/978-3-319-01866-9_3,

disease spreading. Predictions are then made for evaluating consequences and suggesting action plans to prevent identified risks and act in advance [1]. Infectious disease spreading is considered as a system made of several entities (host, vector, pathogen agent, risk factors, etc.) whose interactions lead to event emergence (new disease cases, water infestation, etc.) at different spatial (regional, continental, etc.) and temporal (seasonal) levels. Infectious disease spreading, considering its evolution and emergence properties, is characterized as complex systems and can therefore be modeled as a process [2].

After presenting the ontology-based architecture of an epidemiological monitoring system, we focus on building an ontology of infectious disease spreading process that is then extended for schistosomiasis. Although Infectious Disease Ontology (IDO) exists and provides formal representation of domain knowledge, the spreading dynamics is not explicitly taken into account. In our proposal, we clearly model the underlying spreading process below the "process ontology". The added value of such an ontology in epidemiological monitoring systems is presented in [3, 4]. Indeed, the simulations performed within the monitoring system for risk prediction are based on numerical models that are hardly usable when data collection is difficult or not possible at all. The purpose of using process ontology is to allow reproducing the possible behaviors of the process of infectious disease spreading from the abstract description of its internal sub-processes. Therefore, reasoning on this "process ontology" will allow predicting possible state or process occurrences, or explain the causes of process or state occurrences without constraining them by the availability in real time of numerical data [5]. In the epidemiological monitoring context, this qualitative reasoning approach is useful in identifying causes of the infectious disease emergence and spreading.

The design of the monitoring system and the usage of the process ontology for prediction are not addressed in this chapter. We have essentially focused on the study of the concepts belonging to the infectious disease domain and their relationships within the spreading process. We have presented a multi-level analysis of the global spreading process dynamics taking into account dependencies between biological, clinical and epidemiological levels of granularity. Rather than giving a complete formal and axiomatic ontology, we have emphasized on designing reusable and extendable process ontology for infectious disease spreading that can be customized to any other infectious disease. The usage of this process ontology within the epidemiological monitoring system is illustrated throughout the observation explanation and decision making functionalities.

The chapter is organized as follows: we present in Sect. 3.2 the different ontological modeling approaches, the ontology-driven monitoring system of disease spreading in Sect. 3.3 and the infectious disease spreading process in Sect. 3.4 Then, after presenting the design of the schistosomiasis ontology in Sect. 3.5, we present the foundations of the modeling of process ontology in Sect. 3.6. We then describe the process ontology of the infectious disease spreading in Sect. 3.7 and detail its extension to the spread of schistosomiasis in Sect. 3.8. Finally, we conclude and discuss in Sect. 3.9.

3.2 The Ontological Modeling Approaches

There are two main approaches for building ontology: an object-centered approach for domain ontology and a process-centered approach for process ontology [6].

Building domain ontology is essentially centered on the building a model describing domain entities and their relationships regardless of their evolution over time. Such ontology provides a domain terminology to facilitate communication between domain actors, and formally makes explicit the domain knowledge [7]. This guarantees semantic reasoning on domain knowledge and ensures interoperability between applications of the same domain.

Building a process ontology enables the study of the way domain entities are involved in the dynamics of processes in which they participate. Process ontology essentially focuses on describing how processes unfold over space and time, and their relationships with involved objects. The process occurrences, usually called events, are also studied as well as the effect of these occurrences on the state of other entities such as objects.

It is important to note that, in the context of the infectious disease monitoring, it is necessary to model process ontology in conjunction with domain ontology as well as the relationships between each other. On one hand, we have to build a terminology and a model to facilitate data integration and application interoperability, on the other hand, we have to design a model underlining sub-processes within the spreading process in order to perform qualitative simulations for control and planning. The domain ontology building mainly consists in extending the Infectious Disease Ontology Core (IDO-Core) for the considered infectious disease such as schistosomiasis. The IDO ontology is detailed in Sect. 3.5.

3.3 Ontology-Driven Monitoring System of Disease Spreading

Monitoring is the continuous surveillance of a phenomenon in order to anticipate and control its evolution. In epidemiological monitoring, the monitored phenomenon is the spreading of a disease among populations over space and time. Disease spreading is characterized by the evolution of the number of infected persons in a population. The risk of being contaminated is drastically increased by the presence of risk factors which influence the disease spreading. These risk factors may be related to environmental, human or behavioral characteristics, among others. Thus, epidemiological monitoring consists in identifying and controlling these risk factors. When an event related to a risk factor is detected, all data required to analyze its impact upon the disease spreading is collected. The analysis consists in handling numerical simulations in order to predict precisely the consequences that may stem from the occurrence of an event constituting a risk factor.

Quantitative risk prediction is driven by numerical simulations using numerical models. These models built from descriptive surveys, help in explaining the dynamics of disease spreading and validating assumptions. However, they are rarely used for prediction and decision-support purposes in monitoring context. Indeed, simulation models are either designed for very limited targets, or require input data that is difficult to acquire in real-time simulations. In addition, the many ways used to represent these models (regular differential equations, agent-based models, etc.) restricts model composition and interoperability that is needed to answer to complex queries. Furthermore, monitoring process brings together many practitioners working in different levels of granularity and thus, have different perspectives (biology, clinic and epidemiology) and understanding on the same phenomenon. To communicate, collaborate and share the knowledge, different organizations and actors involved need common vocabulary. To overcome these issues, an ontology-based qualitative approach is proposed for epidemiological monitoring system showing how an ontology fits into classical monitoring systems.

The architecture of the proposed monitoring system is based on an epidemiological monitoring ontology (Fig. 3.1). Monitoring allows detecting the events related to disease spreading risk factors. The *Monitoring Request & Result Manager* (*MRRM*) supports user's requests analysis and results formatting and interpretation. The simulation models used by the *Quantitative Simulation Manager* are stored in the *Numerical Simulation Model Library*. If simulation models or data (from the *Data Collection System*) are not available, a qualitative simulation (*Qualitative Simulation Module*) is performed based on ontological reasoning using process ontology models. Otherwise, the requests are processed by numerical simulations (*Quantitative Simulation Module*). In this case, the MRRM supervises the implementation of simulations by guiding the selection and composition of numerical simulation models. If a disease spreading risk is identified at the end of the analysis process, an alert is triggered. Decisions are generated and evaluated by a quantitative or qualitative approach to prevent identified risks.

Epidemiological monitoring includes knowledge related to the domain (pathogen, host, disease, monitoring resources such as simulation models, etc.) that are modeled by classical *domain ontology*, and knowledge related to dynamic aspects corresponding to the different underlying processes (parasite life cycle, disease spreading, monitoring phases, etc.) that are modeled by the so-called *process ontology*. In this chapter, we will focus on modeling a spreading process ontology of the monitored disease which is here an infectious disease.

3.4 The Process of Infectious Disease Spreading

In this section, we study the process of infectious disease spreading in order to identify entities and their relationships for the ontology building. We first present the standard model of the process of infectious disease spreading. The

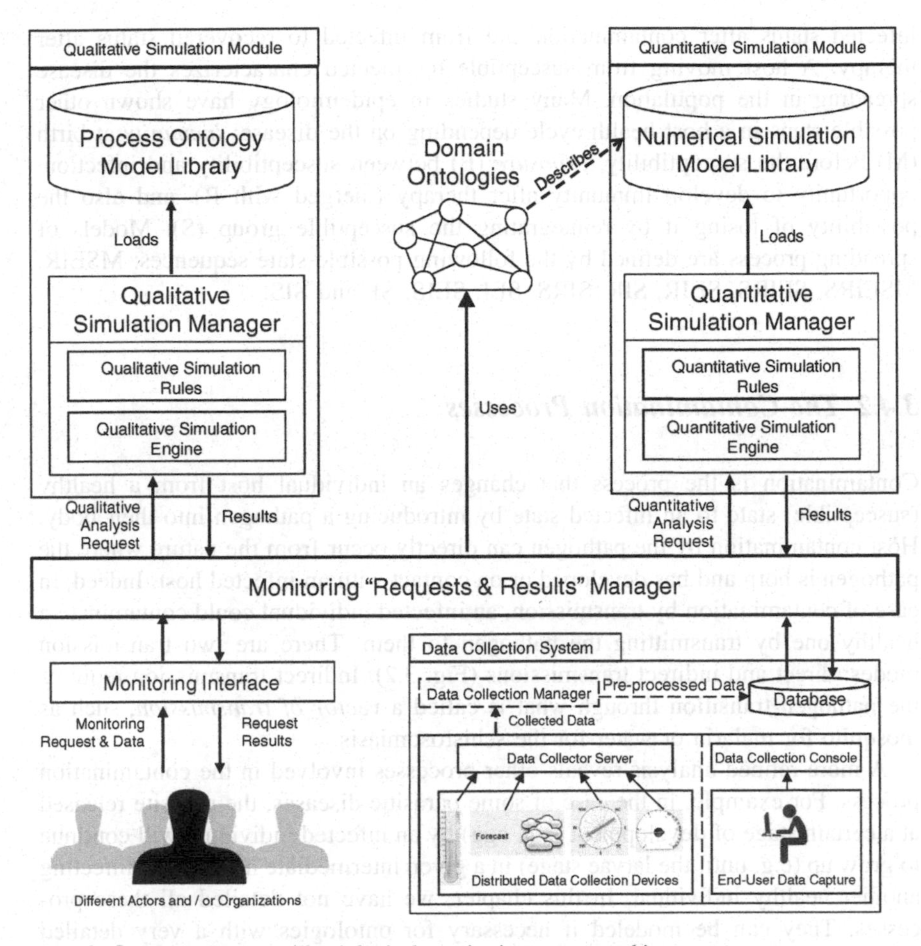

Fig. 3.1 Ontology-driven epidemiological monitoring system architecture

contamination process model is then presented. This contamination process increases the number of infected hosts in the population every time it occurs. We then describe the role played by risk factors in the spreading process.

3.4.1 *Standard Model of the Infectious Disease Spreading Process*

In epidemiology, there is a standard mathematical model representing the process of infectious disease spreading at the population level [8]. It is a compartmental model, dividing a population into three host groups according to their health state: *Susceptible*, *Infected* and *Recovered* (SIR). A host can change from susceptible to infected status after contamination and from infected to recovered status after therapy. A host moving from susceptible to infected characterizes the disease spreading in the population. Many studies in epidemiology have shown other possible steps in a host health cycle depending on the disease: *immunity* at birth (M) before the susceptibility, *exposure* (E) between susceptibility and infection, opportunity to develop immunity after therapy (merged with R), and also the possibility of losing it by reintegrating the susceptible group (S). Models of spreading process are defined by the following possible state sequences: MSEIR, MSEIRS, SEIRS, SEIR, SIR, SIRS, SEI, SEIS, SI, and SIS.

3.4.2 The Contamination Processes

Contamination is the process that changes an individual host from a healthy (susceptible) state to an infected state by introducing a pathogen into their body. Host contamination by the pathogen can directly occur from the nature where the pathogen is born and has developed or on contact with an infected host. Indeed, in case of contamination by transmission, an infected individual could contaminate a healthy one by transmitting the pathogen to them. There are two transmission modes: direct and indirect transmissions (Fig. 3.2). Indirect transmission requires the pathogen transition through what is called a *vector of transmission*, such as mosquito for malaria or water for the schistosomiasis.

A more refined analysis reveals other processes involved in the contamination process. For example, in the case of some parasitic diseases, the parasite released at a certain stage of development (e.g. egg) by an infected individual will continue to grow up (e.g. until the larvae stage) in a given intermediate host before infecting another healthy individual. In this chapter, we have not detailed all these processes. They can be modeled if necessary for ontologies with a very detailed granularity.

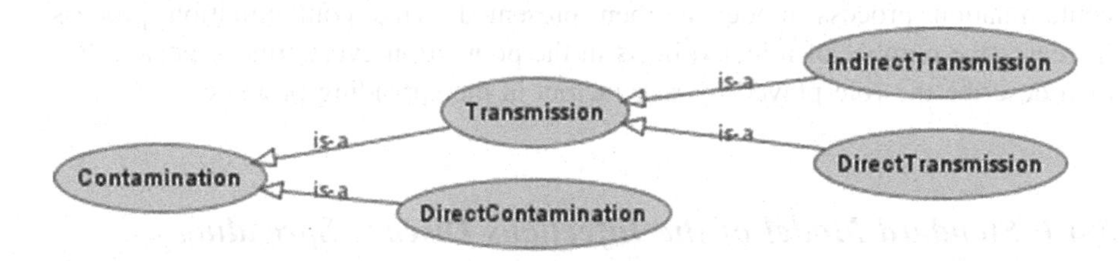

Fig. 3.2 The contamination modes of infectious disease

3.4.3 The Risk Factors in the Infectious Disease Spreading

Risk factors are all the elements, including *pathogen reservoir* [9], that have an impact on the disease spreading. They can be related to biology (infection, susceptibility), economic activities (agriculture, fishing, etc.), environment (water, climate, etc.), logistics (hospitals, dams, etc.), behavior (lack of hygiene), etc. Risk factors could either be events that trigger the occurrence of other events or changes of some object states (e.g. rain could change the status of a water point from dry to filled), or states promoting an event occurrence (e.g. a water point infested with parasite could contaminate a host individual). In this regard, risk factors are closely related to events and object states. They play a causal role in the infectious disease spreading. However, causality in epidemiology is probabilistic [10]. Therefore, modeling risk factors in the process ontology of disease spreading has to take in account probabilistic theories and models.

3.4.4 Summary

The entities considered in our models are mainly *spreading*, *contamination*, *population*, *host* individuals, *states* of host individuals and their *transitions*, *pathogens*, transmission *vectors* and *risk factors*. Categorizations of these entities and their relationships are studied during the design step of the ontology of infectious disease spreading in Sect. 3.7. Theory and model about probabilistic causality of risk factors are not addressed here.

3.5 Design of the Schistosomiasis Ontology

In this section, we present the framework used to build the schistosomiasis ontology (IDOSCHISTO) as an extension of the Infection Disease Ontology. This framework is based on a classical approach in the ontological engineering domain. IDOSHISTO stands for *Infectious Disease Ontology for SCHISTOsomiasis* and is organized in two modules: domain ontology and process ontology. The building of IDOSCHISTO aims at supporting the monitoring of schistosomiasis. The conceptual framework is detailed in Sect. 3.5.1. The ontology language used to formalize the spreading process ontology module is presented in Sect. 3.5.2.

3.5.1 Conceptual Framework

The conceptual framework considered here for building a specific infectious disease ontology is organized in three layers: specific domain ontologies, core ontology and foundational ontology. In the layer of the specific domain ontologies, we distinguish the modeling of domain knowledge (domain ontologies) and process knowledge (process ontologies) of the infectious disease (Fig. 3.3). This distinction is also considered in the core and foundational ontologies by providing a classification of domain entities within top level categories such as *occurrents* and *continuants* [11] (Sect. 3.6.3).

3.5.1.1 Domain Specific Layer

The domain specific layer models the specific aspects of the infectious disease such as spreading, treatment, symptom, diagnosis, vaccine, risk factors, transmission, etc. In this chapter, we focus on modeling the processes of the infectious disease spreading. Two specific domain ontologies are then proposed (Fig. 3.3): the spreading domain ontology and the spreading process ontology. The spreading domain ontology has modeled domain entities involved in the process of the infectious disease spreading and their links between each other. Based on core and foundational ontologies, it assumes the ontological categorization of domain entities in these ontologies. This guaranties its interoperability with the other ontologies built on the same framework and using the same core and foundational ontologies. The process ontology of spreading focuses on representing the unfolding of dynamic entities instead of their categorization only.

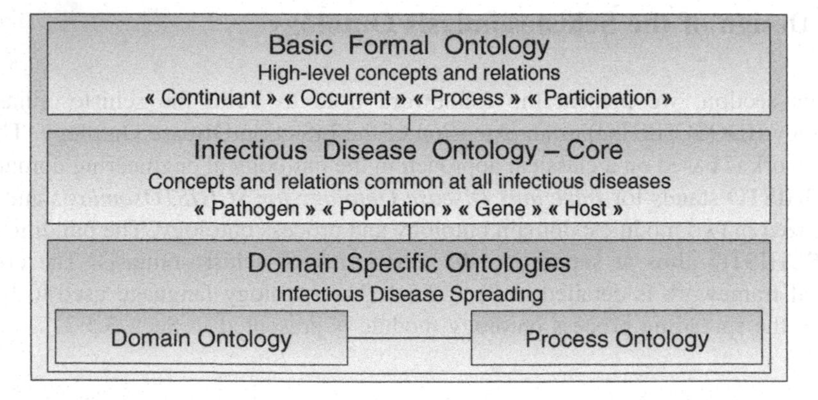

Fig. 3.3 Modeling framework of the infectious disease spreading ontology

3.5.1.2 Core Layer

The core ontology layer, as suggested in ontology building methodologies, contains common general concepts of a domain and their relevant relationships[12, 13]. In the infectious disease domain, an Infectious Disease Core Ontology (IDO-Core) has been developed and which covers its general concepts (e.g. pathogen, infectious agent, gene, cell, organ, organism, population, host, vector organism, human, etc.) and their relationships. These concepts are relevant to both biomedical and clinical perspectives [14].

IDO is developed and maintained by a consortium of experts belonging to the ontological engineering and infectious disease field. Moreover, several ontologies (for Brucellosis, Dengue fever, infective endocarditis, influenza, malaria and other vector-borne diseases, Staphylococcus aureus, tuberculosis) have already been designed as extensions of the Infectious Disease Ontology (IDO). They provide a formal representation of specific disease domain knowledge to support interoperability and reasoning capabilities. These ontologies, usually called domain ontologies, focus on describing domain entities and their relationships regardless of the way they unfold, and this specifically for the *occurrent* entities.

In order to guarantee the interoperability and reusability of the schistosomiasis ontology within the infectious disease domain, we reuse IDO-Core as core ontology in the framework (Fig. 3.3). It is important to notice right now that IDO is already reusing BFO [14] as foundational ontology.

3.5.1.3 Foundational Layer

The foundational ontologies describe very general concepts and relations through all domain knowledge. They support building core and domain ontologies and ensure interoperability between applications. However, these foundational ontologies are based on different philosophical principles such as, for instance, realism (real world) for BFO while DOLCE is descriptive (Natural language and human common-sense). It implies that selecting one of them to reuse in domain ontology building depends on various criteria such as the application purpose (e.g. information retrieval, data integration, etc.) and the domain (e.g. Biomedical, Linguistics, Environment, etc.).

The foundational ontology selection is based on three main criteria: (1) the reuse of a foundational ontology by the core ontology of the framework, namely IDO–Core, (2) the consistency of the categorization of the process concept in the foundational ontology according to the characterization of the process of the infectious disease spreading, and (3) the richness of their relations for describing process dynamics and availability of axioms for automatic reasoning capabilities.

If we prioritize the interoperability of the schistosomiasis ontology in the infectious disease domain, the first criterion constrains us in using BFO as foundational ontology in the framework (Fig. 3.3). Although we aim at building an interoperable ontology within the infectious disease domain, we give advantage to its use for supporting monitoring activities. Therefore, we have used the ONSET tool [15] to test if this choice fits the application purpose of our ontology. The following criteria are specified for the different steps of the ONSET tool. In the first step, the categories Ontological Commitment, Representation Language, Subject Domain and Applications were equally scaled at 3. The Software Engineering category is put at 0, meaning that it is ignored in the foundational ontology selection process. For the following step consisting in specifying details in the different categories, we choose the following values:

- Ontological Commitment: Universal/Classes/Concepts, realist, multiplicative, actualism, concrete entities, no usage of properties and their values, mereology and model of space–time were not necessary, temporal aspects and different granularity levels needed, one-layered architecture and no situations and situoids usage.
- Representation Language: OWL DL.
- Subject Domain: Medical Informatics or Biomedical.
- Applications: Ontology Driven Information Systems and Data Integration.

ONSET proposed both DOLCE and BFO as possible foundational ontologies for our purpose according to the criteria we provided above. The reasons for which DOLCE and BFO were selected are that they have been used in biomedical domain and applied to database integration. Moreover, DOLCE has also been applied to ontology driven information system while ONSET says that BFO has not.

Though BFO has not been applied to ontology driven information system, we stick to this choice because of its fitting with the first criterion about the core ontology and the fact that it appears in the ONSET results.

BFO is narrowly focused on the task of providing a genuine upper ontology [16] which can be used in support of domain ontologies developed for scientific research, as for example in biomedicine within the framework of the Open Biomedical Ontology (OBO) Foundry [14].

3.5.2 Formalization of the Infectious Disease Spreading Process

This section presents the formal specification language used to describe the process of the infectious disease spreading. This formalization aims at providing a representation of the process dynamics to the epidemiological monitoring system so as to perform qualitative simulations. Qualitative simulations on this process ontology allows prediction, explanation of event occurrences related to the infectious disease spreading and decision making for control and prevention of the disease spreading.

The existing process ontology languages[1] are essentially designed for process specification in manufacturing and business sectors. In this chapter, we are using PSL[2] [17]. PSL is a set of logical terms that enables representing possible sequences of a set of activities, their transitions, their occurrence conditions, the effects of their occurrences on the state of the involved entity. The logical terms are specified in an ontology that provides a formal description of the components and their relationships that make up a process. The primitive concepts of PSL are *activities, activity occurrences, objects* and *time points*. To formalize the processes of the infectious disease spreading, we have been using the PSL concepts of *activity* and *activity occurrence* for respectively modeling processes and events. The two latter concepts are described in the next section according to their characterization within the infectious disease domain.

PSL is organized in two categories: (1) "Core Theories" which describe a set of primitive concepts (*object, activity, activityoccurrence, timepoint*), relations (*occurrence_of, participates_in, between, before, exists_at, is_occurring_at*), constants (*inf+, inf-*) and functions (*begin of, end of*), and (2) "Definitional Extensions" extending primitive elements of the language for defining complex concepts and relations. PSL axioms are available in Common Logic Interchange Format[3] (CLIF). The PSL primitives used in our models are detailed in http://www.mel.nist.gov/psl/. The PSL representation supports automated reasoning on processes. Therefore, it can be used within the monitoring system for process planning, process modeling, process control, simulation, etc.

3.6 Ontological Modeling of Processes

In the next sections, we explore the definition of the process of infectious disease spreading as well as the categorization of processes in the chosen foundational ontology, namely BFO. Then, we present the definition and classification of *events, objects* and *states* in BFO. These three concepts are indispensable in process modeling. Finally, we will present the existing relations for process modeling by focusing essentially on those that are relevant for the process of infectious disease spreading.

Although there are numbers of process definitions, this section only addresses those that are relevant to infectious disease spreading. More precisely, we consider definitions within complex systems, infectious disease domain and BFO.

[1] Cyc, DDPO, oXPDL, m3po et m3pl, PSL, etc.

[2] The PSL was developed at the National Institute of Standards and Technology (NIST), and is approved as an international standard in the document ISO 18629.

[3] http://philebus.tamu.edu/cl/

3.6.1 What is a Process?

The design of infectious disease spreading process is fundamentally based on Le Moigne's definition. He states that we can talk about process when there is, over *time*, the position change in a referential *Space-Form* of a collection of any *Products* identifiable by their shape [...]. However, process modeling in the domain of ontology engineering is based on formal ontology. BFO definition of process comes also from this domain: A *processual* entity is an occurrent [span:Occurrent] that exists in time by occurring or happening, has temporal parts and always involves and depends on some entity. Although we have the term *process* in BFO hierarchy, its definition is closer to an event than a process type.

Le Moigne's definition of process highlights three main concepts: *time*, *change* (of form) and *space* while BFO mainly focuses on the temporal aspect. Aligning these two definitions, space and objects can be considered as involved in the processual entities and the change happens on the object and process states. Therefore, processes will be modeled in the next part of this chapter as an entity unfolding over time and space, involving objects, and that could undergo changes. These process properties that change over time are called temporal properties. Referring to the concept of *time window* as defined in [18], the change occurs in a temporal property (of the process) from a time window to another (e.g. a disease spreading which may be slow in a time window and fast in the next one).

3.6.2 Processes and Associated Concepts

Modeling of process dynamics requires the consideration of other concepts such as events, objects and states. We study in this section the relations between processes and these concepts as well as their definition and classification in BFO.

3.6.2.1 Processes Versus Events

Ontologically speaking, events are defined as inherently countable occurrences [19]. This occurrence has a temporality; it is either instantaneous or extended within a defined interval of time [*Tinit*, *Tfinal*]. The event properties are immutable. This means that an event which has already occurred does not change over time. However, in available BFO 1.1 version, the term *event* does not appear explicitly whereas it is classified under the "Instantaneous temporal boundaries" category in the SPAN ontology presented in [16]. This latter categorization restricts them to instantaneous events. Extended events are closer to the definition of processes (not *processual* entities) in BFO: A process entity [span:ProcessualEntity] that is a maximally connected spatiotemporal whole and has bona fide beginnings and endings corresponding to real discontinuities.

To avoid any confusion between event definition in BFO and events as a process occurrence in the modeling spirit of the infectious disease spreading, we will consider concepts given under the processual entity of BFO as the type level. And when operationalizing the process of infectious disease spreading, the primitive PSL *activity_occurrence* is used to model events at the instance level, and *activity* as process type like processual entities in BFO.

3.6.2.2 Processes Versus Objects

Objects are entities without temporal parts. Objects can be either tangible in a physical point of view or abstract in a mental or social point of view. An object persists (by enduring) and its state (values of its properties) can undergo a change. These changes result from *internal* or *external process* occurrences. This internal or external distinction of processes to object [18] is due to the granular structure of the real world. We detail this granular aspect later in this document (Sect. 3.7.2). For example, the object *person* has internal processes (e.g. infection or digestion) and external processes (e.g. disease spreading or contamination processes). More generally, the object structure is maintained by internal and external processes on the one hand; and on the other hand, the object itself is involved in external processes of its environment. For instance, an infected *person* is involved in the spreading of an infectious disease by transmitting pathogen to a healthy individual.

In BFO 1.1, an object is defined as "a material entity [snap:MaterialEntity] that is spatially extended, maximally self-connected and self-contained (the parts of a substance are not separated from each other by spatial gaps) and possesses an internal unity. The identity of substantial object [snap:Object] entities is independent of that of other entities and can be maintained through time."

For PSL, an object is "Anything that is not a time point or an activity". More precisely: "an OBJECT is intuitively a concrete or abstract thing that can participate in an ACTIVITY. The most typical examples of OBJECTs are ordinary, tangible things, such as people, chairs, car bodies, [...]. OBJECTs can come into existence (e.g., be created) and go out of existence (e.g., be "used up" as a resource) at certain points in time. [...]." [20]

These two points of view on objects are compatible and do not raise any problem in modeling of objects participation in the process of infectious disease spreading. It is important for the categorization of objects in BFO and their definition in PSL that they fit the objects as they are considered in infectious disease domain.

3.6.2.3 Processes Versus States

In the process modeling of infectious disease spreading, representation of states inhering in objects or processes is crucial. Indeed, for the control and prevention of the disease spreading, it is necessary to capture the state changes for determining entities such as risk factors. Moreover, a state may also provide a potential of change [19]. For example, the health status of a person and the infection rate of a population are states respectively inhering in host individuals and population collections. Speed of the disease spreading can be fast or slow, depending on the number of infected persons in the population or the presence of risk factors. PSL provides primitives for state modeling but the concept state does not explicitly appear in the BFO 1.1 hierarchy.

3.6.3 Categorization of Process, Event, Object and State in BFO

In foundational ontologies, there is a general distinction between *continuants* (or *endurants)* and *occurrents* (or *perdurants)* [11, 21]. *Endurant* concepts correspond to entities without temporal part whereas *perdurant* concepts correspond to entities occurring over time with different phases. BFO distinguishes the *occurrent* and *continuant* entities at the top of its hierarchy:

- Continuant: an entity [bfo:Entity] that exists in full at any time in which it exists at all, persists through time while maintaining its identity and has no temporal parts. Examples: a heart, a person, the color of a tomato, the mass of a cloud, the disposition of blood to coagulate, etc.
- Occurrent: an entity [bfo:Entity] that has temporal parts and that happens, unfolds or develops through time. Examples: the life of an organism, the most interesting part of Van Gogh's life, the spatiotemporal region occupied by the development of a cancer tumor, etc.

This distinction allows modelers to classify real world entities (such as objects, processes, events and states) in these two categories.

Based on definitions given in the previous sections, we classify objects as *continuants* and events as *occurrents*. An event is generally considered as an instance of a process in a specific interval of time and does not change. We also consider a state as *occurrent* although it doesn't involve dynamics. The two main questions are: (1) in which category processes could be classified? (2) Does this classification respect Le Moigne's definition on which the infectious disease spreading is based? Le Moigne characterizes processes through the four dimensions of Space–Time (x, y, z, t). Referring to the principle of quadridimensionnalism [22], processes are *occurrents*. In the same way as in BFO, processes are occurrent entities.

Objects are continuants but they are closer to *independant_continuants* concepts. The term object already exists under the latter category but its definition could restrict the modeling of the involved object in the process of the infectious disease spreading. States are not explicitly mentioned in the BFO concepts. They are modeled as values of the *dependant_continuants* and *temporal properties* respectively associated with the objects and processes. It should be noted, however, that states are modeled here at the *instance level* and not as *concept type*. It is the same for the events that are represented as instances of the processes. This allows us to represent processes with PSL as *activities* and events as *activity_occurrences*. These adjustments do not violate in any way the principle of BFO classification, but simply allow us to find a compromise to classify the processes of the infectious disease spreading with the BFO and represent their dynamics with PSL.

3.6.4 Relations Between Process, Event, Object and State

Several relations are proposed for ontological modeling. Here we pay attention to the relations used to represent process dynamics. All of these conceptual relations are not available in PSL and in the Relation Ontology (RO) [23] used with BFO concepts.

- *Participation relation*: a *continuant* participates in an *occurrent* [16]. Participation relation could need to be specialized for a very specific domain modeling.
- *Mereological relation*: For instance, a process is made of sub-processes.
- *Internal and external relations*: a process is internal or external to an object [18].
- Causal relations [24]:
  - *causal relation*: an event causes another event
  - *allow, prevent*: a state allows an event
  - *initiate, terminate*: an event initiates a state
  - *enable, disable, perpetuate*: a state enables another state
- *Enaction relation*: The enaction was introduced in [18] to characterize the link between objects and processes in which they *actively* participate: an *object* enacts an *occurrent*. This relation is essentially interesting because of the complex nature (dynamism) of spreading process and could be used to model interactions between entities of different granularity levels.

RO-Bridge links BFO concepts to relations defined in RO. However, these relations are too general to be automatically used for a specific domain application. Some of these relations are specialized and complemented with PSL relations for modeling the process of infectious disease spreading.

3.7 An Ontology for Infectious Disease Spreading

The building of ontology for the infectious disease spreading is based on the framework described in Sect. 3.5.1. The domain ontology and process ontology are both addressed in this section but we emphasize on the latter. In Sect. 3.7.1 a general model of infectious disease spreading is proposed for the domain ontology.

Then, a multi-dimensional analysis for the process of infectious disease spreading is provided in Sect. 3.7.2. The formalization of the spreading process is presented in Sect. 3.8.4 with the extension of the general model for schistosomiasis.

3.7.1 A General Model of the Spread of Infectious Disease

The spread of Infectious disease occurs in a host population and is influenced by risk factors. A population is a collection of individuals; it is therefore modeled as a continuant [25]. According to the complex nature of the infectious disease spreading and its spatio-temporal properties, it is considered as a process in this context. In compliance with BFO classification of processes, the spreading is an occurrent entity. Risk factors are even more difficult to be linked to the spreading. In our model, a risk factor is considered either as an event (hand shaking) or as a state inhering to an object (a host being infected). For instance, if an event *rain* takes place out of the scope of the domain of the disease spreading, this event could have a negative effect on the disease spreading. The spreading process and the raining cannot be linked by an *internal/external* relation because spreading process and raining occurrence are both occurrents. We cannot use the *enaction* relation [18] to say that the rain enacts the disease spreading, because the subject that enacts has to be an object. For the time being, we use the general causal relation [26] between risk factor and spreading process.

The contamination process causes the spreading process because its occurrence implies an increasing number of infected individuals (also called the *disease prevalence*) in the population. This process involves the host and the pathogen.

The model of Fig. 3.4 defines the general process of infectious disease spreading. Of course, entities and relationships are to be specialized by any model that extends this model for a particular infectious disease.

This general model is aligned to the core and foundational entities of the framework (Sect. 3.5.1). Figure 3.5 highlights the hierarchical relations between the general model entities and the IDO and BFO entities. The alignments are done manually and concern the following entities: Spreading (IDOSCHISTO_Spread),

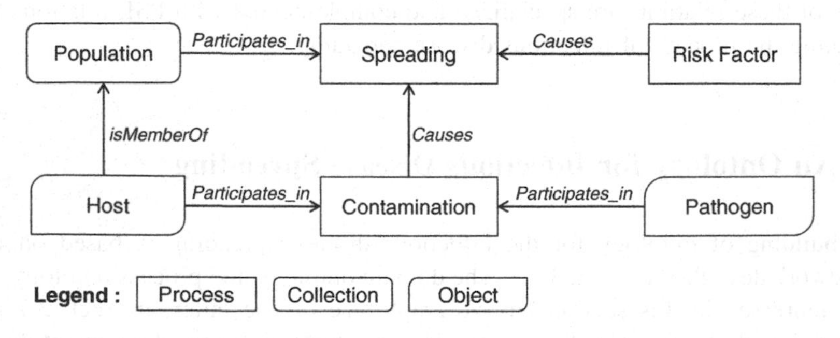

Fig. 3.4 A general model of infectious disease spreading

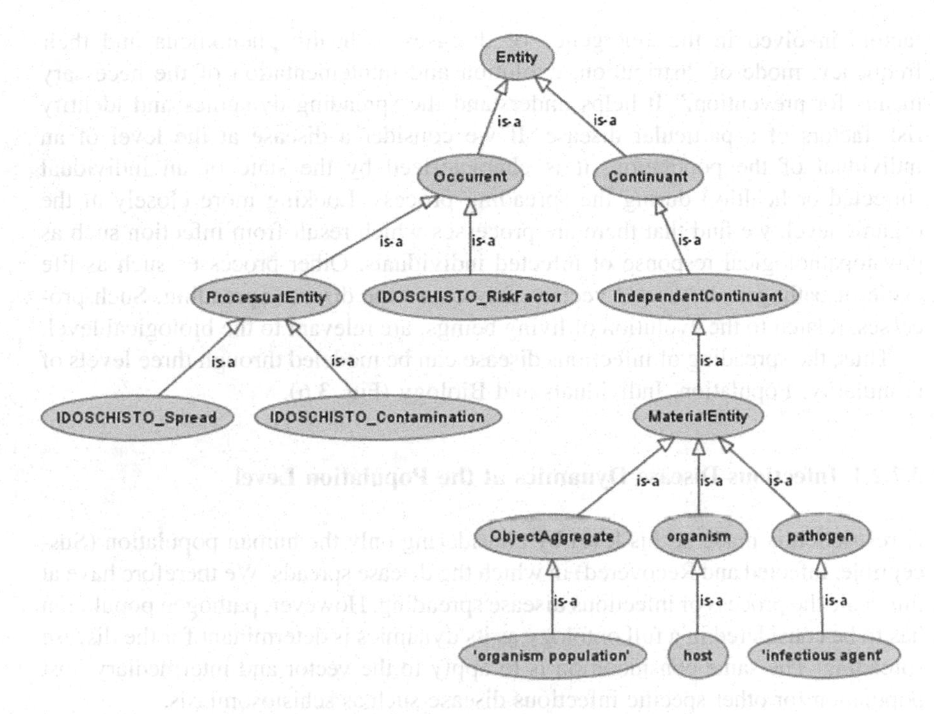

Fig. 3.5 Alignment of the spreading general model to IDO and BFO

Contamination (IDOSCHISTO_Contamination), Risk factors (IDOSCHISTO_RiskFactor), Population ('organism population'), Host (host), and Pathogen ('infectious agent'). The concepts preceded by IDO belong to the IDO-Core ontology and those without prefix belong to the BFO foundational ontology. One can notice that some concepts such as the population already existing in the core and foundational ontology are just imported and don't need to be specialized or redefined.

3.7.2 *Analysis of the Dynamics of the Infectious Disease Spreading*

The analysis provided here can be adapted to any infectious disease and more specifically a parasitic disease.

The spreading of infectious disease in an epidemiological perspective is characterized by the mechanism of the disease spreading in a population. Epidemiology is defined as[4]: "a particular scientific discipline that studies the different

[4] Recommendations "ethics and best practices in epidemiology", 1998, France.

factors involved in the emergence of diseases or health phenomena and their frequency, mode of distribution, evolution and implementation of the necessary means for prevention." It helps understand the spreading dynamics and identify risk factors of a particular disease. If we consider a disease at the level of an individual of the population, it is characterized by the state of an individual (infected or healthy) during the spreading process. Looking more closely at the organic level, we find that there are processes which result from infection such as physiopathological response of infected individuals. Other processes such as life cycle of pathogen, hosts and vectors may cause the disease spreading. Such processes, related to the evolution of living beings, are relevant to the biological level.

Thus, the spreading of infectious disease can be modeled through three levels of granularity: Population, Individuals and Biology (Fig. 3.6).

3.7.2.1 Infectious Disease Dynamics at the Population Level

A restriction is made at this level by considering only the human population (Susceptible, Infected and Recovered) in which the disease spreads. We therefore have at this scale the process of infectious disease spreading. However, pathogen population has to be considered in a full ontology as its dynamics is determinant for the disease spreading. The same consideration is to apply to the vector and intermediary host population for other specific infectious disease such as schistosomiasis.

3.7.2.2 Infectious Disease Dynamics at the Individual Level

The individual level is made of individuals among which we can distinguish human, pathogen, intermediary host, vector, etc. Processes resulting from the interaction between two individuals of different populations such as the contamination process are modeled at this level.

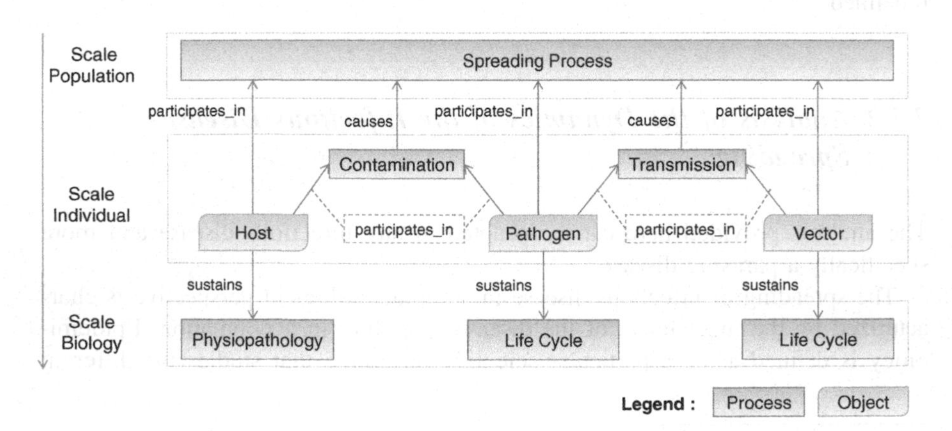

Fig. 3.6 Multidimensional model of the infectious disease spreading dynamics

3.7.2.3 Infectious Disease Dynamics at the Biological Level

The biological level concerns living beings organism with processes such as the life cycle of the pathogen, the pathological response of the infected person, etc. At this scale, it is more difficult to talk about objects (cells, organs, genes, etc.) as it is close to clinical and biological aspects, which are not addressed in this chapter. However, such processes involve more perceptible objects such as human, pathogen and vector. These processes are modeled with the internal relations to the human, vector and pathogen and their occurrences affect the structure of these objects.

This analysis reveals that the spreading process, as the macro-process of the infectious disease, emerges from the interactions between entities at the individual level and process occurrences at the biological level. Applying this to the schistosomiasis, we have therefore proposed a multi-scale model of its spreading process by integrating epidemiological, individual and biological levels in a same model (Fig. 3.6).

3.8 Use Case: Schistosomiasis Spreading in Senegal

This section briefly describes schistosomiasis domain knowledge and emphasizes on the process dynamics.

3.8.1 Knowledge Domain for Schistosomiasis

Schistosomiasis (or bilharzia) is a parasitic disease that affects mammals and poses serious public health challenges in Senegal. In this work, we model the spread in human population. The parasite responsible thrives in water and humans. The dense water network and permanent fresh water points [27] keep prevalence[5] high even during the dry season. The contamination process involves mollusks as intermediary hosts. The mollusks act as a development environment for the parasite. Risk factors cause the contamination process at the individual scale and the spreading at the population scale. Water is the vector of transmission of schistosomiasis. Others in the risk factor category include climate change, rainfall, agriculture and fishing activities (Fig. 3.7).

[5] In epidemiology, the prevalence or prevalence proportion is the proportion of a population found to have a disease. It is arrived at by comparing the number of people found to have the disease with the total number of people studied, and is usually expressed as a fraction, as a percentage or as the number of cases per 10,000 or 100,000 people.

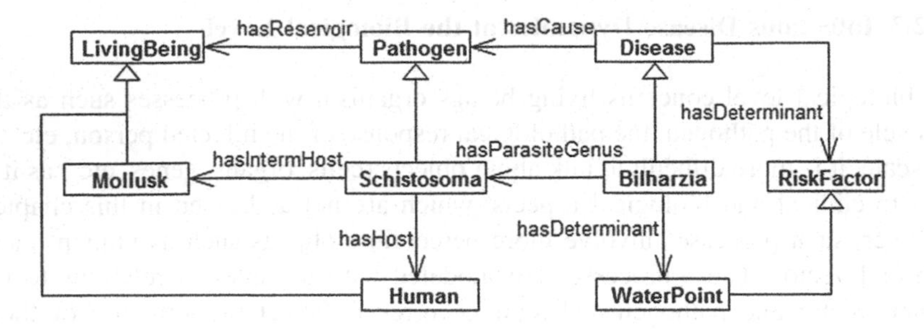

Fig. 3.7 A simplified schistosomiasis domain knowledge model

3.8.2 Spreading Process of Schistosomiasis

The contamination mode of Schistosomiasis is based on an indirect transmission. An infected person contaminates water by urine or feces containing parasite eggs. These eggs hatches in fresh water given the right conditions (temperature of 25–30°C, sunshine, neutral pH) and then, larvae are released. These larvae are housed in gastropod mollusks and evolve to the stage of cercaria before being released again into the water. On contact with water, a healthy person is infected with skin cercaria penetration. From the moment, a pathological reaction is triggered in the organism of the infected individual. The human bodies are contaminated by Cercariae. These cercariae grow to adulthood and mating occurs in the human body. Females lay eggs which are then released by urine or feces and the cycle continues. Here, we just show the human health status and the life cycle of the parasite:

- Schistosomiasis inherits the standard SIR model presented in Sect. 3.4.1. Furthermore, an Exposed (E) state is included between Susceptible and Infected states (SEIR). It is also possible for a recovered person to lose his immunity against the disease; thus, the complete spreading model for schistosomiasis is SEIRS. Figure 3.8 shows the human health status and all transitions. These transition processes are parts of the contamination process at the individual level and parts of the spreading process at the population level.
- The life cycle of the parasite follows several phases: egg, miracidium, sporocyst, cercaria, schistosomula and adult. Each of these phases—taking place itself in an interval of time and in an environment—represents a state of the pathogen in its life cycle. Transition from one state to another is subject to certain conditions (time or event) as showed in Fig. 3.9.

3.8.3 Extension of the Multidimensional Spreading Model for Schistosomiasis

We propose an extension of the multidimensional model (Fig. 3.6) for schistosomiasis. According to the description given in the previous section, in the case of

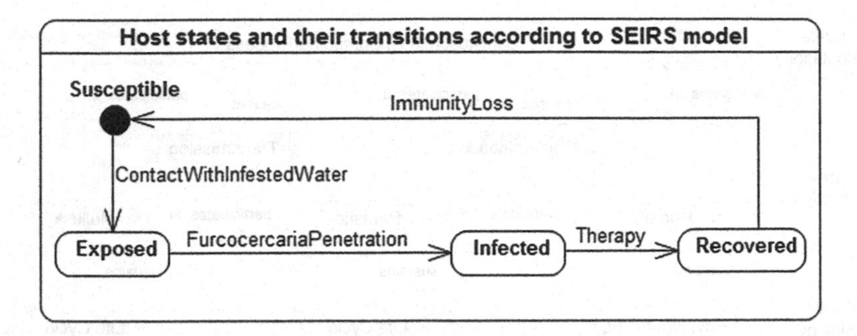

Fig. 3.8 Human health status and their transition

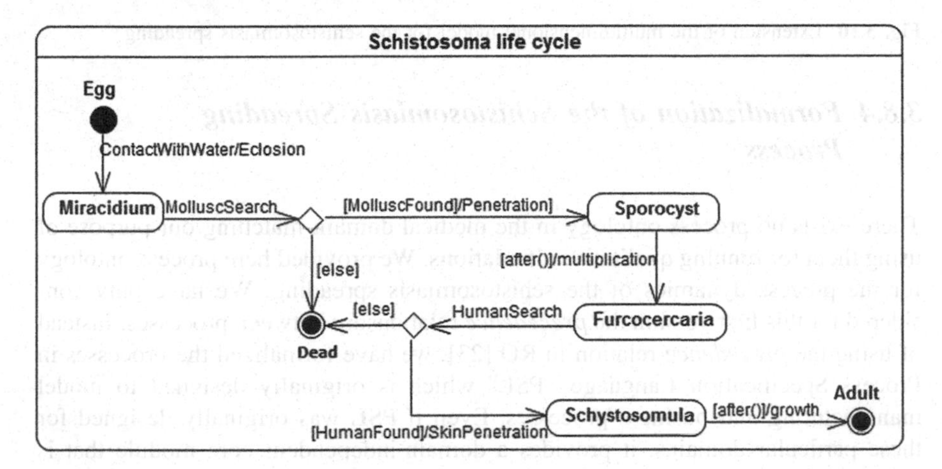

Fig. 3.9 The life cycle of the schistosomiasis parasite

schistosomiasis we have a global dynamics (disease spreading), a set of populations (humans or hosts, pathogens or parasites, and intermediary hosts or mollusks) and a set of biological processes within individuals of the different populations (pathological response in humans, life cycles of the parasite and mollusks). Figure 3.10 shows an extract of this extension (due to a lack of space, we cannot illustrate the whole model extension for schistosomiasis). For example, as shown in Fig. 3.10, the contamination process involves a human and a parasite at the cercaria stage.

To make this extension operational within the monitoring system for the schistosomiasis and support reasoning capabilities, the spreading process for the schistosomiasis is formalized with the Process Specification Language (PSL). An example and its usage are given in the next section for explanation and decision making.

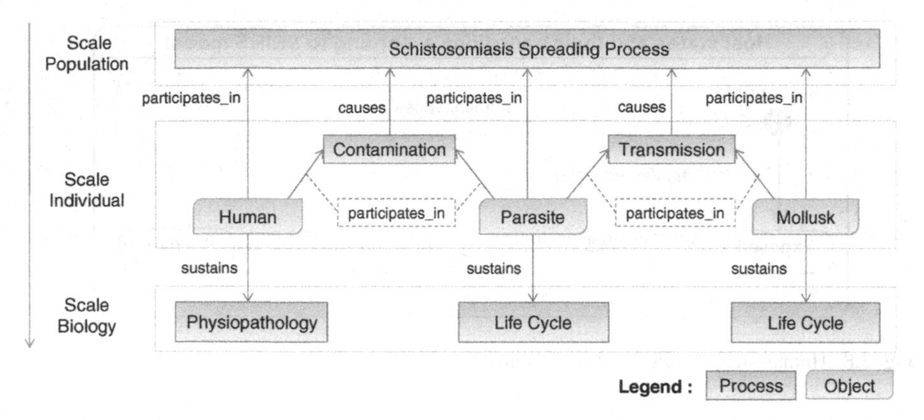

Fig. 3.10 Extension of the multidimensional model for the schistosomiasis spreading

3.8.4 Formalization of the Schistosomiasis Spreading Process

There exists no process ontology in the medical domain matching our purpose of using them for running qualitative simulations. We provided here process ontology for the process dynamics of the schistosomiasis spreading. We have only considered for this first version the *precedence* relationship between processes. Instead of using the *precedence* relation in RO [23], we have formalized the processes in Process Specification Language—PSL[6] which is originally designed to model manufacturing and business processes. Even if PSL was originally designed for these particular domains, it provides a domain-independent core module that is complete enough for our actual purpose and generic enough to let us extend it if needed. As an example, we give below the relation between the *Exposition* and *Contamination* processes (individual level) based on the SEIRS model already presented in this section.

```
;; The occurrence of a contamination is preceded by an exposition
(forall (?occContamination)
  (implies (and (occurrence_of ?occContamination Contamination)
                (subactivity_occurrence ?occContamination ?occSpread))
    (exists (?occExposition)
      (and (occurrence_of ?occExposition Exposition)
           (subactivity_occurrence ?occExposition ?occSpread)
           (earlier ?occExposition ?occContamination)))))
```

[6] http://www.mel.nist.gov/psl/

Reasoning on process ontologies can help interpreting the origin of an observation using preconditions and precedence relationships. It could be useful to identify causes of an infectious disease emergence or spreading. For instance, the occurrence of a new infected host is the result of the occurrence of a contamination process. Since every contamination is preceded by an exposition, according to the SEIRS model above, this means that there exists at least a water point (as it is the source of contamination) in the area of the infected host infested by the schistosomiasis parasite. Therefore, several decisions could be taken depending on whether the host lives in the same area where he was diagnosed (block access to water points and carry out sanitization) or comes from outside (trigger an alert to the public health organization of the area where he comes from).

3.9 Conclusion and Discussion

In this chapter we have presented the results of the analysis and design of schistosomiasis ontology within the context of epidemiological monitoring. The building of the schistosomiasis ontology is based on a framework which is reusing the Infectious Disease Ontology Core (IDO-Core) and the Basic Formal Ontology as a foundational ontology. In the framework, we have distinguished the domain ontology and the process ontology at the domain specific layer. We emphasized on the process ontology building. A general model of the process of infectious disease spreading is proposed. This spreading process model is based on the standard mathematical SIR model of infectious disease spreading. During the analysis of infectious disease spreading process, we identified three dimensions according to the considered granularity level: population, individual and biological level. Each of them was made of a set of processes. The interactions and relationships between processes and objects of different scales are highlighted in the multidimensional model. This multidimensional model could be extended to any type of infectious disease as it is based on high-level concepts and relations that are common to all infectious diseases. An extension of the model was applied to the spreading of schistosomiasis in Senegal. Using such a model to predict and control the spread of infectious disease provides a solid challenge for epidemiological monitoring systems. It is also important to keep schistosomiasis ontology modular so that it could be partially reused for specific purpose such as in clinical or biological contexts.

Looking ahead, a complete axiomatization and formalization of the process ontology need to be provided for more powerful and complete reasoning capabilities. In terms of dynamic systems modeling, the specification of concepts and relationships within complex systems are still immature. For instance, probability or multi-level relations were not well studied and formalized. It is also necessary to study and build theories and technics for using process ontologies to make prediction when numerical data or simulation models are not available to perform numerical simulations within a monitoring system [3, 4]. For the case of schistosomiasis spreading in Senegal, formal models of both domain and process ontologies need to be implemented in a real-time monitoring system to help public health authorities prevent and control the disease spreading.

Process modeling is a key interest across several disciplines. Two main modeling approaches have been developed: numerical and qualitative. Numerical modeling approaches are used to describe physical characteristics of processes in numerical models. These models are built from descriptive surveys and help in understanding the dynamics of disease spreading and validate assumptions. They are then used in monitoring systems for prediction, explanation of observed phenomena and decision-making by performing numerical simulation. While qualitative approaches, such as ontological approach, consider processes as concepts (abstraction). They provide qualitative models for qualitative simulations. However, these two approaches are not antagonistic, but complementary. Another point is that the process categorization does not win unanimous support. Although processes are considered as *occurrents* in foundational ontologies, this classification was discussed by some philosophers. Galton [6] and Stout [28] attempted to consider a process as a *continuant* allowing that it might undergo change. Stout [28] further distinguished between dynamic continuants—processes—and static continuants—physical objects. Galton [6] provided two ontologies: HIST (the *occurrents*) containing events and EXP (the *continuants*) containing processes and objects.

References

1. Buton, F.: De l'expertise scientifique à l'intelligence épidémiologique : l'activité de veille sanitaire. Genèses **4**(65), 71–91 (2006)
2. Le Moigne, J.-L.: La modélisation des systèmes complexes. Dunod, Paris (1990)
3. Camara, G., Despres, S., Djedidi, R., Lo, M.: Modélisation ontologique de processus dans le domaine de la veille épidémiologique. In: Conférence Reconnaissance des Formes et Intelligence Artificielle, Lyon, 24–27 Jan 2012
4. Camara, G., Despres, S., Djedidi, R., Lo, M.: Towards an ontology for an Epidemiological monitoring system. In: Rothkrantz, L., Ristvej, J., Franco, Z. (eds.) Proceedings of the 9th International Conference ISCRAM, Vancouver, Canada, (2012)
5. Forbus, K.D.: Qualitative Reasoning. The Computer Science and Engineering Handbook, pp. 715–733 (1997)
6. Galton, A.: Experience and history: processes and their relation to events. J. Logic Comput. **18**(3), 323–340 (2008)
7. Studer, R., Richard Benjamins, V., Fensel, D.: Knowledge engineering: principles and methods. Data Knowl. Eng. **25**(1–2), 161–197 (1998)
8. Kermack, W.O., McKendrick, A.G.: A contribution to the mathematical theory of epidemics. Proc. R. Soc. Lond. A. Contain. Pap. Math. Phys. Character (1905–1934) **115**(772), 700–721 (1927)
9. Santana, F., Schober, D., Medeiros, Z., Freitas, F., Schulz, S.: Ontology patterns for tabular representations of biomedical knowledge on neglected tropical diseases **27**(13), i349–i356 (2011)
10. Fagot-Largeault, A.: Épidémiologie et causalité. In: Valleron, A.-J. (ed.) L'épidémiologie humaine. Conditions de son développement en France et rôle des mathématiques, pp. 237–245. EDP sciences, Paris (2006)

11. Simons, P., Melia, J.: Continuants and occurrents. Proceedings of the Aristotelian Society, Supplementary Volumes, **74**, 59–75, 77–92 (2000)
12. Despres, S., Szulman, S.: Merging of legal micro-ontologies from European directives. Artif. Intell. Law **15**(2), 187–200 (2007). doi:10.1007/s10506-007-9028-2
13. Scherp, A., Saathoff, C., Franz, T., and Staab, S.: Designing core ontologies. Applied Ontology 6(3), 177–221 (2011). doi:10.3233/AO-2011–0096
14. Grenon, P., Smith, B., Goldberg, L.: Biodynamic ontology: applying BFO in the biomedical domain. In: Studies in Health Technology and Informatics. Ontologies in Medicine, vol. 102, pp. 20–38. IOS Press (2004)
15. Khan, Z., Keet, C.M.: ONSET: Automated foundational ontology selection and explanation. In: Teije, A., Völker, J., Handschuh, S., Stuckenschmidt, H., d'Acquin, M., Nikolov, A., Aussenac-Gilles, N., Hernandez, N. (eds.) Proceedings of the 18th International Conference on Knowledge Engineering and Knowledge Management (EKAW'12), Galway City, October 2012. Lecture Notes in Computer Science, vol. 7603, pp. 237–251. Springer, Berlin (2012)
16. Grenon, P., Smith, B.: SNAP and SPAN: towards dynamic spatial ontology. Spat. Cogn. Comput. **4**(1), 69–103 (2004)
17. Gruninger, M.: Using the PSL ontology. In: Staab, S., Studer, R. (eds.) International Handbooks on Information Systems. Handbook of Ontologies, pp. 419–431. Springer, Berlin (2009)
18. Galton, A., Mizoguchi, R.: The water falls but the waterfall does not fall: new perspectives on objects, processes and events. Appl. Ontology **4**(2), 71–107 (2009)
19. Mourelatos, A.P.D.: Events, processes and events. Linguist. Philos. **2**(3), 415–434 (1978)
20. Schlenoff, C., Gruninger, M., Tissot, F., Valois, J., Lubell, J., Lee, J.: The process specification language (PSL): overview and version 1.0 specification. NISTIR 6459, National Institute of Standards and Technology, Gaithersburg, MD (2000)
21. Lewis, D.: On the Plurality of Worlds. Blackwell Publishers, Oxford (1986)
22. Sider, T.: Four-dimensionalism: an ontology of persistence and time. Oxford University Press, Oxford (2001)
23. Smith B, Ceusters W, Klagges B, KöhlerJ, Kumar A, Lomax J, Mungall C, Neuhaus F, Rector AL, Rosse C (2005) Relations in biomedical ontologies. Genome Biol. **6**, R46
24. Galton, A., Worboys, M.F.: Processes and events in dynamic geospatial networks. In: Rodrıguez, M.A., Cruz, I.F., Egenhofer, M.J., Levashkin, S. (eds.) GeoS'05 Proceedings of the First international conference on GeoSpatial Semantics, Mexico City, Nov 2005. Lecture Notes in Computer Science, vol. 3799, p. 45. Springer, Berlin (2005)
25. Wood, Z., Galton, A.: A taxonomy of collective phenomena. Appl. Ontology **4**(3–4), 267–292 (2009)
26. Galton, A.: States, processes and events, and the ontology of causal relations. In: Formal Ontology in Information Systems. Frontiers in Artificial Intelligence and Applications, vol. 239, pp. 279–292. IOS Press (2012)
27. Talla, I., Kongs, A., Verlé, P.: Preliminary study of the prevalence of human schistosomiasis in Richard-Toll (the Senegal river basin). Trans. R. Soc. Trop. Med. Hyg. **86**(2), 182–191 (1992)
28. Stout, R.: The life of a process. In: Debrock, G. (ed.) Process Pragmatism: Essays on a Quiet Philosophical Revolution, pp. 145–158. Rodopi, Amsterdam (2003)

Part II
Knowledge-Based Systems

Chapter 4
A Context-Centered Architecture for Intelligent Assistant Systems

Patrick Brézillon

Abstract We propose a conceptual framework for implementing intelligent assistant systems (IASs) that (1) work on experience base instead of knowledge base, and (2) deal with the decision-making process and not the result only. Considering experts' experience instead of domain knowledge supposes to have a uniform representation of elements of knowledge, reasoning and contexts. We propose Contextual Graphs (CxG) as such a formalism of representation. A contextual graph is an (micro-) experience base with a task realization on which IASs have to work. This opens a challenge on a new type of simulation, namely a CxG-based simulation with real-time management of context and actions. We are developing such an IAS for supporting anatomo-cyto-pathologists that analyze digital image of slides as part of breast cancer diagnosis. An example illustrates this application.

4.1 Introduction

Generally, support systems are designed and developed for helping actors to realize their task. The support concerns data processing or contribution in problem solving. In this chapter, actors are experts that make critical decision. For example, the surgeon will rely on the anatomo-cyto-pathologist's decision to operate or not a woman for a breast cancer. Such experts rely on a highly compiled experience because they generally act under temporal pressure and are very concerned by the consequences of their decision. Conversely to domain knowledge, expert knowledge, which results of a contextualization process, appears as chunks of contextual knowledge. Thus, decision-making must be considered through its processing as

P. Brézillon (✉)
LIP6, University Pierre and Marie Curie (UPMC), Paris, France
e-mail: patrick.brezillon@lip6.fr

C. Faucher and L. C. Jain (eds.), *Innovations in Intelligent Machines-4*,
Studies in Computational Intelligence 514, DOI: 10.1007/978-3-319-01866-9_4,

well as its result. The two challenges for designing efficient support systems then are: (1) modeling the contextualization process, and (2) exploiting bases of experiences rather than usual bases of knowledge.

The reuse of a contextualized experience is never direct because each context of a decision-making is unique, and thus any experience must be revised to be efficient in another context. Thus, an actor has to (1) identify how the initial experience was contextualized, (2) isolate the reusable part of the experience, and (3) apply the process of contextualization in the new context. Brézillon [8] presents a process of contextualization-decontextualization-recontextualization in the domain of scientific workflows (SWFs). SWFs are used in many applications, namely in physics, climate modeling, medicine, drug discovery process, etc. For example, Fan et al. [14] show in virtual screening that a researcher looks in a repository for finding a SWF that was designed in a working context close to his working context (phase of contextualization). In a second step, the researcher identifies the contextual elements in the SWF independently of their instantiation and obtains a SWF model (phase of decontextualization). Finally, the researcher identifies how the contextual elements of the SWF must be instantiated in his working context and will check different SWF instances (phase of recontextualization). Thus, the experience acquired by the researcher—as a decision maker—during this process relies on context management. Representing such an experience supposes a formalism providing a uniform representation of elements of knowledge, reasoning and contexts. We extend this idea in the domain of decision-making support systems.

At the strategic level, the executive board of an enterprise designs robust procedures (decontextualization phase) from practices developed on the field (contextualization phase), and, after, the same actors develop practices by tailoring procedures to their specific needs and working context (recontextualization phase). In decision-making, the phase of contextualization and decontextualization on the one hand, and the phase of recontextualization on the other hand are considered separately by the head of the organization and by the actor facing decision-making respectively. The head of the organization tries to design and develop procedures that can be applied in a class of problems, while an actor develops a practice for adapting a procedure to the specific working context in which his decision-making is. The head proposes pairs {problem, solution} for representing task realization as procedures, while the actor builds his experience with triples {problem, context, solution} for representing his activities. As a consequence, one cannot represent procedures and practices in the same way. Solving a particular problem (e.g. diagnosing a patient) involves the creation of context-specific models. "Context-specific" refers to a particular case, setting, or scenario.

Thus experience reuse by a support system supposes to be able to manage the process of contextualization, decontextualization and recontextualization. This implies, first, to use a formalism allowing a uniform representation of knowledge, reasoning and context, and, second, to have support systems with powerful functions for processing such a representation.

Hereafter, the chapter is organized as follows. Section 4.2 discusses the need to make context explicit for representing experience. Section 4.3 presents the specificity of CxG-based simulation as the main tool for exploiting experience bases. Section 4.4 gives an overview of the functions that a support system must possess for managing experience bases. Section 4.5 relates our approach with other works. Section 4.6 ends this chapter with a conclusion.

4.2 Representation of Experience Within Contexts

4.2.1 Making Context Explicit in the Representation

The effective application of a procedure on a specific problem supposes to account for the working context in which the problem must be solved. This leads to establish a practice that is tailored to that specific context. The essence is to understand and model how work actually goes done (i.e. the practice), not what is supposed to happen (i.e. the procedure). A practice is the way in which actors adapt a procedure relying on their preferences, the particularities of the task to realize, the situation where the task is realized and the local environment where resources are available. The resulting practice expresses actors' activity, while a procedure corresponds to a task model. As a consequence, there are as many practices (or activities) as actors and contexts.

Context contains elements about the actor, the task realization, the situation and the local environment where available resources are. Context depends on the actor's focus and cannot be considered in an abstract way. The context constrains what must be done in the current focus [10], but, conversely, the focus determines what is contextual knowledge and external knowledge at a given moment. Addressing the focus supposes to select elements from the contextual knowledge and to assemble, organize and structure these contextual elements in a chunk of contextual knowledge, which becomes a part of the problem solving at the considered focus. Therefore the focus and its context are intertwined, and the frontier between external and contextual knowledge is porous and evolves with the progress of the focus. At each step of the focus, a sub-set of the contextual knowledge is proceduralized for addressing the current focus. This "proceduralized context" is built from a part of the contextual knowledge that is invoked, assembled, organized, structured and situated according to the given focus. Once exploited in the focus, the chunk of contextual knowledge can be reused as any other pieces of contextual knowledge.

4.2.2 Consequences for Decision Making

The problem of decision-making reuses can be explained in the light of the different units of analysis (different perspectives) for modeling human behaviors:

Table 4.1 A classification of opposed terms

Level \ Context	Decontextualized	Contextualized
Strategic	Logic of functioning	Logic of use
Tactical	Task	Activity
Operational	Procedure	Practice

procedures and practices [4], task and activity [13, 17], logic of functioning and logic of use [20], etc. Indeed, in one perspective (procedure, task, logic of functioning), one considers only the task, while according to the other perspective one considers the actor, the task, the situation in which the actor realizes the task, and the local environment with its resources available. Brézillon [9] proposes an interpretation of this dichotomy in a referential {decisional levels (policy, strategy, tactics, operation), contexts} as illustrated in Table 4.1. Links with the decisional levels are discussed in [7] and results are applied in (5; [14]). This chapter discusses the dichotomy in terms of context.

Procedure is the (formal) translation of a task realization at operational level. The translation takes into account task realization and the constraints imposed by the logic of functioning at the strategic level. Conversely, a practice is the expression of an activity lead by an actor accomplishing a task in a particular situation with the available resources in the local environment. Actor's experience appears as the accumulation of practices based on activities lead in logic of use. Experience can be discussed in terms of activities and, concretely represented in terms of practices.

A decision-making is an activity that starts with the analysis of the working context (identification of the relevant contextual elements and their instantiations) to have a picture of the problem as complete as possible before any action. Brézillon [8] speaks of a two-step decision-making. The instantiated contextual elements are then assembled, organized and structured in a proceduralized context that allows the actor to make his decision and continue his activity. Then, operating a decision consists of the assembling and execution of actions in a sequence. Indeed, these steps constitute a unique process.

The distinction between contextual elements and instances is important for the reuse of experience because a difference between a past context and the working context can be either a difference on either a contextual element (e.g. a contextual element only exists in one context) or on an instance (e.g. a contextual element takes different values in the two contexts). For example, most of car accidents occur on the way between home and work, when drivers act automatically based on past experience of the way, say, for crossing a crossroad. An accident occurs when the driver does not pay attention to the specificity of the driving context, such as a new contextual element (e.g. oil poured from a truck on the pavement) or a change in the instantiation of a contextual element (e.g. a car does not stop at the red traffic light). Experience management supposes to consider explicitly the working context as well as the decision-making process.

4.2.3 The Contextual-Graphs Formalism

The Contextual-Graphs (CxG) formalism [4] proposes a representation of a task realization as a combination of diagnosis and actions. Diagnosis is represented by contextual elements. A contextual element corresponds to an element of the nature that must be analyzed. The value taken by the contextual element when the focus is on it—its instantiation—is taken into account as long as the situation is under the analysis. Afterwards, the instantiation of the contextual element does not matter in the line of reasoning that can be merged again with the other lines of reasoning (practices) corresponding to the other known values of the contextual element. For example, the calculation of an integral (task realization) can be done by several methods depending on the form of the integrand (the contextual element): integration by parts, using trigonometric variables, etc. (the possible values). The form of the integrand identified (the instantiation), the integral is computed and the type of integrand does not matter any more after.

The CxG formalism gives a uniform representation of elements of knowledge, reasoning and context. Contextual graphs are acyclic due to the time-directed representation and warranty of algorithm termination. Each contextual graph has exactly one root and one end node because the decision making process starts in a state of affairs and ends in another state of affairs (generally with different solutions on the paths) and the branches express only different contextual-dependent ways to achieve this goal. If a contextual graph represents a task realization, paths in the graph represent the different practices developed by actors for realizing the task. There are as many paths as practices known by the system. Actor's experience corresponds to all the practices developed by the actor as represented in the contextual graph in a way that is structured by contextual elements.

Among the four elements of a contextual graph, we discuss here of action, contextual element and activity (see [4], for a complete presentation). An action is the building block of contextual graphs at the chosen granularity of the representation. A contextual element is a pair of nodes, a contextual node and a recombination node; a contextual node has one input and N outputs (branches) corresponding to the N known values of the contextual element (leading to different alternative methods). The recombination node is [N, 1] and represents the moment at which the value of the contextual element does not matter anymore. Sources of contextual elements are very diversified : the actor, the task to realized, the situation in which the task is realized by the actor, and the local environment where are resources for task realization. An activity is a sub graph in a contextual graph that is identified as a work unit by actors (i.e. in the sense discussed in Table 4.1) because it appears on different paths of a contextual graph or in different contextual graphs (i.e. different task realizations).

Figure 4.1 gives a contextual-graph representation of a task realization. Circles (1 and 2) represent contextual elements (CEs) with exclusive values V1.1 or V1.2, V2.1 or V2.2. Square boxes represent action (e.g. a component in a SWF). There are three paths representing three different practices for realizing the task.

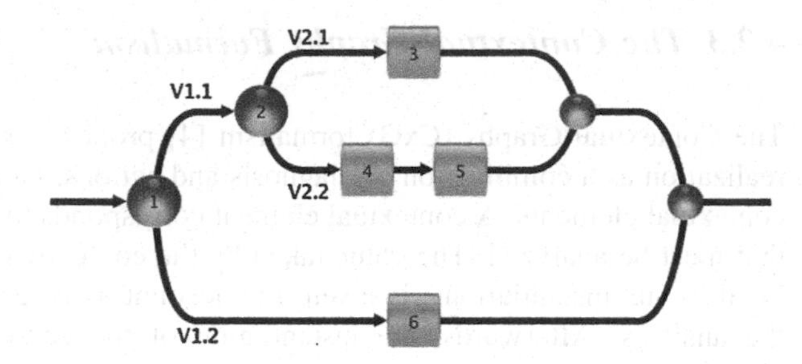

Fig. 4.1 Representation of a task realization as a contextual graph

For example, in a working context where CE-1 is instantiated to V1.2, actor's activity corresponds to the execution of the action A6. CE instantiation only matters when the focus arrives on this contextual element during practice development. Because the instantiation of CE-1 is V1.2, the other contextual element CE-2 and its instantiation are not considered for the practice development. This means that CE instantiation needs to be done only in real time conditions. We call working context the set of all the contextual elements of a contextual graph (CE-1 and CE-2 and their values in Fig. 4.1) and their instantiations (e.g. V1.2 in Fig. 4.1), i.e. the chosen values at the moment of the practice development.

4.2.4 Discussion

Our context model is different from other models that propose, for instance, context as a layer between data and applications or as a middleware or ontology found in the literature. In our model, context is intimately linked to knowledge and reasoning. Contextual elements structure experiences differently of (1) knowledge bases used by expert systems, which were represented in a flat way because context is not represented explicitly, and (2) knowledge organization in an ontology where links between concepts depend only on the domain (is-a, kind-of, etc.) while elements in our context model concern the actor, the task, the situation and the local environment. Moreover, contextual elements have a heterogeneous nature not necessarily linked to the domain (e.g. a choice may be made because the actor is in a hurry). This implies that for using an experience base, a support system must have some specific functions.

A contextual graph allows the incremental enrichment of experiences by the refinement of existing practices. The introduction of a new practice generally corresponds to a contextual element that was not considered explicitly up to now because having always the same value, but that has a different value in the working context at hand. Thus, this contextual element is introduced in the contextual graph with the value implicitly considered up to now, and the value taken in the working context and the action(s) corresponding to this new instantiation.

The notion of chunk of knowledge a la Schank [21] has a clear implementation in CxG formalism as an ordered series of instantiated contextual elements (the proceduralized context). In Fig. 4.1, the proceduralized context of Action 3 is given by CE-1 with the instance V1.1 followed by CE-2 with the value V2.1. Each proceduralized context is specific of an item. For example, the proceduralized context of actions 3 and 4–5 differs from the proceduralized context by the value V2.2 of CE-2, i.e. the instantiation of CE-2. Actions 3 and 4–5 have respectively the proceduralized contexts CE-1(V1.1)-CE-2(V2.1) and CE-1(V1.1)-CE-2(V2.2). This illustrates the two-steps in the decision-making process (first, selection of a contextual element, and, second, the instantiation of this contextual element). The proceduralized context of a practice evolves dynamically during its development by addition at the contextual node or removing at the recombination node of a contextual element.

The representation of the proceduralized context as an ordered sequence of instantiated contextual elements leads to the generation of relevant explanation. The CxG formalism provides a structured representation of actors' experience such as a graph of practices. As a consequence, a support system needs powerful tools for working on such an input.

4.3 Exploitation of Experience Bases

4.3.1 Introduction

Expert systems used a flat base of rules (atoms of human reasoning) and a base of facts, and their reasoning was represented as a series of rules (the reasoning trace) that fired according to heuristics not always clearly stated and control knowledge often implicitly coded in the inference engine (e.g. check the "less expensive rule" first). Nevertheless, expert systems were considered as "model based" and introduce a means of modeling processes qualitatively [11].

Conversely, CxG formalism provides a structured base of experiences described as practices. Each path in a contextual graph (i.e. equivalent to a series of independent rules in an expert system) corresponds to a practice effectively used by a human actor in the working context. Thus, if the expert system builds its own reasoning by assembling a sequence of fired rules according to its heuristics, the support system develops practices effectively developed by human actors by building the proceduralized context jointly with the development of the practice. Thus, the management of an experience base supposes the building of a context-specific model. The support system must be able to play two roles: the role of a CxG Manager for acquiring incrementally new knowledge and learning new practice for enriching the experience base, and the role of a CxG simulator for developing practices. Several general tasks are associated with these roles, namely:

1. Browsing a contextual graph for analyzing the different practices, alternatives, etc. An example is the reading of a recipe and its variants in a cookbook (a recipe being comparable to a task realization).
2. Simulating the practice development in a given context. In the recipe example, the "chef" (with the available ingredients in the kitchen) follows a recipe to prepare a dish according to an initial context (his taste, guests, social importance of the meal, available ingredients, equipment, etc.).
3. Practice learning because all the practices cannot be known in advance (the number of working contexts is large) and the more the support system will be used, the more it will learn and be more efficient afterwards. In the recipe example, this concerns the notes written by the "chef" to adjust the recipe according to the observed results (cooking, too salty or not, guests' feedback, etc.).
4. Explanation generation for exploiting the properties of the experience base because the structure of a chunk of contextual knowledge—the proceduralized context—is known [6]. In the recipe example, cooking separately meat and vegetables may be explained by the fact that one guest is vegetarian.

4.3.2 Contextual-Graph Management

Design and development of the representation of a task realization in a domain as a contextual graph belong to the human actor. This is one of the roles of a support system as a CxG Manager to support the actor in this work. The two main functions of the CxG Manager are (a) the edition of a contextual graph, and (b) the browsing of the contextual graph. The support system can use these two main functions in other roles such as learner or explainer.

The CxG_Platform [9] contains an editor with the usual functions for managing a contextual graph and managing data. The piece of software is available at cxg.fr under GNU license and a screen copy is given Fig. 4.4. It is an interface used by an actor wishing to edit a contextual graph, to read practices for selecting the best one in his working context, to browse alternatives of a practice, to explore a contextual graph at different granularity (by representing an activity by an item or by the contextual graph representing this activity), to analyze (contextual) information attached to each item (date of creation, comments, etc.). The software is written in Java, and contextual graphs are stored as XML files to be reused by other software. Note that an activity, being itself a contextual graph, is stored as an independent XML file. Design and development of the software is user-centered for an easy use by non-specialists in computer science and mathematics. The two specific functions are the incremental knowledge acquisition and the possibility to link an item in the CxG to an external document (Word, PDF, etc.), to run an external piece of software, etc.

Browsing is quite different from simulation with respect to time in the decision-making process. A CxG Browser works at a tactical level where time is not considered explicitly (e.g. duration of an action and loop on routine action do not matter) because browsing concerns more practices in a comparative way rather than the effective development of a practice in a specific context (e.g. comparison of a given recipe in different cookbooks like at an operational level). The focus is more on the realizability of the task than its effective realization (i.e. the first step in the decision making process).

Browsing a contextual graph is a kind of qualitative simulation of experiences for two reasons. Firstly, observing only the reasoning contained in the contextual graph (e.g. to know that a particular action must be executed, but without concern about the effective execution of the action). This concerns only a description of the experience base, the presentation of all the practices known by the system. It is important for learning and explanation capabilities. Secondly, looking what are the relevant contextual elements intervening, their values taken in different working contexts, and the actions to assemble for developing the different practices. In our application for the subway in Paris [18], operators used CxG-based browsing to replay how a colleague solved a given incident, look for alternatives, analyzing the working contexts, and even may study new practices. For developing a practice, the CxG simulator takes into account (1) time, and (2) the qualitative effect of an action execution (e.g. its duration that may constrain the start of another action).

4.3.3 Simulation in Contextual Graphs

4.3.3.1 Introduction

A CxG simulator manages the instantiations of the contextual elements. Thus, a CxG-based simulation is, on the one hand, at the tactical level with an experience base containing all the practices developed for realizing a given task, and, on the other hand, at an operational level for developing a particular practice in a specific context. Time dependency appears because (1) an unpredicted event may modify the instantiation of a contextual element and thus the CxG-based simulation itself, and (2) the execution of an action may impact practice development in different ways independently of what is concerned by the execution of the action. For example, the execution of an action has a duration that may influence the reasoning at the tactical level, or may result in a change of instantiation of a contextual element. Action (or components in SWFs) management was not considered in expert systems except in a hard way like rule packets [2].

A CxG-based simulation presents the progressive building of a practice in the working context. It is a kind of quantitative simulation during which the system mimics actor's behavior. Based on the working context, the CxG simulator develops the reasoning described in the contextual graph and takes into account the activities and their results. However, there are differences between a CxG-based simulation and a model-based simulation.

4.3.3.2 Recall on Model-Based Simulation

Usually, a simulation allows to describe the evolution of a (formal) model from a set of initial conditions. The model expresses a statement about a real system that is based on formalized concepts and hypothesis about the functioning of the real system. Such a model is given by a structure that is specified by parameters that appear in the relationships between variables (a usual formalism of representation is differential equations). A model-based simulation gives a description of the evolution of the variables with respect to an independent variable, generally time, given a set of values for the parameters and a set of initial conditions for the variables (similar to our working context). The evolution of some variables is then compared to temporal observations of the real system. (We will not discuss here the aspect time-based or real time of the representation.)

In a formal model, time appears through the evolution of the variables in the model structure and relationships between variables (e.g. $y(t)$ in a model expressed in the formalism of differential equations like $dy/dt = -a.y + b$). The working context in a model-based simulation (initial conditions and parameter values) is constant during all the simulation: The initial conditions $y(0)$ specify the initial state of the model and the parameter values generally are not modified during the simulation (the values of $y(t)$ are not in the initial state).

At a qualitative level, the (formal) model is a representation of the statements that translates in a formal language hypothesis and formalized concepts needed about observations of the real system. "Browsing" a model is exploiting its mathematical properties for predicting variables evolution (number and stability of steady states, self-oscillations, exponential decreasing curve, etc.) for different sets of parameter values that verify some constraints such as the conditions to have an unstable steady state.

At a quantitative level, model-based simulation is used to find the best set of parameter values and initial conditions describing a set of real-world observations (generally by optimization methods). Here, the formal model is used for the prediction of any behavior of the real system in other contexts, assimilating this context to constraints and initial conditions. In a model-based simulation, the working context is used for the starting state only, variables evolving after, during the whole model-based simulation.

Formal models address the evolution of a system, and the corresponding trajectory is unique because the model structure is unique (parameter values are constant during the whole simulation). Thus, a model-based simulation relies on {model, parameters, initial conditions of variables} where model structure, parameter values and initial conditions are fix during the whole simulation.

4.3.3.3 CxG-Based Simulation

A contextual graph contains the different practices known for accomplishing a given task in different contexts. For a given context, the particular practice that is developed appears like the model structure corresponding to the working context. Thus, practice development (i.e. the model structure) supposes to join the building and the application of the practice. Then, a contextual graph appears as a structure of practices organized by context-specific models that correspond to all the working contexts already met by actors.

At the quantitative level, a CxG simulator needs to know the instantiations of the contextual elements needed, and the effects of action execution. The execution of an action may modify the instantiation of a contextual element. In the recipe example, the "chef" may decide to replace pepper that is missing by paprika if he is in a hurry or, otherwise, s/he goes out and buys pepper. The change of working context (change of instantiation of a contextual element) leads the CxG simulator to consider another practice. This may leads to stop the simulation (e.g. the required resource is no more available) with three options. First, the simulation must be restarted in the new working context and a new practice will be developed. Second, the change of working context corresponds to a routine action in the practice development that must be executed several times by the CxG simulator. There is a need of a clustering phase to group together those executed contexts that are instances of the same defined context. The general approach to clustering is broken down into two major steps, templatization and clustering. There is a third option, where the change of working context concerns a contextual element not yet met during the current practice development. Then, the simulation can be pursued because there is no divergence between the practice under development and the future practice.

In a contextual graph, time appears implicitly through the directed acyclic organization of the contextual graph. In a CxG-based simulation, the practice is built progressively as a context-specific model based on the working context and the execution of an action may put into question the simulation.

4.3.3.4 Discussion

Table 4.2 gives a comparison of model-based simulation and CxG-based simulation according to seven characteristics.

The dichotomy between the model-based approach and the CxG-based approach can also be discussed on workflows and scientific workflows. Workflows are presented as a way to automate decision-making processes in the engineering area. A workflow is the automation of a task realization, in whole or part, during which documents, information or tasks are passed from one actor to another one for action, according to a set of procedural rules (see [22]). Scientific workflows are workflows applied for scientific task realization, which are closer to the human decision making than engineering. Scientific workflows are concerned with the automation of scientific processes in which tasks are structured based on their control and data dependencies.

The dynamics of the SWF building is described at a coarse granularity level and there is no real direct support for implementing SWFs because too many variants

Table 4.2 Comparison of model- and CxG-based simulations

	Model-based	CxG-based
Goal	Represent the behavior of a real system	Represent a task realization on the real system
Real system	An internal viewpoint	An external viewpoint
Tactical level	A model structure	A graph of model structures (practices)
Operational level	Simulation from an initial state	Simulation and building of a context-specific model
Working context	Initial values of variables and parameters (constant during the simulation)	Contextual elements and instantiations (may vary during the simulation)
Simulation	Evolution of model variables	Building and use of a model specific of the working context
Type of support	Prediction, interpretation of deviation (real-system centered)	Task realization on the real system (use-centered)

can be developed according to the large number of possible contexts. The decision-making process begins by assembling, organizing, structuring contextual elements, and concludes on the building of an action sequence that is applied.

Introducing CxG in SWF [8] leads to a representation similar to the contextual graph in Fig. 4.2 by replacing "action" by SWF components, where a component represents a sub-task to execute.

4.3.4 *Simulation Management*

4.3.4.1 Introduction

Task realization supposes the building and the use of a practice according to the context at hand. Finding the best practice consists of the progressive assembling of

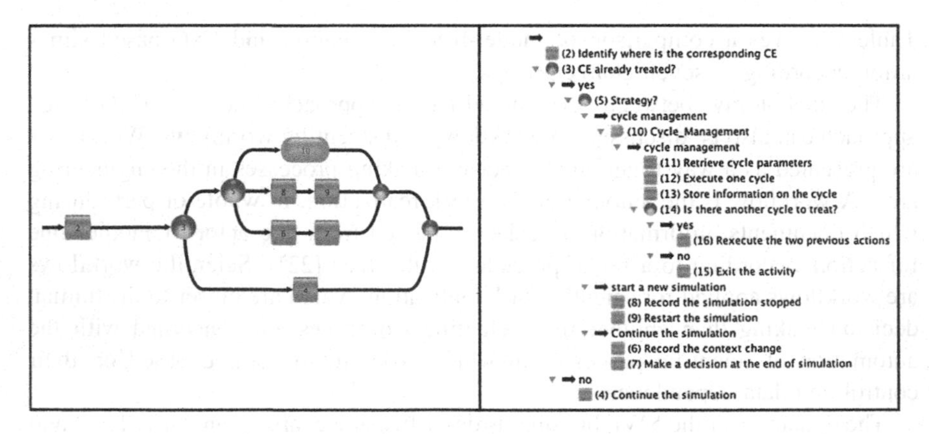

Fig. 4.2 A representation of context management as a contextual graph

the components that are relevant in the working context. The practice built is the best solution for the working context. Because the practice is built at the same time it is used, all that occurs during this process may impact. The main point here is actions (or components in SWF) because even at an operational level, they may impact the practice development: duration, etc. The double aspect building and use of a practice comes from the actor's activity that is gained in his experience.

Because a CxG simulator works on a base of experiences like an (human) actor, it is possible to represent the «expertise» of the CxG simulator also in a contextual graph, the «personal experience base» of the support system. This observation is coming from the fact that if the support system must deal with expert's activity, we must represent support systems' activity in the different modules of management instead of a simple description of the corresponding task. For example, the weak explanatory power of first expert systems comes from a modeling of explanation generation in terms of task description, not like an explainer using contextual knowledge.

4.3.4.2 Working Context Management

The working context has two parts, a static part with the list of the contextual elements in the contextual graph and their known values, and a dynamic part with the instances, i.e. the value taken by a contextual element for the problem solving at simulation time.

A contextual element allows the management of alternatives for a part of the task realization by different methods (or actions in CxG or SWF components). An alternative corresponds to a given value taken by the contextual element in the working context (i.e. its instantiation) and, thus, the choice of a particular method that has been previously used with success by an actor. A contextual element, CE°, has as many (qualitative or quantitative) values as known alternatives:

$$\text{Value}(\text{CE}^\circ) = \text{V1}^\circ, \text{V2}^\circ, \text{V3}^\circ, \text{etc.}$$

The CxG simulator progressively develops the practice by using the working context. Arriving to a contextual element, the CxG simulator looks for its instantiation in the working context to select the right path to follow and the action to execute. The instantiation can be known prior the practice development or provided by the actor to the system during the practice development or found by the CxG simulator in the local environment. In the example given Fig. 4.1, the list of contextual elements will be:

- Contextual element CE-1
- Values: V1.1, V1.2
- Instantiation: V1.2
- Contextual element CE-2
- Values: V2.1, V2.2
- Instantiation: N/A

During a CxG-based simulation, the instantiation of contextual elements may be altered by either an external event or an internal event. The external event corresponds to an unpredicted event, i.e. not represented in the contextual graph. For example, an external resource stops to be available like an ingredient needed in the recipe that is outdated. An internal event occurs as the result of an action execution. For example, the "chef" decides to prepare for more persons than guests to keep a part of the preparation for someone that will come the day after.

The alteration of the instantiation implies a change of the working context. The CxG simulator has two options for reacting to a change of the working context. First, the altered instantiation concerns a contextual element already crossed, the CxG simulator must decide (1) to stop the development of the current practice and re-start the simulation; (2) to redo the part of the practice concerned (e.g. for a routine action); or (3) to finish the development of the practice at hand and then analyze the need for a new simulation in the new working context. The CxG simulator will have to interact with the actor to make a decision on the strategy to apply. Second, the altered instantiation concerns a contextual element not yet crossed by the focus, and the CxG simulator can continue its simulation to progress in the contextual graph because this change of instantiation does not affect the part of the practice already built. The management of the working context also can be represented in the CxG formalism and the different options are represented in Fig. 4.2.

When monitoring a simulation, the CxG simulator that arrives to a contextual element must check the status of the instantiation and makes a real-time decision making on what to do next. This is a part of the "personal experience base" of the CxG simulator.

At a contextual element with an unknown instantiation, the CxG simulator enters a phase of interaction with the actor to have this instantiation. The actor either provides the instantiation of the contextual element, or may wish to study which value of the contextual element is the better for the working context. The actor then enters a phase of interaction with the support system to browse the different alternatives before selecting one and asking the CxG simulator to resume the simulation. This shows that the working context is intimately associated with the simulation.

Working-context management is a key task of the CxG simulator for managing its interaction with actors, information received from external sources, the impact of action on the practice development, and the simulation itself.

4.3.4.3 Action Management

A contextual graph represents the task realization at a tactical level (because the contextual graph contains all the practices used for the task realization at the operational level. An action (or an activity) is executed at the operational level (e.g. execution of an external program or a service). The way in which an action is executed does not matter at the tactical level, but there are some consequences that may impact the practice development as a side effect at the tactical level. The most

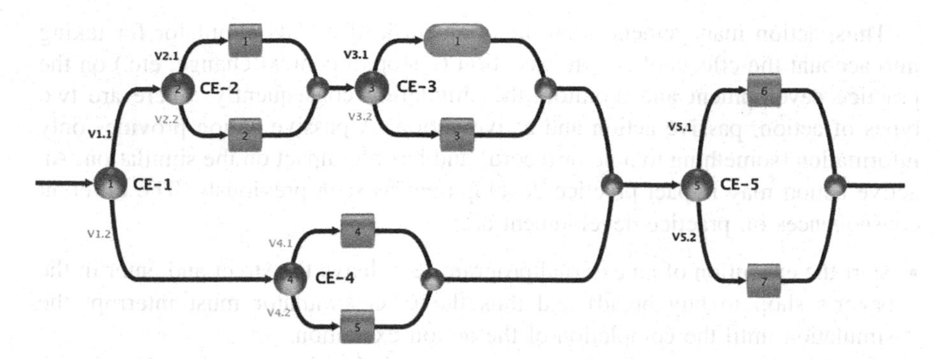

Fig. 4.3 Consequences of the execution of the activity 1 on the contextual graph

obvious consequence is the duration of the action execution that may delay the practice development. Other consequences may be indirect like a change of the instantiation of a contextual element. Figure 4.3 gives a theoretical example for discussing the different situations that may be encountered.

Consider that the CxG simulator reaches the Activity-1 (the oval on Fig. 4.3) by crossing successively CE-1, which has the value V.1.1, CE-2 with the value V2.1, execute Action-1, and crosses CE-3 with the value V3.1. The proceduralized context of Activity-1 is represented by the ordered series CE-1(V1.1), CE-2(V2.1), CE-3(V3.1).

Now, suppose that the activity realization implies the change of an instantiation. Different situations must be discussed:

1. Activity-1 modifies the instantiation of CE-2 to V2.2. This contextual element has been already crossed, and this part of the practice can be accomplished by either action 1 or action 2. Thus, there is no effect on the practice development except if Action 2 itself modifies the instantiation of a contextual element in the contextual graph.
2. Activity-1 provides an instantiation for CE-4. This contextual element is on a branch that will be not considered by the CxG simulator (the branches of a contextual element are exclusive), and this will not affect the practice development.
3. Activity-1 provides the instantiation of CE-4 to V5.2. This contextual element has not been crossed yet, and the change of instantiation does not matter for the part of the practice already developed. If the instantiation of CE-5 was not given initially, the CxG simulator records the instantiation for when it will need it. If the instantiation V5.1 of CE-5 was given initially, there is a conflict, and the CxG simulator needs to alert the actor for making a decision on the instantiation.
4. Activity-1 modifies the instantiation of either CE-1 or CE-3. This means that Activity-1 will not be crossed in the new working context. It is a situation of conflict that needs a decision from the actor for continuing the simulation of the practice development or for starting a new simulation (and then a new practice development) in which Activity-1 will not be considered.

Thus, action management is an important task of a CxG simulator for taking into account the effects of action execution (a stop, a context change, etc.) on the practice development and monitors the simulation consequently. There are two types of action, passive action and active action. A passive action provides only information (something to note or record) and has no impact on the simulation. An active action may impact practice development as seen previously. The different consequences on practice development are:

- Start the execution of an external program (e.g. leave the street and enter in the baker's shop to buy bread) and thus the CxG simulator must interrupt the simulation until the completion of the action execution,
- Abandon the practice development because it is worth continuing the task realization (e.g. going to the hairdresser, abandon because too many persons are already waiting),
- Wait during all the action-execution duration that constrain the progress of the simulation (e.g. stop at the station to fill the car tank or execute a repetitive routine several times),

Actions are domain dependent, but some consequences of their execution can be described by domain-independent features such as duration of the execution, dependency to other actions, to other actors, etc. A possible modeling of action management in a contextual graph is represented in Fig. 4.4.

The outputs of the action-management module (e.g. the alerts to the CxG simulator) are recorded in a base of external facts that will be exploited by the CxG simulator, which will decide to end the simulation or to call for the context-management module. This supposes to transfer the information from the module action management to the module working-context management

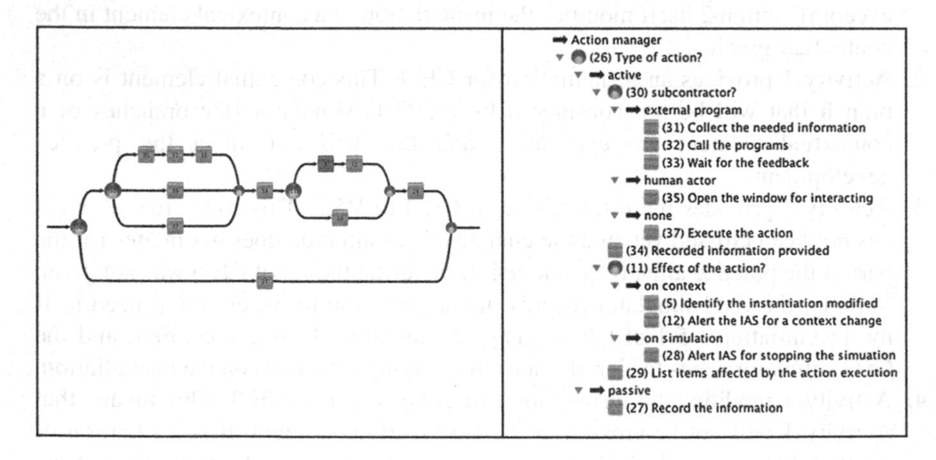

Fig. 4.4 A representation of the module "action management"

4.3.4.4 An Architecture for a Support System

The support system works on a base of actors' experiences, and thus belongs to the category of intelligent systems. However, the actor is an expert in his domain, and the support system only assists actors. An intelligent assistant system (IAS) is an agent that is designed to assist a user that is an expert in his domain. The goal of the IAS is to help intelligently the expert in his decision-making process, not to provide an additional expertise in the domain. In the previous sections, we pointed out the need for an IAS to be able to tackle working-context management and action management. This implies that the IAS too must have a "personal experience base".

The IAS must play different roles with the user, the two main roles being collaborator and observer. Collaboration has three more specific roles, namely CxG manager, practice learner and explainer. Different tasks like CxG management, and CxG-based simulation are associated with these roles.

The modules *working-context management* and *action management*, which are discussed in the previous sections, are part of the "personal experience base" of the IAS and belong to the IAS management module. The personal-experience base also contains knowledge (also described in a contextual graph) about coordination of its different modules (action, working context) and a module for interaction management with the actor. Figure 4.5 gives an idea of such a personal experience base.

Figure 4.6 gives the general architecture of an IAS. In addition to the modules already discussed, there is a base of facts that allows the transfer of some data or information between the different external applications and services triggered by

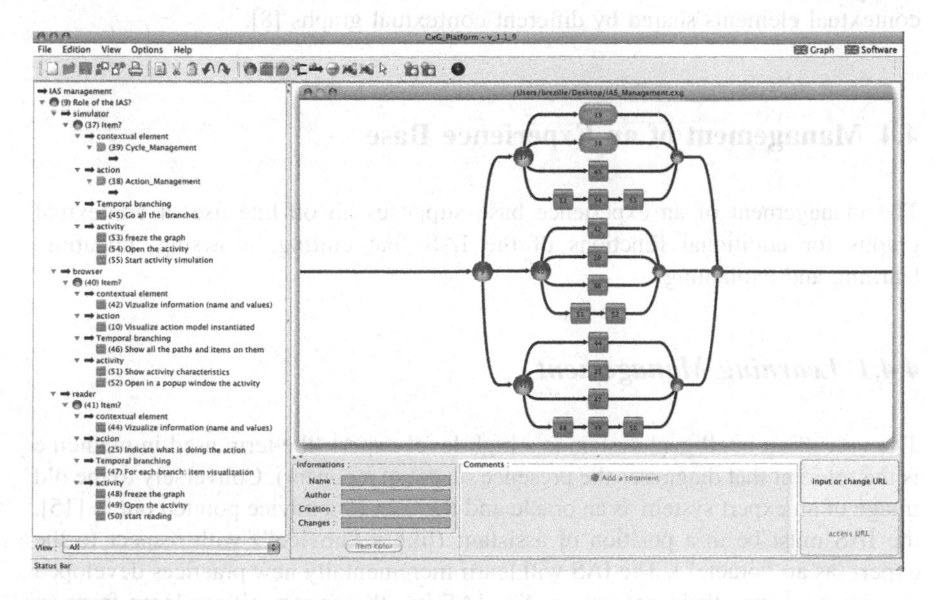

Fig. 4.5 A CxG_representation of the IAS management

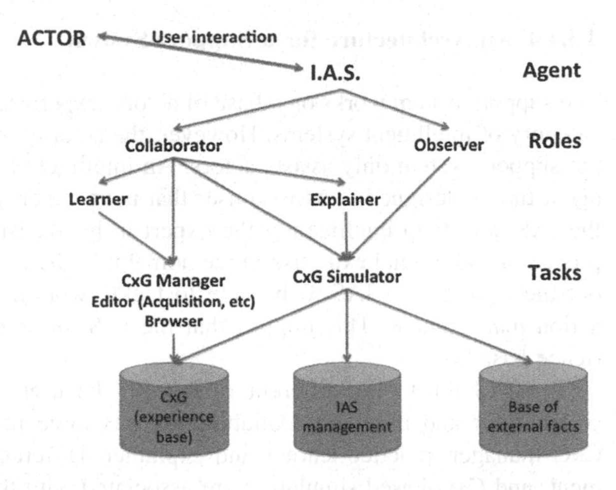

Fig. 4.6 Proposal of architecture for the IAS

an action in the contextual graph. The IAS manages this fact base because some information of the operational level may be relevant for it (e.g. the change of instantiation of a contextual element). Domain expertise is contained uniquely in the CxG experience base, and data and information of the domain are contained in the base of facts. Thus this architecture may be applied in different domains.

Thanks to the CxG formalism, an IAS can manage these different contextual graphs. For example, a contextual graph may contain activities, themselves described as individual contextual graphs. A future extension will be to manage the interaction among collaborators accomplishing each a particular task interacting with the other tasks. In terms of contextual graphs, this means to identify contextual elements shared by different contextual graphs [8].

4.4 Management of an Experience Base

The management of an experience base supposes an off-line use of contextual graphs for additional functions of the IAS like editing, browsing, updating, learning and explaining.

4.4.1 Learning Management

The user of an intelligent system is a high level expert (the term used in medicine is the referent that diagnoses the presence or not of a cancer). Conversely to the old image of an expert system as an oracle and the user as a novice pointed out in [15], the IAS must be in a position of assistant (like a "novice") with respect to the expert (as an "oracle"). The IAS will learn incrementally new practices developed by experts during their interaction. The IAS has the opportunity to learn from an expert when it fails to follow the practice that the expert is developing. Generally, the IAS expects that the expert will execute an action A1, when the expert executed action A2. The conflict arises because the expert considers a contextual element that was not considered by the IAS. The reason is often because this contextual element kept the same value during the whole task realization and, in the context at hand, the expert has identified a different instantiation, and, thus, decided to execute the new action A2. The IAS acquires the new piece of knowledge (A2) and the new contextual element with its two values (the old one and the new one) leading to execute either A1 or A2, and, thus, learns a new practice in a kind of practice-based learning and enriches its experience base.

There is an eventual more drastic change of the experience base when the expert decides that it is not a simple action that is concerned but a sub graph. For example, consider the action 1 "Take water" in coffee preparation (see Fig. 4.7a). Implicitly, the actor considers that he is speaking of running water, i.e. this contextual element "Type of water" does not appear in the representation. Now, suppose that the same actor is in a hurry a morning and decides to use "hot (running) water". An IAS that will observe the actor's behavior will fail to follow the actor reasoning because it does not know the difference between "cold water" and "hot water". Then, the actor will have to provide the IAS (see Fig. 4.7b) with the contextual element CE1 "In a hurry?" with the two values, namely "No" for the previous action 1 "Take (cold) water" and the value "Yes" for the new action 2 "Take hot water".

Now, suppose that the IAS helps another actor in the same task of coffee preparation, and that this new actor only uses mineral water instead of running water to prepare his coffee. Then, when the IAS will ask him "Are you in a hurry?", and the actor will have to explain that this question may be relevant for running water, but not for mineral water in a bottle. The IAS will need to add a new contextual element CE2 "Type of water?" with the value "running water" for the previous practices (toward CE1) and the value "mineral water" for a new action 3 "Take your bottle of water" (see Fig. 4.7c).

The example presents two types of learning for the IAS. In the first learning, the IAS learns by assimilation of a new practice (refinement of an existing practice). It is a practice-based learning. In the second learning, the structure of the experience base is modified for integrating a new method for preparing coffee (i.e. use of mineral water versus running water). It is a procedure-based learning and the IAS must learn by accommodation.

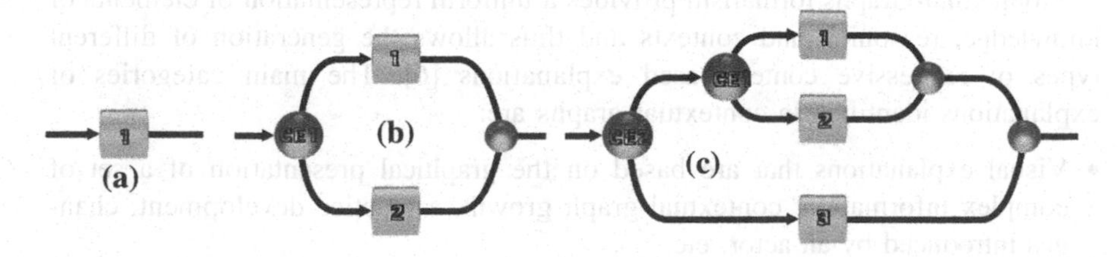

Fig. 4.7 The three steps of water management in coffee preparation with (a) "take water", (b) take cold or hot water, and (c) take either running water or mineral water.

The IAS may be a trainer for actors not quite familiar with practices for realizing a task as described in a contextual graph. The training here consists in explaining the elements followed during a practice development, especially contextual elements and their possible instantiations.

4.4.2 Explanation Management

More than twenty years ago, Artificial Intelligence was considered as the science of explanation [16], but few concrete results were obtained at that time (e.g. see PRCD-GDR [19], in French). There are two main reasons. Firstly, the user was considered as a novice that has to accept what the expert (as an oracle) said, while the user is an expert in his domain with at least a level of expertise as high as any expert system. Secondly, the knowledge considered was either technical knowledge or heuristics. In the former option, contextualization was not present, and in the latter option the contextualization was present but in a highly compiled expression (a "chunk of knowledge", [21]). Moreover, an expert system must understand the user's question and then build the answer jointly with the user.

Trying to imitate a human reasoning, the expert system (ES) presented the trace of its reasoning like a sequence of fired rules that was supposed to be an explanation of the way in which the expert system reached a conclusion. It was right, but (1) ES reasoning was built from "atoms" of the expert reasoning, and (2) explanations were generated at the implementation level because it was not possible to explain heuristics provided by human experts without additional knowledge.

In Contextual Graphs, the explicit representation of context at the same level of knowledge and reasoning provides a new insight on explanation generation. In a previous section, we showed that a proceduralized context is attached to each item in a contextual graph. A proceduralized context is a subset of contextual knowledge that is assembled, organized and structured to be used in the current focus (i.e. a kind of chunk of contextual knowledge). Its representation is an ordered sequence of instantiated contextual elements that can be used for explanation generation. A contextual graph representing a base of experiences, the IAS applies a human expert's reasoning, and not an "automated reasoning" constrained by control knowledge hidden in the inference engine (e.g. fire the first possible rule).

Contextual-Graphs formalism provides a uniform representation of elements of knowledge, reasoning and contexts and thus allows the generation of different types of expressive context-based explanations [6]. The main categories of explanations identified in contextual graphs are:

- Visual explanations that are based on the graphical presentation of a set of complex information: contextual-graph growth, a practice development, changes introduced by an actor, etc.

- Dynamic explanations that describe the progress of a practice development, the chronology of a practice creation, the anticipation and presentation of alternatives at different granularity levels.
- User-based explanations that detail practice parts unknown by the actor and sum up parts that were developed by the actor.
- Context-based explanations that present, say, why the same activity is «good» in one context and «bad» in another one.
- Micro- and macro-explanations that use activity representation either as an entity (for a macro-explanation) or a sub graph of activities (for a micro-explanation).
- Real-time explanations that lead the IAS to acquire incrementally new knowledge and learn practice, and guide the actor to follow the progress of a practice development.

These different types of explanation can be combined in different ways such as visual and dynamic explanations for presenting future alternatives and abandoned options.

4.4.3 Discussion

The contextual-graph representation puts in the front stage different interesting findings:

- A contextual graph is the representation of a task realization by one or different actors having the same role. Actors realizing the same task with different roles must be represented in different contextual graphs. For example, when a physician and a computer engineer work on bio-image analysis, they have different views that must be represented in different contextual graphs.
- A contextual graph gives a representation of a task realization at a given level of granularity at which actions are the building blocks of the representation. Thus the effective realization of the task (e.g. interaction between components of a workflow) does not matter. However, some aspects of the actions may have some effects on the representation. For example, an action execution may modify the instantiation of a contextual element and thus the practice development.
- For interacting intelligently with an actor that is an expert in his domain, the IAS, on the one hand, adheres to the expert's viewpoint, and, on the other hand, makes explicit needed tools that are required for the management of context, actions, the contextual graph, learning, explanations, etc. We are working on such an architecture for IASs (see Fig. 4.6). The key point here is that all the domain knowledge and expertise are only in the experience base, and thus the IAS architecture may be reused in a domain-independent way.

4.5 Related Works

In a classical case-based reasoning (CBR) scenario, a case consists of a problem description and a solution. A case contains a set of (structured) information entities, and optional artifacts. Structured information is represented as attribute—value pairs), while the optional meta-information contains unstructured textual information. Atzmueller [1] uses stored cases (experiences) for selecting an appropriate task and method, reusing those stored task-configurations that are similar to a (partially) defined characterization. The process of capturing and reusing complex task-experiences is led in four main steps: Experience Retrieval, Task Instantiation, Task Evaluation and Deployment, and Experience Maintenance. Thus, a case is recalled as a whole and its characterization is then adapted to the context at hand. In the Contextual-Graphs approach the practice, the equivalent of the case, is identified jointly with its use. The main difference here is that cases are represented in a flat way, while practices are organized in a Contextual-Graphs representation. In the CBR, the approach is "result-oriented" while in Contextual Graphs, the approach is "reasoning-oriented".

Clancey [12] proposed that solving a particular problem (e.g. diagnosing a patient) involves creating situation-specific models. "Situation-specific" refers to a particular case, setting, or scenario. "Situation-specific" is not "situated cognition" that refers to how people are conceiving and thus coordinating their identity, values, and activities in an ongoing process enabled by high order consciousness. In the CxG approach, context concerns an actor accomplishing a task in a particular situation in a specific local environment. A practice development is associated with the progressive building of a "context-specific model". The "situation-specific model" is embedded in the problem solving as a static description fixed initially and filled progressively during the problem solving. Conversely, the context-specific model (i.e. the proceduralized context) is built in parallel with the practice development with the movement of contextual elements entering and leaving the proceduralized context.

A model-based simulation is a top-down (deductive) modeling, while a CxG-based simulation corresponds to a bottom-up (inductive) modeling. In a model-based simulation, the whole working context is defined at the start of the simulation and stays constant during the simulation, while in a CxG-based simulation, the working context evolves during practice development. A formal model is given initially (its structure is confronted to observations), while a practice (the contextualized model of a task realization) is built progressively from the contextual graph and evolves with its working context. In that sense CxG-based simulation is a different type of simulation. The behavior of a CxG simulator is comparable to the usual model-based simulator's behavior, supposing that (1) contextual elements in the contextual graph can be compared to the parameters in the formal model (a change of parameter values impacts on the model behavior like a change of instantiation modifies the practice developed), and (2) variables in a model-based simulation are related to the result of the progressive building of the practice corresponding to the working context.

These approaches can also be discussed with respect to decisional levels: Case-based reasoning approach is at an operational level and model-based simulation at a tactical level. An IAS plays the role of a CxGG Browser at the tactical (qualitative) level and of a CxG simulator at the operational (quantitative) level. The CxG Browser allows working on the experience base. Because we are not developing a practice like in a simulation, it is a tool at the strategic or tactical level. The CxG simulator is a tool at the tactical level or the operational level: taking into account the specificity of the working context to find the best practices.

4.6 Conclusion

Our goal is to develop a decision support system to a user that has a high level of expertise in a domain not well known or too complex. Users' expertise is highly compiled like chunks of contextual knowledge built mainly by experience. Such an expertise is generally used in a decision-making process leading to a critical and definitive decision. In the project MICO, the expert is an anatomo-cyto-pathologist that analyzes digital slides (coming from biopsies) to diagnose if a patient in a surgery has or not a breast cancer.

The consequences are:

(1) The decision support system must behave as an intelligent assistant, following what the expert is doing, how he is doing it, anticipating potential needs. This supposes that the system possesses a representation of the experts' reasoning. The IAS must be an excellent secretary, fixing alone all the simple problems of human experts, and preparing a complete folder on complex situations letting experts make their decision.
(2) The decision support system must work from the practices developed by the experts with all the contextual elements used by the expert during practice development. The line of reasoning of the system is drawn from lines of experts' reasoning described in an experience base, which gives a user-centered representation of the domain [3].
(3) The decision support system must be able to develop the decision-making process in real time to analyze the association diagnosis and action built by experts during their reasoning [10]. Indeed, the system simultaneously develops the decision-making process and its context-specific model.
(4) The decision-making process being highly contextual, the decision support system must benefit from its interaction with the expert to learn new practices by acquiring incrementally the missing knowledge, and thus enriching its experience base.
(5) Making context explicit in the experience base leads to the possibility to generate relevant explanations for:

- Presenting the rationale behind a practice with alternatives abandoned;
- Training (future) experts on the different practices developed;
- Facilitating experience sharing among experts in a kind of dynamic corporate memory;
- Allowing a first step towards the certification of their protocol.

(6) The main tool of an intelligent assistant system is the CxG-based simulator, thanks to a uniform representation of elements of knowledge, reasoning and contexts. Its originality is to build at the same time the practice and its application. Indeed the CxG-based simulator is the key element of an IAS for real-time decision making because it is possible to account for unpredicted events, thanks to an explicit modeling of context as contextual elements covering the user, the task realization, the working situation, the local environment with its available resources. All the items are interdependent and also time-dependent. Thus, intelligent assistant systems cover a more general problematic than context-aware applications. This seems to us the key point for mobile decision making because the instantiations of contextual elements are taken into account at the moment it is necessary.

Acknowledgments This work is supported by grants from ANR TecSan for the MICO project (ANR-10-TECS-015), and we thank partners (IPAL, TRIBVN, UIMAP team of Service d'Anatomie Cytologie Pathologie at La Pitié, Thalès, Agfa) for the fruitful discussions.

References

1. Atzmueller, M.: Experience management with task-configurations and task-patterns for descriptive data mining. KESE, http://ceur-ws.org/Vol-282/02-AtzmuellerM-KESE-Paper-CRC.pdf (2007)
2. Brézillon, P.: Interpretation and rule packets in an expert system. Expert systems theory and applications (IASTED). Acta Press, Anaheim (1989)
3. Brézillon, P.: Focusing on context in human-centered computing. IEEE Intell. Syst. **18**(3), 62–66 (2003)
4. Brézillon, P: Context modeling: Task model and model of practices. In: Kokinov, B., et al. (eds.) Modeling and Using Context. Springer, Hedelberg (2007)
5. Brézillon, J., Brézillon P., Tijus, C.: 3C-Drive: New model for driver's auto evaluation. In: Kiyoki, Y., Tokuda, T. (eds.) Proceeding of the 18th European-Japanese Conference on Information Modelling and Knowledge Bases (EJC2008). Waki Print Pia, Kanagawa, Japan, 2–6 June (2008)
6. Brézillon, P.: Explaining for contextualizing and contextualizing for explaining. In: Proceedings of 3rd International Workshop on Explanation-Aware Computing ExaCt 2008, Patras, Greece, CEUR Workshop Proceedings, ISSN 1613-0073, online CEUR-WS.org/Vol-391/00010001.pdf, 21–22 July 2008
7. Brézillon, P.: A contextual methodology for modeling real-time decision making support. In: Burstein, F., Brezillon, P., Zaslavsky, A. (eds.) Supporting real-time decision making: The role of context in decision support on the move, annals of information systems, series decision support systems (2010)

8. Brézillon, P.: Contextualization of scientific workflows. In Beigl, M., et al. (eds.) Modeling and Using Context (CONTEXT-11). Springer, Heidelberg (2011)
9. Brézillon, P.: Modeling activity management instead of task realization. In: Ana, R., Frada B. (eds.) Fusing DSS into the Fabric of the Context. IOS Press, Amsterdam (2012)
10. Brézillon, P., Pomerol, J.C.: Contextual knowledge sharing and cooperation in intelligent assistant systems. Le Travail Humain **62**(3), 223–246 (1999)
11. Clancey, W.J.: Viewing knowledge bases as qualitative models. IEEE Expert: Intell. Syst. Appl. **4**(2), 9–23 (1989)
12. Clancey, W.J.: Model construction operators. Artif. Intell. J. **53**, 1–115 (1992)
13. Clancey, W.J.: Simulating activities: relating motives, deliberation, and attentive coordination. Cognitive Syst. Res. **3**(3), 471–499 (2002)
14. Fan, X., Zhang, R., Li, L., Brézillon, P: Contextualizing workflow in cooperative design. In: Proceedings of the 15th International Conference on Computer Supported Cooperative Work in Design (CSCWD-11), pp. 17–22, Lausanne, Switzerland, 8–10 June 2011
15. Karsenty, L., Brézillon, P.: Cooperative problem solving and explanation. Expert Syst. Appl. **8**(4), 445–462 (1995)
16. Kodratoff, Y.: Is artificial intelligence a subfield of computer science or is artificial intelligence the science of explanation? In: Bratko, I., Lavrac, N. (eds.) Progress in Machine Learning, pp. 91–106. Sigma Press, Cheshire (1987)
17. Leplat, J., Hoc, J.M.: Tâche et activité dans l'analyse psychologique des situations. Cahiers de Psychologie Cognitive **3**, 49–63 (1983)
18. Pomerol, J.C., Brézillon, P., Pasquier, L.: Operational knowledge representation for practical decision making. J. Manage. Inf. Syst. **18**(4), 101–116 (2002)
19. PRC-GDR: Actes des 3e journées nationales PRC-GDR IA organisées par le CNRS (1990)
20. Richard, J.F.: Logique du fonctionnement et logique de l'utilisation. Rapport de Recherche INRIA no **202**, 1983 (1983)
21. Schank, R.C.: Dynamic Memory, a Theory of Learning in Computers and People. University Press, Cambridge (1982)
22. Workflow Coalition: http://www.wfmc.org/standards/docs/tc003v11.pdf (last access: 03/07/13)

Chapter 5
Knowledge Engineering or Digital Humanities?

Territorial Intelligence, a Case in Point

Francis Rousseaux, Pierre Saurel and Jean Petit

Abstract Knowledge Engineering (KE) usually deals with representation and visualization challenges, sometimes socio or bio inspired, collective aspects being quite often taken into account. Nevertheless with knowledge-based Territorial Intelligence, KE is faced with natively situated know-how, distributed hope and network-centered emerging organizations, as far as this domain aims at providing tools to support and develop our local and territorial communities. Furthermore knowledge-based Territorial Intelligence has to cope with its own paradoxes and success, to challenge its sustainable existence: as a matter of fact, thanks to big data and its digital tools, people may have thought that they where living in a global village, territories-independent, practicing a perpetual nomadism. So they now require participation for defining their collective policies and social perspectives, leading to their common sustainable development. How knowledge-based Territorial Intelligence will manage to make available efficient solutions to support and develop our original way to collectively inhabit places and earth? That is the question we try to present throughout some technical and scientific aspects along this dedicated chapter.

F. Rousseaux · J. Petit
CReSTIC, University of Reims Champagne-Ardenne, Reims, France
e-mail: francis.rousseaux@univ-reims.fr

J. Petit
e-mail: jean.petit@etudiant.univ-reims.fr

P. Saurel (✉)
Sciences Normes Décision, University of Paris-Sorbonne, Paris, France
e-mail: pierre.saurel@paris-sorbonne.fr

C. Faucher and L. C. Jain (eds.), *Innovations in Intelligent Machines-4*,
Studies in Computational Intelligence 514, DOI: 10.1007/978-3-319-01866-9_5,

5.1 Introduction to Digital Humanities

Compared to Classical information Systems, Knowledge-Based Systems are much more dependent on paradigm shifts [1, 2] that shape their fields of application[3, 4, 5]. Fields of application modify the types of data integrated into the information system [6], but also relations between data and thus the structure and shape of the databases. More radically these fields may need a representation inconsistent [7] or less compatible [8] with the usual computerized centralized representations of knowledge [9, 10].

When an intervention in the field of culture and the Humanities—literature, history, geography, philosophy, politics, theology, music, visual and graphic art—is at stake, it now seems [11] that digitalization and computation have an impact on the humanities, that culture impacts Computer Science, with the computational field opening onto the cultural, and the humanities opening onto the computational field. This is what can be called the modern-day notion of digital humanities, which more and more researchers and application designers are considering, especially when they work on e-Learning or Serious Games. At least two modes of knowledge representation confront each other—that of knowledge transmission on the one hand and that usually linked to digital knowledge representation on the other hand.

> This phenomenon sheds light on the strategic position of the Chinese Academy of Science, unveiled by one of its prominent members [12] when he states: “Although the answers to the computational dimensions of culture are not clear, we must foresee them because we simply cannot afford not to see their consequences [….] I am hopeful and optimistic, and believe this could be the beginning of a new area in computing that would seamlessly integrate information technology with social sciences in a connected world.”

Researchers David Radouin and Stéphane Vandamme pointed this out not long ago, at a seminar[1] on the humanities:

> From the 19th century, we have inherited a clear cut separation between the Humanities and Sciences, corresponding to a growing specialization and disciplining. The modern-day university system was built on this disciplinary base, while at the same time calling for its trespassing, for the sake of educating complete individuals. How to think such a project today? […] Do we have to acknowledge the difference between the two different cultures, while trying not to break their unity—as, precisely, forms of cultures—for all that? Is it a matter of reinventing a dialogue between two distinct entities or of questioning the nature of this distinctiveness? Can we, do we have to reactivate ancient forms of connection, or on the contrary, acknowledge a profound evolution in the two terms inviting to new connections? Here are some of the questions that immediately come to mind for those who claim to be interested in the humanities.

As Knowledge Engineering means to serve digital humanities [13] and provide them with innovative applications [14, 15, 16, 17], it was expected that the methods be put to the test of new connections, even of drastic restructurings [18, 19, 20]. This confrontation is obviously bi-directional, as knowledge engineering and its models

[1] http://www.institutdeshumanites.fr/?q=seminaire/seance-du-23-mars-2012

[21, 22, 23] change the way humanities see themselves, and humanities, with their specificities and cognitive capacity [24, 25, 26, 27], widen the scope and modes of representation in Knowledge Engineering and drive the classic digital modes of representation into a corner [28] and eventuality to some of their limits [29].

This chapter focuses on the case of land planning which encompasses geography, town planning, anthropology, and political sciences, and which raises new questions to the engineering of territorial knowledge, prone to make both this key field of artificial intelligence and our own vision of the configuration of territories evolve in today's digital world.

5.1.1 Knowledge Engineering Applied to Territories

According to André Corboz, the broadest and most general definition of a territory [30] "[...] is a space socially constructed in a given time and place, by a given society. Space, place, society: three useful terms, but yet unstable terms that have to be discussed one after the other. Nevertheless one can not but notice that this definition admits a fixed relationship between a specific geographical area on the one hand and a group socially determined which inhabits it on the other hand. There is a one to one correspondence between an area and its occupier—this is not surprising since this definition was conceived in the 19th century, at a time of ardent nationalistic fervor. The two complementary aspects are the border (as a defense against the outside world) and appropriation (of the area thus protected). Or else, to reduce the definition to its minimal form: the territory, in its materiality is an area surrounded by a fence, occupied by a unique and homogeneous society. This definition is not only static, but it is also anachronistic. Our societies are no longer homogeneous, they are multicultural; they comprise groups whose systems of values are really contrasted, even incompatible at times. If that was the case, one would have to admit that there is one territory per group. But the very idea of area does not hold anymore when one tries to think in terms of planning. One has to move on to the notion of network: on the one hand there are networks of directions, frequencies, key hubs, transshipping points, markers, and thresholds; these are long distance highways and airports. But on the other hand there is also a network as far as decision-making structures are concerned: [...] the decisions to intervene are taken away from the places where the intervention is to take place, and sometimes very far away from them. There is no perimeter in a network, endpoints at most. The mutation of the territory into a network thus rejects the very notion of a continuous limit which becomes pointless. The network is admittedly a more subtle reality than the territory: it cannot be easily represented nor measured. It is an elastic reality, even if based upon heavy equipment which deeply modifies the backdrop of the territory: [....] the territory is divided and distributed differently, depending on the project. [....] Perceiving the territory as a network enables to feed the imagination on the territory from another angle, from other criteria which introduce the notions of time and flow. Up to now, the territory was a shape and a stretch of land which bore a name; the territory was semanticized; it could be the topic of speeches; it had an internal distribution; it was essentially static. Yet, the network—or rather networks—are also a way of fathoming the same area; a territory cannot exist without an imagination of the territory."

Almost everything has been said about the recent shift that occurred in our conception of the territory, linked to the need to reconsider the very act of planning these territories. Two dimensions clearly appear here: first the physical space, enclosed or not, and secondly the decision space that does not necessarily overlap the enclosure of the physical space. These two dimensions are not the only ones that potentially structure the territory but the evolution of their separation is enough to induce a shift in the traditional paradigm of land planning.

5.1.2 Shift in the Traditional Paradigm of Land Planning

Local authorities currently face a brutal shift in the classical paradigm of land planning [31], the founding principles of their traditional organization and governance [32, 33]. If territories now comprise networks, local authorities are also part of a worldwide network, so that the discrete values of time and space which used to prevail are fading away to allow a spatiotemporal continuum to step forward. The interactions between the different levels are more and more numerous and more and more frequent making the decisions taken locally and with no impact on other scales less and less effective.

We assume that the profound causes of this shift are to be sought in the advent of the digital in the territories which has drastically changed representations and social uses on the one hand [34, 35, 36], and in the now strong and steady demand for dialogue and community participation in the decision processes and in territorial policies on the other hand [37, 38], with a growing will to implement sustainable development. But we have to admit that the second reason is indirectly linked to the first one. Indeed it is the advent of specific digital technologies that makes possible and supports consultation and participation at the different territorial levels. It is now clear that the digital revolution is triggering a radical transformation in local authorities and traditional land planning.

The networks of stakeholders involved in the decision are themselves organized digitally [39, 40] so that the representation of knowledge relating to the territories and the decision has to consider at least three coupled and interdependent information systems: the network of places and their supports and infrastructures, the decision-making network, and the network of stakeholders (particularly citizens) impacted by the decisions [41, 42, 43].

5.1.2.1 Internet and Digital Revolution

If the massive advent of digital technology in the territories has led to a shift in the classical approaches and representations used in land planning activities, it is first and foremost because the digital technology floods the territories with data and applications, and as a result they are required to provide coherent and controllable interpretations. While land planning used to be centered on and punctuated by centralized actions, organized in hierarchical institutions under the principle of subsidiarity, and controlled by stabilized representations thanks to multi-year development plans, it is now compelled to become the dynamic arrangement of a territorial continuum whose topological and cartographic granularity ceases to be its main characteristic, and where data proliferate.

Thus, by opening for example virtual interconnected worlds with no apparent territorial roots, digital technology seems to take part in a movement of deterritorialization; but it can also make massive and dazzling reterritorializations possible, as was the case during the Arab Spring in 2011 in Tunisia and in Egypt. Social network users, non-experts in territorial policies, managed to share spontaneous and tactical information via the Web, which from a strategic point of view, eventually helped and played a significant role in the collapse[2] of a police state ruling the national territory with an iron fist. Simple actions in the digital world triggered a drastic change in a territory, which was not thought to be possible so suddenly.

Land planning management, thanks to territorial engineering, thus becomes territorial intelligence, in the sense of a field of application emerging from knowledge engineering [44, 45, 46, 47], with its specificities and particularities [48, 49]. It involves territorial information systems—correlating, aggregating and merging data which are often geo-localized—it requires from the co-learning organization to set into place territorial observatories, engages co-inhabitants in collaborative networks, co-develops more and more personalized Web services and coproduces common decisions. In this case as in others in Knowledge Engineering (eLearning, Serious Games), one has to imagine, simulate, model, visualize, control, decide and play to experiment and discuss.

However, the decline of radical expansionism, which has given way to sustainable development utopias, increases the questioning as regards the diversity of cultural and anthropological attitudes on the conservation and development of territories.

[2] See the Round Table of June, 21, 2011 at the UNESCO Headquarters http://unesdoc.unesco.org/images/0021/002116/211659f.pdf.

5.1.2.2 Decline of Radical Expansionism: Promises of Sustainable Development

Sustainable development thus appears like an operational fiction which, no doubt has the great merit of gathering and pooling together dynamic forces, but whose scientific foundations are fragile, no matter the way they are broached. Some researchers indeed underline the ambivalence of sustainable development, such as Sauvé [50], Canada Chairholder in environmental education, when she states:

> As in any social construction, the concept of sustainable development emerged from a specific historical context, it is rather topical at the moment and lies at the heart of tensions, it has become a thing taken for granted whose genesis has been forgotten; it serves specific interests while appearing as a consensual value. Its promoters assert its heuristic status (it is a path or rather a bridge toward a new world), but at the same time, they mix up this concept with a universal principle and insist on its institutionalization; from a proposal, we move on to a norm, a requirement and from now on it becomes THE path, THE bridge, and eventually it becomes the destination. The concept of sustainable development corresponds to the social construction of a saving project, a life buoy in the midst of the security crisis which is the hallmark of our present societies, but it seems that we mix up means, meaning and purpose.

The position can even be more radical at times, like the one held by Boileau [51], professor at the Ecole Nationale des Ponts et Chaussées in Paris:

> We have to abandon the sustainable development dyptic, an ambiguous phrase that has been skewed by facts and turned into a useless slogan, and concentrate the action on worthy and lasting coexistence, by institutionally adopting at all costs a pluralist framework of mutual respects for the different civilizations, worships and customs. It means that we have to consider one's ability to respect different views as one of the key values in all doctrines, ours included, and to admit that others also contribute to the constraining consultation which is dictated by the global dimension of the issues. Now that the communist adventure has failed with its progressive and conquering scientism, it is time to acknowledge the responsibility of the neo-liberal logic and set up a pluralistic and collective dimension of the planet urgently.

In any case, we can distinguish at least three distinct approaches to sustainable development, each of them presenting blatant weaknesses:

- A pragmatic sense of sustainable development, widely accepted today, assumes that it is legitimate to act locally while thinking globally, aiming at a globally harmonious conversion and avoiding the trap of local optimization of ancient orders. It is in this very sense that the recommendations emerging from the 1992 Rio Summits and the tools of the Agenda 213 (Agenda 21) are based and implemented by local authorities worldwide. This approach, developed among others by Nobel Prize Winner Ostrom [52] has the merit of making populations aware of the challenges, but there is no evidence that it converges toward a global balance, not to mention the fact that it carries a radical anthropological injustice (addressed more fully below).
- A scientistic sense of sustainable development implies a reexamination of all the activities which mankind can control directly or indirectly, in order to reverse

the huge matrix and minimize their so-called adverse long-term effects. But which detailed and universal criteria should be used? Which exhaustive description of human and non-human activities should be called up? Which specific objectives should be aimed at? It would mean a genuine restructuring of the economy which would grant the advent of a new polymorph and particularly demanding actor: planet Earth, represented among men by some kind of Supreme Court. This approach has a practical sense only at a local level, with no guarantee whatsoever as to the ability of the local combinations to produce a satisfactory global outcome.
- Concerning the idealistic sense, it aims at the possibility of a radical cultural restructuring, which might guide the organization of human activities towards an environmentally friendly and lasting anthropological equity. It is in this trend that we can locate the anthropological research of Descola [53] which shows that naturalism is but an anthropological position among others perfectly identified which he describes, along with Latour [54, 55], as particular universalism. It would imply to include a real diversity in the cultural anthropological positioning. We would thus have the specification of a meaning both critical and pragmatic for sustainable development, trying to answer the following questions: Which universal values should be chosen to back-up a policy of heritage for natural and cultural goods? How to respect the various ways of being into the world and define customs that could be agreed upon by all its occupiers? Is a relative universalism—and no longer a particular one—possible?

5.1.2.3 Suspicion as Regards Public Policies: Social Demand for Dialogue and Participation

There is a general consensus among observers to claim that social demand for dialogue and community participation is steadily growing in the territories [56, 57, 58] at a time when suspicion as regards public policies is more and more blatant.

With what success on the ground? When Loïc Blondiaux and Jean-Michel Fourniau undertake to assess researches on public participation in democracies [59]—whether they relate to the impact on the decision-making process, the transformation of individuals, the structural and substantial effects of participation, the importance of the conflict, the influence of the positive [60], the institutionalization of participation and its legal codification, the professionalization of participation or else the redefining of expertise—they observed that:

> Community participation in negotiations and public debates as well as in the expertise and decision process is at the heart of the mutations concerning public action that have occurred in Western democracies over the last decades. An increase in the number of stakeholders taking part in the decision-making process, creation of new spaces for participation more open to ordinary citizens and civil-society associations, consecration on the political and legal level of a participation imperative; all these are among the many elements showing an evolution of our democracies and public modes of action, not only on

> the local, regional and national level, but also on the international level. Participation has become an issue way beyond the background of Western democracies, reaching countries structured around authoritarian decision-making processes. [....] It may be a way of organizing the information, or else the agreement of the population on given projects and policies, or, on the contrary of organizing power sharing as regards deliberation and decision between the governors and the governed. Participation can enhance understanding and agreement or, on the opposite, be a way to express diversity and conflict within a democracy.

As for the process of mass digitization at work in the territories and elsewhere, we may wonder if it constitutes an incentive to community participation. Does it provide tools for consultation? If participation is indeed a key element of digital culture [61, 62, 63], alternative media derived from it tend to weigh on the failures of the classic representative system, rather than grasp a community voice expressing normative claims thanks to digital tools. Thus, according to Monnoyer-Smith [64]:

> The increasingly widespread use of the Internet and of peer to peer practice, the development of social networks, and the creative appropriation of the web cause deep and lasting changes in the relationship between citizens and their representatives and renew the forms of political mediation. However, the expectations expressed by some political scientists for a new era of active involvement that would curb the steady erosion of an electorate staying away from the ballot box and would rejuvenate democratic life have led, for a large part, to disillusion. These expectations were most likely widely based on an erroneous analysis of the causes of communities' lack of interest, but they also reveal how difficult it is for research to tackle the question of the use of technologies other than in a deterministic perspective, if only to criticize it in fine.

5.2 Knowledge Engineering for Land Planning: Researchers' Spontaneous Mobilization

Knowledge Engineering spontaneously rallied the field of spatial management, following Information and Communication Technologies (ICT) which had opened the way to lay the foundations for modern territorial Engineering. Metadata are standardized to encode the content of geo-referenced data that complete the information systems. New field ontologies arise [65, 66], along with reusable inference patterns [67], capable of steering territorial engineering to the higher level of knowledge processing. But we are still a long way away from the paradigm of digital humanities, capable of thinking a participatory intelligence of territorial dynamics and real ecoumenal arrangement.

We will show how the taking into account of real cases characterized by new expectations of emerging stakeholders can bring about this change of perception.

5.2.1 *Knowledge Engineering Applied to the Classical Approach of Spatial Management*

In a few decades' time, the incredible creativity with which the ICT have broached engineering and then the humanities will probably come as a surprise, prompted by the researchers' and developers' desire to devise new tools for new uses and applications, without always thinking about the strong expectations of future users.

That was the case when the stakeholders of the computerization of our societies decided to provide tools for spatial management. It is only after a first techno-orientated phase led by the pioneers of information systems that a first draft for a more activist approach to territorial intelligence came to life, claiming the hybrid culture of digital humanities.

5.2.1.1 A Techno-Scientific Approach Urged by the ICT

In the 1980s, local authorities started to express their need for IT tools, whether generic or dedicated to their role as public space developers, and created a niche that appealed first to Information Technology Consulting societies, and later to software package developers. This is how ICT and knowledge engineering, more or less integrated in job-orientated software packages, structured a new business field in a context of computer science development, but still a long way away from the paradigm of digital humanities.

That was the time of the emergence of Geographical Information Systems (GIS) which extended classical data bases to cover territorial data bases, including geo-referenced data. Geographical information systems also incorporated geographical base maps, urban and architectural plans, drawings and photographs. These systems were used by town councils and local authorities to picture their urban heritage and come up with town-planning projects thanks to their ability to support thematic management views and land reattribution simulation.

Knowledge representation and structuring thus appear as relational databases whose stored content may be enriched by their multi-media form.

Very quickly, these systems are endowed with models of geo-localized knowledge representations, electronic chart display systems, maps and pictures, as well as supports for simulation, constraint optimization, and reasoning stemming from artificial intelligence. Such devices help, for instance, in the deploying of a wireless network by using intervisibility calculation functionalities. Such systems, coupled with document management apparatus, also have practical application in economic intelligence or else for decision support in economic and environmental crisis management provided time-based representations are included [68]. The rapid development of the Internet (Web) has increased these systems functionalities, now able to interoperate with information-seeking tools, content-based browsing tools, and web-content searching tools.

The use of these GIS coupled with web-content retrieval and extraction tools to assist correlation, data-aggregation or geo-referenced data fusion operations soon becomes widespread and leads to what might be called a territorial web. Following this, there is an increase in the availability of maps, pictures, geo-referenced plans, so much so that, thanks to the implementation of metadata adapted to cooperative applications, territorialized applications for institutions as well as for private individuals equipped with smart phones are on the rise. There is a boom in the number of data and territorial applications which constantly offer new uses and new services to communities or to individuals, ICT –driven and led by knowledge engineering eager to decipher promising valuation area, whose key-words are: visualization, simulation, decision support, struggle to overcome big data, use of virtual reality and serious games [69], not to mention the Internet of Things and Cloud Computing.

It is only in the early 2000s that the possibility of a digital humanities approach arose in the spatial management careers. That was typically the case with the innovative concept of Territorial Intelligence.

5.2.1.2 Outline of a Paradigmatic Inversion: Territorial Intelligence Approach

The concept of Territorial Intelligence is at the crossroads of the concepts of territory, knowledge society, and sustainable development. It ceases to be inherently defined by the technologies it uses. It is explicitly part of the digital humanities, since it is finalized by business activity and citizens' ambitions, and cannot be reduced to the technical tools employed. In Europe for example, the concept was developed by the French teams working on the European project European Network for Territorial Intelligence[3] [70]. Territorial Intelligence means:

> [...] the body of interdisciplinary knowledge which contributes to the understanding of territorial structures and dynamics on the one hand, and whose aim is to become a tool for the stakeholders of the territories' sustainable development on the other hand. Ibid.

If the territory is no longer defined as a physical space, we understand that the intelligence is that of the territorial community, a construction of the stakeholders and at the same time a corporate citizen:

> By involving both the stakeholders and the territorial community in the pooling of data and in their cooperative use, territorial intelligence improves, through a process at once active, iterative and prospective, their understanding of the territory's structure and dynamics and their collective mastery of territorial development [71].

The reference to the community is related to the concept of social capital which is vital for community development, a concept deeply rooted in the Anglo-Saxon world whose aim is to promote local development. According to the World Bank[4]:

[3] http://www.territorial-intelligence.eu

[4] World Bank Poverty Net website (http://web.worldbank.org), keyword: poverty.

> Social capital is defined as all the conventions and social relations rooted in the structures of society and which enables the members of the community to coordinate their actions so as to reach their goals.

But the reference to sustainable development as the key orientation for territorial intelligence also leads to a global approach and to a participatory governance based on the notion of partnership. Sustainable development offers a comprehensive approach taking into account economic, social, environmental and cultural objectives, without being reduced to short term economic and/or financial aspects. Sustainable development simultaneously promotes the governance' decentralized tendencies and offers participatory and partnership-based methods to implement sustainable development actions.

Politically, sustainable development introduces a plurality of viewpoints and the constructivist nature of a common world in the making. As such, community development involves community participation and the partnership of stakeholders, which argues for the use of packages facilitating seamless sharing (in the sense of radical seamlessness as a tool for management and cyber-democracy) and open sharing of data (in the sense of open-data), e-participation and cooperative management of partnership projects. In short, territorial intelligence combines knowledge, action and participation so as to stimulate innovation to advance sustainable development, based on ethical principles which claim to be high.

However, territorial intelligence, as with all social systems, faces the question of which stakeholders matter—and consequently which facts to take into account in the agenda of the reality publicly discussed—on the one hand, and of which packages are suitable to produce collective choices on the other hand. These two obstacles lead to question once again the very notion of territory with its usual focal points linked to the stakeholders' strategies: if expertise, science or so-called proven facts cannot alone help us decide between the different viewpoints inherent to pluralist universes, and if participatory methods of animation alone may not be sufficient to produce agreements and understanding, we may need to adopt mapping tools to follow the territorial arrangements and innovative metrics so as to assess the attachments [72] between the objects that comprise them.

> Territorial intelligence is a way for researchers, stakeholders and territorial community to acquire a better knowledge of the territory, but also to establish better control over its development. The appropriation of information and communication technologies, of data themselves is an essential step so that the stakeholders get into a learning process enabling them to act in a relevant and effective way. Territorial intelligence is especially useful to help territorial stakeholders to project, define, apply and assess the policies and actions for a sustainable territorial development [70].[5]

[5] This definition emerged from the experience born out of the spreading of the Catalyse method in Europe (http://www.territorial-intelligence.eu/catalyse/), which, since 1989, has offered tools for territorial diagnosis, assessment and observation for the stakeholder partnerships eager to increase, conduct and assess sustainable development projects within their territories.

Territorial intelligence does not only consider knowledge and information as tools, even when referring to multi-criteria and spatial analytical tools, but as an essential vector of development in a knowledge society. It does not view the territory as a firm or a market, but mainly as an area of cooperation that does not exclude the competitive sector, just like what collective intelligence does:

> Collective intelligence refers to the results drawn from collaboration and knowledge sharing, and from competition between many individuals… It can be viewed as a kind of network activated by the recent evolution of information technologies […]. Territorial intelligence aims to be the discipline whose object is territorial sustainable development in a knowledge society, and whose territorial community is the subject. Its goal is to stimulate a dynamics of sustainable development at the level of territories based on the combination of economic, social, environmental and cultural objectives, on the interaction between knowledge and action, on information sharing, on consultation in project preparations and on cooperation in the way actions are implemented and assessed. On the one hand it gathers and produces interdisciplinary knowledge useful to grasp the dynamics and the territorial systems; on the other hand it hopes to become a tool for territorial sustainable developers Ibid.

The distinctive features of territorial intelligence are summed up along that same vein by Philippe Dumas, a professor of computer science at the University of Sud-Toulon, when he states:

> Intelligence as a cognitive process and a way to organize information, and the territory as a space where meaningful relationships can develop.

5.2.2 Presentation of Our Study Fields

It is time to build on specific cases recently dealt with through our own interdisciplinary investigations, under the research group Territorial Assemblages (Agencements territoriaux) which we created not long ago [73, 74, 75].

5.2.2.1 Will La Vallée Scientifique de la Bièvre Join the Paris-Saclay Cluster Within Le Grand-Paris?

We are here interested in the recent attempt, led by a group of elected representatives, to promote the territory of La Vallée Scientifique de la Bièvre within Le Grand Paris, as a reaction to the creation of the neighboring Paris-Saclay Cluster, referred to by its promoters as future ecosystem of growth intending to attract the Ile-de-France and national economy and to become an international hub/cluster for knowledge economy. The different positioning of the two projects is still a very topical issue.

Le Grand Paris is a project that aims to transform the Paris conurbation into a great European capital and world metropolis of the 21st century, so that it is in symbiosis with its environment, like the first five of the kind which are New-York,

London, Tokyo, Shanghai and Hong Kong. The first bill of law for Le Grand Paris was adopted on May, 27, 2010, following a vote in the French Senate. The senators approved the conclusions of the joint committee by 179 votes to 153, opening the path to the construction of a double automated metro loop around Paris. The creation of this new 130 km-long transport artery around the capital will connect nine economic hubs of the area—Plaine-Commune, Roissy, Orly, Saclay, La Défense, Champs-sur-Marne, Evry, Seine-Oise and Montfermeil-Clichy-sous-Bois—through project agreements for future stations and will provide suburb-to-suburb links. It will cost around 21.4 billion euros in investment according to the government's calculations, and will position the Ile de France region among the first four World Cities, along with New York, London and Tokyo, according to Christian Blanc, former secretary of State for the development of the Capital Region.

While some critics view Le Grand Paris project as political gamesmanship mixed with substantive issues, others see it as a strong political and financial commitment for a priority area.

In connection with Le Grand Paris project, the Campus Plan and the Grenelle Environment Round Table, an Operation of National Interest (ONI) plans to turn of Le Plateau de Saclay into a world-class research and innovation center, a territory with high scientific and technological potentiality to boost the national economic growth. A perimeter of ONI covering a large part of the plateau was defined by Decree of the State Council in March 2009, ensuring consistency between spatial planning and the protection of agricultural areas. The project will receive an exceptional investment of one billion euros made possible thanks to a National Loan.

As the local authority representatives for the territories separating Saclay from the Capital city feared to witness the relocation to the Paris-Saclay Cluster of research institutes settled in their municipalities or departments, they in turn came up with the idea of pooling their effort to mobilize an exploratory territory called La Vallée Scientifique de la Bièvre (VSB) which they hoped to promote within the political and organizational dynamics of Le Grand Paris. A few years before, they had created the VSB, an intercommunity informal group conceived as a think tank, without realizing they would have to bring it to the forefront of the strategy for the Ile-the France region (Fig. 5.1).

If it is clear that La Vallée Scientifique de la Bièvre (VSB) has not been conceived to operate the geographic junction between two non-contiguous territories (Paris-Saclay Cluster and Paris—and Le Grand Paris) following the model of a valley linking a plateau to a town, this candidate territory is nonetheless compelled to position itself so as to be able to demonstrate its complementarity with the Paris-Saclay Cluster. The VSB will take the form of an Urban Campus.

By studying the documents produced by the VSB Round Table (the fourth Round Table was held in Fontenay-aux-Roses and led to the production of a benchmark plan for the planning and development of the VSB), the difficulties met by the subscribers to develop their line of arguments become obvious. To reach general agreement and establish the VSB as a key territory for Le Grand Paris, the

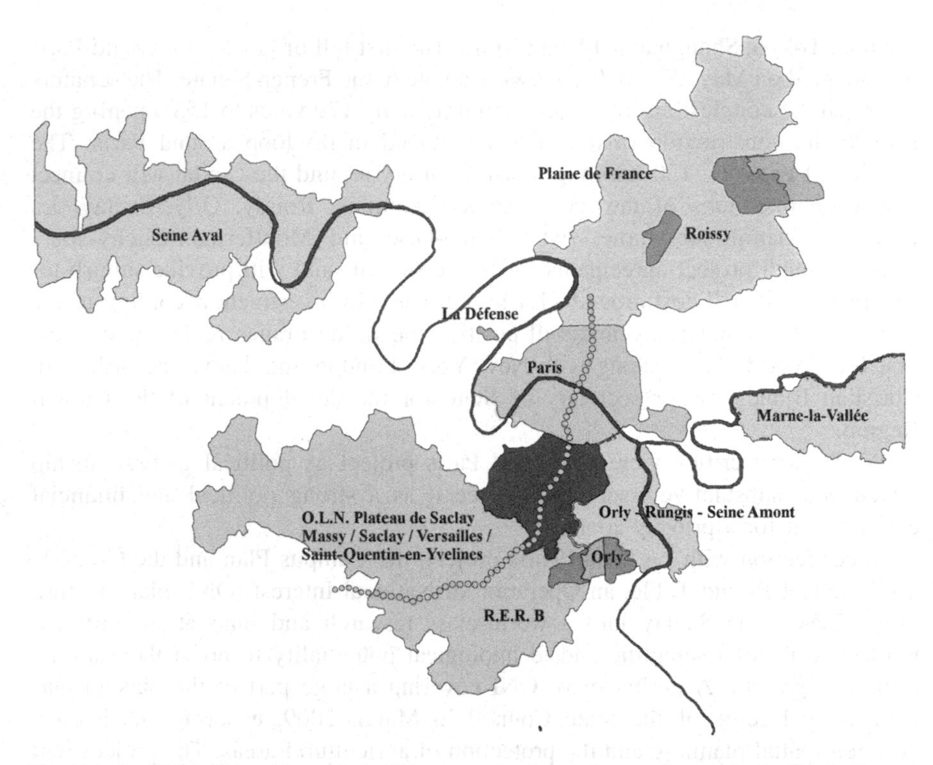

Fig. 5.1 La Vallée Scientifique de la Bièvre among the project territories in the Ile de France region

plan was to offer disruptive categories and model representations adapted to the intricacy of the areas and potential stakeholders, which would open onto readable and convincing public policies (Fig. 5.2).

How to operate the shift of a sector-specific approach, refocus the debate and unquestionably demonstrate the added value of the VSB project within Le Grand Paris? For the time being, a proposal for cooperation has been made to our research group Territorial Assemblages (Agencements Territoriaux) by a representative of the VSB which consists of two parts:

1. Think through the scope of description categories mobilized to present the territorial projects for the Paris-Saclay Cluster and La Vallée Scientifique de la Bièvre, and offer a mode of comparison;
2. Think about controllable means to develop the VSB exploratory territory and make it gain recognition within Le Grand Paris, and to conceive a convincing enough demonstration so that Le Grand Paris take the VSB into consideration and even integrate it in its project.

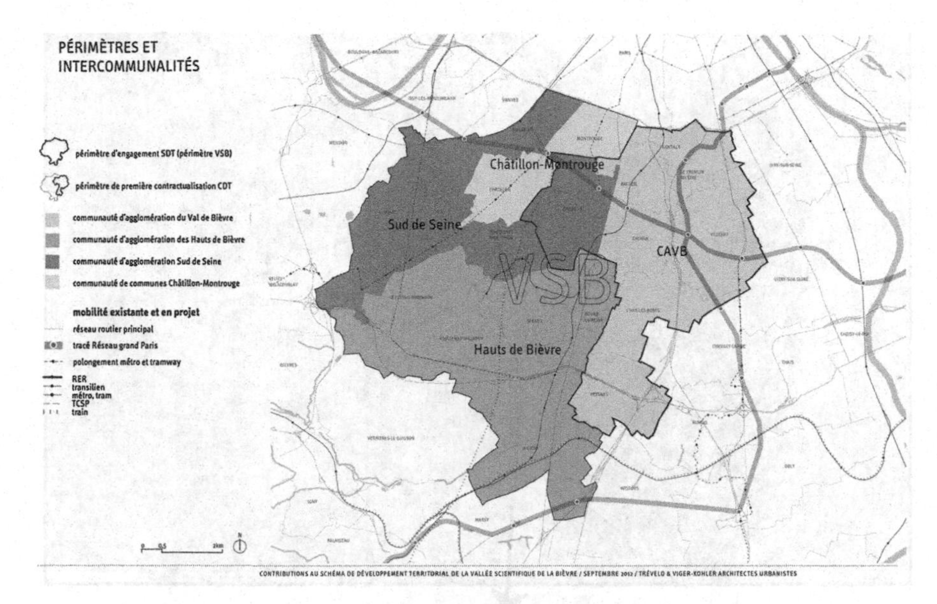

Fig. 5.2 The project territory of La Vallée Scientifique de la Bièvre

5.2.2.2 Taonaba in the French West Indies: A Constructive Criticism on Sustainable Development?

Taonaba is the name given to the eco-tourism planning project for Belle-Plaine canal on Les Abymes territory in Guadeloupe. The municipality's main idea is to create a Mangrove Center. Indeed, the site where the project is located is typical of a coastal wetland area, remarkable both for its ecological diversity recognized as having national and international importance, and for the large stretch of its swamp forest. Moreover, there is an agricultural area adjacent to it, next to an interesting historical heritage—the remains of Belle-Plaine sugar plantation. The will to preserve and develop all these assets guided the Taonaba project[6] (Fig. 5.3).

The key idea of the project was to improve the ecosystems present on Les Abymes coastal areas in a logic based on sustainable development, by synergizing three different tools:

- Eco-tourism development so as to be the driver of tourist activity in Les Abymes region by highlighting Les Abymes territory and its natural environment;
- Ecological conservation through education in environmental protection and a better understanding of the ecosystems (agricultural areas and coastal wetland areas);

[6] http://www.ville-abymes.fr/IMG/pdf/presentation_taonaba_synthetique_2.pdf

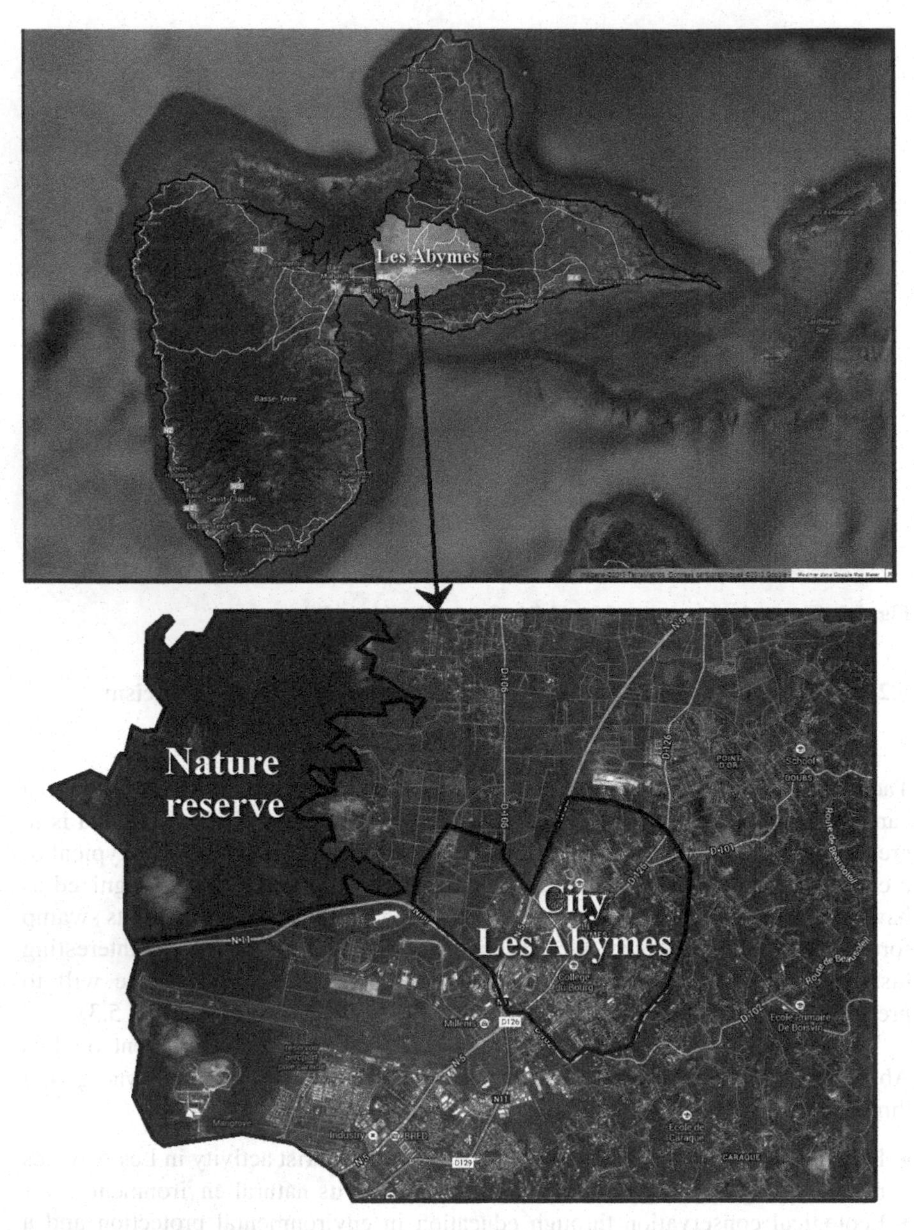

Fig. 5.3 Taonaba's mangrove center location on Les Abymes territory in Guadeloupe. Original picture after a google map satellite view

- Local development (social well-being): by fostering the development of employment–generating activities for the local stakeholders, and by creating a space vital for oxygenation at the city's doorstep (the green lungs of the city).

Three potential axes of activities on location and three types of target audience:

- A science research center on wetlands such as mangrove swamps aimed at students and researchers;
- An educational provision, directed mainly at school groups, which allows an entertaining and yet scientific discovery of these ecosystems;
- A mass-market tourist offer, for local visitors and tourists alike, offering a wide range of indoors and outdoors activities, at the same time pedagogical, entertaining and innovative compared to the services already available in Guadeloupe.

The territorial approach—territory perceived as having a "geographical, economic, cultural or social cohesion at the level of a living or employment area"; approach aimed to express "the common economic, cultural and social interests of the territory's residents" and to allow "the study and implementation of development projects"—is directly inspired by the creative process at work in the development of 'administrative counties' known as Pasqua's 1995's law on spatial plannig and development of the territory (LOADT) modified by Voynet's 1999's law. It is based on:

- A mission: rally together all the stakeholders, users and residents of the area around a common and coherent project;
- An economic interest: develop around the Mangrove Center ecotourism and agri-tourism activities compatible with the main development of Belle-Plaine canal;
- Resources: a territorial Charter, but also territorial contracts between the local authorities and the territory's stakeholders ensuring a legal framework and financial support for the activities developed, thanks to the development of partnerships (Fig. 5.4).

The process was implemented as follow:

1. Carry out a diagnostic study of the territory to reveal the issues at stake.

   For Belle-Plaine, the defined issues are:

- Manage conflicts and mobilize the territory's stakeholders;
- Maintain and develop agriculture;
- Control urban development;
- Act for the Conservation of the natural and historical resources and the prevention of pollutions;
- Help in the structuring of a territorial identity.

2. Determine the development axes for the concerned territory in collaboration with all the stakeholders involved.

   For the Belle-Plaine case, the identified axes are:

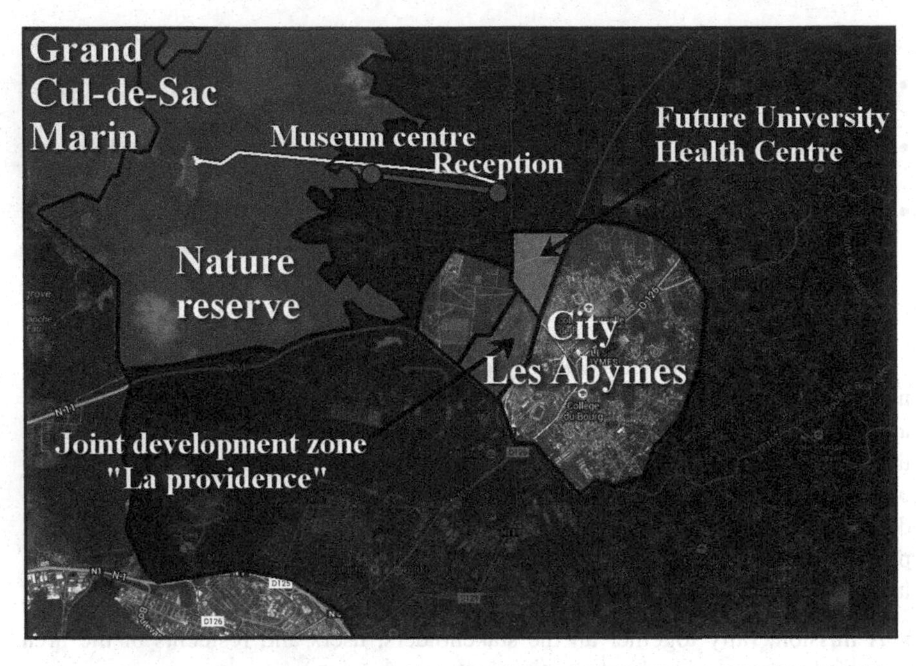

Fig. 5.4 Les Abymes mangrove swamps and marshlands areas within Le Grand Cul-de-Sac Marin in Guadeloupe. Original picture after a google map satellite view

- Use ecotourism as a driver of tourism development;
- Evolve towards the development of an integrated and sustainable agriculture;
- Create frameworks for the development of an innovative and well-suited town-planning;
- Characterize and define the specific identity of the territory.

3. Draw up a territory charter in close cooperation with all the stakeholders (yet to come in the case of Belle-Plaine).
4. Build an agenda for actions corresponding to the development axes (yet to come in the case of Belle-Plaine) (Fig. 5.5).

From a practical point of view, the method used was as follow:

- Consultation meetings with the different types of stakeholders (farm managers, resident owners, project managers, businesses);
- Attempt at having the town services involved in the project through the appointment of a TAONABA contact person within each department dealing with one or several issues of the project;
- Development of partnerships with the University of the Antilles and Guyana, the National Park, and the Coastal Conservation Authority;
- Fieldwork for an accurate identification of the different users of the areas;
- Cost assessment and fund- seeking for the action program;

Fig. 5.5 Bird's eye view of the mangrove center and of the future the water sports center. *Source*: http://www.ville-abymes.fr/IMG/pdf/presentation_taonaba_synthetique_2.pdf

- Launching of a call for eco-tourism and agri-tourism projects for the Belle-Plaine area (action that will be carried out by an external consulting firm);
- Establishing thematic working groups: agri-urban project, land-use planning, territorial charter, economic development;
- Concrete targeted actions: dealing with the access road, networking of actors and project managers, setting up of informative billboards on the progress of work, participation of local residents in some phases of the procedure (e.g. logo meeting, trail meeting etc.).

Les Abymes services' idea is to pursue the implementation of a sustainable development and participatory consultation approach throughout the Taonoba project life cycle, in the new context of creation of a community of municipalities comprising the town from Les Abymes to Pointe-à-Pitre, called Cap Excellence (http://www.capexcellence.net).

Cap Excellence intends to become the driving force of social, economic and cultural activity in Guadeloupe, and seeks the best way to pool the resources together and drive the economies of scale necessary at a time when national financial support becomes scarce. The slogan put forward is "Cap Excellence aims to build a territorial project based on values of sustainable development and social cohesion".

The Cap Excellence territory, encompassing the territories of Pointe-à-Pitre and of Les Abymes, turns out to be a strange cultural bi-pole made up of the conceptual coming together of Taonaba and the ACTe Memorial,[7] likely to make Cap excellence utterly original, and thus to gain substantial visibility. The tension might be seen as resulting from an opposition between nature (Taonaba) and culture (ACTe Memorial): but in fact it results entirely from the reversal of this opposition, through information sharing and mutual influence, which continues to deplete the Nature/Culture conjecture (Fig. 5.6).

[7] http://www.cr-guadeloupe.fr/upload/documents/Macte12P.pdf

Fig. 5.6 Future ACTe memorial site in Pointe-à-Pitre in Guadeloupe. Original image from the brochure available at: http://www.cr-guadeloupe.fr/upload/documents/Macte12P.pdf

The ACTe is indeed a memorial project centered on the painful issue of slavery, a very sensitive question in Guadeloupe. One thing worth mentioning at this point is the plant-like look of the ACTe Memorial architecture which echoes the mangrove swamp, when Taonaba (a Tainos First Nations place name) became a haven for runaway slaves (Maroons). It is also interesting to notice the tension between the sacred aspect and the desire to visit, valid both for the mangrove swamp seen as a sanctuary and for the sacred memorial. All this refers the notion of heterotopy, notion on which we will return in depth.

The political leaders of the Cap Excellence Corporation have asked our research group Territorial Assemblages (Agencements territoriaux) to help draft this Taonaba/ACTe Memorial bi-pole through innovative thinking.

5.2.3 Back to Conventional Spatial Management Assumptions: How to Overcome Them?

Spatial management, in its conventional sense, implies globally stable territories that only catastrophes can destabilize, apart from the periods corresponding precisely to so-called wished for and assumed planning actions carried out through planning and project management techniques. The categories that govern this constant virtual stability are most of the time not fully thought out, placed under the high spatial and temporal protection of abstract Topos and Chronos, and then developed on discrete scales corresponding implicitly to the hierarchy of territorial powers that rule over them. Added to this is the modern-day triumph of a naturalistic cosmology which, in Western or Westernized societies, validates an irreconcilable duality between Nature and Culture.

These assumptions must be overcome absolutely to raise territorial intelligence to the status of digital humanities.

5.2.3.1 Territories Virtual Stability Assumption

As many other municipalities, Les Abymes recently launched its local Agenda 21 (http://www.ville-abymes.fr/spip.php?article30). The elected representatives considered that the implementation of the Territorial Climate and Energy Plan (TCEP) and of the Agenda 21 would allow the sustainable development of the territory given the political stakeholders' great willingness to be part of a participatory eco-responsible approach. The connection of the Agenda 21 with the Local Town Planning (LTP) is under way.

According to the principles of sustainable development adopted at the Earth Summit in Rio, an Agenda 21 program has to address the economic, social, cultural and environmental aspects harmoniously, and promote the broadest possible participation in the population and civil society stakeholders (associations, businesses, administrations). These programs are strongly regulated by benchmark laws. In France, the Grenelle Environment Round Table in 2007 provided an opportunity to elaborate a long-term roadmap regarding the environment and sustainable development. A National Strategy for Sustainable Development (NSSD) called Toward a Fair and Equitable Green Economy was adopted on July, 27, 2010. It set the French policy as regards sustainable development for the period 2010–2013. A frame of reference for the assessment of the local Agenda 21 was built and an Observatory established. Sustainable Development Indicators[8] were recently developed by the Observatory of the Territories managed by the DATAR.

Drawing up an Agenda 21 is a participatory process generally drafted over a 3–4 year period. The key steps in its composition are the territorial diagnosis and the writing of the Action Plan. We would like to highlight the fact that local sustainable development processes, and particularly the method used to draft the local Agenda 21, are consistent with today's conceptions of what the management of public policies should be [76]. And yet these ≪ conceptions ≫ are rather at odds with the ≪ institutional arrangements ≫ intending to manage common resources, as described by Ostrom [77]. The Agenda 21 are basically scheduled action plans which place the local political stakeholders in a central position to

[8] http://www.developpement-durable.gouv.fr/-Le-referentiel-pour-l-evaluation-.html, http://observatoire-territoires-durables.org/, http://www.territoires.gouv.fr/observatoire-des-territoires/fr/node

coordinate the approach, and place it in the footstep of a strategic planning model which, as is well-known, is being increasingly criticized [78].

Thus, even when dealing with a process said to be participatory, the territories respecting the notion of sustainable development are supposed to be virtually stable in that the elected representatives ruling over them resort to project management and planning methods mobilizing functional services structured around functional areas: transport, housing, education, health, culture. However, "The territory, over-impressed as it is with the marks and readings from the past, looks more like a palimpsest. In order to set up new equipment and to use some of the land more efficiently, it is often necessary to modify its substance irreversibly. But the territory is neither a disposable packaging nor a consumer product which can be replaced. Each territory is unique, hence the necessity to recycle, to scratch once again the old text men have imprinted on this irreplaceable material which is the soil, so as to write a new one addressing today's needs, which, in turn, will be discarded. Thus some regions, dealt with too harshly or in an improper way, present holes, like an over-scratched parchment. In the language of the territory, these holes are called deserts" [30].

The vocation of the territory is to induce a planning dynamics from its inhabitants, as they unfold their projects and their vision of living together. That is how the competitive attractiveness of the territories appears, generating migration flows difficult to reconcile with a conception of the territories as developable and virtually stable entities.

This vision has, of course, an impact on the representation and structuring of the information system. Even though the traditional tools of digital representation of data are adapted to the conception of a stable territory—with for example a conventional structure of a relational database, be it enriched with and composed of metadata—this structuring reaches its limits if the representation is also to return an evolution dynamics of the database structure.

5.2.3.2 How to Outsmart Territorial, Spatial and Temporal Scales, Prone to Spontaneously Organize Powers and Institutions in a Hierarchical Way?

According to Foucault [79], "our own time might be the time of space". We would be in the epoch of simultaneity, of juxtaposition, the epoch of the near and the far, of the "side-by-side", of the dispersed. We would be at a time when the world is no longer seen as a long life that would develop through time, but rather as a network that connects points and intersects with its own skein. One could argue that some of the ideological conflicts which feed today's polemics are waged between the pious descendants of time and the tenacious inhabitants of space. However, it should be noted that the notion of space which seems to be today at the heart of our concerns, our theory and our systems is not an innovation; space itself, in Western cultures, is endowed with a history, and one cannot ignore this inevitable "interweaving of time and space" [31].

So as to retrace briefly this history of space, we could say with Michel Foucault that in the Middle Ages it was a hierarchical set of places: "sacred spaces and secular spaces, protected spaces and open, defenseless spaces, urban spaces and rural spaces—so much for the daily life of men; in the cosmological theory, there were supra-celestial spaces opposed to the celestial space; and the celestial space was in turn opposed to the terrestrial one; there were spaces where things were put because they had been violently discarded and then spaces, on the contrary, where things would find their natural ground and stability. It was all this hierarchy, this opposition, this intersection of places which could roughly be called medieval space, a space of localizations.

From now on, we would be at a time when space appears to us as connections of sites, marked by concerns which fundamentally affect space, undoubtedly far more than time; time appearing probably only as one of the various distributive operations that are possible for the elements that are spread out in space.

Yet, despite all the technologies that invest it, despite all the networks of knowledge, which makes it possible to determine or formalize it, the modern space may not be fully desacralized—unlike time probably which was indeed desacralized in the 19th century.

Although there was indeed some theoretical desecration of space (the starting point of which was Galileo's work), we may not have reached the point of a practical desecration of space. Our lives may still be ordered around some infrangible oppositions, which the institutions and practice have not yet dare damage; oppositions which we take for granted: for example, between the public and the private sphere, between the family and the social space, between the cultural and the useful space, between the recreational and the work space; all these oppositions are still nurtured by "a faint presence of the sacred" [31].

Traces of this faint presence of the sacred in space can be detected in the law. Different legal regimes are enforced in places categorized differently depending on their particular use: public places—private places including digital spaces where the written words do not benefit from the same protection (freedom of the press) and will not have the same consequences (libel).

But digital technology, in so far as it allows a spatial continuum and a time continuum to be operationalized, desecrates the representations and discrete scales of time and space, and thus breaks down the assumption that territorial power has to be hierarchical and correspond legitimately to the intertwining of the spatial and temporal discrete scales.

The notion of heterotopy was introduced by Michel Foucault in a text he presented at the Cercle d'études architecturales (Architectural study group) in Tunisia in 1967[9]:

[9] (http://foucault.info/documents/heteroTopia/foucault.heteroTopia.fr.html). The article is published in ≪ Dits et écrits ≫ , ≪ Des espaces autres ≫ in Architecture, Mouvement, Continuité, n°5, octobre 1984, pp. 46–49.

> There are also, probably the in every culture, every civilization, real places, actual places, places that are shaped in the very fabric of society, and which are kinds of counter-sites, kinds of utopias actually achieved where the real sites, all the other real sites that can be found within a culture are simultaneously represented, questioned and inverted. Places of this kind are not part of any place, even though it may be possible to indicate their location. The places, because they are intrinsically different from the sites they reflect and speak about, I shall call them, by way of contrast with utopias, heterotopies.

Thus hierarchy gives way to heterarchies, just to the extent that topology gives way to heterotopies. Accordingly, in his speech at a seminar on sustainable cities organized by Cap Excellence, the Deputy Mayor of Cachan showed, through the real example of La Vallée Scientifique de la Bièvre, how he managed to simulate stabilized conquest situations without the depletion of resources inherent to conquests. The basic idea is as follow: when a territory wishes to promote a project, it starts by informing the neighboring territories organized in an informal and not permanently established association—as was the case with La Vallée Scientifique de la Bièvre in its early stages. As a result, a conquest is simulated which encourages the adoption of a common viewpoint, by promoting one's own views or by moving away from them for the sake of adopting better ones. If abulia could prevail when alone, it now becomes necessary to show, explain, present arguments, change so as to keep this conquest with all one's might—the conquest, which in this specific case is a virtual one, has not exhausted any of the resources. Indeed, according to Montesquieu [80]:

> Conquests are easy to make because they are made with all one's forces; they are difficult to preserve because they are defended with only a part of one's forces.

Once the alchemy of the conquering plunge has taken place, the coherences identified and the concessions made, it remains to loosen the grip of the virtual conquest and come back to the territory per se. For once, there are no wounded soldiers caught up by the enemy in this retreat, because it is also a virtual one. Even failure is not a defeat. The hierarchy has given way to the heterarchies.

5.2.3.3 Duality Nature/Culture: How to Fight the Hegemony of Naturalistic Cosmology?

Cap Excellence is an excellent way to combine, at the level of an enlarged territory, cultural reflection and reflections on Taonaba (tourism, biodiversity). The question of preservation, of how it impacts Taonaba and the ACTe Memorial, is still to be dealt with. How to agree on a policy for cultural and natural heritage? How to respect the various ways of being into the world and define customs that could be agreed upon by all its occupiers? All the more so if we want to avoid the simplistic naturalistic view. It seems to be the case if the Guadelupian identity is consistent with Edouard Glissant's prospective views [81].

The Cap Excellence Territory- Guadeloupe's main territory- has to build an identity beyond Nature and Culture, but in their mutual instruction. Taonaba is a specular echo of the ACTe Memorial and vice versa. "For naturalism is just one of many ways to configure the world, that is to contrive some identifications by allotting attributes to existing beings, ascribing, starting from the available options, to an unspecified alter a physicality and an interiority comparable to or differing from those found in any human experiences. So that identification can go down four ontological routes. Either most existing entities are supposed to share a similar interiority whilst being different in body, and we have animism, as found among peoples of the Amazonian basin, the Northern reaches of North America and Siberia and some parts of Southern Asia and Melanesia. Or humans alone experience the privilege of interiority whilst being connected to the non-human continuum by their materiality and we have naturalism—Europe from the classical age. Or some humans and non-humans share, within a given framework, the same physical and moral properties generated by a prototype, whilst being wholly distinguishable from other classes of the same type and we have totemism—chiefly to be found among Australia's Aborigines. Or all the world's elements are ontologically distinct from one another, thence the necessity to find stable correspondences between them and we have analogism –China, Renaissance Europe, West Africa, the indigenous peoples of the Andes and Central-America. Yet, it can be shown that not only each of these modes of identification foreshadows a kind of community more specifically adapted to the pooling in a common destiny of types of entities it distinguishes—each ontology creating a distinctive sociology- but also that the ontological boundaries impact on the definition and on the attributes of the subject, therefore that each ontology fosters an epistemology and a theory of action adapted to the problems it has to solve. In other words, the problem we face is as follow: how can a naturalistic epistemology, bearer of universalistic values, amend so as to accept non-naturalistic epistemologies?" [82][10] (Fig. 5.7).

However, naturalism alone is accountable for radical expansionism, and also for sustainable development regulation. In that respect, sustainable development could be criticized, especially in cultures (Guadeloupe) claiming inherent anthropological mix (Fig. 5.8).

> For naturalism recognizes the signs of otherness in the discontinuity of the spirits, as opposed to animism for example, which reads them in the discontinuity of the bodies. Is different from me the man who speaks another language, believes in other values, thinks along different lines, has another vision of the world. As such he is no longer my exact fellow creature since the "collective representations" he adheres to and which influence his actions are poles apart from mine. Strange habits, enigmatic or disgusting practices are then explained by the fact that those indulging in them cannot help believing (thinking, picturing, imagining, judging, guessing…) that this is the way they have to proceed so as to reach a specific goal. It is a question of 'mentalities', and if they are allegedly knowable up to a point by the traces they leave in public expressions, it is however, impossible to understand their functioning in depth, because I cannot completely creep into the mind of one of my fellow creature, no matter how close. From that perspective, it is easy to understand that radical otherness lies on the side to those who are deprived of mind or who do not know how to use it: the savage in the past, the mentally-ill today, and above all, the multitude of non-humans: animals, objects, plants, stones, clouds, all this material chaos

[10] Translated by Janet Lloyd.

Interiority \ Body	Same	Different
Same The face of otherness	**Totemism** Aboriginal Australia *Absence of inheritance of class*	**Naturalism** Europe from the classical age *Discontinuity of the spirits*
Different The face of otherness	**Animism** Amazonia, Northern part of North America, Northern Siberia, Melanesia, part of Southeast Asia *Discontinuity of the bodies*	**Analogism** China, Renaissance Europe, Western Africa, the Andes and Mesoamerica *Absence of analogy*

Fig. 5.7 The different cosmologies, depending on the way to consider the other's interiority versus physicality [82]

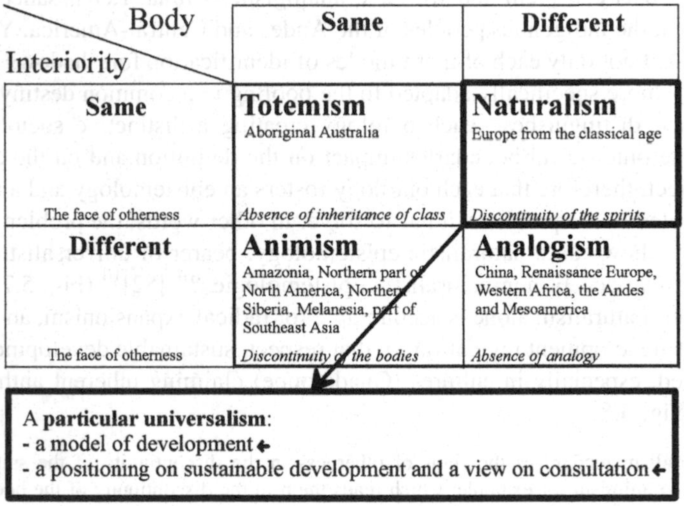

Interiority \ Body	Same	Different
Same The face of otherness	**Totemism** Aboriginal Australia *Absence of inheritance of class*	**Naturalism** Europe from the classical age *Discontinuity of the spirits*
Different The face of otherness	**Animism** Amazonia, Northern part of North America, Northern Siberia, Melanesia, part of Southeast Asia *Discontinuity of the bodies*	**Analogism** China, Renaissance Europe, Western Africa, the Andes and Mesoamerica *Absence of analogy*

A particular universalism:
- a model of development
- a positioning on sustainable development and a view on consultation

Fig. 5.8 Naturalistic cosmology would be based on a particular universalism [82]

whose reality is repetitive whilst man, in his great wisdom, strives to determine their composition and operation rules.

How, then to escape the dilemma of naturalism, this far too predictable oscillation between the monistic hope of natural universalism and the pluralistic temptation of cultural relativism? Most specifically how to step back from the comforting thought that our culture would be the only one to have gained a true understanding of nature whilst other cultures would only have access to representations—rough representations, yet worth considering for benevolent spirits, false and pernicious because of their contagiousness for the positivists? This epistemological regime, which Latour calls "distinctive universalism" establishes the development of anthropology and legitimize its success, so much so that it is difficult to imagine a patient leaving his mental hospital without risking ostracism and facing a sterile wandering mesmerized by the mirages of singularities [82].

5.3 Towards a Participatory Intelligence of Territorial Dynamics

How to allow the advent of a genuine participatory intelligence of territorial dynamics? We may have to come back to an antepredicative sense of topological space and chronological time, close to Khôra and Kairos, through ecoumenal assemblages, concept developed by Augustin Berque and to which we will return later.

This question relates directly to the representation of knowledge and its structuring but also to the evolution dynamics of its representation and the interactions between the knowledge representations different stakeholders come up with. These representations are at least of three types already identified and pointed: places and infrastructure, decision-making structures and finally the stakeholders involved in the decisions which are likely to take part in the decision-making processes.

Since knowledge structuring requires separate representations taking into account several categories and the evolution and interaction dynamics, the traditional tools of knowledge representation are outdated, that is why new tools have to be developed.

As part of this deconstruction process, the question of incommensurability arises: how to arrange contents to reduce their incommensurability to its commensurable portion, or make arrangements that make them commensurable, thus allowing the interoperability of information systems that will derive from it.

5.3.1 Territorial Intelligence: Providing Tools for the Deconstruction of Radical Naturalism

In what follows, we present three action-researches led by our research group Territorial Assemblages (Agencements territoriaux) on the areas of La Vallée Scientifique de la Bièvre and Cap Excellence. All three try to provide tools for a deconstruction of radical naturalism, which currently prevails in Western or Westernized cultures and modify the representation at stake and the structuring of the information system.

5.3.1.1 Heterotopies and Heterarchies : Weakening the Arbitrariness of the Temporal and Spatial Scales Disruptions

The first part of our research on La Vallée Scientifique de la Bièvre, is built upon the heterotopies as developed by Michel Foucault. We criticize the discretized spatial and temporal scales supported by hierarchies of power that create gaps between the discrete levels, and introduce complexes of heterogeneous and fringed spaces.

We drew inspiration from Walter Benjamin in his book The Arcades Project, from Georges Pérec in his research on intimate space organization [83], Gaston Bachelard and his phenomenological investigations, from Berque and his Ecoumenes [84], from Deleuze [85] in his Thousand Plateaus, eventually it is Michel Foucault's thinking that has held our attention. We mobilize the heterotopies conceived by Michel Foucault as early as 1967—with in the background Deleuze's notions of assemblage function and Foucault's concept of power apparatus—to provide a perspective on La Vallée Scientifique de la Bièvre within Le Grand Paris.

Here are the main principles Michel Foucault outlined in his heterotopology. We may reformulate these principles as follows.

- First principle: In any given society, one or several heterotopies are created.
- Second principle: Any society can remove or reorganize an existing heterotopy. It can also organize a heterotopy that did not exist before.
- Third principle: A heterotopy can juxtapose several incompatible spaces in one given place.
- Fourth Principle: Heterotopies open to heterochronies.
- Fifth principle: Heterotopies have a system of opening and closing that isolate them from their surroundings.
- Sixth principle: Heterotopies have a function as regards the remaining space.

A heterotopological reading of the Paris-Saclay Cluster territories and La Vallée Scientifique de la Bièvre territories allows to position the VSB project as complementary to the Plateau de Saclay. At first glance, the supporting documents for the two projects are rather similar. Some try to make sense of what is by building on the gains and their inner dynamics, while others try to give meaning to what the cluster will stand for—scientific supremacy, excellence, blending of functions, emergence of a new identity exceeding the mere juxtapositions of structures. In both cases, there is a striking lack of homogeneity in the arguments put forward. No sooner have the effects of a first argument been sketched out that a second one is broached before the full implications of the first are grasped. The intricacy of the territories under scrutiny is obviously not compatible with the type of description, and the methodological tools are still lacking.

a. The VSB project struggles to step out of the tracks of Paris-Saclay Cluster Project

The methodological approaches driving the creation of the Paris-Saclay Cluster territory and La Vallée Scientifique de la Bièvre territory differ from one another first and foremost in terms of what has not been fully thought out or, to put it differently, as regards the implicit hypotheses on which they are based. In fact, the Plateau de Saclay dynamics rests on axioms that should not be taken for granted and need to be supported, and whose criticism may not be politically well-advised, at a time when the country as a whole fears for its rank in the world. The VSB creators discreetly refrain from referring to those axioms, but do not criticize them openly, nor run the risk of putting forward some others.

The Table 5.1 summarizes the main implicit hypotheses present in the ONI Saclay Plateau Project, which we try to verbalize by picking out the phrases from which they emanate in the texts:

The territories of Paris-Saclay Cluster and La Vallée Scientifique de la Bièvre share key characteristics:

- Both territories offer a coherent functional and spatial diversity in the making; they are project-territories that can only be understood in their dynamics, always in the backdrop of global competitiveness and necessary economic growth;
- Both territorial projects try to find reference points to their respective advantage and assessment tools for their own dynamics.

b. Differential positioning of the VSB project compared to Paris-Saclay Cluster

The positioning differences are presented in the comparative Table 5.2:

The foreshadowing of Paris-Saclay Cluster is based on symbolic figures like the figure of the local showcase territory that has worldwide impacts, and on methods such as imitation/adaptation of practices implemented where "they work", postulating a cause and effect relationship. Development of large clusters would lead to virtuous practices—spontaneous coordination, resource pooling and sharing, result exchanges, interactions and interdisciplinarity—which in turn would lead to the accelerated ripening of the fruit of these practices (realization of economical potentialities, innovative breakthrough, creation of dynamic companies and, appeal to the best in the world).

These symbolic figures are not called forth in the VSB, they are not criticized either for that matter, yet not the slightest substitutive figure is called forth in their

Table 5.1 Main implicit hypotheses in Paris-Saclay Cluster

Phrases present in the text of the ONI cluster Paris-Saclay project	Implicit hypotheses
Actualize potentialities by promoting development plans	Structural development would mechanically lead to the realizations of economic potentialities
Coordinate and rally around joint actions, mutualize collective equipment	Geographic proximity would lead to spontaneous coordination
Accelerate the ripening of industrial processing of scientific breakthrough and business development	The densification of potentialities would mechanically bring about the acceleration of the ripening processes
Promote strong interactions, technology transfer, and interdisciplinary	The cluster organization would automatically prompt a boom in interactions
Encourage, at the interfaces of traditional disciplines, the emergence of future technological breakthrough and scientific revolutions	The contact of very specialized and complementary research works would naturally produce scientific revolutions and scientific shifts at the interfaces

Table 5.2 Different features of Paris-Saclay cluster and the VSB

Projects' characteristics	Paris-Saclay cluster	La Vallée Scientifique de la Bièvre
Benchmark	Operation of national interest (ONI)	Reference plan, population involvement
Type of structure	Functional cluster, innovative campus, synergy and simultaneity	Urban campus with emphasis on spatiality, intermediate integration between Paris university districts and a Saclay–like cluster
Development horizon	Speed, acceleration, then progressive deploying	Slow speed; progression
Implementation regime	Shift, exception and exemplariness, pro-active approach	Continuity, in keeping with a historical and geographical tradition
Model of integration	Interconnection through fast networks (transport, communication) of the whole Le Grand Paris area, overture, and energy integration	Overture (sharing, global, sustainable)
Rhetorics	Youth, positive announcement effect	Maturity, regularity, age of the installations
Inspiration	Innovative approach and benchmark of the best clusters worldwide	Singularity (its own scale, its own way)
Impulsion	Pro-active, massive investment	Participatory, confidence
Fields of activity	Economic and social specialization, restricted thematic areas, knowledge economy	Diversity
Topology	Densely populated concentric zone, linked to the city through fast intercity connections	Part of the southern Ile-de-France cone for innovation, intermediate link
Ecology	Environment (plateau = metaphor of the platform and of intervisibility); city life and proximity to nature are reconciled; farming activities maintained; natural water resources and architectural heritage preserved	Living place (valley = metaphor of a hierarchy and intervisibility)
Local/global	Ecosystem of growth to boost the Ile–de-France region and the nation's economy; expect global positive effects from a local perspective, aim to be one of the world's emblematic place, an international hub for knowledge economy	Deal with the negative effect of globalization at a local level (dual trend—co-variant yet not linked—for people to settle within the territory while not working there, and, in a globalized job market, for executives to live away from the territory)
Spatial positioning	Compact development on 7,700 ha to save space, higher density and opening-up are thought out simultaneously	Variety

(continued)

Table 5.2 (continued)

Projects' characteristics	Paris-Saclay cluster	La Vallée Scientifique de la Bièvre
Visibility	Showcase, emblem, seamless and attractive offer, reference internationally recognized on the résumé of researchers and students alike	Discrete 'distinction'
Model of growth	Continuous innovation in knowledge economy, symbolized by the 'knowledge triangle'—education, research and innovation—	Local growth through redistribution, try to 'be a local metropolis', that is to say ward off the curse of gentrification and individualism
Points of comparison	Comparable to the most successful similar clusters worldwide with the emergence of a synthesis, of a new model	Comparable the other territories comprising Le Grand Paris
Assessment	Move up in the Shanghai Ranking, creation of wealth and innovative companies, high-performance, quality of life and functional diversity	"Time will tell", quality of life in the long-term for the users.
Governance	Heterogeneous federation (49 municipalities, 23 higher education and research institutes, 3 competitive/competitiveness clusters)	Homogeneous federation (mayors)

stead, probably because it is not that easy to produce categories as elementary and powerful, whose radical criticism would be the target of serious accusations (pessimism, fatalism, lack of enthusiasm and spirit, even defeatism).

c. A possible heterotopic approach of the VSB

Paris-Saclay Cluster is somewhat like a television show: it is a showcase, visible worldwide, and appealing as is the case for any place where one can gain visibility. It is obviously the hybridization of a crisis heterotopy with a heterotopy of deviation (coming with the first principle), since the proposed solution to solve the ongoing economic crisis is a radical inversion of the set of deviation, by which all that is not conspicuously in the showcase is presented as deviant. Its rhythm is synchronous with its counterparts the new showcase territory competes with. Its pace may be even faster, in so far as the project asserts itself as mimetic in its goals—even more so than in its construction—(second principle).

Paris-Saclay Cluster is dense (scientific density and compact development but with an open lay-out, “protected and yet open”) and optimized. As such it radiates throughout the world like a hologram, thus generalizing the third principle of Foucault’s heterotopology. It is based on the principle of acceleration (fourth principle) and of the centrifugation of the slow, which is at the heart of its selection process (fifth principle).

Paris-Saclay Cluster is a heterotopy of compensation—maybe combined with a heterotopy of illusion (sixth principle), intended to be an allegorical embodiment of a post-crisis situation, provided the stakeholders, who in fine will have to determine its operational content, commit themselves to this fiction in the long term, or else it will be no more than a mere utopia.

In addition, La Vallée Scientifique de la Bièvre presents a categorical consistency coupled with sensitivity differentiators, mainly characterized by a positioning on the fringe of the showcase, marked by an unapologetic discretion—discretion which, however, has been damaged by the very creation of the VSB and accounts for the present lack of consistency in matter of communication policy and public awareness.

By probing further into its positions, one may wonder how the VSB could complete its proposals so as to make them indispensable and indisputable. La Vallée Scientifique de la Bièvre territory could work on the unthoughts of the Paris-Saclay Cluster ONI by experimenting them on a full scale, away from the spotlights (discretion), with more freedom on the choice of its speed (precision of the slow motion), away from the need for challenge simultaneity (demand for diversity), and above all at the very level of the functional operations and the stakeholders operational participations (involvement, confidence).

Thus at the level of these two contiguous territories, the heterotopy embodied by Paris-Saclay Cluster would gain strength, credibility, sustainability, and would widen its percolation surface.

While keeping its own imagination and working on a territorial identity that would preserve its other assets, the VSB can support change and capitalize on it for

Table 5.3 The five operational fictions to master

	Fictions that need to be checked
1	Structural development would *mechanically* lead to the realizations of economic potentialities
2	Geographic proximity would *mechanically* lead to spontaneous coordination
3	The densification of potentialities would *mechanically* bring about the acceleration of the ripening processes
4	The cluster organization would *automatically* prompt a boom in interactions
5	The contact of very specialized and complementary research works would *naturally* produce scientific revolutions and scientific breakthrough

its own economic and social development. In other words, the VSB could be thought out as a heterotopy preventing Paris-Saclay Cluster from turning into a utopia, by working on its unthoughts, studying in depth the realistic outcomes of its opportunity requirements and practical modalities of implementation. This would mean ensuring that the five main fictions given in Table 5.3 meet controlled conditions of implementation.

5.3.1.2 Assemblage Theory: Weakening the Naturalist Assumptions and Modeling

The second approach to our research on La Vallée Scientifique de la Bièvre, is built upon the assemblage theory [85, 86], coupled with the theory of simplicial complexes [87–89].

Both territorial projects under study present a list of proposals, backed by agents, which have impacts over time. At the beginning, the relational definition of the proposals is empty, then it changes when the agents get connected. The similarities they demonstrate are revealed through the extraction of the lists of agents and proposals. The proposals may be networks of cohesive facts, but what is at stake here is to try and understand the representativeness of the agents (who cogenerate the proposals) and the links between these agents and the proposals. The idea of democracy resurfaces with the possible identification of backings, strong and weak links, cohesive elements (hard to disconnect) and also of the territorial or local effects or impacts. Eventually, the structure construction and the identification of trajectories within the arrangements [86, 90] should make it possible to optimize the network of proposals by assessing the strength or weakness of the links that bind them together, spot the agents that have to be convinced, those whose influence should be strengthen or weaken, order the proposals depending on relational distances, increase or reduce the distance between two proposals so as to alter the relational density which determines the strength of the link (Fig. 5.9).

The aim is to promote the making of a common world, an optimized view of the *best* proposals to implement. The best-supported proposals must achieve the most desired effects, but a significant place must be left to the proposals or effects that, though not backed by the majority, can turn out to be key (cf. representativeness of marginal viewpoints). Similarly, the identification of *gaps* within the arrangement allows to detecting the absence of key or intermediary agents who could back a proposal and make it more influential.

Several web content searches were carried out as part of a first experiment. The results from the query "Paris-Saclay Cluster" performed in September 2011 on Google™ are presented below by way of illustration. The ranking proposed by this search engine makes it possible for an analyst to identify the dynamic and heterogeneous entities of a given situation. We analyzed the first five pages of results provided by Google™ so as to extract the agents and proposals of the Paris-Saclay Cluster. We present an excerpt from the results in the Table 5.4 in which the proposals are bolded.

Table 5.5 shows the links existing between the agents and the proposals. Some are quite expected, while others are less so [73]. Besides, even if it is only an excerpt from the results returned by Google™, we can already see a certain variety in the entities and their links. The agents involved in several proposals connect these proposals and potentially modify the form and dynamics of the global arrangement.

Let's take the example of the proposal "setting up of a R&D site in the Palaiseau district". It emerges from the interlinking of four agents: "EDF", "R&D", "Paris-Saclay Cluster" and "Palaiseau District". Moreover, it can be positioned on a map (Palaiseau district of the Cluster territory). Another example is the following proposal: "be exemplary as regards energy savings and energy efficiency". It emerges from the interlinking of the agents "Paris-Saclay Cluster" and "Ecole Polytechnique district" which can also be positioned on a map. The two clusters formed by "setting up of a R&D site in the Palaiseau district" and "be exemplary as regards energy savings and energy efficiency", peopled by the interconnected

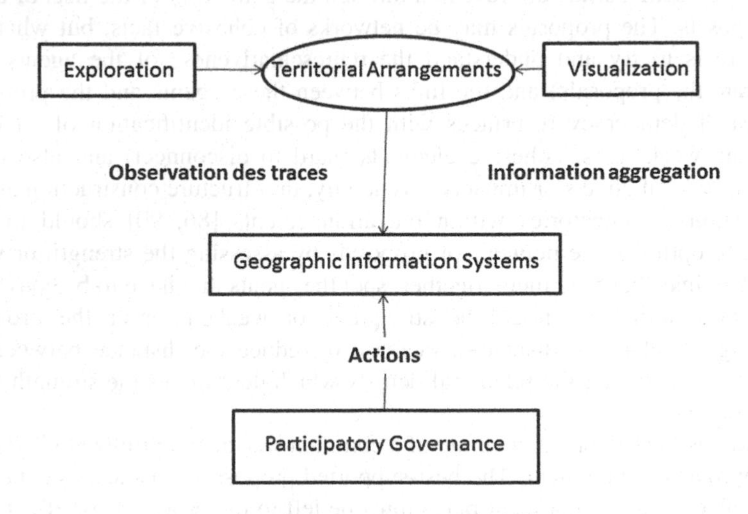

Fig. 5.9 Architecture of a system of territorial participatory governance

Table 5.4 Excerpt from the results of the query performed on Google™ (2011 09 30)

EDF confirms **the setting up of its main R&D site** within Paris-Saclay cluster, in the Palaiseau district
After a long gestation period, Paris-Saclay Campus **was given a crucial boost in 2007 thanks to operation campus launched by the French president**
The Ecole Polythecnique district, as well as Paris-Saclay cluster as a whole, has to **be exemplary as regards energy savings and energy efficiency**
Paris-Saclay cluster is created to **capitalize on interactions between higher education, research and industry and to contribute to the creation of innovative start-ups and growth in general**
Our municipality (Jouy en Josas), located 17 km from Paris, is now **part of the 49 municipalities selected in the draft bill on the creation of the public body Paris-Saclay**
Amendment to provision of title V relating to the creation of Paris-Saclay cluster—art.22—composition of the governing council for the future Paris-Saclay public body
In July 2008, the French minister for higher education and research asked the different stakeholders to **commit themselves and respond to the remarks of the assessment committee on operation campus**
The preliminary mission for the Paris-Saclay public body plans to **have a study conducted on the strategic planning of the Paris-Saclay cluster**
Renovating Paris-Sud university is listed among the important elements in this file (Paris-Saclay campus)
The Building of the first intermediary road section from the Ecole Polytechnique to the CEA, then to Saint Quentin **is planned for 2013**

agents, share one common feature, "Paris-Saclay Cluster", and thus are strongly linked. In mathematical terms, "setting up of a R&D site in the Palaiseau district" and "be exemplary as regards energy savings and energy efficiency" form two simplices with a common apex. All the project proposals, agents and links, together form a simplicial complex (Fig. 5.10).

The territorial arrangements modeled by simplicial complexes allow the emergence of proposals for the analysis of the agents and help assessing their effects. Using these techniques for scanning and visualization, it is possible to identify:

- Their representativeness (number of stakeholders per proposal and effect);
- Their influence (center/periphery radar);
- The minorities (amount of deviation between minority and majority/the chance for the minority to have its way).

All this helps plan the optimization of the network thus composed to reach a common and well-balanced view of the strongest proposals by, among other things, identifying a missing key agent or, on the contrary an agent that have to be discarded, increasing or reducing the strength of some links, and so on. As an example, why would not the "building of the first intermediary road section" (transport) from the Ecole Polytechnique to the CEA, then to Saint-Quentin receive the support from the Ministry of Transport or else transport companies or citizens associations?

Table 5.5 Table of the connections between agents and proposals

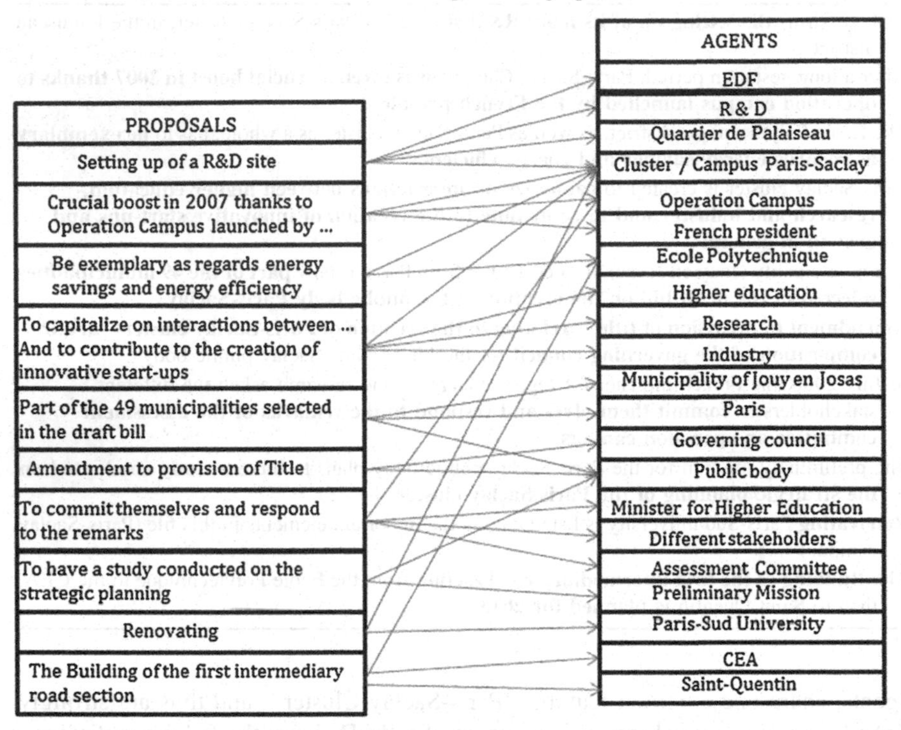

In the case of territorial intelligence, arrangement excluded, we would have:

- The assessment and the prioritization of the services to provide or already existing;
- The identification and ranking of the needs;
- The arbitration as regard the spatial localization of the services/the balance between supply and demand.

Thanks to the modeling of territorial arrangements, it is possible to picture the networks of stakeholders, proposals and effects. But the reading and analysis of their mathematic representation seems abstract for most of these projects' stakeholders. The spatial projection of the arrangements and the use of indicators—social, economic, cultural, etc.—makes the results look more concrete and give a careful interpretation of them. In this context, a GIS (Geographical Information System) can turn out to be a key element in territorial intelligence. It allows to geo-localize the data for a given territory (services, needs, etc.) which can be coupled with marketing data or territorial indicators (social, economic, etc., for example average household income data). This, coupled with a territorial intelligence-like approach permits to envision the development of new services or the restructuring of old services within a framework of sustainable development, of partnerships and

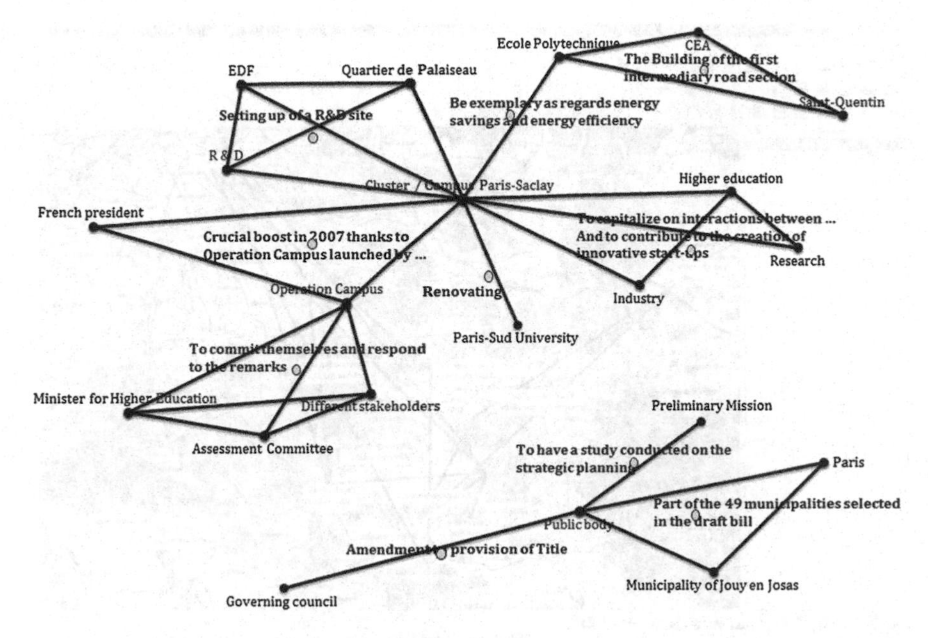

Fig. 5.10 Simplicial complex associated to the query

citizens involvement in the decision making process. By analogy, these maps could make the projects more readable and operational. They could offer a more concrete depiction of the arrangements once both the agents and proposals are "positioned" on the geographic area.

The localization of the stakeholders and proposals can be interpreted differently depending on:

- Origin and date of the web page (URI, meta-data, key words etc. promised by the semantic web whose aim is to upgrade web data);
- Origin and date of the hit itself (i.e.: creation of an apartment block in a given place);
- Absence of time and place for all which is abstract.

The analysis focuses on the hits clearly geo-localized on the digital record (in green in the figure below), or not geo-localized but easy to infer (in blue in the figure below), or else on hits too abstracts or uncertain (in italics in the above figure). The calculation function for territorial arrangement was integrated to the multi-purpose Geographical Information System Quantum™ (http://www.qgis.org/) as well as the resources from the OpenStreetMap project whose aim is to create free world maps under free license with the help of the GPS and other free data (http://www.openstreetmap.org/).

Figure 5.11 shows the spatial projection of geo-localized elements (in green), and inferred elements (in blue) of the territorial arrangement for Paris-Saclay

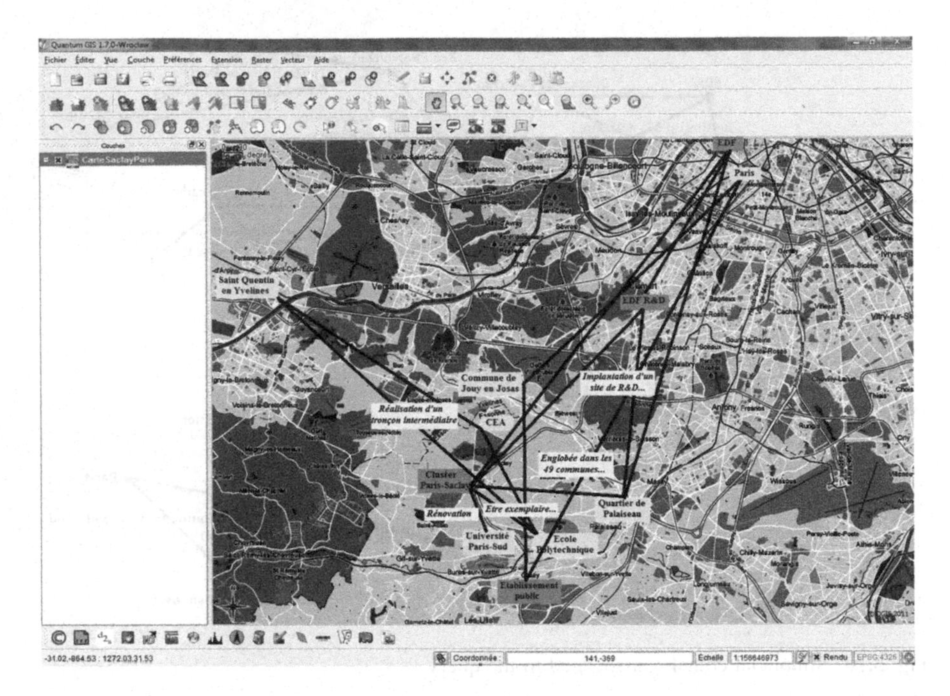

Fig. 5.11 Screenshot of an excerpt from the Paris-Saclay cluster arrangement in the GIS quantum™

Cluster. The proposals emerging from the conglomeration of stakeholders (supporters or opponents) are shown in yellow. Finally, it should be noted that e-participation could allow the inferred elements (in blue) and the abstract elements (in italics in the chart of proposals and agents) to be submitted to the community or to territorial project managers so as to dispel ambiguities, localize some of them, or else remove the "false agents" from the list.

Modeling the arrangements by simplicial complexes on the research project Taonaba-ACTe Memoral in Guadeloupe, is a way test the scalability of the approach.

5.3.1.3 Consultation as Driver of a Pluralistic Approach of Territorial Planning

A territory's spatial planning reflects the vision of the world of those who implemented it. Considering the crises and controversies arising from the inconsistencies on the Guadeloupe territory, and taking into account cross-cultural issues, we will define and propose an alternative territorial approach based on cooperation and respect for the various visions of the world.

The first aim of our research on Territorial Intelligence is to prove that it is possible to link a territory's spatial planning to its vision of the world. Then, by showing that there are not one but some territories, we will show that resorting to inadequate territorial approaches may trigger crises. The need to preserve the cohesion of these pluralist territories led up to develop an alternative approach based on dialogue.

a. A particular territorial planning after universalisms

Our investigation starts with a remark by Augustin Berque who writes in "Ecoumene" [84]:

> Our cities stand for what we are us. They pertain to our very being. Their shapes are the face of our medial body. Suffice it to see, throughout the world, the great sensitivity with which these shapes express the social structures and their evolutions.

From this mesological assertion, the author concludes that "ontology lacks in geography and geography lacks in ontology" [84]. It thus seems essential to combine ontology (as the study of the being as such) and geography as a first step. We will manage this merger so as to build a model compatible with both disciplines through the parallel study of Descola's work on universalism and the key features of the different land-use planning projects.

In his book ≪ Beyond Nature and Culture ≫ , Descola explains that Man structures the world around his experience, following a logical two-step process, namely identification and relations/connections. If ontology and geography are co-dependent as Berque implies, then the structure of certain sites bear traces of it.

The study covers the ACTe Memorial and Taonaba, both located in the French department of Guadeloupe in Overseas France, more precisely the towns of Point-à-Pitre and Les Abymes whose common desire to preserve and enhance the cultural—for one—and natural—for the other—heritage led to the establishment of two real territorial projects (Figs. 5.12, 5.13).

Identification, the first step of the process defined by Descola, consists in assigning symbols related to interiority ("mind, soul, consciousness", [53]) and physicality ("external shape, substance, physiological, perceptual and sensory-motor processes", [53]), two parameters humans are endowed with and which they project onto all existing beings. A double dichotomy—interiority/physicality, difference/similarity—occurs and allows to determine the relative place of all existing beings whose aggregation form a "collective". These various positions are thus marked by continuities or discontinuities which are typical of the related ontological pattern.

We can notice, when analyzing Western perception, how a specific set of qualities is assigned to all existing beings, but:

> Humans are the only ones to have the privilege of interiority while being connected to the continuum of non-human by their physical characteristics [82].

At the end of this identification process, we get a dual ontological pattern composed of two collectives, Humans and Non-Humans, which reflects the distinction between what is called Nature and Culture. If we stick to our initial reasoning, we should find the characteristics of this particular pattern in the ACTe Memorial and Taonaba projects (Fig. 5.14) (Table 5.6).

Fig. 5.12 ACTe memorial, future Caribean slave trade and slave life center. Original image from the brochure available at: http://www.cr-guadeloupe.fr/upload/documents/Macte12P.pdf

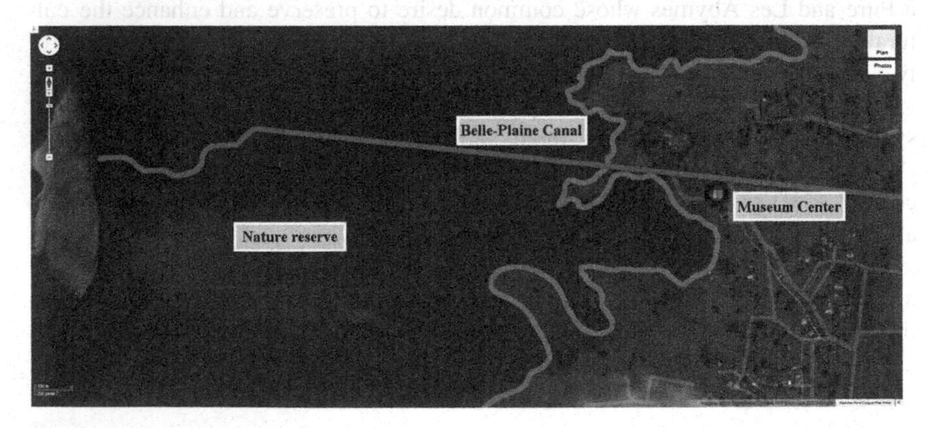

Fig. 5.13 Taonaba, ecotourism at the natural reserve of Grand Cul-de-Sac Marin. Original picture after a google map satellite view

And indeed it is easy to see that the ACTe Memorial consists of a cultural building (same name) on the one hand and a park (Morne Mémoire) on the other. These two entities are consistent with the Western ontological pattern. A similar duality is to be found in the morphology of the Taonaba project, where a cultural

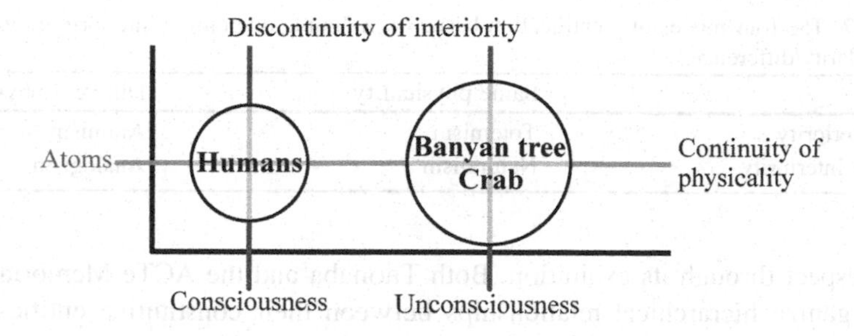

Fig. 5.14 Ontological pattern of the western vision: continuity as regards physicality, and discontinuity as regards interiority between the humans (*H*) and non humans (banyan tree, crab)

Table 5.6 Table identifying the western view on existing beings' features depending on interiority and physicality

	Humans	Dichotomy	Non-humans (banyan tree, crab…)
Physicality	Atoms	Similarities	Atoms
Interiority	Consciousness	Differences	None

center (the museum center) is built on the outskirts of the nature reserve of Grand Cul-de-Sac Marin.

Scale does not seem to be a limiting factor in the spatial expression of this duality, since Taonaba and the ACTe Memorial resort to structures different in size and express this specificity in themselves, the ACTe Memorial being dedicated to culture, while Taonaba is dedicated to nature.

This nature/culture opposition is for many the only and universal paradigm, while in fact it is a particular paradigm among four possibilities called "naturalism" (Table 5.7).

The second step in Descola's method is made of "relations". Beside the intrinsic properties of the existing beings extrinsic relations are added. They fall into two groups:

- Some potentially reversible between two equivalent terms (gift, exchange and predation) situated at the "same ontological level" [53];
- Some univocal based on connexity (genetic, temporal or spatial) between non-equivalent terms (production, transmission and protection) "linking several ontological levels" [53].

According to Descola, any relation can be broken down into results of these atomic relations. Thus the relations nature enjoys with culture in the ACTe Memorial project reveals a relation of production, that is to say "the imposition of form upon inert matter [by] an individualized intentional agent using a model of the object he originates".[11] Nature is indeed shaped as a park where visitors will have the opportunity to wander and meditate as in "philosophical gardens".[12] As for Taonaba, it endorses a relation of protection, "the non-reversible domination of the protector over the one who benefits from that protection" [53], because nature

Table 5.7 The four modes of identification based on a double dichotomy interiority/physicality and similarity/difference

	Same physicality	Different physicality
Same interiority	Totemism	Animism
Different interiority	Naturalism	Analogism

gains respect through its exhibition. Both Taonaba and the ACTe Memorial projects organize hierarchical relationships between their constitutive entities with culture ruling over nature.

We have shown that these two sites reflect naturalism, a specific anthropological and mesological paradigm which implies a conception of the world divided between nature and culture, and postulates a man endowed with a consciousness opposed to all the other existing beings devoid of consciousness. Beside this intrinsic division, naturalism has affinities with extrinsic hierarchical relations because they are formed between non-equivalent existing beings. These relations are noticeable on the ground in the communication routes: a footbridge for the ACTe Memorial and the Belle-Plaine canal for Taonaba. We think, although this is impossible to prove with only two entities, that reversible relations—expressing relationships between equivalent structures—will not imply an order in their layout and their connections, and that univocal relations—hierarchizing the entities-will orchestrate their respective roles and relationships. It is possible, if we follow this two-step logical process—identification and relations—to link the development of a territory with a vision of the world on an ontological graph (Fig. 5.15).

b. Critical Review on the Different Land Management Approaches

We highlighted, in the first step of our study, the intrinsic issue of the land management process, namely determine how to convey the communities' worldviews in a given area. If the areas are singular, the territories are plural because they embody a geographic, economic, cultural, or social cohesion at different scales. Since there can be no more than a management for a given area, the process leading to implement its organization is of paramount importance for the territories' coherence.

Focusing on our case study, we can distinguish several visions of the world associated to specific territories of Guadeloupe, which we will call "ontological territories":

- A French department in Overseas France, hosting the Western domineering vision;
- A geographic territory claiming an inherent anthropological mix designated as Creole;

[11] Descola http://www.college-de-france.fr/media/philippe-descola/UPL35675_descola_cours0304.pdf.

[12] http://www.cr-guadeloupe.fr/upload/documents/Macte12P.pdf

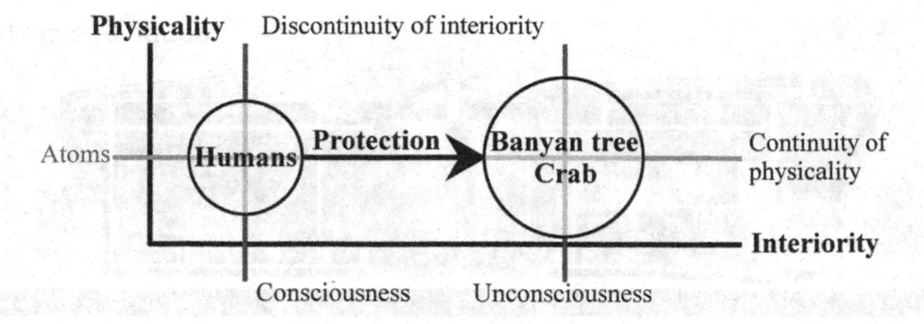

Fig. 5.15 Ontological graph functional for Taonaba's land management, western vision with continuity of physicality and discontinuity of interiority between the human (*H*) and non-human (banyan tree, crab) collectives and relation of protection between humans and non-humans

- More specific areas, such as Belle-Plaine included in the Taonaba project, where people of slave and Maroon descent identify with a cultural territory in its own right.

We will thus analyze the Taonaba and ACTe Memorial projects in depth and assess their success in dealing with these ontological territories.

Considered by its contemporaries "as an "act" likely to instill and produce a new culture", the ACTe Memorial project echoes the quest of an identity for the Creole territory in Guadeloupe. However, it should be noted that the project complies with the dual nature/culture pattern. This initial dual conception is nevertheless toned down by the use of the banyan tree analogy as it incorporates an analogist vision of the world. This analogist vision is conveyed by the tree, both in the architecture of the Memorial, "literally rooted in the land of Guadeloupe as it represents roots in the proper sense, while being visible by all since these roots are those of the banyan tree enclosing ruins as it thrives, thus protecting them from destruction"[13] and in the Morne Mémoire Park with the "landscaping of an old preserved banyan tree" (Fig. 5.16).

As such, the ACTe Memorial project, designed by the "Atelier d'architecture (BMC Jean-Michel Mocka-Celestine and Pascal Berthelot) and Atelier Gold/Marton (Marton and Michael Fabien Gold)" agencies, is a pretty conclusive test of hybrid planning. If the result is interesting, the contest that led to select this major territorial project is a risky method to use. Indeed, the development depends on the personal worldview of the architectural teams, and there is no way to know if this result is truly representative of the various national and regional territorial expectations.

Unlike the ACTe Memorial project in which the designers could freely decide on its development, the Taonaba project is directly inspired by the creative process at work in the development of 'administrative counties' (LOADDT Pasqua 1995 and Voynet Act n° 99–503 of 25 June 1999).

[13] The ACTe Memorial, a foundation for the Guadelupian society http://www.lecourrierdelarchitecte.com/article_714, on 2010/10/31.

Fig. 5.16 Memorial act cross, from the website "Le courier de l'architecte" at http://www.lecourrierdelarchitecte.com/article_715

The Voynet Act has had two major consequences as regards territorial approach. First it ensures the coherence of the French territory through sustainable development, which is a variant of the naturalistic concept of the world. It also includes participatory consultation as part of the process.

We expect that all the actors involved should take part in the dialogue. However, we notice that "the project is initially developed without taking into account the population living on the Belle-Plaine area" [91] and that "(1) the city of Les Abymes, (2) the semi-public company in charge of land use planning in Guadeloupe (3) the architect and contractor of the project" [91] alone are included in the process. Similarly, if we analyze the Western approach in planning this project, we realize that the Maroons are located in the natural zone and that the other actors are directly connected to its cultural space. It thus seems logical and coherent with the characteristics of the model of land management not to mobilize the zone of non-humans since only "people" can take part in the process. And yet, once again, if you restrict the definition of a "people" to the collective of Humans alone, it means you disregard all the other cosmologies which ascribe to non-humans interiorities similar to ours. It is thus impossible to reach a real dialogue if it is coupled with a spatial planning approach based on a particular vision of the world, as is the case in some research works in Sustainability Science such as the "participatory processes of co-construction of policy-making,"[14] in which it is wrongly assumed that everyone shares the same worldview.

Using this particular Western approach has another consequence, directly linked to the plurality and coherence of the territories. If the Western vision allows for a continuity in the territory of metropolitan France, it leads to discontinuity for the Maroon and Guadeloupe territories. This ontological incompatibility of the territories, inherent to the particular vision of this approach, leads to a phenomenon of "desynchronization of the territories", meaning that on ontological level, the territory is no longer synchronous with its population. This is, in our view, the starting point of a crisis for the Taonaba project. "Dwellers feel displaced and

[14] Reims University International Sustainability Science Research Center, http://www.univ-reims.fr/site/laboratoire-labellise/habiter-ea-2076/les-axes-de-recherche/axe-amenagement-urbanisme,11231,23890.html.

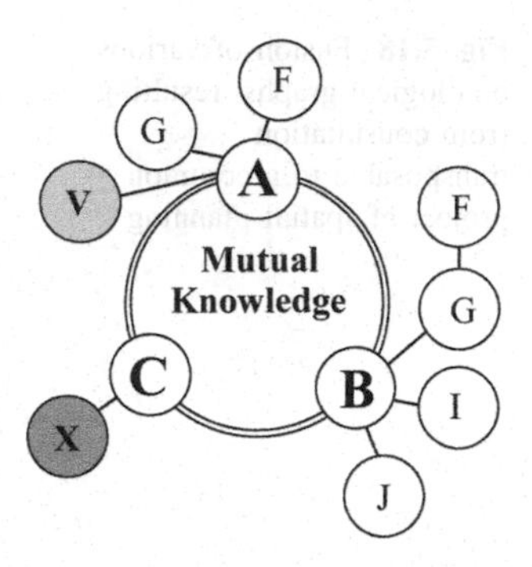

Fig. 5.17 Graph of mutual recognition: reciprocal knowledge as driver of new relations between the ontological graphs (models) of existing beings

dispossessed of a space they have appropriated for a long time; vandalism (destruction of equipment, recurring thefts) occurs regularly" [91].

As we have seen the development of a country is linked to a specific conception of the world. Yet, this development is expressed in an area that can be shared by several ontological territories. It results from the approaches related to the ACTe Memmorial and Taonaba projects that a risky approach may lead to better results than a particular approach including a consensus-building process. How, therefore, to set the basis of a controlled territorial approach and avoid the pitfall of a particular vision?

c. Using Relative Universalisms for Territorial Management

According to us, Descola provides some light on the question in his analysis of relative universalisms "with relative as in "relative pronoun," that is, making a connection" [82]. Their goals are, in Descola's words, to respect "the diversity of the states of the world" [82]. By studying the possibility of establishing relations between different universalisms, we hope to create new possibilities for territorial management.

If we consider Alain Le Pichon's analysis in mutual anthropology, mutual knowledge is "the art of discovering and producing a concise network of 'relations of relations'" [92]. This, in theory, allows us to create relations between several ontological models because they are in themselves a set of relations. "Mutual knowledge is built step after step and develops through mutual acceptance and recognition of the other's models" [92]. Le Pichon illustrates this acceptance and recognition with the visual play of anamorphosis, in which, by adjusting the arrangement of "these mirrors of hard, distorting glass, which is the way a given culture looks onto another, [we get] the common field of mutual knowledge" [92] (Fig. 5.17).

> In the game of mutual knowledge, players, partners have to gradually adjust their respective positions until they find the right arrangement and harmony that allow for the emergence of mutual recognition [92].

In this example:

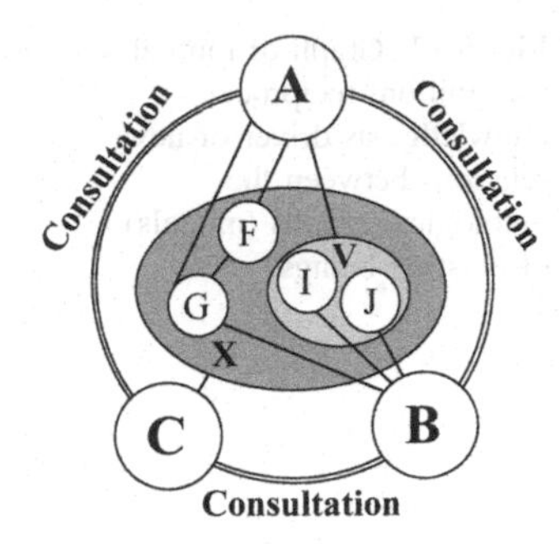

Fig. 5.18 Fusion of various ontological graphs, resulting from consultation transposable to a common project of spatial planning

- A, B and C are humans;
- X is a collective including the collectives V, G and F;
- V is a collective including the collectives I and J;
- The blue links are undefined relations.

Consultation, which implies the will of several people to reach agreements (to adjust with each other) for a joint project in which they could mutually identify with, would be nothing but the expression of mutual knowledge in a single object. Consultation could thus lead to 'harmonia'[15] in a really pluralist Spatial Planning project.

Consultation, by merging all the visions of the world associated to territories rallied around a common spatial management, could theoretically generate a place where all these visions would merge (Fig. 5.18).

Is it possible then, using this method, that all the different territories acknowledge one another? Because, at the end of the day, if planning is the result of all stakeholders, it does not match up any, as Michel Foucault's reasoning makes us understand.

> [Land Planning] functions as a heterotopy in this respect: it makes this place that I occupy at the moment when I look at myself [through its organization], at once absolutely real, connected with all the space that surrounds it, and absolutely unreal, since in order to be perceived it has to pass through this virtual point which is [my vision of the world] [79].

Thus including in a single project all the different visions of the territories at stake is not a problem in itself, since each stakeholder will see a coherent planning that matches their vision of the world. Consultation in territorial planning allows the creation of heterotopic places that ensures the coherence of the territories thanks to a pluralist territorial planning.

Our research on Territorial Intelligence was built around an ontological approach of the territory. It is possible, as we have seen, to link a territory's spatial management to a way of perceiving the world but the basis for an alternative approached both controlled and neutral have to be set. Dialogue has turned out to be the driver of this new open-minded territorial approach as it allows the

[15] From the Greek, meaning "arrangement".

expression of the territories' ontological plurality thanks to the development of heterotopic places.

5.3.2 Social Construction of Human Territories

According to Di Meo [93], the territory reflects—beyond its strictest definition of an administrative and political entity—an "appropriation of space simultaneously economic, ideological and political (hence social) by groups who present a particular image of themselves, their history and their singularity". In this highly subjective context, the characterization and understanding of the perception of a same territory by the stakeholders are difficult indeed, but nevertheless particularly interesting from the standpoint of spatial planning and public territorial policy process. The search for related information including groups of stakeholders resorting to the same territorial rhetoric represents a multidisciplinary scientific problem.

It is high time we came back to the central political philosophy which enables us to think through the question of the territory without oversimplifying it.

5.3.2.1 Ecumenes, Traces, Repetitions, Reminiscences, the Distance Problem

The spatial and temporal scales, discretized by the former territorial managers now look artificial and arbitrary. They give way to original choreographies which are baffling for the political authorities: Chronos and Topos surrender to Khôra -the existential place in so far as it is not a mappable place -, Ecoumene (see below) and Kairos—the opportune and decisive moment, which cannot be reduced to a chronological moment—the physical ways of being into the world in one's singularity. The traces, reminiscences, repetitions, problems of distance are once again the basic features of a territory which condition remembrances, the combining of traces and living together. In short, the territory is the condition of possibility of culture. Indeed, culture is the memory of culture, the revival of culture. Culture is what makes the world work: the steps of the donkey charting the Greek Khôra driven by the interplay of light and shade, slopes, grass, a spring, prevailing winds, sea swell. It is all of this, the earth and the sun. The path is repeated, sameness is charted, engramming the route. Traces become collective, collectable, coupled with signages, signposts, landmarks, tags, directions, junctions. Then, spaces are enclosed (heterotopy of the museum, the garden, the cemetery) which form special and separate places, whether specialized or not (territorial facilities such as hospitals and cultural centers), themselves symbolized as cultural attractors (cultural venues).

> The geographicity of the being is indeed nothing but the relation by which the stretch of land is so little alien to the thinking thing that it pertains to its very being. This relation is

> inextricably geographical and ontological. I call this Ecumene, returning the old Greek word oikoumenê its feminine form, making it both the earth and mankind; that by which the earth is human and mankind is terrestrial. The ecumene is the whole set and the condition of human environments, in their very humaneness, but no less in their physicality and ecology. This is what the Ecumene is, the human being's dwelling place (oikos). The ecumene is the bond of mankind with the earth, bond which is simultaneously ecological, technical, and symbolic [84].

Perec [83] perhaps best knew how to make us feel the intimacy between the question of territory and that of the mnemonic trace, through rituals:

> I would like there to exist places stable, motionless, intangible, untouched and almost untouchable, immutable, rooted; places that would be references, points of departure, sources. My native country, the cradle of my family, the house where I was born, the tree I would have seen grow (which my father would have planted on the day I was born), the attic of my youth filled with intact memories… Such places do not exist and it's because they do not exist that space becomes a question, ceases to be an evidence, ceases to be incorporated, ceases to be appropriated. Space is a doubt. I must continually mark it, designate it, it never belongs to me, it's never given to me, I must conquer it. My spaces are fragile: time is going to wear them away, to destroy them. Nothing will resemble what was any longer, my memories will betray me, oblivion will infiltrate my memory, I shall look at a few old yellowing photographs with broken edges without recognizing them. Space melts like sand running through one's fingers. Time bears it away and leaves me only shapeless shreds. To write: to try meticulously to retain something, to cause something to survive, to wrest a few precise scraps from the void as it grows, to leave somewhere a furrow, a trace, a mark or a few signs.[16]

5.3.2.2 Violence Versus Mediation

American anthropologist Keeley [94] is specialized in the study of prehistoric wars. He has shown that if violence contrasts with mediation—immediateness making it impossible to deter confrontation—this state of affairs is as old as the hills:

> After exploring war before civilization in search of something less terrible than the wars we know, we merely arrive where we started with an all-too-familiar catalog of deaths, rapes, pillage, destruction, and terror. This is a brutal reality that modern Westerners seem very loathe to accept. They seem always tempted to flee it by imagining that our world is the best of all possible ones or that life was better when the human world was far simpler. During this century, anthropologists have struggled with such complacent and nostalgic impulses, even in themselves. Their ambition was and is to explore the human condition at all times and in all places, to enlarge the narrow view of it that the written records of civilized life provide and to, in every sense, "arrive where we started and know the place for the first time". But these goals and the raw subject matter of anthropology—the origins of humans and their various cultures, social life before cities, states, and historical records—are in every culture but our own the province of mythology. Myths are a consequence of many impulses and serve many purposes, but chief among these are didactic and moralizing ones […]. "The facts recovered by ethnographers and archaeologists indicate unequivocally that primitive and prehistoric warfare was just as terrible and effective as the historic and civilized version." Ibid.

[16] Translated by John Sturrock.

According to him, this scientific reality is difficult to admit because it is at odds with the great Western myths:

> Even today, most views concerning prehistoric (and tribal) war and peace reflect two ancient and enduring myths: progress and the golden age. The myth of progress depicts the original state of mankind as ignorant, miserable, brutal, and violent. Any artificial complexities introduced by human invention or helpful gods have only served to increase human bliss, comfort, and peace, lifting humans out of their ugly and hurtful state of nature. The contradictory myth avers that civilized humans have fallen from grace—from a simple and primeval happiness, a peaceful golden age. All the accretions of progress merely multiply violence and suffering; civilization is the sorry condition that our sinfulness, greed, and technological hubris have earned us. In the modern period, these ancient mythic themes were elaborated by Hobbes and Rousseau into enduring philosophical attitudes towards primitive and prehistoric peoples. Ibid.

This issue relates directly to the question of living space and to Deleuze's concept of becoming-animal, spreading in a deterritorializing territory and whose keys are to be found in the Lebenswelt, as specified by Von Uexküll [95] among others.

5.3.3 Animal Territory, Borders: Khôra and Kairos

> The nomad has a territory; he follows customary paths; he goes from one point to another; he is not ignorant of points (water points, dwelling points, assembly points, etc.). But the question is what in nomad life is a principle and what is only a consequence. To begin with, although the points determine paths, they are strictly subordinated to the paths they determine, the reverse of what happens with the sedentary. The water point is reached only in order to be left behind; every point is a relay and exists only as a relay. A path is always between two points, but the in-between has taken on all the consistency and enjoys both an autonomy and a direction of its own. The life of the nomad is the intermezzo. Even the elements of his dwelling are conceived in terms of the trajectory that is forever mobilizing them. The nomad is not at all the same as the migrant; for the migrant goes principally from one point to another, even if the second point is uncertain, unforeseen, or not well localized. But the nomad goes from point to point only as a consequence and as a factual necessity; in principle, points for him are relays along a trajectory. Nomads and migrants can mix in many ways, or form a common aggregate; their causes and conditions are no less distinct for that (for example, those who joined Mohammed at Medina had a choice between a nomadic or Bedouin pledge, and a pledge of hegira or emigration). Second, even though the nomadic trajectory may follow trails or customary routes, it does not fulfill the function of the sedentary road, which is to parcel out a closed space to people, assigning each person a share and regulating the communication between shares. The nomadic trajectory does the opposite: it distributes people (or animals) in an open space, one that is indefinite and non-communicating. The nomos came to designate the law, but that was originally because it was distribution, a mode of distribution. It is a very special kind of distribution, one without division into shares, in a space without borders or enclosure. The nomos is the consistency of a fuzzy aggregate: it is in this sense that it stands in opposition to the law or the polls, as the backcountry, a mountainside, or the vague expanse around a city ("either nomos or polis"). Therefore, and this is the third point, there is a significant difference between the spaces: sedentary space is striated, by walls, enclosures, and roads between enclosures, while nomad space is smooth, marked only by "traits" that are effaced and displaced with the trajectory. [...] With the nomad, on

the contrary, it is deterritorialization that constitutes the relation to the earth, to such a degree that the nomad reterritorializes on deterritorialization itself. It is the earth that deterritorializes itself, in a way that provides the nomad with a territory.[...] The variability, the polyvocality of directions, is an essential feature of smooth spaces of the rhizome type, and it alters their cartography" (Treatise on Nomadology, A Thousand plateaus, [85][17]).

The territory concretizes, in the words of Simondon [96]: first overloaded with many contingencies—spatial and temporal like the notions of border and enclosure, but also cognitive like memory which, we assume, is inscribed in places and turns out to be inscribed in human and non-human relationships as well—these contingencies slowly fade away partly under the influence of digital technology, and the territory gradually radicalizes its essence, by deterritorializing all which is not consubstantial with it. The territory soon emerges as the condition of possibility of remembrance, of reminiscence, of traces, of living together and collective resilience, and goes well beyond mere spatial and temporal contingencies.

5.4 Conclusions: A Well Thought out Knowledge Engineering for Digital Humanities Integrating Differentiated Cosmologies

The territorial intelligence we outline resort to complex systems.

Complex systems are systems with a large number of differentiated entities which interact in complex ways: nonlinear interactions, feedback loops, memory of past interactions. They are characterized by the emergence, at a global level, of new properties unobservable at the level of constituent entities. The local level generates organized forms emerging at the global level, which in turn influences the local level (this is the notion of immergence). Local and global interactions can be combined in the description of their dynamics. In human societies, entities can be agents highly sophisticated themselves, endowed with cognitive, representation and intention faculties, able to develop strategic behavior taking into account the strategies of others crossways.

Complex systems are structured on several levels of organization, composed of heterogeneous entities that may themselves be complex. They cover both physical and natural systems, from the cell to the ecosphere, as well as sophisticated artificial systems—more and more inspired by natural systems men surround themselves with. The foundations for the science of complex systems (Institut des systèmes complexes de Paris Ile-de-France (http://www.iscpif.fr/AAP2013)) are both ambitious (reconstruction, modeling and simulation of the systems' dynamics at different scales of observation) and rigorous (confrontation to measures at these different scales).

[17] Translated by Brian Massumi.

Understanding complex systems requires their modeling. The models thus constructed are doubly restrained by the usual science rules: they must provide a reconstruction of observable inputs and must be as uncluttered as possible. The reconstruction in itself raises difficult problems—known as inverse problems—: given a phenomenological corpus, which modeling of the entities and interactions are compatible with the corpus. Which are, among the compatible modeling, the simplest? These inverse problems are further complicated when the reconstructions and their associated modeling generate more than one level of emergence. This complication culminates in social sciences and humanities models, in which agents both model and picture the system in which they are themselves included. With the rapid growing mass of more and more sophisticated data, the reconstruction processes tend to form a large class of inverse problems common to all disciplines.

5.4.1 Scientific Challenges of Knowledge Engineering for an Enactive Intelligence of Territorial Dynamics

Human interventions in complex systems often produce counter-productive effects, at odds the intentions they stem from. They are highly dependent on how the stakeholders model these complex systems. Complex-system engineering thus comes down to the first class of inverse problems which is their modeling and reconstruction. But it also refers to a new class of difficult problems: finding among the possible actions, those whose consequences are most desirable, or most qualitatively viable over a certain period of time. Once again, this is a large class of inverse problems which raises conventional questions of global control (acting on global interactions), but also new questions and paradoxes on distributed control (acting on local interactions). Is it possible to organize self-organization, to plan emergence? These issues are even more intricate when engineering covers several levels of organization that is the control of systems of systems.

5.4.1.1 Tools for a Collective Intelligence

In a recent e-book, Noubel [97] writes that Collective Intelligence is the study and optimization of emergent properties—be there internal-subjective or external-objective—of the groups, so as to increase their ability to live and evolve fully. It would thus invent tools for a universal governance (global, local, interdisciplinary, cross-cultural…) while also developing skills useful and immediate for today's organizations through an ethic of collaboration. Collective intelligence is the ability of a group of people to work together to express their own future and achieve it in a complex background.

He advocates collective intelligence -which he distinguishes from its archetypes he refers to as original collective intelligence, pyramidal collective intelligence and collective intelligence swarm- so as to better assert that in the present situation when an economy of abundance (vs. scarcity), of contribution (vs. production) and functionalities (vs. equipment) is possible, it is time to create the holoptic systems needed to operationalize this collective intelligence.

According to him, a holoptic apparatus is a physical or virtual space whose architecture is intentionally conceived to give its stakeholders the faculty to see and perceive the whole of what unfolds in it. "Holopticism is the means by which any participant perceives, in real time, the manifestation of other members of the group (horizontal axis) as well as the superior emerging organization (vertical axis). Thus a sports team works in a holoptical situation because each player perceives what the other players are doing, and each player perceives the emerging figure of the team. Each player then reacts accordingly, which in itself modifies the global pattern, and so on. In this case, the holoptical architecture is organically defined by the 3D space in which our basic organic senses communicate. The opposite of holopticism is panopticism. It consists of a spatial architecture organized so that all information converges toward a central point, while it is partially—and even totally—inaccessible to the others. Video surveillance systems, banks, intelligence services, and jails are examples of panoptical-based environments. This type of organization occurs sometimes in physical space, and sometimes as a result of the way information is distributed. In most companies information systems are a hybrid mix of panoptical and holoptical. While these may offer a certain level of transparency, it is still true that access rights diminish at lower levels in the hierarchy. Information systems in most companies still very much reflect such hierarchies. Absolute holopticism is a necessary but not sufficient condition for the emergence of original collective intelligence. This is also the case for global Collective Intelligence environments. From a technical perspective such artificial spaces can be built for communities having many participants by inventing knowledge and exchange spaces that are accessible and available to everyone in real time, do not overwhelm people with too much information, but provide each one with 'angled' artificially synthesized information (offering an angle, a pertinent point of view that fits with the individual user's situation, and not generalist views), allow materialization (as a perceptible object for our senses, even if virtual), namely the visualization and circulation of objects-link destined to organize the convergence and the synchronization of the community[18]" [97].

Still according to Noubel, once the holoptic visualization tools are more accomplished, it will be easier for an individual to know how to make his individual interests and those of the community converge. This evolution will have a strong impact on the economy as it evolves from a context of swarm intelligence (everyone does the same thing without knowing where it leads) to one of a collective intelligence (everyone builds their benefits depending on the information the community returns).

[18] Translated by Franck Baylin.

A question remains: what allows us to know whether an action is beneficial or not for oneself as well as for the community? Apart from extreme and evident cases, and despite our best intentions, forecasting the outcome remains an act of faith to say the least, a form of research into equilibriums. But one thing seems certain. We make better predictions when collective experience is solicited, precise holistic evaluation methodologies are implemented (including qualitative and quantitative metrics), and actions are clearly accepted and supported by the community. Today, actions undertaken in public life that go through a preliminary evaluation based on these three steps are rare. Products are launched on the market, companies are created, policies are conducted, and social actions are initiated without the moral and ethical consent of the public and the citizens and without applying any methodologies that evaluate the advantages for and threats to the community. Let's anticipate that collective information and evaluation systems will one day be at the disposal of all who wish to weigh, support and invest in projects that are estimated to be beneficial for the community. From an entrepreneurial point of view, this will not only be a guarantee of sustainability but also a source of enhanced upstream support from the marketplace. For the public, this is a guarantee of more safety. For investors, it is a way to bet in an ethical and social dimension on sustainable development, the very foundation of the economy. These financial bets on the future can be rewarded in proportion to the precision of the estimates and the risk over time. Ibid.

Noubel's vision is in perfect synergy with the theory of complex systems mentioned above.

5.4.1.2 Modeling an Economy of Functionalities and Public Web Services

Web-Content Retrieval and Extraction is a key issue for territorial intelligence so as to provide efficient and convenient Web services. How to and why use Web content mining?

All the discovered traces could allow, beyond the use of official documents as participatory systems, to reconstruct the list of stakeholders involved in a process. According to Soulier [75], Web content mining is used to collect traces of the action and provides the data for its analysis. Simplicial complexes can be the archetypal mathematical support for this analysis, which must not be taken as a form of data analysis in the usual sense. It is a complementary and enriching form of visualization, which is not subjected to the same constraints as regard data-gathering nor to the same axioms.

It should be noted that the representation by simplicial complexes provides multidimensional networks. They are no longer graphs, or even hypergraphs. A geometric perspective presents the arrangements obtained as a "collage" of polyhedra of all sizes, like Robert Rauschenberg's Combines.[19] Their contacts (the 13 possible connections between attributes) can form chains of adjacencies. Not only is the notion of path used in graphs generalized, but a whole set of quantitative and qualitative data on the structure becomes available [88]. Thus, the separate parts, more or less closely connected, and the length the paths ahead, even the closures or "missing" parts are indicators of participation.

Thus the experimentation of web content mining for La Vallée Scientifique de la Bièvre or the Taonaba site in Guadeloupe aims to preparee a full-scale capture. Illustrative examples would give way to demonstrative examples meant to assess the performativity of these multidimensional networks. The term full-scale should not be understood as an underlying desire for the absolute or exhaustiveness.

In short, the experimentation of Web content mining is used to initiate a specification for a possible resort to automation, and to clarify how the semantic interpretation of the mathematical model can impact the search and vice versa. The most difficult problem during a search is to discriminate between substances and attributes. In the absence of a predefined nomenclature, the hits surface from targets that range from a blog entry to a thematic collection of documents, the entire archives of an institutional website or even a dedicated journal.

Thus the experimentation raises the following typical questions:

- What benefits can be drawn from a quantified assessment of a full-scale search?
- What are the advantages in using multiple search engines and/or several types of search engines?
- What are the effects of searches extended over long periods of time, in terms of updates, archiving, evolution, and stability?

5.4.2 Digital Humanities Integrating Differentiated Cosmologies: A New Eldorado for Knowledge Engineering?

If computer science and digital technologies started their meteoric career in the industry, even before seducing the public, the explanation probably lies in the combination of two distinct reasons. First, engineering, much more than the humanities, is based on formal and computable models which are particularly well adapted to be operationalized on Turing machines. Secondly, the very culture of engineers is, even before being a scientific culture, the quest for empirical result, open to experimentation, provided it leads to economies of scale.

Yet, the possibility for digital humanities was there from the start of computer science as Turing's work on artificial intelligence and his imitation game [98] shows. The digital has not only conquered engineering, but also people's everyday lives. Digital humanities still has to take off. This will, no doubt, be the revolution of the twenty-first century.

Digital humanities are in need for knowledge engineering to become scientific disciplines, but the reverse is perhaps even more blatant. The epistemologies

[19] "The objects I use are most of the time trapped in their ordinary banality. No quest for rareness. In New York, it is impossible to walk the streets without seeing a tire, a can, a cardboard. I only take them and make them their own world…".

underlying today's knowledge engineering are often sketchy and therefore without nuance, and even at times radically hegemonic—as field researchers did not get a chance to ponder over ontological differentiation, engrossed as they were by the technical challenges their new discipline posed them. Thus, digital humanities integrating differentiated cosmologies, capable of structuring a cultural dialogue within a globalized humanity, will most likely be a real Eldorado for second-generation knowledge engineering.

Territorial intelligence is a prototype of this new knowledge engineering. One of the main obstacles it has to deal with is the confrontation of the different possible representations of the territory involved in this territorial intelligence: representation through places and structures, representation stemming from the structuring body, the ruling body, and the decision makers and eventually representation by the actors and agents impacted by the decision.

It is only after a comprehensive confrontation of these different representations has taken place that structuring will be effective and the information system fully operating.

A representation that would be but a digitization of the territories falls prey to the expectations of the information systems it faces. Interactions between these systems are real and effective. A preliminary conception of this confrontation and of the dynamics that bear it is necessary to ensure a representation that would be not immediately or too quickly challenged. This condition is essential to make it acceptable.

Providing tools to make the confrontations between the three main forms of identified representations operating would, on the contrary, ensure their projections onto each plane, and particularly onto the plane of the territory, and complement the traditional information systems based only on the plane of the territory.

Acknowledgments Thanks to Eddie Soulier, Jacky Legrand, Florie Bugeaud, Houda Neffati and Philippe Calvez for their scientific collaboration and support during that research. Thanks to Catherine Gerber for her support during the English translation process. Thanks to the CPER AidCrisis Project and the French Ministry of Research that supported most of related research. Thanks to the mayor of Les Abymes for helping us collecting the local data.

References

1. Kuhn, T.S.: La Structure des révolutions scientifiques. Flammarion (Champs), Paris (1983)
2. Helmer, O., Rescher, N.: On the epistemology of the inexact sciences. Manage. Sci. (October), 25–52 (1959)
3. Bailey, A.D. Jr., Whinston, A.B., Zacarias, P.T.: Knowledge representation theory and the design of auditable office information systems. J. Inf. Syst. 1–28 (1989) (Spring)
4. Bhimani, A., Roberts, H.: Management accounting and knowledge management: in search of intelligibility. Manage. Account. Res. 1–4 (2004) (March)
5. Birkinshaw, J., Sheehan, T.: Managing the knowledge life cycle. MIT Sloan Manage. Rev. (Fall), 75–83 (2002)

6. Boland Jr, R.J., Singh, J., Salipante, P., Aram, J.D., Fay, S.Y., Kanawattanachai, P.: Knowledge representations and knowledge transfer. Acad. Manage. J **44**(2), 393–417 (2001)
7. Bonner, S.: Experience effects in auditing: the role of task-specific knowledge. Account. Rev. (January), 72–92 (1990)
8. Brown, J.S., Duguid, P.: Creativity versus structure: a useful tension. MIT Sloan Manage. Rev. (Summer), 93–94 (2001)
9. Chen, A.N.K., Hwang, Y., Raghu, T.S.: Knowledge life cycle, knowledge inventory, and knowledge acquisition strategies. Decis. Sci. **41**(1), 21–47 (2010)
10. Argyris, C.: Organizational learning and management information systems. Acc. Organ. Soc. **2**(2), 113–123 (1977)
11. Meunier, J.G.: Humanités numériques, enjeux et méthodes, Texto!, Volume XVII n°1&2, Coordonné par Jean-Louis Vaxelaire (2012)
12. Wang, F.-Y.: Is culture computable? IEEE Intell. Syst. **24**(2), 2–3 (2009)
13. Ahrens, T.: Talking accounting: an ethnography of management knowledge in British and German brewers. Acc. Organ. Soc. **22**(7), 617–637 (1997)
14. Ericsson, K.A., Prietula, M.J., Cokely, E.T.: The Making of an Expert. Harvard Business Review, Hoboken (July–August), pp. 114–121 (2007)
15. Gibbins, M., Qu, S.Q.: Eliciting experts' context knowledge with theory-based experiential questionnaires. Behav. Res. Account. **17**, 71–88 (2005)
16. Hansen, M.T.: The search-transfer problem: the role of weak ties in sharing knowledge across organization subunits. Adm. Sci. Q. **44**(1), 82–111 (1999)
17. Hislop, D.: Knowledge Management in Organizations. Oxford University Press, USA (2009)
18. Dixon, N.M.: Common Knowledge: How Companies Thrive by Sharing What They Know. Harvard Business School Press, Boston (2000)
19. Easterby-Smith, M., Lyles, M.A.,(eds.): The Blackwell Handbook of Organizational Learning and Knowledge Management (Blackwell Handbooks in Management). Blackwell Publishers, Hoboken (2003)
20. Majchrzak, A., Logan, D., McCurdy, R., Kirchmer, M.: Four keys to managing emergence. MIT Sloan Manage. Rev. (Winter), 14–18. (Continuous discourse with potential participants, continuous updating of knowledge maps, blurring the boundaries between participants inside and outside the organization, and governing through reputation networks), (2006)
21. Bukowitz, W., Petrash, G.: Visualizing, measuring and managing knowledge. Res. Technol. Manage. (July/August), 24–31 (1997)
22. Coyne, K.P., Clifford, P.G., Dye, R.: Breakthrough thinking from inside the box. Harvard Bus. Rev. (December), 70–78 (2007)
23. Dalkir, K., Liebowitz, J.: Knowledge Management in Theory and Practice. The MIT Press, Cambridge (2011)
24. Amabile, T.M., Barsade, S.G., Mueller, J.S., Staw, B.M.: Affect and creativity at work. Adm. Sci. Q. **50**(3), 367–403 (2005)
25. Govindarajan, V., Trimble, C.: Strategic innovation and the science of learning. MIT Sloan Manage. Rev. (Winter), 67–75 (2004)
26. Hammer, M., Leonard, D., Davenport, T.: Why don't we know more about knowledge. MIT Sloan Manage. Rev. (Summer), 14–18 (2004)
27. Schulz, M.: The uncertain relevance of newness: organizational learning and knowledge flows. Acad. Manage. J. **44**(4), 661–681 (2001)
28. Alvesson, M.: Review: the politics of management knowledge by Stewart Clegg and Gill Palmer. Adm. Sci. Q. **43**(4), 938–942 (1998)
29. Dilnutt, R.: Knowledge management in practice: three contemporary case studies. Int. J. Account. Inf. Syst. **3**(2), 75–81 (2002)
30. Corboz, A., Tironu, G.: L'espace et le détour, Entretiens et essais sur le territoire, la ville, la complexité et les doutes, Editions hepia, l'Age d'Homme, Lausanne, (2009)
31. Malone, J.D.: Shooting the past: an instructional case for knowledge management. J. Inf. Syst. (Fall), 41–49 (2003)

32. Mizruchi, M.S., Fein, L.C.: The social construction of organizational knowledge: a study of the uses of coercive, mimetic, and normative isomorphism. Adm. Sci. Q. **44**(4), 653–683 (1999)
33. Nag, R., Corley, K.G., Gioia, D.A.: The intersection of organizational identity, knowledge, and practice: attempting strategic change via knowledge grafting. Acad. Manage. J **50**(4), 821–847 (2007)
34. Mullin, R.: Knowledge management: a cultural evolution. J. Bus. Strategy (September/ October), 56–61 (1996)
35. O'Dell, C., Grayson, C.J.: If only we knew what we know: Identification and transfer of internal best practices. Calif. Manage. Rev. (Spring), 154–174 (1998)
36. O'Leary, D.E., Selfridge, P.: Knowledge management for best practices. Commun. ACM (November) (2000)
37. Simonin, B.L.: The importance of collaborative know-how: an empirical test of the learning organization. Acad. Manage. J. **40**(5), 1150–1174 (1997)
38. Singh, J.: Collaborative networks as determinants of knowledge diffusion patterns. Manage. Sci. (May), 756–770 (2005)
39. Dyer, J.H., Nobeoka, K.: Creating and managing a high-performance knowledge-sharing network: the Toyota case. Strateg. Manag. J. **21**, 345–367 (2000)
40. Bunderson, J.S.: Recognizing and utilizing expertise in work groups: a status characteristics perspective. Adm. Sci. Q. **48**(4), 557–591 (2003)
41. George, J., George, A.: An integrative approach to planning and control using a stakeholder-based knowledge management system. J. Appl. Manage. Account. Res. (Winter), 1–20 (2011)
42. Kogut, B.: The network as knowledge: generative rules and the emergence of structure. Strateg. Manag. J. **21**, 405–425 (2000)
43. Wenger, E., McDermott, R., Snyder, W.M.: Cultivating Communities of Practice: A Guide to Managing Knowledge. Harvard Business School Publishing, Cambridge (2002)
44. Becerra-Fernandez, I., Sabherwa, R.: Knowledge Management: Systems and Processes. M. E. Sharpe, Armonk (2010)
45. Geisler, E., Wickramasinghe, N.: Principles of Knowledge Management: Theory, Practice and Cases. M. E. Sharp, Canada (2009)
46. Atwood, C.G.: Knowledge Management Basics. (ASTD Training Basics Series). ASTD Press, Alexandria (2009)
47. Awad, E.M., Ghaziri, H.: Knowledge Management: Updated 2nd Edition. International Technology Group, LTD (2010)
48. Elias, N., Wrigh, A.: Using knowledge management systems to manage knowledge resource risks. Adv. Manage. Account. **15**, 195–227 (2006)
49. Leitner, K., Warden, C.: Managing and reporting knowledge-based resources and processes in research organisations: specifics, lessons learned and perspectives. Manage. Account. Res. (March), 33–51 (2004)
50. Sauvé, L.: Being Here Together. In McKenzie, M., Hart, P., Heesoon, B., Jickling, B. Fields of green: restorying culture, environment, and education pp. 325–335. Hampton Press, New Jersey (2009)
51. Bouleau, N.: Risk and Meaning: Adversaries in Art, Science and Philosophy. Springer, Berlin (2011)
52. Ostrom, E., Hess, C.: Understanding Knowledge as a Commons: From Theory to Practice. The MIT Press, Cambridge (2006)
53. Descola, P.: Par-delà nature et culture. Gallimard, Paris (2005)
54. Latour, B.: Cogitamus: Six lettres sur les humanités scientifiques. La Découverte, Paris (2010)
55. Latour, B.: Avoir ou ne pas avoir de réseau: That's the question, in Madeleine Akrich et al. (sous la direction de) Débordements. Mélanges offerts à Michel Callon, Presses de l'Ecole des Mines, pp. 257–268 (2010b)
56. Arnstein, S.: A ladder of citizen participation. J. Am. Inst. Plann. **35**(4), 216–224 (1969)

57. De Jouvenel, B.: Arcadie, Essais sur le mieux-vivre, Tel gallimard (2002)
58. Renard, J.: Territoires, territorialité, territorialisation, Controverses et perspectives. Norois, n°210/2009, Dir. M. Vanier, pp. 109–110 (2009)
59. Blondiaux, L., Fourniau, JM.: Un bilan des recherches sur la participation du public en démocratie : beaucoup de bruit pour rien ? Revue de sciences sociales sur la démocratie et la citoyenneté Participations, n°1/2011, Edts de Boeck, (2011)
60. Deleuze, G.: Qu'est-ce qu'un dispositif ?, Rencontre internationale ≪ Michel Foucault philosophe ≫ , Paris, 9/11 Janvier 1988, Seuil (1989)
61. Callon, M., Lascoumes, P., Barthes, Y.: Agir dans un mode incertain. Essai sur la démocratie technique, La couleur des idées, Seuil (2001)
62. Sgard, A.: Marie-José Fortin et Véronique Peyrache-Gadeau, Le paysage en politique, Dossier ≪ Paysage et développement durable ≫ . Revue électronique Développement durable & territoires, **1**(2), Septembre (2010)
63. Lévy, J., Lussault, M.: Dictionnaire de la géographie et de l'espace des sociétés, Belin, Paris (2009)
64. Monnoyer-Smith, L.: La participation en ligne, révélateur d'une évolution des pratiques politiques?, Revue de sciences sociales sur la démocratie et la citoyenneté Participations, n°1/ 2011, Edts de Boeck (2011)
65. O'Leary, D.E.: Using AI in knowledge management: knowledge bases and ontologies. IEEE Int. Syst. (May–June), 34–39 (1998)
66. Wright, W.F., Jindanuwat, N., Todd, J.: Computational models as a knowledge management tool: a process model of the critical judgments made during audit planning. J. Inf. Syst. (Spring), 67–94 (2004)
67. Yayavaram, S., Ahuja, G.: Decomposability in knowledge structures and its impact on the usefulness of inventions and knowledge-base malleability. Adm. Sci. Q. **53**(2), 333–362 (2008)
68. Rousseaux, F.: Contribution à une méthodologie d'acquisition des connaissances pour l'ingénierie des Systèmes d'Information et de Communication: l'exemple de CHEOPS pour l'aide à la gestion de crises collectives à caractère géopolitique, Habilitation à Diriger des Recherches, Université Paris 6, 22 février (1995)
69. Daniell, K.A., White, I., Ferrand, N., Coad, P., Ribarova, IS., Rougier, J-E., Hare, M., Popova, A., Jones, NA., Burn, S., Perez, P.: Co-engineering participatory water management processes: theory and insights from australian and bulgarian interventions. Ecol. Soc. **15**(4), art. 11, IF 3:3 (2010)
70. Girardot, JJ.: Principes, Méthodes et Outils d'Intelligence Territoriale—Évaluation participative et Observation coopérative, Séminaire européen de la Direction Générale de l'Action Sociale du Portugal ≪ Conhecer melhor para agir melhor ≫ , EVORA (Portugal), DGAS 7–17, 3–5 mai (2000)
71. Girardot, J.J.: Intelligence territoriale et participation, Actes des 3ème rencontres ≪ TIC & Territoire : quels développements ? ≫ , Enic et Cies, ISDM, n° 16, 13 p., Lille, 14 mai (2004)
72. Hennion, A.: Vous avez dit attachements ?, in Akrich M., Barthe Y., Muniesa F., Mustar P. (eds.), Débordements Mélanges offerts à Michel Callon, Paris, Presses des Mines, pp. 179–190 (2010)
73. Soulier, E., Legrand, J., Bugeaud, F., Rousseaux, F., Neffati, H., Saurel, P.: Territorial Participative Agencement: a Case Study of Vallée Scientifique de la Bièvre as Multidimensional Network. In: Proceedings of Workshop "Les effets de la participation", Congrès Démocratie & Participation, 18–21 octobre, Paris (2011)
74. Rousseaux, F., Soulier, E., Saurel, P., Neffati, H.: Agencement multi-échelle de territoires à valeur ajoutée numérique, Politiques publiques, Systèmes complexes, Bourcier, D., Boulet, R., Mazzega, P. (eds.), Hermann, pp. 169–193 (2012)
75. Soulier, E., Rousseaux, F., Neffati, H., Legrand, J., Bugeaud, F., Saurel, P., Calvez, P.: Performativité organisationnelle des outils d'agencements territoriaux: vers une intelligence

territoriale à base d'engagements ? In: Proceedings of Conférence internationale ACFAS, Organisations, performativité et engagement, Montréal, pp. 150/173, 7–11 mai 2012
76. Muller, P.: Les politiques publiques. Presses universitaires de France, France (2009)
77. Ostrom, E.: Understanding Institutional Diversity. Princeton University Press, Princeton (2005)
78. Mintzberg, H.: Grandeur et décadence de la planification stratégique. Dunod, Paris (1994)
79. Foucault, M.: Dits et écrits 1984, Des espaces autres (conférence au Cercle d'études architecturales, 14 mars 1967), in Architecture, Mouvement, Continuité, n°5, pp. 46–49, octobre (1984)
80. Montesquieu, C.-L.: Considérations sur les causes de la grandeur des Romains et de leur décadence. Gallimard, Paris (2008)
81. Glissant, E.: Introduction à une poétique du divers. Gallimard, Paris (1996)
82. Descola, P.: Anthropologie de la nature, exposé à l'Unesco le 16 juin (2006)
83. Perec, G.: Espèces d'espace. Galilée, Paris (2000)
84. Berque, A.: Écoumène, introduction à l'étude des milieux humains. Belin, Paris (2000)
85. Deleuze, G., Guattari, F.: Mille Plateaux. Minuit, Paris (1980)
86. Igor K.: Diagramme et agencement chez Gilles Deleuze : L'élaboration du concept de diagramme au contact de Foucault, Filozofija i društvo **20**(3), 97–124 (2009)
87. Atkin, R.: Mathematical Structure in Human Affairs. Heinemann, London (1974)
88. Atkin, R.: Combinatorial Connectivities in Social Systems. Birkhäuser Verlag, Basel (1977)
89. Atkin, R.: Multidimensional Man: Can Man Live in 3-dimensional Space? Penguin Books, City of Westminster (1981)
90. Huet, F., Gkouskou Giannakou, P., Hugues Choplin, Charles Lenay, Entre territoire et apprentissage, les dynamiques d'agencement, De Boeck Université, Revue internationale de projectique 2008/1, pp. 55–67 (2008)
91. Lahaye, N.: Evaluation de la participation et graphe d'influence pour une gouvernance participative en éco-tourisme. Le cas du projet écotouristique Taonaba, en Guadeloupe, XLVème colloque de l'ASRDLF http://asrdlf2008.uqar.qc.ca/Papiers%20en%20ligne/LAHAYE-evaluation.pdf, Rimouski, Canada (2008)
92. Le Pichon, A.: Stratégies transculturelles pour un monde multipolaire, Alliage n°55–56, pp. 17–28, http://revel.unice.fr/alliage/index.html?id=3586 mars (2004)
93. Di Méo, G.: Géographie sociale et territoires. Paris, Nathan (1998)
94. Keeley, L.H.: War before Civilization. Oxford University Press, Oxford (1996)
95. Von Uexküll, J.: Mondes animaux et mondes humains. Denoël (1956)
96. Simondon, G.: Du mode d'existence des objets techniques. Aubier (2012)
97. Noubel, J.F.: Intelligence collective, la révolution invisible, The Transitioner, http://www.thetransitioner.org/wikifr/tiki-index.php?page=La+révolution+invisible (2007)
98. Turing, A.: Computing Machinery and Intelligence. Mind, **59**(36), 433–460 (1950)

Chapter 6
Integrating Knowledge Engineering with Knowledge Discovery in Database: TOM4D and TOM4L

Laura Pomponio and Marc Le Goc

Abstract Knowledge Engineering (KE) provides resources to build a conceptual model from experts' knowledge which is sometimes deficient to interpret the input data flow coming from a concrete process. On the other hand, data mining techniques in a process of Knowledge Discovery in Databases (KDD) can be used in order to obtain representative patterns of data which could allow to improve the model to be constructed. However, interpreting these patterns is difficult due to the gap which exists between the expert's conceptual universe and that of the process instrumentation. This chapter proposes then a global approach which combines KE with KDD in order to allow the construction of Knowledge Models for Knowledge Based Systems from expert knowledge and knowledge discovered in data. This approach is grounded in the Theory of Timed Observations on which both a KE methodology and a KDD process are based, so that the resulting models can be compared.

6.1 Introduction

A Knowledge Based System (KBS) carries out a set of knowledge intensive tasks for the purpose of putting in practice problem-solving capabilities, comparable to those of a domain expert, from an input data flow produced by a process.

In particular, a knowledge intensive task requires, by construction, a *Knowledge Model* in order to interpret the input data flow according to the task to be achieved, to identify an eventual problem to be solved and to produce a solution to this one.

L. Pomponio (✉) · M. Le Goc
Laboratoire de Sciences de l'Information et des Systèmes (LSIS), UMR CNRS 7296, Aix-Marseille University, Domaine universitaire de Saint Jérôme, Avenue Escadrille Normandie Niemen, 13397 Marseille, France
e-mail: laura.pomponio@lsis.org

M. Le Goc
e-mail: marc.legoc@lsis.org

C. Faucher and L. C. Jain (eds.), *Innovations in Intelligent Machines-4*, Studies in Computational Intelligence 514, DOI: 10.1007/978-3-319-01866-9_6,

The Knowledge Engineering (KE) discipline provides methods, techniques and tools which facilitate and improve the modelling task of expert knowledge. In this field of study, most approaches model separately expert knowledge regarding the expert's reasoning mechanisms from expert knowledge specific to the domain of interest. Thus, a model of the expert's knowledge, called *Expert Model* (or *Knowledge Model*), obtained through this discipline will be generally made up of a model describing how the expert reasons about the process (a conceptual model of the expert's reasoning tasks) and of a representation of the knowledge used in the involved reasoning (a conceptual model of the domain knowledge). This latter is derived from the *Process Model* utilized by the expert in order to formulate his own knowledge. Knowledge Engineering allows then to establish a back and forth way between the expert's knowledge and the built *Expert Model* where the validity of this latter can be evaluated. However, two of the main drawbacks with the KE approaches are (1) the cost of knowledge acquisition and modelling process, which is too long for economic domains that use technologies with short life cycles and (2) the validation of the *Expert Model* which is mainly oriented to "case-based".

An interesting alternative to deal with these problems is to resort to the process of Knowledge Discovery in Database (KDD) which uses Data Mining techniques in order to obtain knowledge from data. In this approach, the process data flow is recorded by a program in a database where the data contained in such a database are analysed by means of Data Mining techniques in a KDD process with the purpose of discovering "patterns" of data. An n-ary relation among data can be considered a pattern when this relation has a power of representativeness according to the data contained in a database. This representativeness is related to a form of recurrence within the data; that is to say, an n-ary relation among data of a given set is a pattern, when this relation is ''often" observed in the database. Thereby, a set of patterns is then considered as the observable manifestation of the existence of an underlying model of the process data contained in the database. Nevertheless, establishing the meaning, regarding the expert's semantics, of such a *Data Model* entails a difficult task. One of the reasons for this difficulty is the deep difference between the universe of the process instrumentation, from where the data come, and the conceptual universe of the expert's reasoning where exist scientific theories and theirs underlying hypothesis. As a consequence, the validation of a *Data Model* is an intrinsically difficult task and a lot of work has to be done to constitute a knowledge corpus from a validated *Data Model*.

Thus, in this last decade the idea of combining Knowledge Engineering with Knowledge Discovery in Database emerges with purpose of taking the advantages of both disciplines in order to reduce the construction cost of suitable *Knowledge Models* for Knowledge Based Systems. The main idea is to make possible the cross-validation of an *Expert Model* and a *Data Model*. This aims to define a general perspective, by combining Knowledge Engineering with Knowledge Discovery in Database in a global approach of knowledge creation carried out from experts and knowledge discovered in data. The key point to achieve this is then to find a KE methodology and a KDD process which allow to produce *Expert*

Models and *Data Models* comparable each other by knowledge engineers and easily interpretable by experts.

As far as we know, only the KE methodology and the KDD process which are based on the Theory of Timed Observations [1] allow to compare their models each other. This theory has been established to provide a general mathematical framework for modelling dynamic processes from timed data by combining the Markov Chain Theory, the Poisson Process Theory, the Shannon's Communication Theory [2] and the Logical Theory of Diagnosis [3]. Thus, this theoretical framework provides the principles that allow to define a KE methodology, denominated TOM4D (Timed Observation Modelling For Diagnosis) [4–7], and a KDD process called TOM4L (Timed Observation Mining For Learning) [8–13]. Owing to that, both TOM4D and TOM4L are based on the same theory, the models constructed through both can be easily related and compared to each other.

The purpose of this chapter is to describe the way the Theory of Timed Observations builds a bridge between Knowledge Engineering and Knowledge Discovery in Database. In line with this aim, a global knowledge creation perspective which combines experts' knowledge with knowledge discovered in a database is presented. In order to show how models built through this perspective can be collated and complement each other, the proposed approach is applied to a very simple didactic example of the diagnosis of a vehicle taken from the book by Schreiber et al. [14].

The next section completes this introduction by presenting arguments about the need of a global approach which fuses Knowledge Engineering and Knowledge Discovery in Database. The main concepts of the Theory of Timed Observations are then introduced in order to present the TOM4D KE methodology and the basic principles of the TOM4L KDD process. Next, both TOM4D and TOM4L are applied to the didactic example above mentioned in order to show how the corresponding *Expert Models* and *Data Models* can be compared to each other. Finally, the conclusion section synthesizes this chapter and refers to some applications of our approach of knowledge creation on real world problems.

6.2 Two Knowledge Sources, Two Different Approaches

Creating or capturing knowledge can be originated from psychological and social processes or, alternatively, from data analysis and interpretation. That is to say, the two significant ways to capture knowledge are: synthesis of new knowledge through socialization with experts (a primarily people-driven approach) and discovery by finding interesting patterns through observation and intertwining of data (a primarily data-driven or technology-driven approach) [15].

6.2.1 Knowledge Engineering: A Primarily People-Driven Approach

Considering knowledge as intellectual capital in individuals or groups of them, the creation of new intellectual capital is carried out through combining and exchanging existing knowledge. With this perspective, Nonaka's *knowledge spiral* [16, 17], illustrated in Fig. 6.1, is considered in the literature as a foundational stone in knowledge creation. Nonaka characterizes knowledge creation as a spiralling process of interactions between explicit and tacit knowledge. The former can be articulated, codified, and communicated in symbolic form and/or natural language [18], the latter is highly personal and hard to formalize, making it difficult to communicate or share with others [19]. Each interaction between both existing knowledges gives as result new knowledge. Thus, this process is conceptualized in four phases: Socialization (the sharing of tacit knowledge between individuals), Externalization (the conversion of tacit into explicit knowledge: the articulation of tacit knowledge and its translation into comprehensible forms that can be understood by others), Combination (the conversion of explicit knowledge into new and more complex explicit knowledge) and Internalization (the

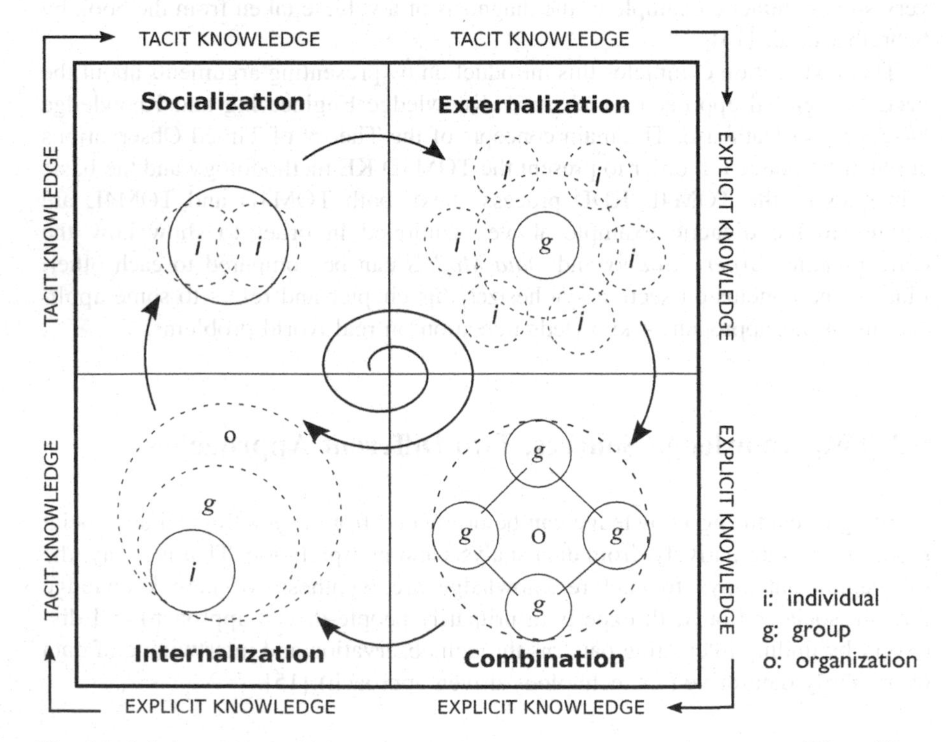

Fig. 6.1 Spiral evolution of knowledge conversion and self-transcending process [20, p. 43]

conversion of explicit knowledge into tacit knowledge: the individuals can broaden, extend and reframe their own tacit knowledge).

The tacit knowledge is, among other things, the knowledge of experts who intuitively know what to do in performing their duties but which is difficult to express because it refers to sub-symbolic skills. Such knowledge is frequently based on intuitive evaluations of sensory inputs of smell, taste, feel, sound or appearance. Eliciting such knowledge can be a major obstacle in attempts to build Knowledge Based Systems (KBSs). Knowledge Engineering (KE) arises then as the need of transforming the art of building KBSs into an engineering discipline [21, 22] providing thus techniques and tools that help to treat with the expert's tacit knowledge and to build KBSs. This discipline motived the development of a number of methodologies and frameworks such as Roles-Limiting Methods and Generic Tasks [23], and later, CommonKADS [14, 24], Protégé [25], MIKE [26, 27], KAMET II [28, 29] and VITAL [30]. In particular, CommonKADS is a KE methodology of great significance which proposes a structured approach in the construction of KBSs. Essentially, it consists in the creation of a collection of models that capture different aspects of the system to be developed, among which is the Knowledge Model (or Expert Model) that describes the knowledge and reasoning requirements of a system, that is, expert knowledge. Other two important modelling frameworks are MIKE and PROTÉGÉ, where the former focuses on executable specifications while the latter exploits the notion of ontology. All these frameworks or methodologies aim, of one or another way, to build a model of the expert's knowledge.

6.2.2 Knowledge Discovery in Database: A Primarily Data-Driven Approach

The traditional method of turning data into knowledge is based on data manual analysis and interpretation. For example, in the health-care industry, specialists periodically analyse trends and changes regarding health in the data. Then, they detail the analysis in a report which becomes the basis for future decision making in the domain of health. However, when data volumes grow exponentially and their manipulation is beyond human capacity, resorting to automatic analysis is absolutely necessary. Thus, computational techniques help to discover meaningful structures and patterns from data.

The field of Knowledge Discovery in Database (KDD) is concerned with the development of methods and techniques for making sense of data. The phrase *knowledge discovery in database* was coined at the first KDD workshop in 1989 [31] to emphasize that knowledge is the end product of a data-driven discovery. Although the terms KDD and Data Mining are often used interchangeably, KDD refers to the overall process of discovering useful knowledge from data, and Data Mining refers to a particular step in the mentioned process [32]. More precisely,

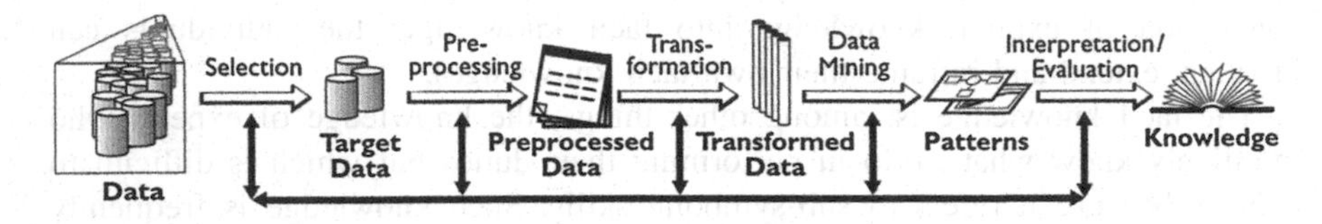

Fig. 6.2 Overview of the steps constituting the KDD process [33, p. 29]

this step consists of the application of specific algorithms in order to extract patterns from data.

The typical KDD process is depicted in Fig. 6.2 and summarized as follows [33]. The starting point is to learn the application domain and its goals. Next, to select a dataset or a subset of variables on which discovery is to be performed. Then preprocessing takes place which involves removing noise, collecting the necessary information to account for noise, deciding on strategies for handling missing data field, etc. The following step is data transformation which includes finding useful features to represent the data, depending on the goal of the task, and to reduce the effective number of variables under consideration or to find invariant representation for the data. After that, data mining is carried out. In general terms, this involves selecting data mining methods and choosing algorithms; and through these ones, searching for patterns of interest. Finally, the mined patterns are interpreted removing those that are redundant or irrelevant, and translating the useful ones into terms understandable by users. This discovered knowledge can be incorporated in systems or simply documented and reported to interested parties.

In a KDD process, finding patterns in data can be carried out through different techniques such as Decision Trees [34], Hidden Markov Chain [35], Neural Networks [36], Bayesian Networks [37], K Nearest-Neighbour [38], SVM [39], etc. All these techniques allow to obtain a model representative of the studied data where this model have to be interpreted and validated by expert knowledge.

6.2.3 The Need of One Integral Approach

The model-building of an observed process can be carried out through KE or KDD. As Fig. 6.3 depicts, given a process about which an expert has knowledge, a model M_e of this process can be constructed from expert knowledge by applying KE techniques. In turn, the process can be observed through sensors by a program which records data describing its evolution. Thus, these data can be analysed by applying data mining techniques in a KDD process in order to obtain a model M_d of the process. In an ideal world, both M_e and M_d would complement each other in order to have a process model M_{PR} more complete and suitable. That is, M_e must be validated with the process data perceived through sensors and M_d must be validated with expert knowledge. Nevertheless, some drawbacks arise. Knowledge Engineering approaches do not address the treatment of knowledge discovered in

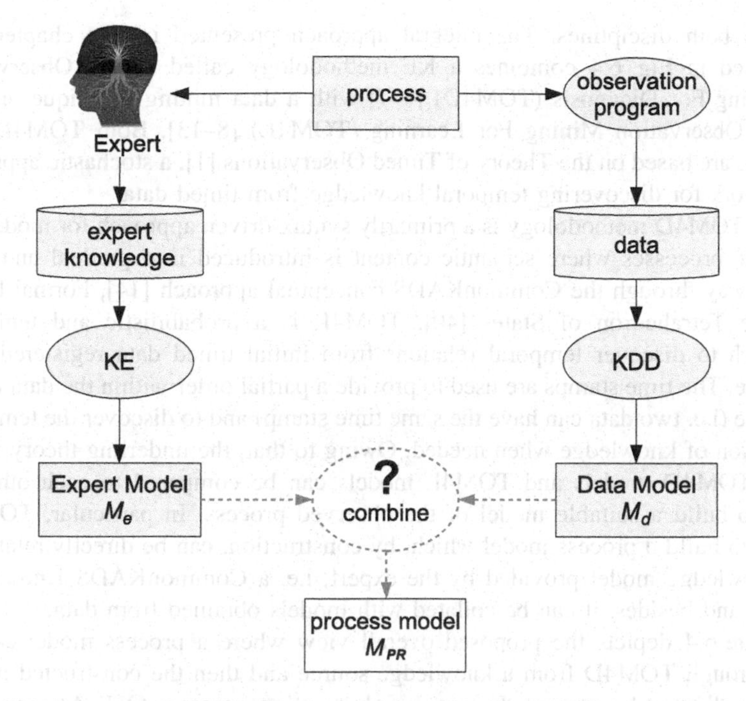

Fig. 6.3 Building a process model from two knowledge sources

databases, that is to say, sometimes the interpretation of discovered patterns is not trivial for an expert. Besides, relating models M_e and M_d obtained through KE and KDD, respectively, proves to be difficult owing to the different theories and the different natures of the representation formalisms used in both disciplines.

As [15] establishes, although capturing knowledge is the central focus of both fields of study, knowledge creation has tended to be approached from one or the other perspective, rather than from a combined perspective. Thus, a holistic view of knowledge creation that combines a people-dominated perspective with a data-driven approach is considered vital. In line with this need, this article proposes to integrate a KE methodology with data mining techniques in a KDD process in order to define a human–machine learning process.

6.3 Two Knowledge Sources, One Integral Approach

Models obtained through Knowledge Engineering (KE) and Knowledge Discovery in Database (KDD) will be able to be related and collated each other, if a bridge between the mentioned areas is established. We believe that fusing KE and KDD into a global approach of learning or knowledge acquisition, nourished with knowledge discovered in data and experts' knowledge, requires a theory on which to base both disciplines. The integral approach presented in this chapter and illustrated in Fig. 6.4 combines a KE methodology called Timed Observation Modelling For Diagnosis (TOM4D) [4–7] with a data mining technique, named Timed Observation Mining For Learning (TOM4L) [8–13]. Both TOM4D and TOM4L are based on the Theory of Timed Observations [1], a stochastic approach framework for discovering temporal knowledge from timed data.

The TOM4D methodology is a primarily syntax-driven approach for modelling dynamic processes where semantic content is introduced in a gradual and controlled way through the CommonKADS conceptual approach [14], Formal Logic and the Tetrahedron of States [40]. TOM4L is a probabilistic and temporal approach to discover temporal relations from initial timed data registered in a database. The time stamps are used to provide a partial order within the data in the database (i.e. two data can have the same time stamp) and to discover the temporal dimension of knowledge when needed. Owing to that, the underling theory is the same, TOM4D models and TOM4L models can be compared to each other in order to build a suitable model of the observed process. In particular, TOM4D allows to build a process model which, by construction, can be directly related to the knowledge model provided by the expert, i.e. a CommonKADS Knowledge Model; and besides, it can be collated with models obtained from data.

Figure 6.4 depicts the proposed overall view where a process model can be built through TOM4D from a knowledge source and then the constructed model can be validated by experts. In turn, an observation program $\Theta(X, \Delta)$ requires a model of the observed process for recording data in respect of the evolution of this one. These data are then analysed by means of TOM4L to produce a process model. This model can be directly related to the TOM4D model built from the expert's knowledge and consequently, it can be either validated by the expert or it

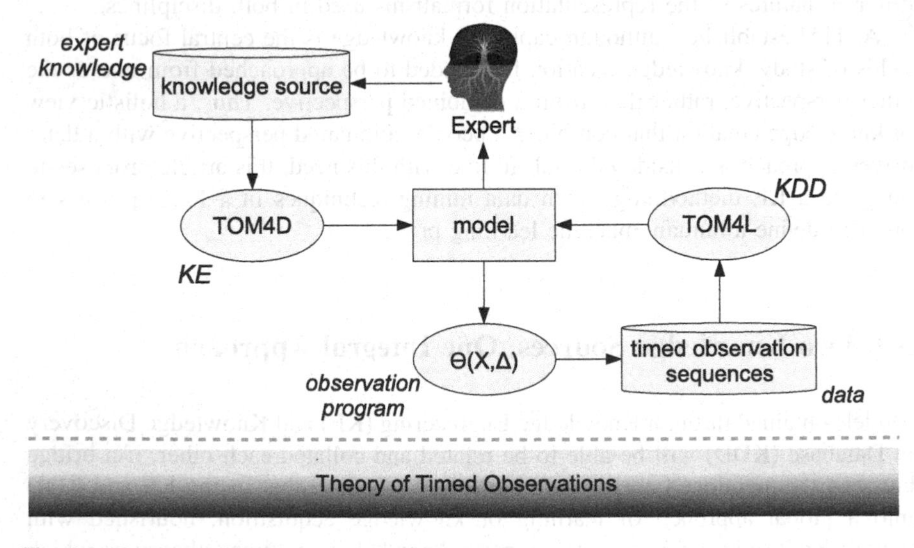

Fig. 6.4 Human-machine learning integral approach

can be utilized as pieces of new knowledge when the learning approach is applied to an unknown process. In this way, the built model can be defined through a back and forth way between experts' knowledge and knowledge discovered in data, establishing thus, an integral human–machine learning approach.

6.4 Introduction to the Theory of Timed Observations

The Theory of Timed Observations (TTO) [1] provides a general framework for modelling dynamic processes from timed data by combining the Markov Chain Theory, the Poisson Process Theory, the Shannon's Communication Theory [2] and the Logical Theory of Diagnosis [3]. The main concepts of the TTO, required in order to introduce the TOM4D KE methodology and the TOM4L KDD process, will be described in this section. These concepts are the notions of *timed observation* and *observation class*.

The Theory of Timed Observations defines a dynamic process as an arbitrarily constituted set $X(t) = \{x_1(t), \ldots, x_n(t)\}$ of n functions $x_i(t)$ of continuous time $t \in \Re$. The set $X(t)$ of functions implicitly defines a set $X = \{x_1, \ldots, x_n\}$ of n variable names x_i. The dynamic process $X(t)$ is monitored by a program $\Theta(X, \Delta)$ which observes the functions $x_i(t)$ of $X(t)$; and then, it establishes, records and informs their evolution over time with a finite set $\Delta = \{\delta_j\}_{j=1,\ldots,m}$ of constants δ_i (i.e. a number or a string). The program $\Theta(X, \Delta)$ usually accounts for the functions progression through messages recorded in a database. These messages can be alarms, warnings or reporting events.

This theory considers a message at time t_k as a *timed observation* (δ, t_k) where δ is a constant value of Δ and t_k is the moment in which the observation occurs. For example, let us suppose that timed data recorded in a database are of the form "yymmdd-hhmmss/message_value" where yymmdd-hhmmss is a time stamp and message_value is a value determined by a monitoring program. The message "080313-132225/TEMPERATURE/very_high" can be represented with a timed observation (δ, t_k) where $t_k = 080313\text{-}132225$ and $\delta =$ /TEMPERATURE/very_high. That is, $(\delta, t_k) =$ (TEMPERATURE/very_high, 080313-132225).

In general terms, a timed observation (δ, t_k) is written by an observer program $\Theta(\{x\}, \{\delta\})$ when a function $x(t)$ of continuous time enters in a specific interval of values. The specification of such an observer program refers to a threshold value $\Psi_j \in \Re$ and two immediately successive values $x(t_{k-1}) \in \Re$ and $x(t_k) \in \Re$ so that,

$$x(t_{k-1}) < \Psi_j \wedge x(t_k) \geq \Psi_j \Rightarrow write((\delta, t_k)). \tag{6.1}$$

In this program, *write(msg)* is a predicate which denotes that the element *msg* is recorded in a memory. For example, Fig. 6.5 illustrates a temperature function $x_i(t)$, where values above Ψ_j are interpreted by an observer program $\Theta(\{x_i\},$ $\{$TEMPERATURE/very_high$\})$ as very high temperature; that is, when $x_i(t) \in [\Psi_j, +\infty)$. Thus, given a sequence of values $w = (x_i(t_1), \ldots, x_i(t_{k-1}),$

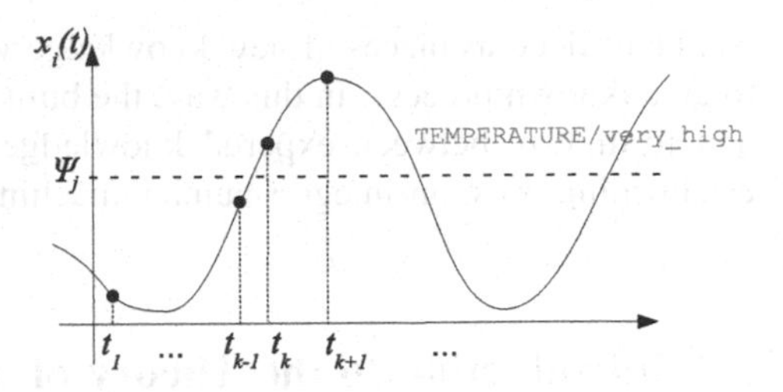

Fig. 6.5 Function of temperature

$x_i(t_k)$, $x_i(t_{k+1})$), the program $\Theta(\{x_i\}$,{TEMPERATURE/very_high}) will write a timed observation (TEMPERATURE/very_high, t_k), which indicates that the function $x_i(t)$ entered the interval $[\Psi_j, +\infty)$ at time t_k.

The Theory of Timed Observations establishes that the existence of a timed observation (δ, t_k), recorded in a database, allows to infer that the mentioned observation has been recorded by an unknown program $\Theta(\{x\}, \{\delta\})$ which implements the abstract logical equation described in (6.2).

$$\forall t_k \in \Gamma, \theta(x, \delta, t_k) \in \Theta \Rightarrow (\delta, t_k) \in \Omega \tag{6.2}$$

This sentence associates the set Θ of all the assignations to a ternary predicate $\theta(x_\theta,\ \delta_\theta,\ t_\theta)$ with the set Ω of all the timed observations carried out by the program $\Theta(\{x\}, \{\delta\})$ (i.e., the database). A timed observation (δ, t_k) is then interpreted as the logical consequence of the assignation of the values x, δ and t_k to a ternary predicate $\theta(x_\theta,\ \delta_\theta,\ t_\theta)$. In other words, this means that the timed observation (δ, t_k) was recorded when the program $\Theta(\{x\}, \{\delta\})$ assigned the values x, δ and t_k to the predicate $\theta(x_\theta,\ \delta_\theta,\ t_\theta)$.

Given the sentences (6.1) and (6.2), the general meaning "**is**" can be always provided to the predicate θ so that the timed observation (δ, t_k) is interpreted as "**at time** t_k, x **is** δ". Considering that x is associated with a function $x(t)$, the meaning "**equal**" can also be attributed to the predicate θ, which leads to the following abuse of language: $\theta(x,\ \delta,\ t_k)$ means "*Equal*$(x,\ \delta,\ t_k)$" (i.e. "$x(t_k) = \delta$"). Consequently, the Theory of Timed Observations considers that a message contained in a database is a timed observation (δ, t_k) written by a program $\Theta(X, \Delta)$ which observes a time function $x(t)$ and implements the abstract Eq. (6.2). In our example, the timed observation (TEMPERATURE/very_high, t_k) indicates that a program $\Theta(x(t), \{\delta\})$, observing a time function $x_i(t)$ and defining implicitly a predicate $\theta(x_\theta,\ \delta_\theta,\ t_\theta)$, has considered $\theta(x_i$, TEMPERATURE/very_high, $t_k)$ true and then it has written the timed observation (TEMPERATURE/very_high, t_k) in the database Ω. This example illustrates the abuse of language frequently carried out, which associates the meaning "$x_i(t_k)$ = very_high" with the interpretation of the function "$x_i(t)$" as a temperature.

According to the Definition 6.1, the interpretation of a timed observation (δ, t_k) is precisely the assigned predicate $\theta(x,\ \delta,\ t_k)$. It is noteworthy that the program $\Theta(\{x\}, \{\delta\})$ could have errors; that is to say, a timed observation (δ, t_k) could have been written in a database although the assertion $\theta(x_i,\ \delta,\ t_k)$ is not really true.

Definition 6.1 Let $X(t) = \{x_i(t)\}_{i=1,\ldots,n}$ be a finite set of time functions; let $X = \{x_i\}_{i=1,\ldots,n}$ be the corresponding finite set of variable names; let $\Delta = \{\delta_j\}_{j=1,\ldots,m}$ be a finite set of constant values; let $\Theta(X, \Delta)$ be a program observing the evolution of the functions of $X(t)$; let $\Gamma = \{t_k\}_{k \in \Re}$ be a set of arbitrary time instants; and let $\theta(x_\theta, \delta_\theta, t_\theta)$ be a predicate implicitly determined by $\Theta(X, \Delta)$. Then,

- a *timed observation* $(\delta, t_k) \in \Delta \times \Gamma$ on $x_i(t)$ is the assignation of values x_i, δ and t_k to the predicate $\theta(x_\theta,\ \delta_\theta,\ t_\theta)$ such that $\theta(x_i,\ \delta,\ t_k)$;
- by definition $o(t_k)$ denotes a timed observation; i.e., $o(t_k) \triangleq (\delta, t_k)$[1] and,
- a finite set $O \subset \Delta \times \Gamma$ of timed observations is disjointly partitioned and ordered in a scenario Ω defined as a set of temporally ordered sequences of timed observations; that is, $\Omega = \{w : \{1,\ldots,n\} \rightarrow O\} | n \in \aleph \wedge \forall i, j \in \{1,\ldots,n\}, i<j, (w(i) = o(t_k) \wedge w(j) = o(t_r) \Rightarrow t_k \leq t_r)\} \wedge \bigcap_{w \in \Omega} \Im(w) = \emptyset \wedge \bigcup_{w \in \Omega} \Im(w) = O$ where $\Im(w)$ denotes the image or range of w, i.e. the observations of the sequence $w \in \Omega$.

Moreover, as follows from that previously explained, timed observations on a particular function implicitly determine a variable, which assumes discrete values and describes the function evolution according to an interpretation of the observer program. That is to say, when Θ considers θ(x_i, TEMPERATURE/very_high, t_k) true and then writes (TEMPERATURE/very_high, t_k), it is implicitly defining a discrete variable which assumes the value TEMPERATURE/very_high. Consequently, a timed observation and the implicit existence of an associated discrete variable enable to define the notion of *observation class*, other important concept in this theory. An observation class associated with a variable x, that assumes values $\delta \in \Delta$, is a set $C_x = \{(x, \delta) \mid \delta \in \Delta\}$. For simplicity reasons, C_x is often defined as a singleton $C_x = \{(x, \delta)\}, \delta \in \Delta$. Thus, this concept establishes the link between a constant $\delta \in \Delta$ and a variable $x \in X$ and then, a timed observation (δ, t_k) is an occurrence of an observation class $C_x = \{(x,\ \delta)\}$. Definition 6.2 specifies this concept.

Definition 6.2 Let $X(t) = \{x_i(t)\}_{i=1,\ldots,n}$ be a set of time functions whose evolutions are observed by a program Θ; let $X = \{x_i\}_{i=1,\ldots,n}$ be a set of discrete variables where each x_i is associated with a time function $x_i(t)$ and its value is determined by an interpretation of Θ about the evolution of $x_i(t)$; and, let $\Delta = \bigcup_{x_i \in X} \Delta_{x_i}$ be such that Δ_{x_i} is a set of values which can be assumed by $x_i \in X$. Then we say that an *observation class* associated with a variable $x_i \in X$ is a set $C_i = \{(x_i, \delta) \mid \delta \in \Delta_{x_i}\}$.

In summary, from a message (TEMPERATURE/very_high, t_k) written in a database, the Theory of Timed Observations allows to consider that there exists a program $\Theta(\{x_i\}$,{TEMPERATURE/very_high}) which wrote the message, by means of observing a time function, maybe unknown for us, noted as $x_i(t)$. This

[1] The symbol $\triangleq$ denotes rewriting or “corresponds to”.

message is then a timed observation (TEMPERATURE/very_high, t_k) indicating that a certain predicate $\theta(x_i,$ TEMPERATURE/very_high, $t_k)$ was assumed true by the program $\Theta(\{x_i\},\{\text{TEMPERATURE/very\_high}\})$. Then, there is tacitly a discrete variable x_i which takes at least the value TEMPERATURE/very_high. Therefore, we can define an observation class $C_i = \{(x_i,$ TEMPERATURE/very_high$)\}$, so that the timed observation (TEMPERATURE/very_high, t_k) is an occurrence of C_i. When knowing that the time function $x_i(t)$ represents the evolution of temperature, it is inferred that (1) x_i denotes a *variable of temperature*, (2) the observation class C_i can then be written as $C_i = \{(\text{very\_high})\}$ denoting that the *temperature is very high* and (3) the timed observation (TEMPERATURE/very_high, t_k) is an occurrence of this class, which means "at time t_k, temperature is very high".

For sake of generality, it is important to note that a predicate $\theta(x_\theta, \delta_\theta, t_\theta)$ is satisfied when the corresponding time function $x_i(t)$ matches against a behavioural model [41]. Such a model can be as simple as the switch of an interrupter or requiring complex techniques, as signal processing techniques for artificial vision.

The TOM4D KE methodology and the TOM4L KDD process are based on these notions of timed observation and observation class, as the next sections describe below.

6.5 TOM4D KE Methodology

TOM4D is a modelling approach for dynamic systems focused on timed observations. The objective of this one is to produce suitable models for dynamic process diagnosis from timed observations and experts' a priori knowledge. This methodology combines then the modelling of the experts' cognitive process, using CommonKADS [14, 24], with a multi-modelling approach for dynamic systems [40, 42]. In addition, TOM4D is a primarily syntax-driven approach [5–7] which resorts to CommonKADS, Formal Logic and the Tetrahedron of States (ToS) [40] as interpretation frameworks and paradigms in order to introduce, in the modelling process, semantic content in a gradual and controlled way.

6.5.1 Multi-Modelling

In this methodology, a system is represented by means of four models, the three models described in the conceptual multi-modelling framework introduced in [43] and a complementary model called *Perception Model* [6].

The models of the multi-modelling framework are *Structural Model* (*SM*), *Behavioural Model* (*BM*) and *Functional Model* (*FM*) which describe different types of knowledge. The *SM* contains knowledge relative to the system components and their structural organization, that is to say, the relations between these ones. The *BM* specifies knowledge about the phenomena which act inside the system in order to transform an input flow into an output flow. Such transformations are measured through the evolution of the values of a set of variables. Thus, these changes in the values define the possible sequences of observation classes that can occur and therefore, the discernible states between them. Finally, the *FM* describes knowledge about the relations among the values that the variables can assume.

For its part, the *Perception Model* (*PM*) contains knowledge about the following elements and aspects of the process: variables and their thresholds, operating goals, and normal and abnormal operating modes.

The relations between the first three models are determined by the notion of *variable* as Fig. 6.6 illustrates. A variable used in a function of the *Functional Model* is associated with a component of the *Structural Model* and, a discrete event of the *Behavioural Model* is the assignment of a value to the variable. Indeed, any specification in these models must be consistent with that one made in the *Perception Model*.

6.5.2 Interpretation Frameworks

CommonKADS [14, 24] is a methodology which offers a structured approach in the development of KBSs by proposing three groups of models. The first group regarding the organizational context and environment, the second one with respect to the conceptual description of the knowledge applied in a task, and the last one concerning the technical aspects of the software artefact.

In particular, the CommonKADS Knowledge Model which belongs to the second group is utilized in our approach. This model describes the types and structures of the knowledge required to accomplish a particular task and thus, it acts as a tool that helps to clarify the structure of a knowledge-intensive information-processing task. This model is developed, in a way that is understandable by humans, as part of the analysis process and therefore, it does not contain any

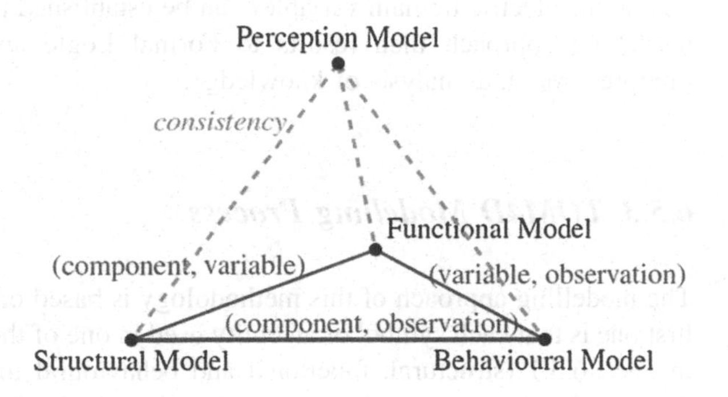

Fig. 6.6 Relations between TOM4D models

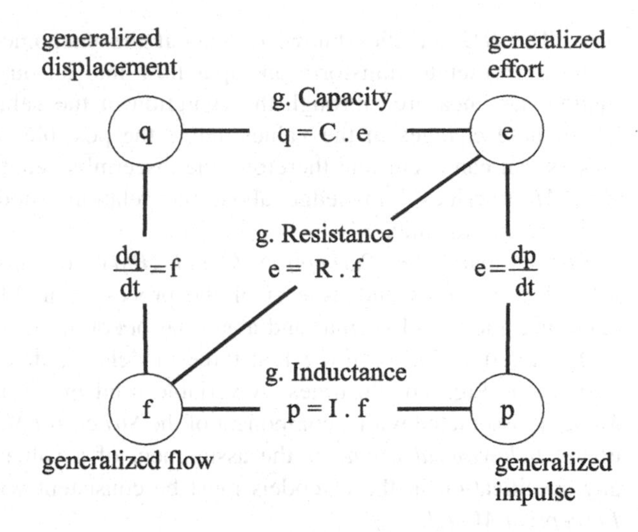

Fig. 6.7 Tetrahedron of states (ToS) (based on [40, p. 1728])

implementation-specific term. Thus, this one is an important vehicle for communication with experts and users about the problem-solving aspects. Consequently, TOM4D uses the aforementioned model as a mean of interpreting and structuring the available knowledge.

Formal logic is also used by the proposed methodology as a resource which provides reasoning mechanisms and gives the possibility of utilizing Reiter's Theory of Diagnosis [44]. In turn, in order to give a physical interpretation to the variables, the Tetrahedron of States (ToS) [40, 45, 46] can be incorporated in the analysis process. The ToS is a framework that describes a set of generalized equations (Fig. 6.7) which are common to a wide variety of physical domains (electromagnetism, fluid dynamics, thermodynamics, etc.). This one allows to map physical variables of a specific domain into four classes of generalized variables (*effort*, *flow*, *impulse* and *displacement*) and to identify the set of relationships among these ones. For example, in the electric domain (Electric ToS), *current* is mapped to generalized *flow*, *electric charge* to generalized *displacement*, *voltage* to generalized *effort* and *magnetic flux* to generalized *impulse*; thus, the relations among the electric domain variables can be established according to the ToS. Our modelling approach then resorts to Formal Logic and ToS as paradigms of interpretation and analysis of knowledge.

6.5.3 TOM4D Modelling Process

The modelling approach of this methodology is based on three principles [7]. The first one is that each symbol of an entity used in one of the three models introduced in Sect. 6.5.1 (structural, functional and behavioural models) denotes a concept that is defined at the level of domain knowledge of a CommonKADS model [14].

This means that the introduction of a symbol that is not associated with an element of the domain knowledge model is prohibited. The second principle is that a variable is always associated with a component or a component aggregate defined in the structural model. The third principle is that a transition between two states is conditioned by the assignment of a new value to a variable. The notion of variable, as aforementioned in Sect. 6.5.1, constitutes thus the common point of the three models.

The modelling process aims to produce a generic model of a system from available knowledge and data, where the three fundamental modelling phases are **knowledge interpretation, process definition** and **generic modelling**. Figure 6.8 illustrates a structure of logical dependences that describe the TOM4D reasoning process for obtaining a model of an observed system. Therefore, how the control flow of the modelling process is carried out, is not part of this structure. The illustrated process, introduced below, gives a general guide in order to understand the principal objectives of this approach. Clearly, the modelling is generally

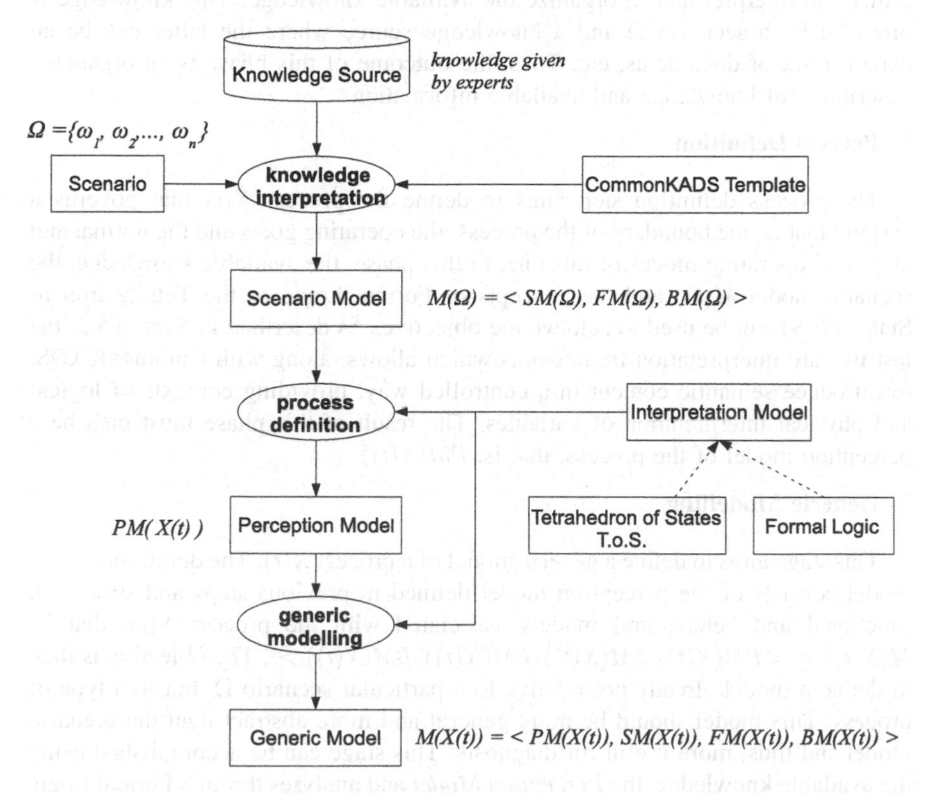

Fig. 6.8 General structure of the TOM4D modelling process

cyclical and each stage can require to return to previous phases with the objective of revising the expert's knowledge, results, ideas, modelling decisions, etc.

1. **Knowledge Interpretation**

The objective of this phase is to define a scenario model. In general terms, a scenario Ω of a system is a set of observations or measures over time on the variables of the system, where these measures describe a certain evolution of the process that drives the system dynamic. Definition 6.1 in Sect. 6.4 introduces the meaning of *scenario* and other concepts such as *timed observation* and *observation class*. In short, a scenario is a set of sequences of timed observations describing partially the behaviour of a process.

The construction of a scenario model $M(\Omega) = <SM(\Omega), FM(\Omega), BM(\Omega)>$ consists of the definition of a structural model $SM(\Omega)$, a functional model $FM(\Omega)$ and a behavioural one $BM(\Omega)$ of Ω.

For the purpose of defining a model $M(\Omega)$, a CommonKADS template is utilized to interpret and to organize the available knowledge. This knowledge is provided by a scenario Ω and a knowledge source where the latter can be an expert, a set of documents, etc. Thus, the outcome of this phase is an organized description of knowledge and available information.

2. **Process Definition**

The process definition step aims to define the process $X(t)$ that governs a system; that is, the boundary of the process, the operating goals and the normal and abnormal operating modes of this one. In this phase, the available knowledge, the scenario model $M(\Omega)$ and the concepts of Formal Logic or the Tetrahedron of States (ToS) can be used to achieve the objective. As described in Sect. 6.5.2, the last two are interpretation frameworks which allows, along with CommonKADS, to introduce semantic content in a controlled way, providing contexts of logical and physical interpretation of variables. The result of this phase must then be a perception model of the process, that is, $PM(X(t))$.

3. **Generic Modelling**

This stage aims to define a generic model of a process $X(t)$. The definition of this model consists of the perception model defined in previous steps and structural, functional and behavioural models associated with the process $X(t)$; that is, $M(X(t)) = <PM(X(t)), SM(X(t)), FM(X(t)), BM(X(t))>$. The objective is then to define a model already not relative to a particular scenario Ω, but to a type of process. This model should be more general and more abstract than the scenario model and thus, more useful for diagnosis. This stage can be accomplished using the available knowledge, the *Perception Model* and analyses through Formal Logic and the ToS.

The results of applying TOM4D to a didactic example will be presented later in order to show how the built TOM4D model can be related to a TOM4L model automatically obtained from data.

6.6 TOM4L KDD Process

TOM4L [12], based on the Theory of Timed Observations [1], is a probabilistic and temporal approach to discover temporal relations for description, diagnosis and prediction from initial timed data Ω registered in a database (i.e. a set of timed observation sequences). The aim is to discover n-ary temporal relations which are representative of the process behaviour which gave rise to Ω.

In particular, the TOM4L approach is implemented by the ElpLab Java software, so that the n-ary temporal relations can be discovered in an automatic way.

6.6.1 Temporal Relations

As described in Sect. 6.4, sequences of timed observations $(\delta, t_k) \in \Delta \times \Gamma$ recorded by a program observing a process allow to establish a set of discrete variables $x \in X$; and consequently, a set C of corresponding observation classes $C_i \in C$. For example, if $C_{1a} = \{(x_i, \delta_a)\}$ is defined as an observation class associated with x_i, then a timed observation (δ_a, t_k) is an occurrence at time t_k of the class C_{1a}. In order to specify that an observation is of a certain class, the symbol '::' is used; e.g., $(\delta_a, t_k) :: C_{1a}$.

TOM4L aims to discover temporal characteristics present in the data that describe the evolution of a process; therefore, detailed descriptions about variables and particular values that these variables can assume are not necessary in this context. In particular, we shall refer to timed observations and observation classes. We recall that the timed observation (δ_a, t_k) can be rewritten as $o(t_k)$ (Definition 6.1); thus, we refer to this observation like $o(t_k)$ and we specify its class with the symbol '::' like $o(t_k) :: C_{1a}$.

A temporal relation between two observation classes describes a temporal constraint between observations of the involved classes. By considering $I = \{[\tau^-, \tau^+] \mid [\tau^-, \tau^+] \subset \Re\}$ a set of time intervals and C a set of observation classes, a temporal relation between two observation classes is a pair $(q, \bar{i})$ where $q \in C \times C$ and $\bar{i} \in I$. Thus, a temporal relation $(q, \bar{i}) = ((C_i, C_j), \ [\tau^-, \tau^+])$ specifies a temporal constraint between timed observations of the observation classes $C_i, C_j \in C$. Figure 6.9 illustrates this relation according to the ElpLab representation.

In particular, two observations verify the aforesaid relation if the elapsed time between an occurrence of C_i and an occurrence of C_j is greater than or equal to τ^- and less than or equal to τ^+. That is to say, two observations $o(t_k), o(t_r) \in \Delta \times \Gamma$ verify the relation $((C_i, C_j), \ [\tau^-, \tau^+])$ if $o(t_k) :: C_i \wedge o(t_r) :: C_j \wedge (t_r - t_k) \in [\tau^-, \tau^+]$.

For its part, an n-ary temporal relation is a sequence m of temporal relations. Thus, a sequence of timed observations verifies an n-ary temporal relation m if the mentioned sequence verifies each temporal relation in m, even if in the middle of

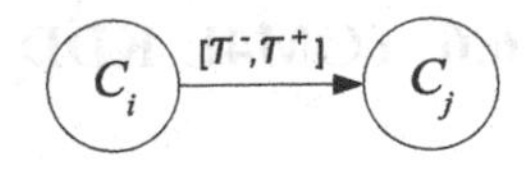

Fig. 6.9 Binary temporal relation $((C_i, C_j),\ [\tau^-, \tau^+])$ between two observation classes

the observation sequence there exist occurrences of classes that are not present in m.

As an example, we consider the observation classes C_{1a}, C_{2b}, C_{3c} and the n-ary temporal relation $m = (((C_{1a}, C_{2b}), [2,5]),\ ((C_{2b}, C_{3c}), [0,4]))$, as illustrated in Fig. 6.10. Besides, we suppose the sequence of timed observations $w = (o(19), o(20), o(22), o(24))$ such that $o(19) :: C_{1a}$, $o(20) :: C_{3c}$, $o(22) :: C_{2b}$ and $o(24) :: C_{3c}$, also illustrated in the figure. In this case, w verifies m owing to the following. Firstly, the class of the first observation coincides with the first class in the n-ary relation (i.e., $o(19) :: C_{1a}$) and the class of the last observation in w coincides with the last class in m (i.e., $o(24) :: C_{3c}$). In addition, the sequence of relations $m = (((C_{1a}, C_{2b}), [2,5]), ((C_{2b}, C_{3c}), [0,4]))$ is verified in w. That is to say, $((C_{1a}, C_{2b}), [2,5])$ specifies that the elapsed time between an occurrence of the observation class C_{1a} and an observation of the class C_{2b} is greater than or equal to 2 and less than or equal to 5. Thus, in w, $o(19)$ and $o(22)$ verify this temporal constraint since that $o(19) :: C_{1a}$, $o(22) :: C_{2b}$, $22 - 19 = 3$ and $2 \leq 3 \leq 5$. In a similar way, $o(22)$ and $o(24)$ verify $((C_{2b}, C_{3c}), [0,4])$. It is noteworthy that between $o(19)$ and $o(22)$, the observation $o(20)$ takes place. However, this does not invalidate that the relation $((C_{1a}, C_{2b}), [2,5])$ is verified, along with the complete n-ary relation, in the sequence of observations w.

In this way, given a set of data describing the behaviour of a process, discovering the n-ary temporal relations that are representative of these data is the central focus in the TOM4L KDD process.

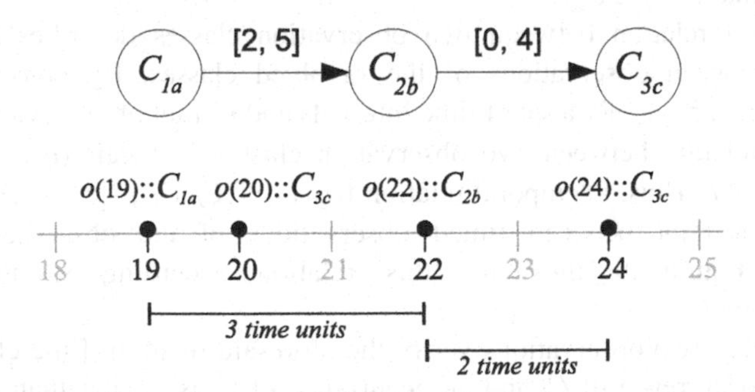

Fig. 6.10 Sequence $w = (o(19), o(20), o(22), o(24))$ of timed observations that satisfies the n-ary temporal relation $m = (((C_{1a}, C_{2b}), [2,5]),\ ((C_{2b}, C_{3c}), [0,4]))$

6.6.2 Stochastic Approach

In TOM4L, the analysis of a sequence w of timed observations consists of finding the more representative sequential relations between observation classes and establishing the temporal constraints in each relation. Thus, the study of the mentioned relations is addressed by resorting the Markov chain theory and the estimation of temporal constraints is dealt with the Poisson process theory. Consequently, in this framework, a sequence w of timed observations has a *stochastic representation* that consists of associating with w a superposition of the Poisson process and a Markov chain.

Given a finite set $O \subset \Delta \times \Gamma$ of timed observations, w is the sequence of all observations in O (i.e., the image of w is equal to O, or $w : \aleph \rightarrow O$ and $\Im(w) = O$) and C is the set of the n classes of observations in w. A stochastic representation of w consists then of a set of matrices reflecting different properties, where the rows and columns refer to the observations classes in C; that is to say, matrices $n \times n$ where the element of row i, column j refers to the sequential relation between the class C_i and the class C_j. We denote by $P(C_j \mid C_i)$ the conditional probability $P(w(k) :: C_j \mid w(k-1) :: C_i)$ of observing an occurrence of C_j having immediately before observed an occurrence of C_i and we denote by $P((C_i\ ,\ C_j))$ the probability $P((w(k-1) :: C_i\ ,\ w(k) :: C_j))$ of observing an occurrence of C_i followed immediately by an occurrence of C_j. Thus, the stochastic representation of w is given by the set of the following matrices. $\mathrm{N} = (N_{ij})_{n \times n}$ is a matrix where each N_{ij} establishes the number of observations of C_i followed immediately by an observation of C_j in w. The matrix $\mathrm{P} = (p_{ij})_{n \times n}$ establishes the transition probabilities between two observation classes, where the value p_{ij} corresponds to $P(C_j \mid C_i)$ and is calculated, based on N, as the rate between the number of the occurrences of C_i followed immediately by an occurrence of C_j and the number of occurrences of C_i followed immediately by an occurrence of any class.

The temporal constraints between two observation classes are calculated by analysing only the two subsequences of w whose observations are of the classes in question. In other words, w is partitioned in a set Ω of sequences w_r, where the observations in each w_r are of a same class C_r. By considering $w_i, w_j \in \Omega$ the subsequences of w whose observations are of the classes C_i and C_j respectively, the temporal constraint $[\tau^-, \tau^+]$ of a relation $((C_i, C_j),\ \ [\tau^-, \tau^+])$ is computed from the average of the elapsed times between an observation of class C_i and the following and first observation of class C_j, when overlapping w_i and w_j.

Based on theses calculations, an algorithm called BJT computes the stochastic representation of a sequence w under study, and an algorithm called BJT4T, based on the mentioned representation and on an abductive reasoning, builds a three of n-ary temporal relations associated with a given observation class C_i, i.e., paths ended in C_i representative of w. Both algorithms belonging to the TOM4L framework are implemented by ELpLab.

6.6.3 BJ-Measure and the Bayesian Networks Building

The BJ-measure [12] is a measure based on information entropy. Considering a superimposition of occurrences of two timed observation classes, this measure allows to evaluate the strength of intertwining of the mentioned superimposition; that is to say, the strength of an oriented binary relation between two observation classes taken from an arbitrarily built set.

Given an ordered binary relation $(C_i,\ C_j)$ between two observation classes, if these classes are independent, the probability of observing an occurrence of C_j at a time t_k is equal to the probability of observing an occurrence of C_j at that time having observed an occurrence of class C_i at the previous time t_{k-1}; that is, $P(C_j \mid C_i) = P(C_j)$. However, according to [12], if the classes are not independent, an occurrence of C_i at a time t_k provides information about an occurrence (or not) of C_j at the subsequent time t_{k+1}. In particular, the interest is in a measure indicating that an occurrence of class C_i at the time t_k increases the probability of observing an occurrence of class C_j at the time t_{k+1}; that is, $P(C_j \mid C_i) \geq P(C_j)$. Thus, the BJ-measure is based on the Kullback–Leibler distance [47] between two probability distributions which can be interpreted as the amount of information lost when a probability distribution is approximated by another distribution. The general idea is then the analysis, on the base of this distance measure, of the distance between $P(C_j \mid C_i)$ and $P(C_j)$ in order to establish if the relation $(C_i,\ C_j)$ is strong or weak.

Consequently, the BJ-measure $BJM(C_i,\ C_j)$ is defined from associating a sequential relation $(C_i,\ C_j)$ with a discrete memoryless communication channel [2] and from using the Kullback–Leibler distance. In particular, the BJ-measure verifies the properties of monotony, dissymmetry, positivity, independence and triangular inequality. Thus, if $BJM(C_i,\ C_j)$ is negative, the relation $(C_i,\ C_j)$ is weak; otherwise it is considered a possibly strong relation, or of interest.

The maximum and minimum values of the BJ-Measure depend on the rate $\tilde{\theta}_{ij}$ between the number of observations of class C_i and the number of observations of class C_j in the studied sequence w. In particular, [12] shows that a sequential relation $(C_i,\ C_j)$ is credible in the sense of the BJ-Measure if and only if $\frac{1}{4} < \tilde{\theta}_{ij} < 4$. This condition allows to select then relations of interest that provide a representative model of the sequence w.

TOM4L proposes an algorithm called Tom4BN [10, 11] to build Naive Bayesian Networks from timed data. Inspired by Cheng et al.'s algorithm [48], Tom4BN uses the properties of monotony, dissymmetry, positivity, independence and triangular inequality of the BJ-Measure to build a Naive Bayesian Network.

The general idea of the Tom4BN algorithm is to remove the sequential relations $(C_i,\ C_j)$ that are not of interest when building, from a set of timed data, the structure of a Naive Bayesian Network associated with a given observation class. For example, if $R_{BN} \subseteq C \times C$ is the set of all binary sequential relations $(C_i,\ C_j)$ with which paths in a Bayesian Network can be built, in principle $R_{BN} = C \times C$;

then, relations (C_i, C_j) where $BJM(C_i, C_j) \leq 0$ or $BJM(C_i, C_j) < BJM(C_j, C_i)$ are removed from R_{BN}. These and other criteria based on the aforesaid properties allow to select the binary sequential relations suitable for building a Bayesian Network from a data set. Consequently, given a goal class, the structure of the Bayesian Network associated with this one is constructed by the aforementioned algorithm from the mentioned criteria.

Based on properties which follow from the discrete memoryless communication channel, the tables of conditional probability for the Bayesian Network are defined from the matrix $N = (N_{ij})_{n \times n}$ established by the stochastic representation. That is to say, are defined the probability $P(C_i)$ of a root node, the probability $P(C_i, C_j)$ for a simple sequential relation and the probabilities for two sequential relations (C_i, C_j) and (C_z, C_j) associated with the same class C_j (i.e., $P(C_j \mid C_i, C_z)$, $P(C_j \mid \neg C_i, C_z)$, $P(C_j \mid C_i, \neg C_z)$, $P(C_j \mid \neg C_i, \neg C_z)$). Thus, from the mentioned definitions, probabilities as for example $P(\neg C_j \mid C_i)$ can be calculated as $P(\neg C_j \mid C_i) = 1 - P(C_j \mid C_i)$; and $P(\neg C_j \mid C_i, \neg C_z)$ as $P(\neg C_j \mid C_i, \neg C_z) = 1 - P(C_j \mid C_i, \neg C_z)$.

Thereby, given a goal class, the Bayesian Network associated with this one can be automatically built through the Tom4BN algorithm from a data set.

6.6.4 Signatures

An n-ary temporal relation m is considered representative of a sequence w of timed observations from evaluating two rates, *anticipation rate* and *coverage rate* [13].

Considering w a sequence of observations, let m be a sequence of temporal relations and let m_s be the sequence resultant of eliminating from m the last binary relation. The anticipation rate T_A of m in w is the rate between the number of subsequences w_j of w that satisfy m (i.e., $w_j \sqsubseteq w \wedge \mathit{satisfies}(w_j, m)$) and the number of subsequences of w that satisfy m_s, as illustrated in Fig. 6.11. That is to say, the percentage of cases in which after observing an instance of m_s, an occurrence of the last relation in m takes place. Clearly, T_A is of great interest in the diagnosis task when allowing to anticipate the occurrence of an observation class, in particular the last class of the model m; i.e., C_i in Fig. 6.11.

In the TOM4L framework, a *signature* [13] is a model m that has certain representativeness in the data, that is, in the sequence w under study. In particular, this representativeness is given when the anticipation rate T_A is above a certain value T_{Amin} (typically, 50 %). In other words, a sequence of temporal relations m is a *signature* in the sequence w of timed observations if and only if the anticipation rate T_A of m in w is greater than or equal to T_{Amin}. Thus, for anticipating the occurrence of an observation class C_i, a signature ending in C_i can be used as predictive model.

In some cases, although the anticipation rate T_A of a model m is above the value established T_{Amin} (i.e. m is a signature), the number of occurrences of m in w is

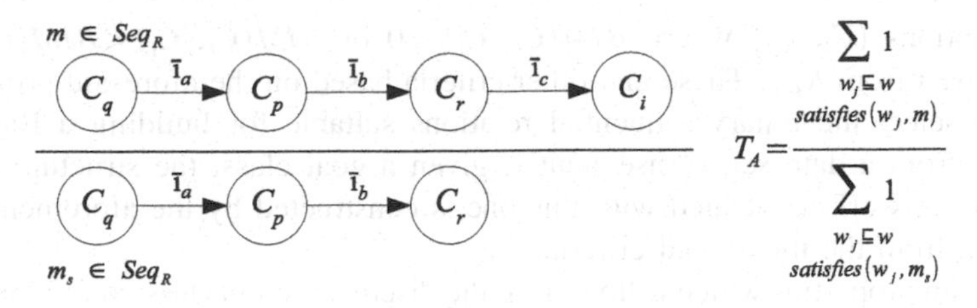

Fig. 6.11 Anticipation rate of m with regard to w (based on [13, p. 47])

low; or put in another way, the number of occurrences of the class C_i to be predicted is low. Therefore, in order to discard signatures where the occurrences of the last class are not significant in w, the coverage rate is established and illustrated in Fig. 6.12. Thus, the coverage rate T_C of m in w is the rate between the number of subsequences of w that satisfy m and the number of occurrences of the last observation class in m.

TOM4L aims, among other things, to discover from a given sequence w, a minimal set of signatures able to predict the maximal number of observations classes defined. That is, to discover a minimal set of temporal relations m whose anticipation rate T_A and coverage rate T_C in w are above the established threshold.

6.6.5 TOM4L Process

The general structure of the TOM4L KDD process is illustrated in Fig. 6.13 and is implemented by the ElpLab Java software, which allows to apply this data mining approach in an automatic way.

As depicted in the figure, stochastic and temporal properties of binary relations are obtained from a stochastic representation which associates a superposition of the Poisson process and a Markov chain with a set Ω of timed observations. A minimal set $R = \{r_j\}_{j=1,\ldots,r}$ of binary temporal relations which satisfies a criterion of interest is then induced from this stochastic representation, where the used criterion of interest is based on the BJ-measure described in Sect. 6.6.3.

From the mentioned minimal set, the TOM4L KDD process allows to compute a Naive Bayesian Network by means of the Tom4BN algorithm and a set of

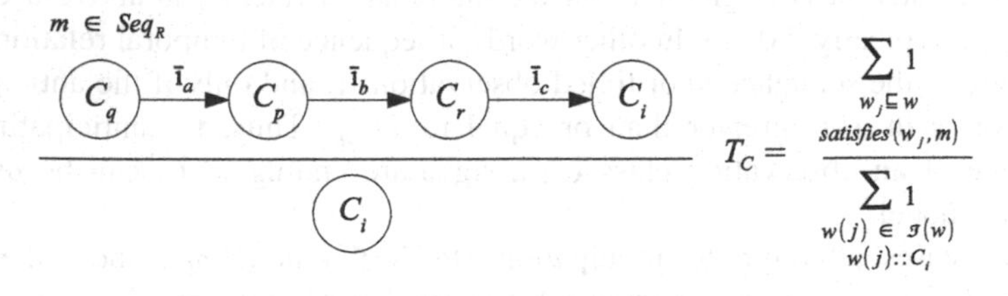

Fig. 6.12 Coverage rate of m with regard to w (based on [13, p. 47])

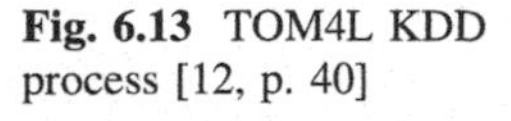
Fig. 6.13 TOM4L KDD process [12, p. 40]

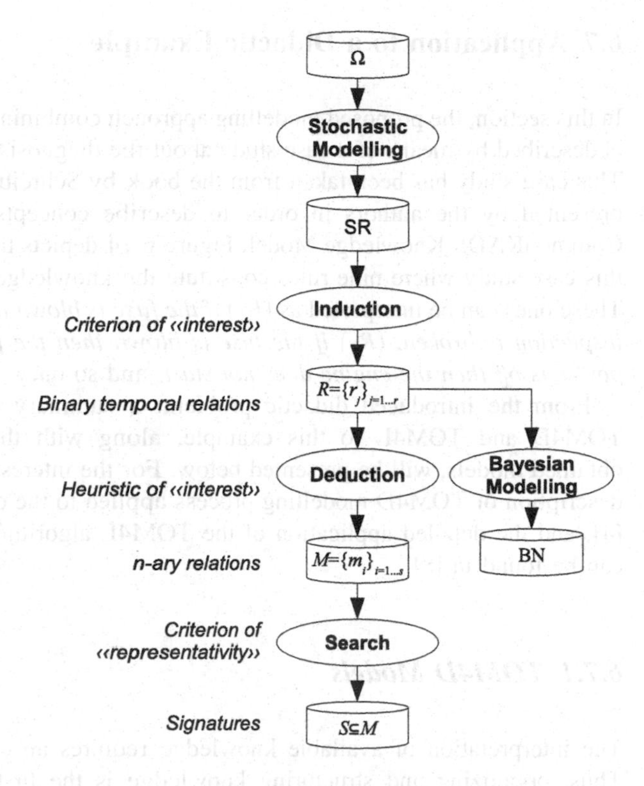

representative n-ary temporal relations. This latter is built through an abductive reasoning which is carried out on R in order to build a minimal set $M = \{m_i\}_{i=1,\ldots,s}$ of n-ary temporal relations m_i which would represent some properties of the process. In particular, the depth of abduction is controlled by heuristics based on the BJ-measure.

In the next stage, an exhaustive search is accomplished to extract from M the minimal set $S \subseteq M$ of n-ary temporal relations which satisfy a criterion of representativeness adapted to temporal relations. These n-ary relations are called signatures and their predictive ability allows to anticipate the occurrence of observable events. Searching for signatures consists in identifying all n-ary temporal relations m_i which finish in a particular observation class C_j (all paths and sub-paths to C_j) whose representativeness in Ω is sufficient. This representativeness is calculated from the coverage and the anticipation rates. The coverage rate of an n-ary relation m_i is the rate between the number of instances of m_i and the number of observation occurrences of class C_j; and, the anticipation rate of an n-ary relation m_i is the rate between the number of instances of the relation m_i and the number of instances of the relation m_i', where m_i' is the result of removing the last observation class C_j of the path m_i.

Models obtained through the TOM4L process, both signatures and Bayesian Networks, can be related to TOM4D models as described in the next section.

6.7 Application to a Didactic Example

In this section, the proposed modelling approach combining TOM4D with TOM4L is described by means of a case study about the diagnosis of problems with a car. This case study has been taken from the book by Schreiber et al. [14] where it is presented by the authors in order to describe concepts and components of a CommonKADS Knowledge Model. Figure 6.14 depicts the domain knowledge of this case study where nine rules constitute the knowledge provided by an expert. These ones can be interpreted as (R_1) *if the fuse is blown then the result of the fuse inspection is broken*, (R_2) *if the fuse is blown then the power is off*, (R_7) *if the power is off then the engine does not start*, and so on.

From the introduced didactic problem, a summary description of applying TOM4D and TOM4L to this example, along with the relation between the obtained models, will be presented below. For the interested reader, the complete description of TOM4D modelling process applied to the example can be found in [4], and the detailed application of the TOM4L algorithms to the same example can be found in [8].

6.7.1 TOM4D Models

The interpretation of available knowledge requires an organization of this one. Thus, organizing and structuring knowledge is the first step in the modelling activity of the TOM4D KE methodology.

6.7.1.1 Organizing Available Knowledge

CommonKADS is an important methodology in terms of modelling experts' knowledge and therefore, it is utilized by TOM4D as a framework of interpretation and organization of knowledge. CommonKADS provides a collection of predefined

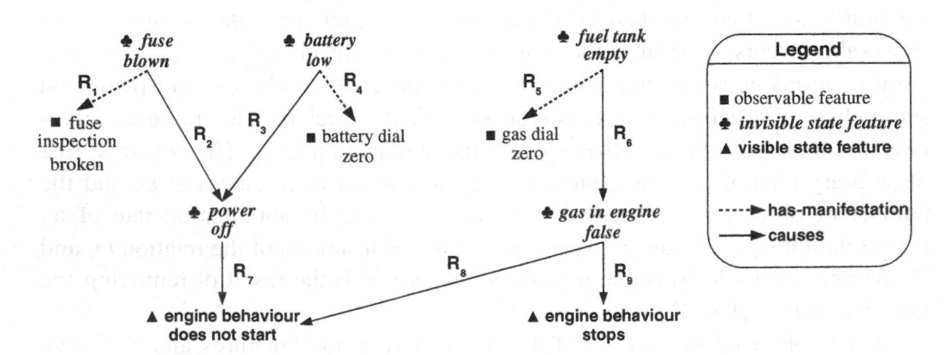

Fig. 6.14 Classification and organization of knowledge pieces

sets of model elements such as task templates and inference catalogues, which detail tasks and inferences typical for resolving a problem of a particular type. These templates also propose a characteristic structure for specifying the domain knowledge from the point of view of the selected type of task. In this case, we shall consider the *diagnosis* template.

The diagnosis template presents a typical domain schema in which each system being diagnosed can be characterized in terms of two types of features: those ones that can be observed and those ones that can represent an internal state of the system. Consequently, as Fig. 6.14 illustrates, the concepts **fuse inspection**, **battery dial** and **gas dial** are considered observable features; and *fuse*, *battery*, *fuel tank*, *power*, *gas in engine* and *engine behaviour* are considered concepts that allow to represent the states of the car. In particular, *engine behaviour* refers to a state which can be perceived in some way; therefore, the last concepts associated with car states can in turn be classified as visible or invisible.

Considering the previous classification, the arrows in Fig. 6.14 show dependences between the knowledge pieces. These dependences are rules which indicate relations between domain concepts. For example, "if there is no gas in engine, engine stops" establishes a causal relation between the concepts "gas in engine" and "engine behavior": gas-in-engine.status=false ⇒ engine-behaviour.status= stops. In this case study, two types of dependences can be observed: rules that indicate that a value assumed by an entity *causes* a certain value in other entity; and rules which establish that a value assumed by an entity *has a particular manifestation* in other entity.

Thus, the previous reasoning, illustrated in Fig. 6.14, describing dependence types and concept types in the specific domain determines the following domain rules specified in (6.3) in the language CLM (Conceptual Modelling Language, [14]).

```
fuse.status = blown HAS-MANIFESTATION                              (R1)
              fuse-inspection.value = broken;
fuse.status = blown CAUSES power.status = off;                     (R2)
battery.status = low CAUSES power.status = off;                    (R3)
battery.status = low HAS-MANIFESTATION                             (R4)
              battery-dial.value = zero;
fuel-tank.status = empty HAS-MANIFESTATION                         (R5)
              gas-dial.value = zero;
fuel-tank.status = empty CAUSES                                    (R6)    (6.3)
              gas-in-engine.status = false;
power.status = off CAUSES                                          (R7)
              engine-behaviour.status = does-not-start;
gas-in-engine.status = false CAUSES                                (R8)
              engine-behaviour.status = does-not-start;
gas-in-engine.status = false CAUSES                                (R9)
              engine-behaviour.status = stops;
```

Considering the aforementioned analysis and the three TOM4D principles introduced in Sect. 6.5.3, the next objective is to define a scenario model $M(\Omega) = < SM(\Omega), FM(\Omega), BM(\Omega) >$ from the given knowledge and a set Ω of sequences of timed measures or observations which describe certain modes of functioning of the car. In a real case, it would be desirable to have a set of timed observations describing the evolution over time of the process under study. In this case, Ω has not been provided; nevertheless, we shall deduce on the basis of the existing domain knowledge a scenario Ω to be assumed.

6.7.1.2 Knowledge Interpretation

The rules in (6.3) represent causal relations which implicitly define the notion of timed sequence of events; thus, from these rules, a set of sequences of timed observations can be assumed, that is, a scenario Ω. Taking in consideration $(R_1)$and (R_2) in (6.3), if the fuse blows at the instant t_0, the *fuse inspection* will result equal to broken at a subsequent moment $t_0 + \Delta t_i$ and the *electric supply* will be off at another moment $t_0 + \Delta t_j$. Affirming the order of sequence between $t_0 + \Delta t_i$ and $t_0 + \Delta t_j$ is not possible from the available information; nevertheless, we assume that all sensors properly work and quickly react, therefore, the order $t_0 + \Delta t_i < t_0 + \Delta t_j$ will be considered. In other words, first the fuse blows, then the *fuse inspection* result is equal to broken and, after that, the *electric supply* is switched off. Analogously, other two assumptions are: first the level of battery falls below the minimum, then the battery-dial is equal to zero and later the *electric supply* is turned off; and besides, first the fuel-tank is empty, then the gas-dial is equal to zero and after that the *gas supply* is empty.

Thus, considering the previous assumptions, it is supposed a scenario Ω of timed observations such that $\Omega = \{w_1, w_2, w_3, w_4\}$ where

$$\begin{aligned}
w_1 =&((blown, t_{10}),\ (broken, t_{10} + \Delta t_{11}),\ (off, t_{10} + \Delta t_{11} + \Delta t_{12}),\\
&(does_not_start, t_{10} + \Delta t_{11} + \Delta t_{12} + \Delta t_{13}))\\
w_2 =&((low, t_{20}),\ (battery_zero, t_{20} + \Delta t_{21}),\ (off, t_{20} + \Delta t_{21} + \Delta t_{22}),\\
&(does_not_start, t_{20} + \Delta t_{21} + \Delta t_{22} + \Delta t_{23}))\\
w_3 =&((empty, t_{30}),\ (gas_zero, t_{30} + \Delta t_{31}),\ (false, t_{30} + \Delta t_{31} + \Delta t_{32})\\
&(does_not_start, t_{30} + \Delta t_{31} + \Delta t_{32} + \Delta t_{33}))\\
w_4 =&((empty, t_{40}),\ (gas_zero, t_{40} + \Delta t_{41}),\ (false, t_{40} + \Delta t_{41} + \Delta t_{42}),\\
&(stop, t_{40} + \Delta t_{41} + \Delta t_{42} + \Delta t_{43}))
\end{aligned} \tag{6.4}$$

From the interpretation of the available knowledge, the concepts fuse, battery, fuel-tank, battery-dial and gas-dial are considered as components of the system. However, the concepts fuse-inspection, power, gas-in-engine and engine-behaviour denote physical entities which are unknown or whose information is insufficient. Consequently, abstract components (or component aggregates) such as *tools that allow fuse inspection*, *electric supply*, *gas supply* and *engine* will be defined to represent these concepts. In addition, the knowledge interpretation from CommonKADS enables to identify the variables of the system such as fuse.status, gas-dial.value, engine-behaviour.status, etc. Thus, these variables and components are defined in (6.5) where the value that, in principle, a variable x_i $(i = 1,\ldots,9)$ can assume, is described in the corresponding set Δ_{x_i} presented in (6.6), denoting ϕ_i an unknown value.

$$\begin{array}{ll} \text{Variables } X = \{x_1,\ldots,x_9\} & \text{Components } COMPS = \{c_1,\ldots,c_9\} \\ x_1 \triangleq \texttt{fuse.status} & c_1 \triangleq \texttt{fuse} \\ x_2 \triangleq \texttt{battery.status} & c_2 \triangleq \texttt{battery} \\ x_3 \triangleq \texttt{fuel-tank.status} & c_3 \triangleq \texttt{fuel-tank} \\ x_4 \triangleq \texttt{fuse-inspection.value} & c_4 \triangleq \texttt{fuseinspectiontools} \\ x_5 \triangleq \texttt{battery-dial.value} & c_5 \triangleq \texttt{battery-dial} \\ x_6 \triangleq \texttt{gas-dial.value} & c_6 \triangleq \texttt{gas-dial} \\ x_7 \triangleq \texttt{power.status} & c_7 \triangleq \texttt{electricsupply} \\ x_8 \triangleq \texttt{gas-in-engine.status} & c_8 \triangleq \texttt{gassupply} \\ x_9 \triangleq \texttt{engine-behaviour.status} & c_9 \triangleq \texttt{engine} \end{array} \tag{6.5}$$

$$\begin{array}{lll} \Delta_{x_1} = \{blown, \phi_1\} & \Delta_{x_4} = \{broken, \phi_4\} & \Delta_{x_7} = \{off, \phi_7\} \\ \Delta_{x_2} = \{low, \phi_2\} & \Delta_{x_5} = \{battery\_zero, \phi_5\} & \Delta_{x_8} = \{false, \phi_8\} \\ \Delta_{x_3} = \{empty, \phi_3\} & \Delta_{x_6} = \{gas\_zero, \phi_6\} & \Delta_{x_9} = \{stops,\ does\_not\_start\} \end{array} \tag{6.6}$$

In the first phase the scenario model $M(\Omega) = <SM(\Omega), FM(\Omega), BM(\Omega)>$ is defined. Although the detailed specification of this model will not be presented, we shall mention the principal points of this one. This model organizes and describes the information and the knowledge available. $SM(\Omega)$ describes the 9 components in (6.5) and the interconnections between them; and $FM(\Omega)$ specifies the relation among the values that the variables can assume through the definition of a set of functions. For example, rule R_5 allows to establish an interconnection between the components c_3 (fuel-tank) and c_6 (gas-dial); and also, the relation between the values of x_3 and x_6 through a function $f_1 : \Delta_{x_3} \rightarrow \Delta_{x_6}$ such that $f_1(empty) = gas\_zero, f_1(\phi_3) = \phi_6$, and where $x_6 = f_1(x_3)$. Besides, $BM(\Omega)$ specifies an initial behavioural model that, because of the 9 existent binary variables, consists of 18 observation classes (e.g., $C_{1,1} = \{(x_1, blown)\}$, $C_{1,2} = \{(x_1, \phi_1)\}$ are observation classes related to x_1) and $2^9 = 512$ characterized states (e.g., a state in which $x_1 = blown,\ x_2 = low,\ x_3 = empty,\ x_4 = \phi_4,\ x_5 = battery\_zero,\ x_6 = gas\_zero,\ x_7 = off,\ x_8 = false,\ x_9 = stops$). However, this model, which describes the available knowledge, is inadequate for analysing or diagnosing behaviour problems. It should be noticed that the existence of only 9 binary components determines 512 discernible states, a number significant with respect to the small number of units. Presumably, certain states in $BM(\Omega)$ are irrelevant for the pursued objectives or, they are meaningless since are impossible physically. Then, the two following stages in the modelling process, illustrated in Fig. 6.8, Sect. 6.5.3, aim to deal with these aspects.

6.7.1.3 Process Definition

In the phase of process definition, the perception model $PM(X(t))$ is defined, where the boundaries and operating constraints such as the set of variables of interest, operating goals, normal and abnormal operating modes are established. After that, in the stage of generic modelling, the objective is to define a model already not of a particular scenario, but a more general model of the car functioning. These two stages resort to the Formal Logic and the Tetrahedron of States in order to carry out a logical and a physical interpretation of the variables as Table 6.1 describes.

From Formal Logic, the components c_i $(i = 1, \ldots, 9)$ in (6.5) can be considered as logical components c_{Bi} described with first order predicate logic where Reiter's diagnosis theory [44] can be applied. Thus, the variables x_i $(i = 1, \ldots, 9)$ can be interpreted as logical variables $\bar{x}_i$ $(i = 1, \ldots, 9)$ which assume values 1 (true) or 0 (false). For example, in Table 6.1, $x_1 = blown$ is logically interpreted as $\bar{x}_1 = 0$ (false).

In principle, the components c_4 (*fuse inspection tools*), c_5 (battery-dial) and c_6 (*gas-dial*) being sensors, they simply replicates the behaviour of the components c_1 (fuse), c_2 (battery) and c_3 (fuel-tank). Consequently, and for reducing the complexity, it is assumed that the former work correctly (i.e. sensors are supposed to never fail) and then they are not necessary in the resultant model. Thus, the logical model of the process is depicted in Fig. 6.15 and the logical relations

Table 6.1 Logical and physical interpretations

Knowledge	Logical interpretation	Physical interpretation	
$x_1 = blown$	$\bar{x}_1 = 0$	$R(t) = \infty$	$(x_1^p = \infty)$
$x_1 = \neg blown$	$\bar{x}_1 = 1$	$R(t) = c_r$	$(x_1^p = c_r)$
$x_2 = low$	$\bar{x}_2 = 0$	$Q(t) = 0$	$(x_2^p = 0)$
$x_2 = \neg low$	$\bar{x}_2 = 1$	$Q(t) \neq 0$	$(x_2^p \neq 0)$
$x_3 = empty$	$\bar{x}_3 = 0$	$V(t) = 0$	$(x_3^p = 0)$
$x_3 = \neg empty$	$\bar{x}_3 = 1$	$V(t) \neq 0$	$(x_3^p \neq 0)$
$x_7 = off$	$\bar{x}_7 = 0$	$U(t) = 0$	$(x_7^p = 0)$
$x_7 = \neg off$	$\bar{x}_7 = 1$	$U(t) \neq 0$	$(x_7^p \neq 0)$
$x_8 = false$	$\bar{x}_8 = 0$	$Qv(t) = 0$	$(x_8^p = 0)$
$x_8 = \neg false$	$\bar{x}_8 = 1$	$Qv(t) \neq 0$	$(x_8^p \neq 0)$
$x_9 = \neg works$	$\bar{x}_9 = 0$	$\alpha.U(t).Qv(t) = 0^{a}$	$(x_9^p = 0)$
$x_9 = works$	$\bar{x}_9 = 1$	$\alpha.U(t).Qv(t) \neq 0$	$(x_9^p \neq 0)$

[a] $\alpha \in \{0, 1\}$ models the car key (off/on) allowing to interpret $x_9 = \neg works$ as the car is stopped (owing to that it is off, there is no voltage or there is no gas). However, we do not have information about α, so we assume that it can not be observed

among the variables are presented in Table 6.2. In the figure, the boxes c_{B7}, c_{B8} and c_{B9} represent logical "AND" components, and the components c_{B1}, c_{B2} and c_{B3} represent boolean value generators. This interpretation allows to specify clearly conditions of normal and abnormal behaviour on the variables and, as mentioned, it allows to resort to Reiter's theory.

Nevertheless, Reiter's theory tacitly assumes that logically *consistent* states correspond to *normal and desired* behaviour, and the *inconsistent* states, denoting a problem with at least one component, coincide with *abnormal and undesired* behaviour. The problem is that this correspondence sometimes is not compatible with a physical interpretation of the variables; thus, a logical model is a strong tool for reasoning but is not sufficient. For example, when observing Fig. 6.15, a state in which $\bar{x}_3 = 0$ and $\bar{x}_8 = 1$ results in an *inconsistent* state and in the mentioned theory, this would indicate that the component c_{B8} does not work. However, in this *inconsistent* state, the fuel tank is empty ($\bar{x}_3 = 0$) and there is gas in the engine ($\bar{x}_8 = 1$); consequently, the mentioned situation can not be associated with the problem of a component. On the contrary, this state is transient and corresponds to *normal* behaviour, although it is not a state of interest for the diagnosis task and it should not be considered. However, in the logical model, it is identified as a state of abnormal behaviour.

The example shows that the logical interpretation of variables required by Reiter's theory must be completed with a physical interpretation. For this purpose, [40] proposes to utilize the Tetrahedron of States (ToS), introduced in Sect. 6.5.2, where the given variables can be mapped to physical variables of the ToS and thus, the relations among them established. In this way, the introduction of semantic content in the physical interpretation of variables is controlled through the ToS framework. In particular, the ToS of hydraulic domain and that one of electric domain, shown respectively in Fig. 6.16a, b, are used in this example.

Each given variable $x_i \in X$ is mapped to a physical variable of the corresponding ToS. For example, using the Hydraulic ToS in Fig. 6.16a, the variable x_3 (fuel tank status) is associated with the gas volume $V(t)$ in the tank, as Table 6.1 specifies where $V(t)$ is also noted as x_3^p. Thus, $x_3 = \mathit{empty}$ which is logically interpreted as $\bar{x}_3 = 0$ (false), is physically interpreted through the ToS as $V(t) = 0$; and, $x_3 = \neg\mathit{empty}$ (or $x_3 = \phi_3$), related to $\bar{x}_3 = 1$ (true), is physically interpreted as $V(t) \neq 0$.

Table 6.2 Logical and physical functional relations

Logical relations	Physical relations	
$\bar{x}_7 = \bar{x}_1 \wedge \bar{x}_2$	$x_7^p = x_1^p \cdot \frac{dx_2^p}{dt}$	$(U(t) = R(t) \cdot \frac{dQ(t)}{dt}$, $(R(t) = \infty \vee Q(t) = 0) \Rightarrow U(t) = 0)$
$\bar{x}_8 = \bar{x}_3$	$x_8^p = \frac{dx_3^p}{dt}$	$(Qv(t) = \frac{dV(t)}{dt}, V(t) = 0 \Rightarrow Qv(t) = 0)$
$\bar{x}_9 = \bar{x}_7 \wedge \bar{x}_8$	$x_9^p = \alpha \cdot x_7^p \cdot x_8^p$	$((U(t) = 0 \vee Qv(t) = 0) \Rightarrow \alpha \cdot U(t) \cdot Qv(t) = 0)$

Thereby, the variables are mapped with physical variables as illustrated in Fig. 6.16 and specified in Table 6.1 where the relations among the variables are established as Table 6.2 presents. Thus, the physical model of the process is illustrated in Fig. 6.17.

This interpretation allows to determine conditions on the variables in order to identify transient states which can be discarded from the model to be built. For example, the states in which $V(t) = 0$ and $Qv(t) \neq 0$ (see Fig. 6.17) can be eliminated from the model; or, what is the same thing, the states in which $\bar{x}_3 = 0 \wedge \bar{x}_8 = 1$.

From this interpretation and a suitable analysis, the transient or physically impossible states can be removed from the model to be built, which results in 21 states of interest.

6.7.1.4 Generic Modelling

From the consideration of the two previous interpretations, a generic model of the process $M(X(t)) = <PM(X(t)), SM(X(t)), FM(X(t)), BM(X(t))>$ is defined. Details of the analysis carried out for establishing this model will not be described. Nevertheless, we shall limit ourselves to present the resultant model of the process formally specified in the TOM4D formalism [4].

In order to facilitate the analysis, the logical variables are used to describe the model; considering always that it is possible to reinterpret them like the variables and components described in (6.5) through Table 6.1. Thereby, observing Fig. 6.15, the models are the following ones.

The perception model $PM(X(t))$ of the process consists of the set X of variables, the set Ψ of threshold values described in Sect. 6.4 which in this case are not present and a set R_q of sentences describing objectives and operating modes. This model is specified as follows:

$$
\begin{aligned}
&PM(X(t)) = <X, \Psi, R_q> \text{ where} \\
&\quad X = \{\bar{x}_1, \bar{x}_2, \bar{x}_3, \bar{x}_7, \bar{x}_8, \bar{x}_9\}, \quad \Delta_{\bar{x}_i} = \{0, 1\}, \quad i = 1, 2, 3, 7, 8, 9 \\
&\quad \Psi = \{\Psi_i\}_{i=1,2,3,7,8,9} \quad \text{(threshold values of the time functions which we do not know)} \\
&\quad R_q = R_{goal} \cup R_n \cup R_{ab} \text{ such that} \\
&\quad R_{goal} \text{ describes the process operating goals } \bar{x}_9 = 1 \\
&\quad R_n \text{ describes the conditions of the normal operating mode :} \\
&\quad (\bar{x}_1 = 1 \wedge \bar{x}_2 = 1 \wedge \bar{x}_3 = 1 \wedge \bar{x}_7 = 1 \wedge \bar{x}_8 = 1 \wedge \bar{x}_9 = 1) \vee \\
&\quad ((((\bar{x}_1 = 0 \vee \bar{x}_2 = 0) \wedge \bar{x}_7 = 0) \vee (\bar{x}_3 = 0 \wedge \bar{x}_8 = 0)) \wedge \bar{x}_9 = 0) \\
&\quad R_{ab} \text{ describes the conditions of the abnormal operating mode :} \\
&\quad (\bar{x}_1 = 1 \wedge \bar{x}_2 = 1 \wedge \bar{x}_7 = 0) \vee (\bar{x}_3 = 1 \wedge \bar{x}_8 = 0) \vee (\bar{x}_7 = 1 \wedge \bar{x}_8 = 1 \wedge \bar{x}_9 = 0)
\end{aligned}
\tag{6.7}
$$

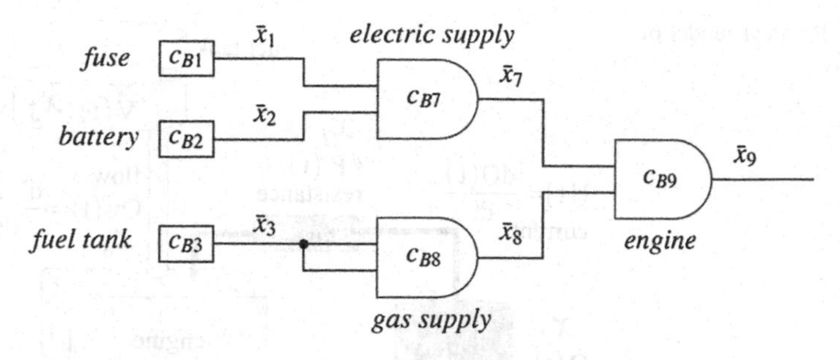

Fig. 6.15 Logical model of the process

The structural model $SM(X(t))$, defined in (6.8), describes the set $COMPS$ of components, the set R_{port} specifying the interconnections between output ports with input ports of components (e.g., $out(c_{B1}) = in_1(c_{B7})$) and the set R_{xport} associating each variable with an output port (e.g., $out(c_{B1}) = \bar{x}_1$).

$$
\begin{aligned}
SM(X(t)) = \; & < COMPS, R_{port}, R_{xport} > \text{ where} \\
& COMPS = \{c_{B1}, c_{B2}, c_{B3}, c_{B7}, c_{B8}, c_{B9}\} \\
& R_{port} = \{out(c_{B1}) = in_1(c_{B7}), out(c_{B2}) = in_2(c_{B7}), out(c_{B7}) = in_1(c_{B9}), \\
& \qquad out(c_{B3}) = in_1(c_{B8}), out(c_{B3}) = in_2(c_{B8}), out(c_{B8}) = in_2(c_{B9})\} \\
& R_{xport} = \{out(c_{B1}) = \bar{x}_1, out(c_{B2}) = \bar{x}_2, out(c_{B3}) = \bar{x}_3, \\
& \qquad out(c_{B7}) = \bar{x}_7, out(c_{B8}) = \bar{x}_8, out(c_{B9}) = \bar{x}_9\}
\end{aligned} \tag{6.8}
$$

The functional model $FM(X(t))$ describes the relations among the values that the variables can assume, as defined in (6.9). This model consists of the set Δ of values belonging to the domain and the image of the functions defined in the set F, the mentioned set F and the set R_f that establishes the relation among the variables (e.g., $\bar{x}_7 = f_{B4}(\bar{x}_1, \bar{x}_2)$).

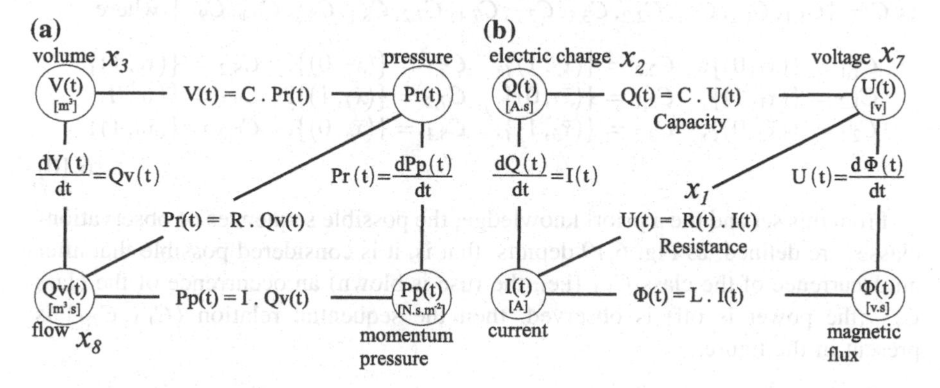

Fig. 6.16 Physical interpretation of variables

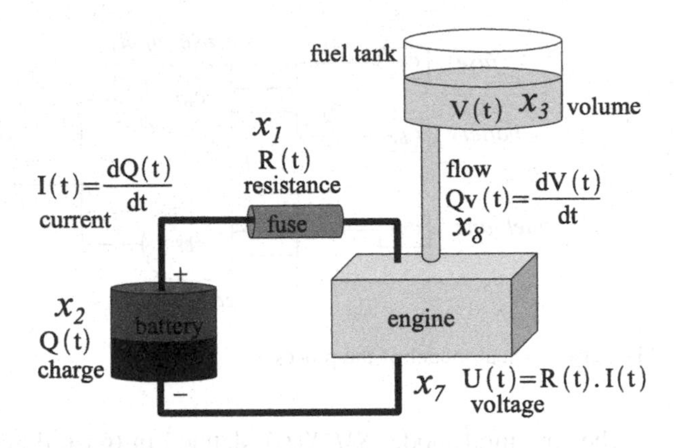

Fig. 6.17 Physical model of the process

$$FM(X(t))$$

$$\begin{aligned}
&\Delta_{\bar{x}_i} = \{0,1\}, \quad i = 1,2,3,7,8,9\\
&F = \{f_{B4}, f_{B5}, f_{B6}\} \text{ with}\\
&f_{B4} : \Delta_{\bar{x}_1} \times \Delta_{\bar{x}_2} \rightarrow \Delta_{\bar{x}_7}\\
&f_{B5} : \Delta_{\bar{x}_3} \rightarrow \Delta_{\bar{x}_8}\\
&f_{B6} : \Delta_{\bar{x}_7} \times \Delta_{\bar{x}_8} \rightarrow \Delta_{\bar{x}_9} \quad \text{and such that}\\
&\quad f_{B4}(y_1,y_2) = and\ (y_1,y_2) \wedge f_{B5}(y) = and(y,y) \wedge\\
&\quad f_{B6}(y_1,y_2) = and\ (y_1,y_2)\\
&R_f = \{\bar{x}_7 = f_{B4}(\bar{x}_1,\bar{x}_2),\ \bar{x}_8 = f_{B5}(\bar{x}_3),\ \bar{x}_9 = f_{B6}(\bar{x}_7,\bar{x}_8)\}
\end{aligned} \tag{6.9}$$

For readability and clarity, we consider to reinterpret from Table 6.1 the logical variables $\bar{x}_i$ $(i = 1,2,3,7,8,9)$ like their corresponding x_i $(i = 1,2,3,7,8,9)$. This reinterpretation then allows us to see the functional model as depicted in Fig. 6.18.

The behavioural model requires the set of observation classes, which is defined as $C = \{C_{1,1}, C_{1,2}, C_{2,1}, C_{2,2}, C_{3,1}, C_{3,2}, C_{7,1}, C_{7,2}, C_{8,1}, C_{8,2}, C_{9,1}, C_{9,2}\}$ where

$$\begin{aligned}
C_{1,1} &= \{(\bar{x}_1,0)\}, & C_{2,2} &= \{(\bar{x}_2,1)\}, & C_{7,1} &= \{(\bar{x}_7,0)\}, & C_{8,2} &= \{(\bar{x}_8,1)\},\\
C_{1,2} &= \{(\bar{x}_1,1)\}, & C_{3,1} &= \{(\bar{x}_3,0)\}, & C_{7,2} &= \{(\bar{x}_7,1)\}, & C_{9,1} &= \{(\bar{x}_9,0)\},\\
C_{2,1} &= \{(\bar{x}_2,0)\}, & C_{3,2} &= \{(\bar{x}_3,1)\}, & C_{8,1} &= \{(\bar{x}_8,0)\}, & C_{9,2} &= \{(\bar{x}_9,1)\}
\end{aligned} \tag{6.10}$$

From this set and the a priori knowledge, the possible sequences of observations classes are defined, as Fig. 6.19 depicts; that is, it is considered possible that after an occurrence of the class $C_{1,1}$ (i.e., the fuse is blown) an occurrence of the class $C_{7,1}$ (the power is off) is observed, then the sequential relation $(C_{1,1}, C_{7,1})$ is present in the figure.

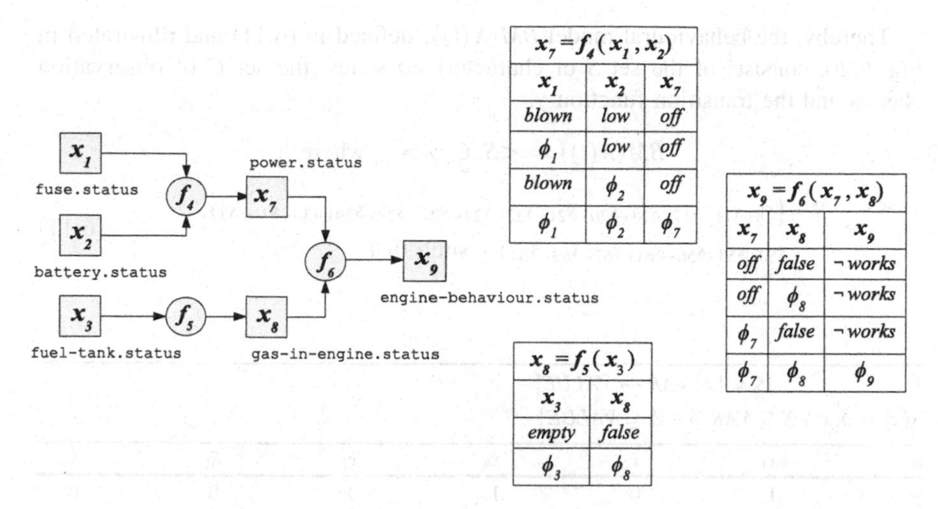

$x_7 = f_4(x_1, x_2)$		
x_1	x_2	x_7
blown	*low*	*off*
ϕ_1	*low*	*off*
blown	ϕ_2	*off*
ϕ_1	ϕ_2	ϕ_7

$x_9 = f_6(x_7, x_8)$		
x_7	x_8	x_9
off	*false*	$\neg$ *works*
off	ϕ_8	$\neg$ *works*
ϕ_7	*false*	$\neg$ *works*
ϕ_7	ϕ_8	ϕ_9

$x_8 = f_5(x_3)$	
x_3	x_8
empty	*false*
ϕ_3	ϕ_8

Fig. 6.18 Functional model

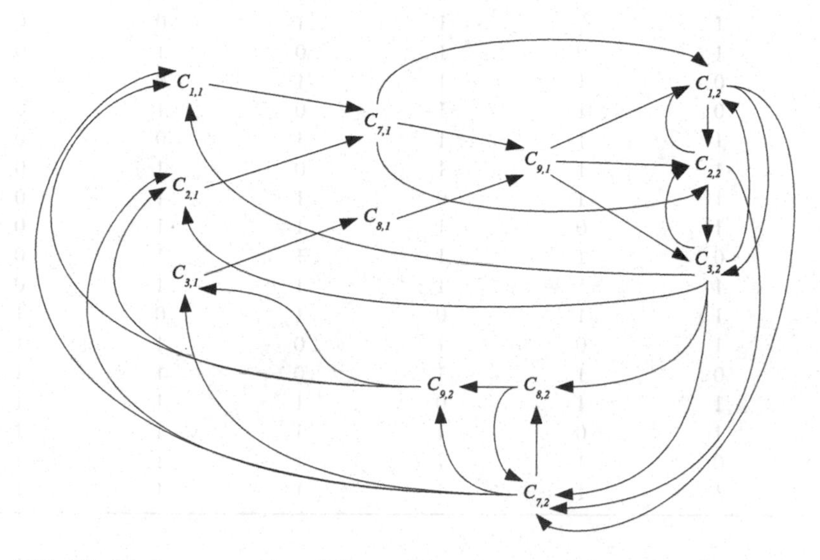

Fig. 6.19 Graphical representation of the possible sequences of observation classes

The occurrence of an observation class entails the assignation of a value to a variable; that is to say, an occurrence of $C_{1,1}$ entails that the value 0 is assumed by $\bar{x}_1$. Consequently, the previous value of $\bar{x}_1$ was not 0. Thus, the possible states between two observation classes can be characterized and established. Recall that only 21 characterized states were considered of interest from the logical and physical interpretations.

Thereby, the behavioural model $BM(X(t))$, defined in (6.11) and illustrated in Fig. 6.20, consists of the set S of characterized states, the set C of observation classes and the transition function γ.

$$BM(X(t)) = <S, C, \gamma> \quad \text{where}$$

$$S = \{s_8, s_{11}, s_{17}, s_{18}, s_{20}, s_{21}, s_{23}, s_{24}, s_{27}, s_{28}, s_{29}, s_{30}, s_{31}, s_{32}, s_{50}, s_{53}, s_{56}, s_{61}, s_{62}, s_{63}, s_{64}\} \text{ such that} \tag{6.11}$$

$S = \{s : VAR \rightarrow VALUE|$
$s(x) = \delta, x \in X \subseteq VAR, \delta \in \Delta \subseteq VALUE\}$

S	$\bar{x}_1$	$\bar{x}_2$	$\bar{x}_3$	$\bar{x}_7$	$\bar{x}_8$	$\bar{x}_9$
s_8	1	0	1	0	0	0
s_{11}	0	1	1	0	0	0
s_{17}	1	1	1	0	0	0
s_{18}	1	1	0	1	0	0
s_{20}	1	0	1	1	0	0
s_{21}	1	0	1	0	1	0
s_{23}	0	1	1	1	0	0
s_{24}	0	1	1	0	1	0
s_{27}	1	1	1	1	0	0
s_{28}	1	1	1	0	1	0
s_{29}	1	1	0	1	1	0
s_{30}	1	0	1	1	1	0
s_{31}	0	1	1	1	1	0
s_{32}	1	1	1	1	1	0
s_{50}	1	1	0	1	0	1
s_{53}	1	0	1	0	1	1
s_{56}	0	1	1	0	1	1
s_{61}	1	1	0	1	1	1
s_{62}	1	0	1	1	1	1
s_{63}	0	1	1	1	1	1
s_{64}	1	1	1	1	1	1

$$C = \{C_{1,1}, C_{1,2}, C_{2,1}, C_{2,2}, C_{3,1}, C_{3,2}, C_{7,1}, C_{7,2}, C_{8,1}, C_{8,2}, C_{9,1}, C_{9,2}\} \text{ where}$$

$$\begin{array}{llll} C_{1,1} = \{(\bar{x}_1, 0)\}, & C_{2,2} = \{(\bar{x}_2, 1)\}, & C_{7,1} = \{(\bar{x}_7, 0)\}, & C_{8,2} = \{(\bar{x}_8, 1)\}, \\ C_{1,2} = \{(\bar{x}_1, 1)\}, & C_{3,1} = \{(\bar{x}_3, 0)\}, & C_{7,2} = \{(\bar{x}_7, 1)\}, & C_{9,1} = \{(\bar{x}_9, 0)\}, \\ C_{2,1} = \{(\bar{x}_2, 0)\}, & C_{3,2} = \{(\bar{x}_3, 1)\}, & C_{8,1} = \{(\bar{x}_8, 0)\}, & C_{9,2} = \{(\bar{x}_9, 1)\} \end{array}$$

$$\gamma : S \times C \rightarrow S \quad \text{such that}$$

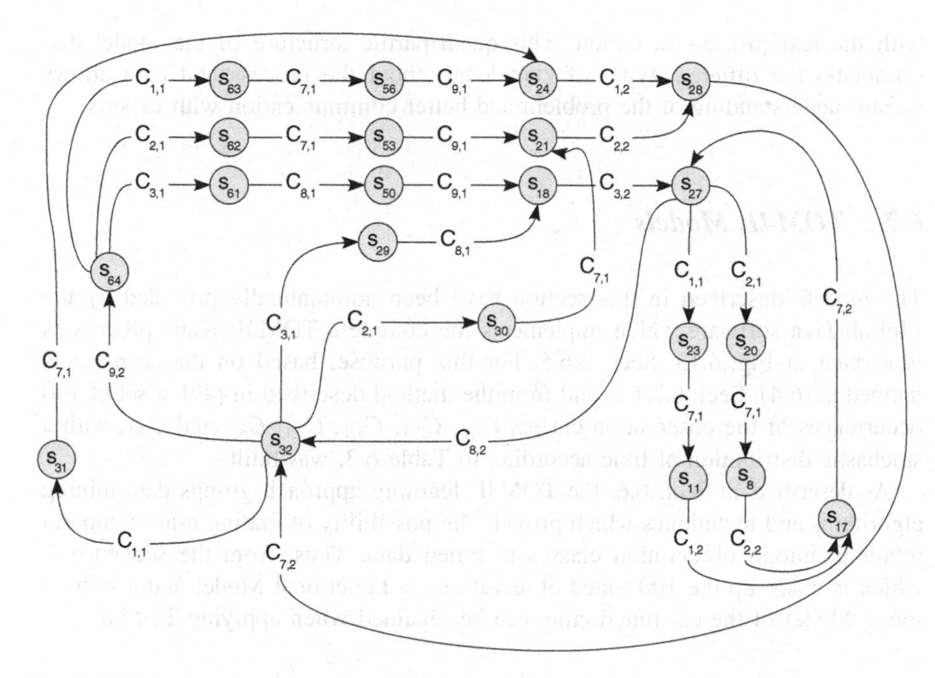

Fig. 6.20 Behavioural model of the process P(t)

$$\begin{array}{llll}
\gamma(s_8, C_{2,2}) = s_{17}, & \gamma(s_{31}, C_{7,1}) = s_{24}, & \gamma(s_{11}, C_{1,2}) = s_{17}, & \gamma(s_{32}, C_{1,1}) = s_{31}, \\
\gamma(s_{17}, C_{7,2}) = s_{27}, & \gamma(s_{32}, C_{9,2}) = s_{64}, & \gamma(s_{18}, C_{3,2}) = s_{27}, & \gamma(s_{32}, C_{3,1}) = s_{29}, \\
\gamma(s_{20}, C_{7,1}) = s_8, & \gamma(s_{32}, C_{2,1}) = s_{30}, & \gamma(s_{21}, C_{2,2}) = s_{28}, & \gamma(s_{50}, C_{9,1}) = s_{18}, \\
\gamma(s_{23}, C_{7,1}) = s_{11}, & \gamma(s_{53}, C_{9,1}) = s_{21}, & \gamma(s_{24}, C_{1,2}) = s_{28}, & \gamma(s_{56}, C_{9,1}) = s_{24}, \\
\gamma(s_{27}, C_{1,1}) = s_{23}, & \gamma(s_{61}, C_{8,1}) = s_{50}, & \gamma(s_{27}, C_{2,1}) = s_{20}, & \gamma(s_{62}, C_{7,1}) = s_{53}, \\
\gamma(s_{27}, C_{8,2}) = s_{32}, & \gamma(s_{63}, C_{71}) = s_{56}, & \gamma(s_{28}, C_{7,2}) = s_{32}, & \gamma(s_{64}, C_{1,1}) = s_{63}, \\
\gamma(s_{29}, C_{8,1}) = s_{18}, & \gamma(s_{64}, C_{2,1}) = s_{62}, & \gamma(s_{30}, C_{7,1}) = s_{21}, & \gamma(s_{64}, C_{3,1}) = s_{61}
\end{array}$$

As a result of this analysis, we consider that the construction of a generic model of the process requires interpretations of the expert's knowledge both in logical and physical terms. These interpretations along with modelling decisions allowed a reduction from 512 to only 21 states physically possible and of interest for diagnosing behaviour problems. The logical model of Fig. 6.15 describes the structure of the expert's diagnosis reasoning and the physical model of Fig. 6.17 provides the diagnosis knowledge required for this reasoning. Thus, both logical and physical models are necessary and complement each other. We believe that these models are, ultimately, those ones "constructed" by experts where, in practice, the combination of these ones simplifies the diagnosis task.

Moreover, the resultant model $M(X(t))$ admits the application of model-based diagnosis techniques and, simultaneously, introduce the dimension of time allowing to model the dynamic of the process in a behavioural model. This model is a crucial element in the supervision of processes since generally it is collated with the real process evolution. This quadripartite structure of the model discriminates the different types of knowledge about the process and then allows greater understanding of the problem and better communication with experts.

6.7.2 TOM4L Models

The models described in this section have been automatically provided by the ElpLab Java software which implements the complete TOM4L KDD process as illustrated in Fig. 6.13, Sect. 6.6.5. For this purpose, based on the scenario Ω defined in (6.4), Sect. 6.7.1.2, and from the method described in [49], a set of 100 occurrences of the observation classes $C_{1,1}$, $C_{2,1}$, $C_{3,1}$, $C_{7,1}$, $C_{8,1}$ and $C_{9,1}$, with a stochastic distribution of time according to Table 6.3, was built.

As described in Sect. 6.6, the TOM4L learning approach groups data mining algorithms and techniques which provide the possibility of finding n-ary temporal relations among observation classes in timed data. Thus, from the sequence w which is made up the 100 timed observations, a Functional Model and a Behavioural Model of the car functioning can be obtained when applying TOM4L.

6.7.2.1 Functional Model

The algorithm Tom4BN [8, 10] which allows to discover naive Bayesian Networks (Sect. 6.6.3) from timed data is applied to the 100 observations of the car example, giving as a result the Bayesian Network shown in Fig. 6.21a.

In this example, classes $C_{i,j}$ are singletons of the form $C_{i,j} = \{(x_i, \delta_j)\}$ and $P(C_{i,j})$, equivalent to $P(x_i = \delta_j)$, is the prior probability of observing an occurrence of the class $C_{i,j} = \{(x_i, \delta_j)\}$ in w. Besides, it should be noted that "$\neg x_i$" refers to any equality except "$x_i = \delta_j$" or, put in another way, "$\neg x_i$" denotes "$x_i = \delta_k \wedge \delta_k \neq \delta_j$".

Thus, this Bayesian Network enables the definition of the Functional Model of Fig. 6.21b, whose functions correspond to those ones of the TOM4D Functional Model (Fig. 6.18, Sect. 6.7.1.4); but unlike the last one, these functions have probabilities associated which provide a certain level of confidence about the established relations among values. For example, the probability of observing that the power is off having observed that the battery is low and the fuse is blown is 0.684; that is, $P(x_7|x_1, x_2) = 0.684$ in Fig. 6.21a. Thus, the level of confidence

Table 6.3 Prior probabilities of the car example [10, p. 76]

$P(\{(x_1, blown)\})$	$P(\{(x_2, low)\})$	$P(\{(x_3, empty)\})$	$P(\{(x_7, off)\})$	$P(\{(x_8, false)\})$	$P(\{(x_9, \neg low)\})$
$P(C_{1,1})$	$P(C_{2,1})$	$P(C_{3,1})$	$P(C_{7,1})$	$P(C_{8,1})$	$P(C_{9,1})$
0.05	0.15	0.3	0.2	0.2	0.1

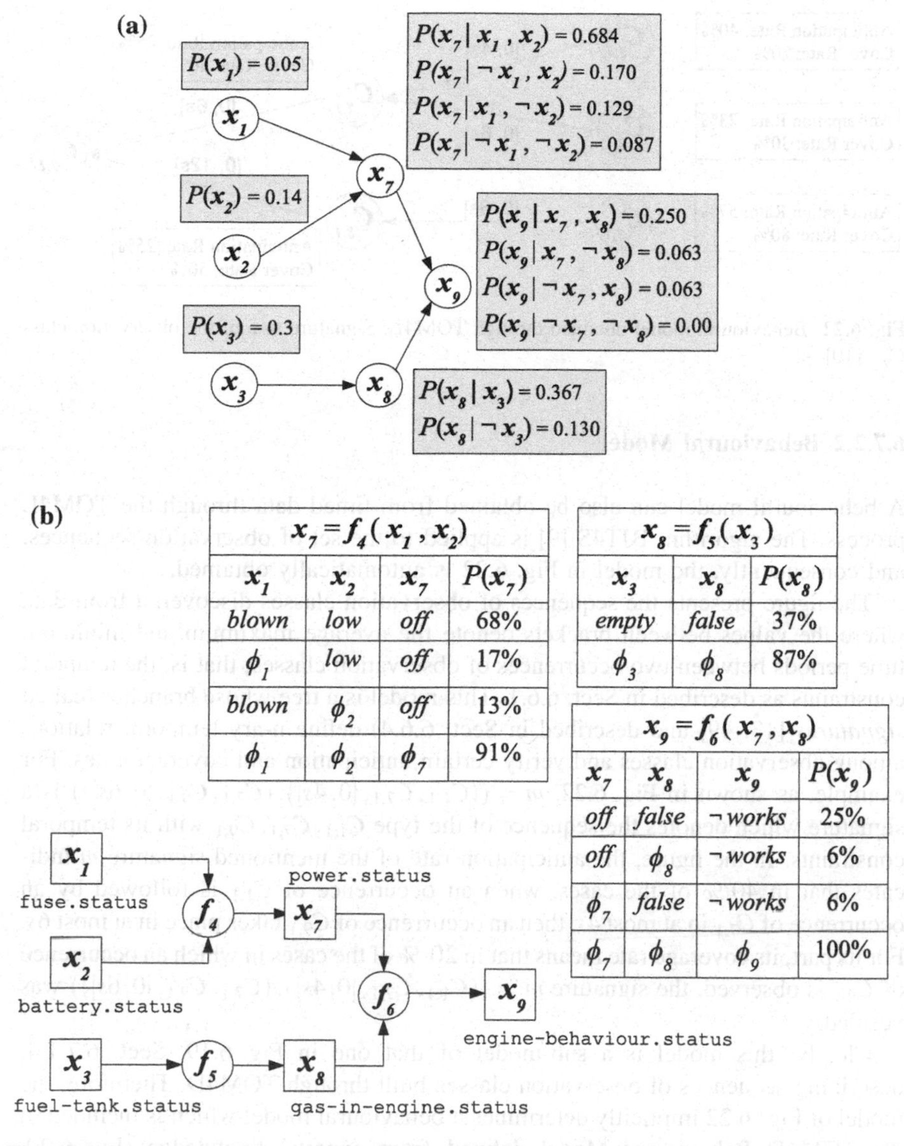

Fig. 6.21 Functional model obtained through TOM4L [10, pp. 81, 83]

when considering $off = f_4(blown, low)$ is approximately 68 % as Fig. 6.21b depicts. Another example is the probability of $\phi_7 = f_4(\phi_1, \phi_2)$, which can be obtained from $P(x_7|\neg x_1, \neg x_2) = 0.087$ when calculating $P(\neg x_7|\neg x_1, \neg x_2) = 1 - P(x_7|\neg x_1, \neg x_2)$.

Hence, the Functional Model with probabilities automatically obtained from data can be compared with the Functional Model defined from experts' knowledge; and thus, both models can be analysed together complementing each other.

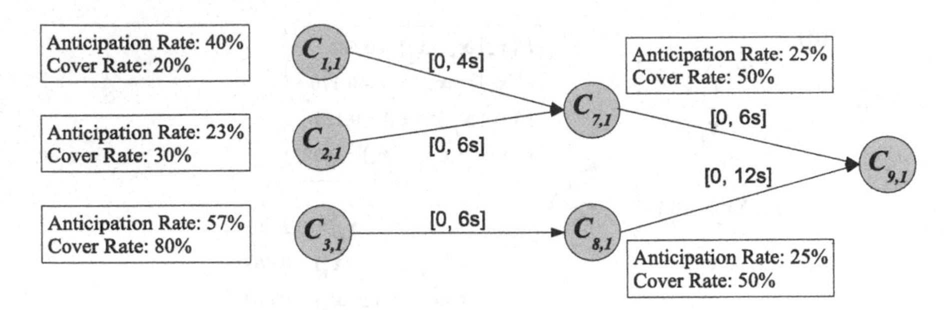

Fig. 6.22 Behavioural model obtained through TOM4L. Signature tree of the observation class $C_{9,1}$ [10]

6.7.2.2 Behavioural Model

A behavioural model can also be obtained from timed data through the TOM4L process. The algorithm BJT4S [9] is applied to the set of observation sequences, and consequently, the model in Fig. 6.22 is automatically obtained.

The figure presents the sequences of observation classes discovered from data where the values between brackets denote the average maximum and minimum time periods between two occurrences of observation classes; that is, the temporal constraints as described in Sect. 6.6.1. This model is a tree whose branches (called *signatures* [13, 49] and described in Sect. 6.6.4) define n-ary temporal relations among observation classes and verify certain anticipation and coverage rates. For example, as shown in Fig. 6.22, $m = ((C_{1,1}, C_{7,1}, [0, 4s]), (C_{7,1}, C_{9,1}, [0, 6s]))$ is a signature which denotes the sequence of the type $C_{1,1}, C_{7,1}, C_{9,1}$ with its temporal constraints. In the figure, the anticipation rate of the mentioned signature m indicates that in 40 % of the cases, when an occurrence of $C_{1,1}$ is followed by an occurrence of $C_{7,1}$ in at most $4s$, then an occurrence of $C_{9,1}$ takes place in at most $6s$. For its part, its coverage rate means that in 20 % of the cases in which an occurrence of $C_{9,1}$ is observed, the signature $m = ((C_{1,1}, C_{7,1}, [0, 4s]), (C_{7,1}, C_{9,1}, [0, 6s]))$ was verified.

Clearly, this model is a sub-model of that one in Fig. 6.19, Sect. 6.7.1.4, describing sequences of observation classes built through TOM4D. Therefore, the model of Fig. 6.22 implicitly determines a behavioural model which is included in the TOM4D Behavioural Model defined from experts' knowledge (Fig. 6.20, Sect. 6.7.1.4). In particular, the model obtained from data provides, in addition, knowledge about temporal constraints between event occurrences. Thus, once again, these models belonging to different disciplines, such as KE and KDD are, can be easily related and compared to each other.

Owing to that, TOM4L models can be related with TOM4D models and the latter are directly related with a CommonKADS conceptual model, the communication with experts about the first one is easier. That is to say, the meaning of the signature $m = ((C_{1,1}, C_{7,1}, [0, 4s]), (C_{7,1}, C_{9,1}, [0, 6s]))$ can be easily explained by saying that in the 40 % of cases, when observing the fuse blown, in at most 4*s* the power is observed off and subsequently, in at most 6*s*, it is observed that the engine does not work. Thus, TOM4D establishes a bridge between experts' knowledge and data, and TOM4L allows automatic learning from these last ones.

6.8 Conclusion

Knowledge acquisition, as a topic of interest in sciences, has been generally addressed from two different perspectives. One approach has been to consider knowledge acquisition as a psychological and social process that consists in the synthesis of new knowledge through socialization with experts. The other approach has been to consider knowledge acquisition as an interpretation and analysis process of data, based on discovering patterns of interest through observation, analysis and intertwining of the data. These two perspectives are, respectively, central issues in Knowledge Engineering (KE) and in Knowledge Discovery in Database (KDD).

Nevertheless, as highlighted by N. Wickramasinghe [15], although knowledge acquisition is the main and central question in both disciplines, the issue has been traditionally approached from one or the other perspective, rather than from an integrative view. We consider then that a whole approach is necessary in order to accelerate the global learning process and even, in extremely complex cases, to provide viability.

Results about probabilistic information and temporal constraints, as well as discovered event sequences which could be unexpected, extend the knowledge about a real process and provide resources to build a more suitable model of this one. However, relating this knowledge to the expert's one is not a trivial task because, generally, the formalisms used by Knowledge Engineering methodologies and by Knowledge Discovery in Database processes to represent knowledge models are different. As a consequence, the comparison between both models can not be *in principle* carried out. We then believe that the main difficulty for relating the mentioned disciplines stems from the lack of a global approach based on a same theory and consequently, from the lack of representation formalisms that can be used in both domains.

Thereby, the central focus of this chapter was the definition of a global human–machine learning process which combines a Knowledge Engineering methodology called TOM4D (i.e. Timed Observation Modelling for Diagnosis) with a Knowledge Discovery in Database process called TOM4L (i.e. Timed Observation Mining for Learning). Thus, with the aim of defining this integral view, the Theory of Timed Observations [1] has been established as a basis for the development of the proposed approach. This theory defines, among other things, the notions of *timed observation* and *observation class*, concepts that enable to specify the traditional notion of discrete event and the Artificial Intelligence notion of alarm (or warning).

This chapter presented then the TOM4D Knowledge Engineering methodology, which allows to build models, by basing on the Theory of Timed Observations, from experts' knowledge. The models built through this methodology are not experts' Knowledge Models but models of the process about which experts have knowledge. By construction, TOM4D models are consistent with and easily relatable to CommonKADS Knowledge Models built from experts' knowledge, CommonKADS being one of the principal KE methodologies. Therefore, models of a process built through TOM4D facilitate the communication with the expert, and thus, the validation of the Knowledge Models. Besides, the chapter introduced the basic elements of the TOM4L Knowledge Discovery in Database process to obtain knowledge from data. The TOM4L process allows to find n-ary temporal relations of observation classes representative of the process that gives rise to data, by using an entropy-based measure called the BJ-measure [9, 12]. In addition, through the aforesaid measure, TOM4L enables to build Bayesian Networks from timed data [8, 10, 11]. Thus, TOM4L models are directly relatable to TOM4D models.

In summary, it was presented a human–machine learning process nourished from experts' knowledge and knowledge discovered in data which, in our opinion, is ultimately a virtuous circle that establishes a positive and corrective feedback to each step. Therefore, a process model which meets the expectation in the knowledge intensive tasks performed by a Knowledge Based System can be built in a more suitable way.

Real world problems have been addressed though this approach. In particular, the security of the dam of Cublize (France), where the resultant models have been validated by the hydraulic dam experts of the French governmental organization (Irstea) which controls the security of hydraulic civil engineering structures in the corresponding country [6, 50]. Moreover, nowadays we are utilizing the presented approach in order to model human behaviour from gerontologists' knowledge and smart environments data, in the context of the GerHome Project of the Centre Scientifique et Technique du Bâtiment (CSTB) of Sophia Antipolis, France [4, 51].

We believe that binding the KE and KDD universes enriches and facilitates the modelling task. Nevertheless, there still exists a difficulty with regard to the discursive and conceptual levels in which each universe is developed. That to say, sometimes, even being able to link the mentioned disciplines, relating models obtained from knowledge discovered in data to models obtained from experts' knowledge is very difficult, because experts' conceptual abstraction level is very high or is far from those concepts at data level. Although this topic has been beyond the scope of the present chapter, we consider of interest to mention that this issue has been addressed by means of a theoretical framework of abstraction levels that we have defined [4, 52, 53], where in each level a KE methodology, like TOM4D, can be combined with a KDD process, like TOM4L, in order to built a set of models linking the data abstraction level (e.g. sensor level) to the expert's conceptual level.

References

1. Le Goc, M.: Notion d'observation pour le diagnostic des processus dynamiques: Application à Sachem et à la découverte de connaissances temporelles. Habilitation à Diriger des Recherches. Université de Droit d'Economie et des Sciences d'Aix-Marseille (2006)
2. Shannon, C.E.: A mathematical theory of communication. Bell Syst. Tech. J. **27**(379–423), 623–656 (1948)
3. Dagues, P.: Théorie logique du diagnostic à base de modèles. Diagnostic, Intelligence Artificielle, et Reconnaissance des Formes, pp. 17–105. Hermes Science Publications, Paris (2001)
4. Pomponio, L.: Definition of a human-machine learning process from timed observations: application to the modelling behaviour of old people at home. Université Aix-Marseille (2012)
5. Pomponio, L., Le Goc, M.: Timed observations modelling for diagnosis methodology: a case study. In: Cordeiro, J.A.M., Virvou, M., Shishkov, B. (eds.) ICSoft 2010—Proceedings of the 5th International Conference on Software and Data Technologies, pp. 504–507. SciTePress, Athens (2010)
6. Le Goc M., Masse E., Curt C.: Modeling processes from timed observations. In: Proceedings of the 3rd International Conference on Software and Data Technologies (ICSoft'08), pp. 249–256 (2008)
7. Le Goc, M., Masse, E.: Towards a multimodeling approach of dynamic systems for diagnosis. In: Proceedings of the 2nd International Conference on Software and Data Technologies (ICSoft'07), pp. 277–282 (2007)
8. Le Goc, M., Ahdab, A.: Learning Bayesian Networks from Timed Observations. LAP LAMBERT Academic Publishing GmbH & Co, KG (2012)
9. Benayadi, N., Le Goc, M.: Mining timed sequences with TOM4L framework. In: Proceedings of the 12th International Conference on Enterprise Information Systems (ICEIS 2010), pp. 111–120 (2010)
10. Ahdab, A., Le Goc, M.: Learning dynamic bayesian networks with the TOM4L process. In: Proceedings of the 5th International Conference on Software and Data Technologies (ICSoft 2010), pp. 353–363 (2010)
11. Ahdab, A.: Contribution à l'apprendissage de réseaux bayésiens à partir de donnèes datées pour le diagnostic des processus dynamiques continus. Université Paul Cézanne, Aix-Marseille (2010)
12. Benayadi, N.: Contribution à la découverte de connaissances à partir de données datées. Université Paul Cézanne, Aix-Marseille III (2010)
13. Bouché, P.: Une approache stochastique de modélisation de séquences d'événements discrets pour le diagnostic des systèmes dynamiques. Université Paul Cézanne, Aix-Marseille III (2005)
14. Schreiber, G., Akkermans, H., Anjewierden, A., et al.: Knowledge Engineering and Management: the CommonKADS Methodology. MIT Press, Cambridge (2000)
15. Wickramasinghe, N.: Knowledge Creation. Encyclopedia of Knowledge Management, pp. 326–335. Idea Group Inc., Hershey (2006)
16. Nonaka, I.: Dynamic theory of organizational knowledge creation. Organ. Sci. **5**, 14–37 (1994)
17. Nonaka, I.: The knowledge-creating company. Harvard Bus. Rev. 96–104 (1991)
18. Alavi, M., Leidner, D.E.: Review: knowledge management and knowledge management systems: conceptual foundations and research issues. MIS Quart **25**, 107–136 (2001)
19. Polanyi, M.: The Tacit Dimension. Doubleday & Company, Inc., NY (1966)
20. Nonaka, I., Konno, N.: The concept of "Ba": building a foundation for knowledge creation. California Manage. Rev. **40**, 40–54 (1998)

21. Feigenbaum, E.A.: The art of artificial intelligence: 1. Themes and case studies of knowledge engineering. In: International Joint Conference on Artificial Intelligence, pp. 1014–1029 (1977)
22. Feigenbaum, E.A.: A personal view of expert systems: looking back and looking ahead. knowledge systems laboratory. Department of Computer Science, Stanford University (1992)
23. Studer, R., Benjamins, V.R., Fensel, D.: Knowledge Engineering: Principles and Methods. Data Knowl. Eng. **25**, 161–197 (1998)
24. Breuker, J., de Velde, W.V.: CommonKADS Library For Expertise Modelling. IOS Press, Amsterdam (1994)
25. Gennari, J.H., Musen, M.A., Fergerson, R.W., et al.: The evolution of protégé: an environment for knowledge-based systems development. Int. J. Hum Comput Stud. **58**, 89–123 (2002)
26. Angele, J., Fensel, D., Landes, D., Studer, R.: Developing knowledge based-systems with MIKE. Autom. Soft. Eng. **5**, 389–418 (1998)
27. Angele, J., Fensel, D., Studer, R.: Domain and task modeling in MIKE. In: Proceedings of the IFIP WG8.1/13.2 Joint Working Conference on Domain Knowledge for Interactive System Design, pp. 8–10 (1996)
28. Cairó, O., Alvarez, J.C.: KAMET II: an extended knowledge-acquisition methodology. In: Palade, V., Howlett, R.J., Jain, L.C. (eds.) Knowledge-Based Intelligent Information and Engineering Systems, pp. 61–67. Springer, London (2003)
29. Cairó, O., Alvarez, J.C.: The KAMET II Methodology: A Modern Approach for Building Diagnosis-Specialized Knowledge-Based Systems ISMIS, pp. 652–656. Springer, London (2003)
30. Motta, E., Stutt, A., O'Hara, K. et al.: VITAL knowledge representation language specification. Human Cognition Research Laboratory of the Open University (1991)
31. Piatetsky-Shapiro, G.: Knowledge discovery in real databases: a report on the IJCAI-89 workshop. IA Mag. **11**, 68–70 (1990)
32. Fayyad, U., Piatetsky-Shapiro, G., Smyth, P.: From data mining to knowledge discovery in databases. IA Mag. **17**, 37–57 (1996)
33. Fayyad, U., Piatetsky-Shapiro, G., Smyth, P.: The KDD process for extracting useful knowledge from volumes of data. Commun. ACM **39**, 29–34 (1996)
34. Quinlan, J.R: C4.5: programs for machine learning. Morgan Kaufmann Publishers Inc., San Francisco (1993)
35. Rabiner L.R. : A tutorial on hidden Markov models and selected applications in speech recognition. In: Proceedings of the IEEE 77, pp. 257 –286 (1989)
36. Michalski, R.S., Carbonell, J.G., Mitchell, T.M.: Machine Learning: An Artificial Intelligence Approach. Morgan Kaufmann, Tioga (1983)
37. Cheng, J., Greiner, R., Kelly, J., et al.: Learning bayesian networks from data: an information-theory based approach. Artif. Intell. **137**, 43–90 (2002)
38. Defays, D.: An efficient algorithm for a complete link method. Comput. J. **20**, 364–366 (1977)
39. Mitchell T.: Machine Learning. McGraw Hill, NY (1977)
40. Chittaro, L., Guida, G., Tasso, C., Toppano, E.: Functional and teleological knowledge in the multimodeling approach for reasoning about physical systems: a case study in diagnosis. IEEE Trans. Sys. Man Cybern. **23**, 1718–1751 (1993)
41. Le Goc, M.: SACHEM, a real-time intelligent diagnosis system based on the discrete event paradigm. Simulation **80**, 591–617 (2004)
42. Chittaro, L., Ranon, R.: Diagnosis of multiple faults with flow-based functional models: the functional diagnosis with efforts and flows approach. Reliab. Eng. Syst. Safety **64**, 137–150 (1999)
43. Zanni, C., Le Goc, M., Frydman, C.: A conceptual framework for the analysis, classification and choice of knowledge-based diagnosis systems. KES—Int. J. Knowl. Based Intell. Eng. Syst. **10**, 113–138 (2006)
44. Reiter, R.: A theory of diagnosis from first principles. Artif. Intell. **32**, 57–95 (1987)

45. Rosenberg, R.C., Karnopp, D.C.: Introduction to Physical System Dynamics. McGraw-Hill, NY (1983)
46. Chittaro, L., Ranon, R.: Augmenting the diagnostic power of flow-based approaches to functional reasoning. In: AAAI-96 Proceedings, pp. 1010–1015 (1996)
47. Kullback, S., Leibler, R.A.: On information and sufficiency. Ann. Math. Stat. **22**, 79–86 (1951)
48. Cheng, J., Bell, D., Liu, W.: Learning bayesian networks from data: an efficient approach based on information theory (1997)
49. Bouché, P., Le Goc, M., Coinu, J.: A global model of sequences of discrete event class occurrences. In: Proceedings of the 10th International Conference on Enterprise Information Systems (ICEIS 2008), pp. 173–180 (2008)
50. Fakhfakh I., Curt C., Le Goc M., Torrès L.: Diagnosis of the Hydraulic Dam Safety based on Multimodelling Approach. Actes du 18ème Congrès de Maîtrise des Risques et de Sûreté de Fonctionnement (2012)
51. Pomponio, L., Le Goc, M., Pascual, E., Anfosso, A.: Discovering models of human's behavior from sensor's data. In: Workshop Proceedings of the 7th International Conference on Intelligent Environments, pp. 17–28. IOS Press, Nottingham, 25–26 July 2011
52. Pomponio, L., Le Goc, M., Anfosso, A., Pascual, E.: Levels of abstraction for behavior modeling in the GerHome project. Int. J. E-Health Med. Commun. **3**, 12–28 (2012)
53. Pomponio, L., Le Goc, M., Pascual, E., Anfosso, A.: Resident's activity at different abstraction levels: proposition of a general theoretical framework. In: The 6th IEEE International Conference on Intelligent Data Acquisition and Advanced Computing Systems: Technology and Applications, IDAACS'2011, pp. 540–545, Prague (2011)

Chapter 7
A CDSS Supporting Clinical Guidelines Integrated and Interoperable Within the Clinical Information System

Bruno Frandji, Dipak Kalra and Marie-Christine Jaulent

Abstract A CDSS (Clinical Decision Support System) aiming to support the exploitation of CG (Clinical Guidelines) by HCP (Health Care Practitioners) has been designed, able to consider the available knowledge about the patient's health stored within the CIS (Clinical Information System), using the CIS native, well-trained functions and ergonomics. Amongst the main methods used, figure rule based decision trees to represent the CG knowledge, concept dictionary bonded to international standard terminological systems for semantic indexing, usage of CIS components as part of the CDSS to ensure the respect of the clinical workflow. The results obtained are threefold: 1) a CG model structure adapted for such CDSS; 2) a semantic interoperability platform populated with SNOMED 3.5 international terminology system between CDSS and Electronic Healthcare Records; 3) a workflow of clinical information systems elements coupled by a rule engine solution allowing authoring CG as decision trees. The semantic interoperability platform is up and running in more than sixty large French healthcare organizations, the CDSS is available for first exploitation experiments.

B. Frandji (✉) · M.-C. Jaulent
INSERM UMR_S 872 Eq. 20, Faculté de Médecine Broussais-Hôtel-Dieu, 15 rue de l'Ecole de médecine 75006 Paris, France
e-mail: Bruno.Frandji@crc.jussieu.fr

M.-C. Jaulent
e-mail: Marie-Christine.Jaulent@crc.jussieu.fr

B. Frandji
MEDASYS, Gif sur Yvette 91130 Paris, France

D. Kalra
Centre for Health Informatics and Multiprofessional Education, University College London, London, UK
e-mail: d.kalra@ucl.ac.uk

C. Faucher and L. C. Jain (eds.), *Innovations in Intelligent Machines-4*,
Studies in Computational Intelligence 514, DOI: 10.1007/978-3-319-01866-9_7,

7.1 Introduction

This chapter presents results of work started from 2004 leading to the creation of a Clinical Decision Support System (CDSS) allowing Health Care Professionals (HCP) to follow Clinical Guidelines (CG).

CDSS are particularly useful in clinical settings when non specialists HCP are required to deal with unusual or complex cases like when a family physician performs the follow-up of a breast cancer after an hospital stay. When based on a CG, the CDSS is able to guide HCP actions and suggest appropriate recommendations [1].

Another benefit from such CDSS is expected in the control of consensual quality insurance procedures and the respect of evidence-based medicine conclusions to fight against adverse medical events. Studies in USA show that the application of CG may lead to significant results such as numbers of legal procedures divided by two and costs divided by five [8].

In addition, the CDSS should contribute to more efficient healthcare, reducing unuseful prescriptions, lengths of stay, hospital costs, sometimes even avoiding specialists involvement. As an example, a study performed on a population of 681 patients undergoing elective endocrine surgery during a 30-months period, shows that length of stay can be divided by two, patients were six times less likely to be admitted to intensive care, hospital costs significantly reduced by applying CG [9].

National institutions and scholarly societies regularly publish new versions or even new CG, in the form of very rich documents full of knowledge and expertise, with many switch cases, sub cases and exceptions. Sometimes, they include annexes presenting semi-formal models of decision trees. Their richness and complexity with many links to subsections make them unhandy to use for HCP in their paper form [11]. Even with strong commitment and personal investment, HCP are often prevented to exploit the CG that they should use during their activities. The inadequate complex paper format problem, added to missing or imprecise information, keeps the best evidence-based CG hardly compatible with professional practices.

Numerous surveys of CG implementation strategies have been published. They show that ICT-Tools are looked for in order to help the usage of CG [10, 15], but with mixed success [7], although the increased effect of computerised CG over non electronic implementation has been highlighted [4].

Several teams have aimed to build CDSS supporting the medical knowledge described within the CG. They mainly use the decision tree paradigm coupled with concepts organised within conceptual systems, sometimes processing web-based Electronic Healthcare Record (EHR). However, since they are proposed as standalone applications without integration within the clinical information system and the clinical workflow, these approaches increase the already heavy burden of HCP with large information input efforts. Therefore it cannot be generalized in practice. Indeed, if guidelines are to produce a large impact, they must be integrated into functions that physicians find useful in the routine care of patients [18].

According to [19], the key success for actual use in practice is patient-centered or data driven suggestions, to take into account existing medical knowledge about the patient. Methods have been set up to address the subsequent needs for tight semantic coupling between the clinical data items referred to within a CG and the patient's data item that are documented within his EHR [14], addressing various challenges such as the granularity of the data items documented within the EHR, the heterogeneity of the EHR data items. International efforts [20] have targeted standards for interoperability between systems, in particular semantic interoperability (the ability of one system to achieve a consistent information of a clinical meaning communicated by another system). Most of the on-going work for indexing the data item from the patient's EHR focuses on the retrospective analysis of healthcare information, including natural language processing techniques in order to try to semantically index the medical information contained in structured and non-structured data, like electronic reports. For example, in the context of the i2b2 challenges,[1] some approaches have been tested to automatically extract medication information from clinical records [6]. More recently, a CDSS based on automated text processing has been developed in the field of oncology for cervical cancer screening [22]. These approaches, although making good progresses, are not yet reliable enough to support CG oriented decision support systems used in routine patient care. Approaches specifically aiming to standardize the structural representation of clinical meaning through the use of archetypes [3] or HL7 templates[2] have been proposed, but research is still ongoing to scale up the development and governance of clinical models and other semantic interoperability artefacts. There is therefore a need for empirical efforts to establish and validate how the semantics of CG and the semantics of the EHR can best be coupled on a large scale with the potential to produce an impact on the actual usage of CG.

7.2 Materials and Methods

We had a succession of challenges to address (semantic interoperability, knowledge representation, inference processing, integration with the CIS…), so we designed an overall stepwise methodology, building upon achieved results to address the next one:

- The first step aimed to build a semantic interoperability platform enabling the CDSS to process patient's medical information maintained by the Clinical Information System (CIS).

[1] https://www.i2b2.org/about/index.html

[2] http://wiki.hl7.org/index.php?title=Templates

- The second thread of work consisted in the definition of a model allowing to transform CG paper based medical knowledge representation into a format suitable for CDSS exploitation.
- During the last phase we conceived the CDSS in such a way that decision trees would be natively integrated with CIS components and able to reuse the available medical knowledge about the patient's health state.

The implementation of the CDSS has been performed by the Research and Development department of MEDASYS®[3] using Microsoft workflow foundation® part of the.net® platform, and the clinical information system DxCare® largely deployed in health institutions in France.

Methodology and models have been created within the INSERM UMR_S 872, team 20, in the framework of the EU projects ≪Semantic Mining≫ (EU-FP6-ICT-2004 n°507505) [2] and ≪DEBUG-IT≫ (EU-FP7-ICT-2007 n°217139) [12].

7.2.1 *Method to Build the Semantic Interoperability Platform*

To build the semantic interoperability platform we have applied a method defined during the ≪Semantic Mining≫ network of excellence project [17]. It is based on a tripod:

- a standardized syntax enabling functional interoperability,
- a semantic indexation engine referencing in a unique way each medical information element of the CIS,
- coupled with a terminology binding mechanism able to link every indexed medical information element with entries from terminology systems.

7.2.1.1 Definition of a Normalized Syntax

To design the model of medical information containers, we projected the HL7 RIM V3 model into the architectural principles for clinical information systems defined within the EU RICHE (EU ESPRIT II N° 2221, 1993) and NUCLEUS projects (EU AIM A2025, 1996) that led to the definition of the CEN-12967 european normative vouchure (HISA 1999) endorsed in 2009 as an ISO-12967 standard.[4] A business layer of abstract objects has been designed and implemented, as shown in Fig. 7.1, structured according to the RIM model. The choice

[3] www.medasys.com

[4] International Organization for Standardization ISO 12967-1 Health informatics—Service architecture, http://www.iso.org, part 1—2009.

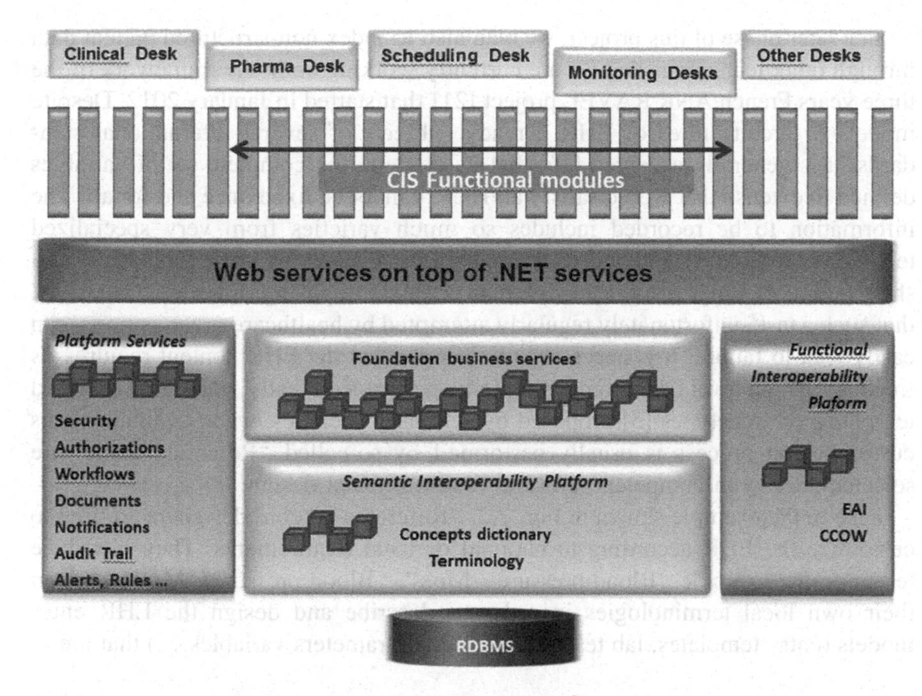

Fig. 7.1 Abstract layer of business services within DxCare® Clinical information System, where client application desks (like the EHR graphical user interface) call functional modules (like the order entry component) that use general services (like authorizations), so-called "foundation" business object services based on the RIM HL7 V3 (like observation object), semantic interoperability platform services (to be discussed below) and functional interoperability services to communicate with external systems

of the RIM has been made by the MEDASYS development team because of the need for functional interoperability with external systems requirements. The business objects have been mapped to the implementation database model so that they are loaded on demand from the database with the actual medical information related to the patient. The so-called "foundation" business layer allows looking for or retrieving the containers of medical information using the RIM normalized syntax (observations, acts, participations, roles, entities…), as shown in Fig. 7.1.

7.2.1.2 Definition of a Medical Information Indexing Engine

In order to set up an indexing mechanism within a clinical information system, we chose a different approach than most of on-going works, since our objectives of information semi-automatic processing within a CDSS and of information reuse in different contexts of the CIS requires an important level of reliability that is not yet provided by non-structured data. Therefore we decided to concentrate on structured information processing.

In a later phase of this project, we plan also to index non-structured patient data through other techniques that we are currently working on in the framework of the three years French ANR RAVEL project [21] that started in January 2012. Despite models or architectures of EHR, already subjects of several international standards, it is generally accepted [16] that its content, the complete set of variables defined in extenso that would satisfy all HCP, cannot be fixed once and for all. The information to be recorded includes so much varieties from very specialized techniques to context and social characteristics, evolves in great proportion in a short period of time, is used for different contexts imposing different constraints, that such a task, unfortunately regularly attempted by healthcare organizations, can easily lead to failure. It is necessary to consider that the EHR content requires its freedom, a high level of customization to be performed locally, guided by standard templates or archetypes [3] proposed by scholarly societies, when available. This customization process is usually performed by so-called "Référents", who are selected HCP with competences in information model design.

Like in the example shown in Fig. 7.2, "Referents" record designers are free to customize the EHR according to national or local requirements. They introduce terms (in the example "Blood pressure—Max", "Blood pressure—Min"...) from their own local terminologies in order to describe and design the EHR entry models (entry templates, lab test models, vital parameters variables....) that are to

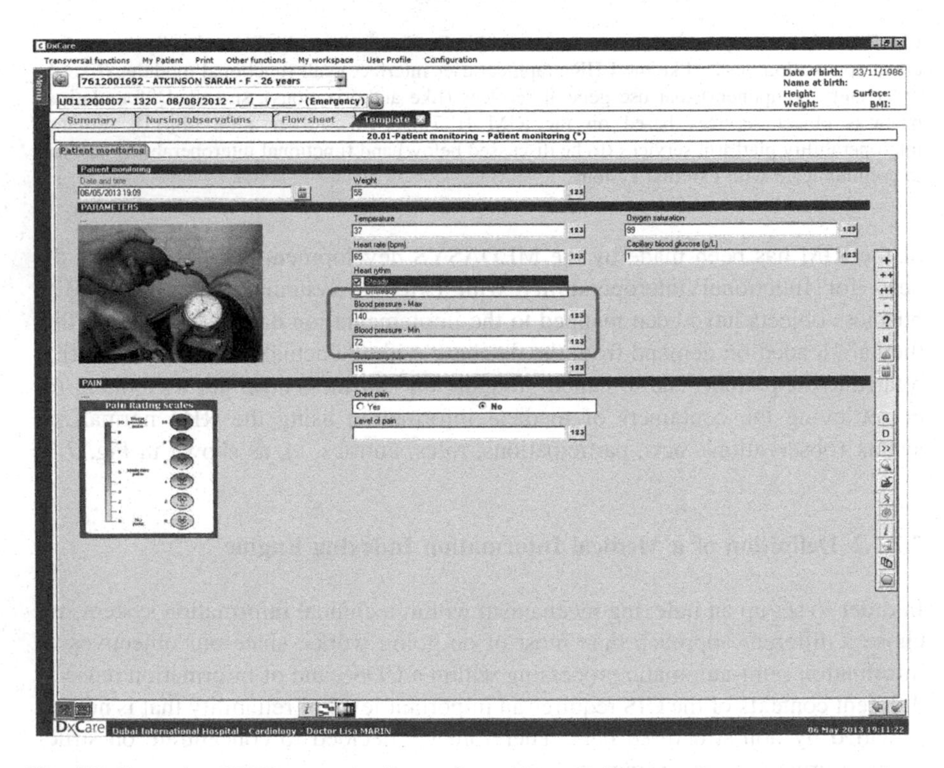

Fig. 7.2 Example of EHR customization performed by ≪Referents≫

be used within their organization or medical specialty. Several terms from these local terminologies may refer to the same medical or social concept (for example "Blood pressure—Max" and "systolic tension"). Some terms refer to concepts that are specializations of others. If this fact was not recognized by the system, information may be duplicated, entered several times, not found, and not reused. A solution to this problem has been proposed by Professor Degoulet who named it "Concepts dictionary" during the design of the "form management module" of the Georges Pompidou European Hospital systems in 1999 [5]. Later, we further refined these basic principles and extended the scope of the dictionary, so we now define a concept as a semantic normalization item which is able to identify and recognize a clinical information element and thereby allowing its reusability by other HCP, other components from the clinical information system and its exploitation by CDSS, medico-economics studies, epidemiological and clinical surveys, as described in Fig. 7.3.

This allows the dictionary of concepts to recognize and reuse the medical entries, thereby avoiding duplication of information entry.

Figure 7.4 summarizes the basic conceptual principles of our approach for semantic indexing.

Since there is no single way to structure an EHR (although archetypes when they are available may provide some guidance), the system allows record designers to define their own models of "observations" (the term "observation" here refers to the syntax used in the RIM HL7 V3), for example "cardiologic observation form", "blood count model", "intensive care vital signs model". Each individual model may be linked to a concept from the dictionary, which is itself bound to one or several standard terminological system. Concepts are organized by structural hierarchical relationships (like "is-a" or "part-of"), able to represent the granularity of the medical information. Currently these structural relationships have been generated from the code structure of the SNOMED 3.5 terminology

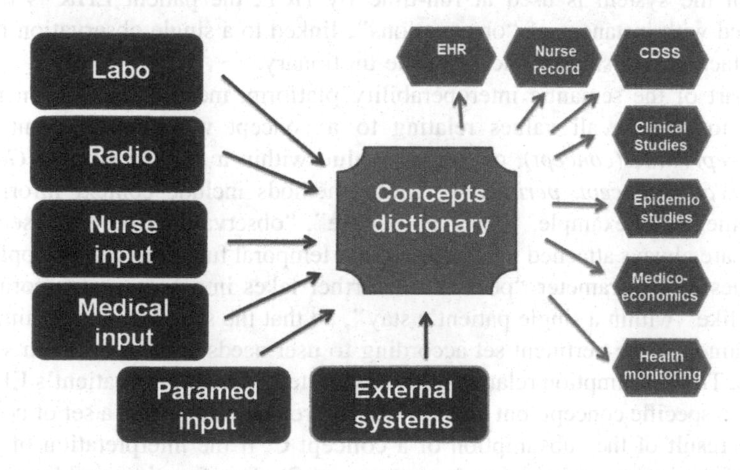

Fig. 7.3 The concepts dictionary as an indexing tool allowing information reuse

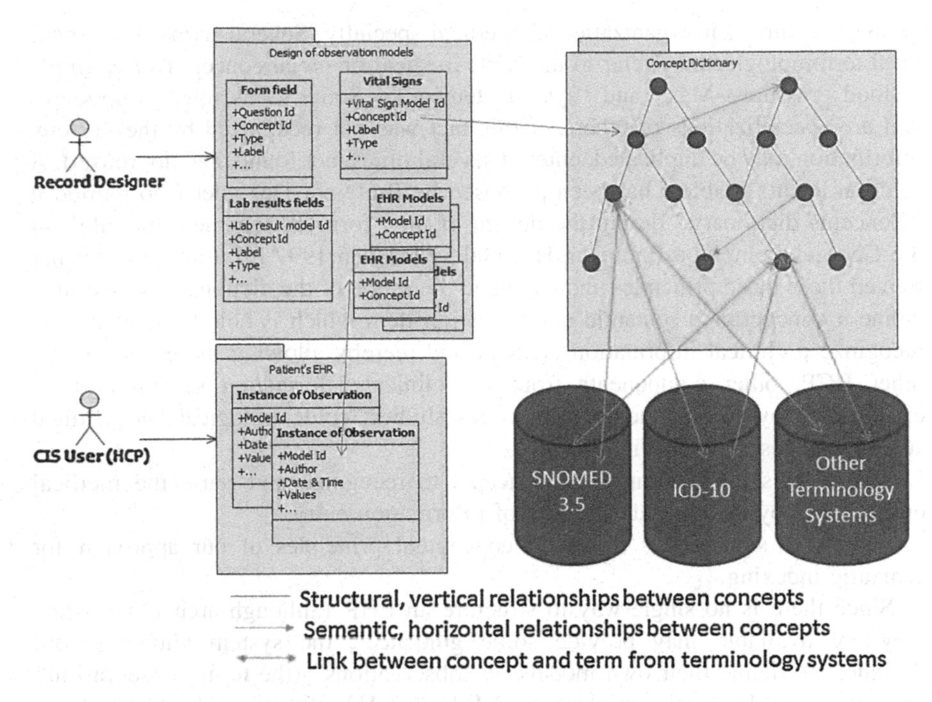

Fig. 7.4 Basic conceptual principles of the semantic interoperability platform

system that has been chosen for France. However, it is foreseen during the RAVEL project to maintain these relationships and to migrate them to reflect more formally designed ontological or terminological systems, such as SNOMED CT. Semantic horizontal relationships complete the definition of the concepts, by linking concepts within a single axis or between several axes.

When the system is used at run-time by HCP, the patient EHR is thereby populated with instances of "observations", linked to a single observation model, itself attached to a single concept of the dictionary.

As part of the semantic interoperability platform, methods have been implemented to retrieve all values relating to a concept within the patient EHR, *GetConceptValues(concept)*, or the last value within a defined context *GetConceptLastValue(concept, perimeter)*. These methods include context information management. For example, "observation date", "observation time", "observation author" are always attached to a value so that temporal functions can be applied on the values. The parameter "perimeter" further takes into account temporal constraints like "within a single patient's stay", so that the set of values obtained can be restrained to the pertinent set according to user needs such as decision support systems. The subsumption relationship is exploited to search the patient's EHR not only for a specific concept, but also for its children (subsumption: a set of concepts S is the result of the subsumption of a concept C, if the interpretation of C is a subset of the interpretation of each concepts of S). This function enables retrieval of relevant information despite different granularities of expression, if the information granularity is correctly expressed by the structural relationships between the concepts. The method *GetSubConceptValues(concept, perimeter)* searches the patient's EHR for the instances of observations indexed by a sub-concept of the selected concept matching the constraints defined in the definition of "perimeter". It is also possible to specify the unit (for example "kg", "g", "mg") in which the value to which the values should be converted so that the result can be processed by inference rules.

One innovation of the "concepts dictionary" is in the indexing method, which is performed on the models of information elements which harbor the interesting characteristic to be patient independent objects. This means that the indexing process is not performed, "a posteriori", after the patient's data has been generated, as it is usual for example with coding tools, but at the system customization step, when defining the entry models elements. For example when defining templates, any entry model (1) is semantically indexed by such a concept (2), as shown by Fig. 7.5.

Record designers, the "Referents", are helped when performing this semantic indexing process by a concept browser, allowing them to select the right concept from the "concepts dictionary", as in Fig. 7.6.

This semantic indexing method, done "a priori" at the EHR customization step, allows "Referents" to perform the indexing definition only once, and leave the actual HCP out of any semantic indexing burden. As a consequence, all information entries which are based on a model are automatically semantically indexed. This covers administrative, physician, nurse, paramedical structured entries as well as lab tests results. The same concept dictionary is delivered to each healthcare organization using DxCare® in order to enable semantic interoperability between healthcare organizations exchanging EHR extracts.

7.2.1.3 Binding of Concepts with Terminology Systems

The third part of the required semantic interoperability platform is the "binding" [17] of the elements of the dictionary of concepts with a standard international terminology system or with a domain ontology that provides the semantic by which one may look for or retrieve medical information. The concepts dictionary is able to index clinical information elements through so called "Concepts pivots" which

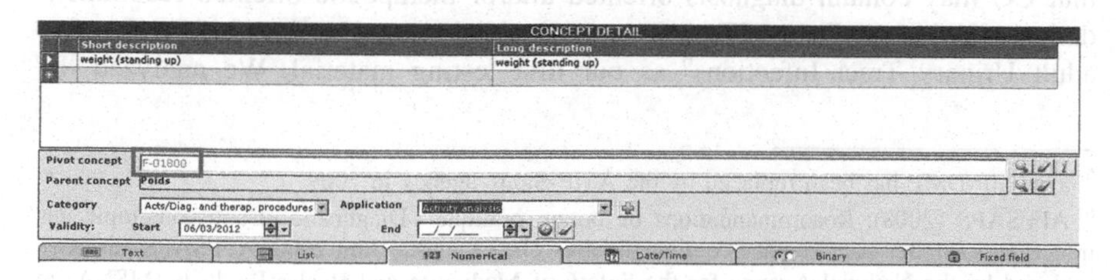

Fig. 7.5 Information element models are semantically indexed by a concept (here F-01800)

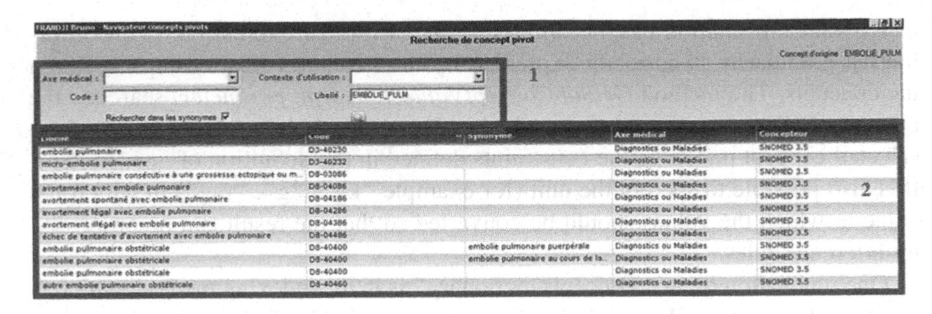

Fig. 7.6 Finding concepts to index information element models. The browser presents a list of concepts (2) corresponding to the users' selection criteria (1). It also permits to browse within a concept's surroundings to navigate to the concept's children, brothers, or through semantic relationships between concepts

themselves may be linked to terminological systems. In France, the GIP-DMP[5] has imposed the usage of the SNOMED 3.5 VF terminological system as the language to describe the contents of the national shared medical records ("Dossier Médical Partagé—DMP"). However the implementation of SNOMED within a clinical system requires prior steps like the RIM-based business layer and the concept dictionary to have already been achieved. Therefore, besides DxCare® customers, no other French healthcare organizations yet claims to exploit the SNOMED 3.5 terminology system integrated within their clinical information system. The concepts dictionary has been populated, using Oracle® database scripts, with the terms, codes, synonyms and semantic relationships contained within this terminological system. "Concept pivots" have been created and "bonded" to the SNOMED terms and codes. Structural relationships and semantic relationships contained within the SNOMED 3.5 files have been mapped between "Concept pivots".

7.2.2 CG Modeling Method

To process CG through a decision support system, it is necessary to capture the comprehensive set of information contained in the textual form through a computerizable model. Our work on the CG modeling method started by recognizing that CG may contain diagnosis oriented and/or therapeutic oriented recommendations. Within the Debug-IT project, we decided to use the French CG related to adult Urinary Tract Infections[6] as our first testing material. We analyzed the

---

[5] The GIP-DMP has been replaced by the ASIP-Santé agency in 2009.

[6] AFFSAPS (2008): Recommandations de bonne pratique—Diagnostic et antibiothérapie des infections urinaires bactériennes communautaires chez l'adulte, Juin 2008. AFFSAPS has been replaced by the National Agency for the Safety of Medicines and Health Products (MSNA) in 2012.

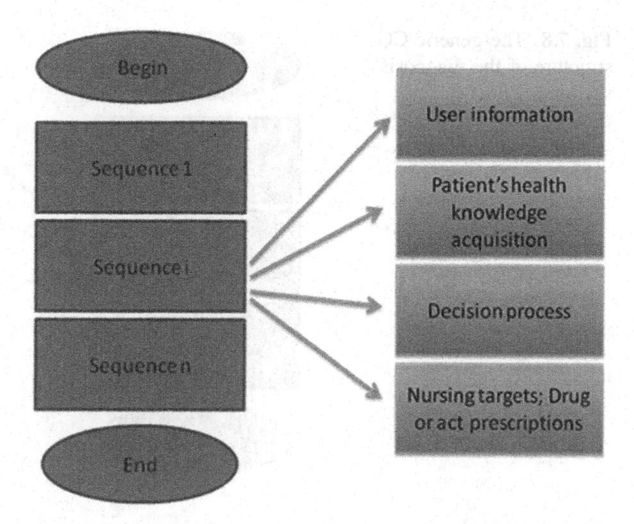

Fig. 7.7 The generic CG structure of the therapeutic section

guideline's therapeutic section and designed decision trees, starting from our experience during the PRESTIGE project, the GLIF formalism and the PRO*forma* "*Plans*, *Decisions*, *Actions* and *Enquiries*". The decision and action steps were composed of about three hundred rules, that we could distribute over various tree branches, using Microsoft® Excel®. To validate these decision trees, partners from the project implemented the rules using several methods like Bayesian networks or Rule Based Fuzzy Cognitive Maps [13]. In addition the CG contains much information for the HCP user to read, such as the level of proof or general information about possible etiologies. We extended the GLIF structure by adding the acquisition of knowledge about the patient's health (such as enquiries in Pro*forma*), which is an implicit step within the CG but is required explicitly by our targeted CDSS. By comparing these validated decision trees to the clinical information system component structure, we could identify a generic pattern adapted for our objectives to exploit the guidelines within a CDSS immerged within the clinical information system: the therapeutic section of the guideline may be seen as a set of sequences, each of them including at least one of four phases,—informing the user,—acquiring knowledge about the patient's health,—reasoning through a decision process,—in order to issue drug or act prescriptions recommendations, as shown in Fig. 7.7. Through the study of other CG addressing different topics (like "Breast Cancer"), we slightly refined the model since nurse-oriented clinical pathways often present a section called "expected outcomes". This important characteristic is dealt with through the positioning of "nursing targets" in addition to drug or act prescriptions:

We proceeded with the same method for the diagnosis section and found that it could be satisfied by a very similar structure, i.e. a set of sequences, each of them including at least one of four phases,—informing the user,—acquiring knowledge about the patient's health,—reasoning through a decision process—in order to issue diagnosis oriented act requests recommendations, as shown in Fig. 7.8.

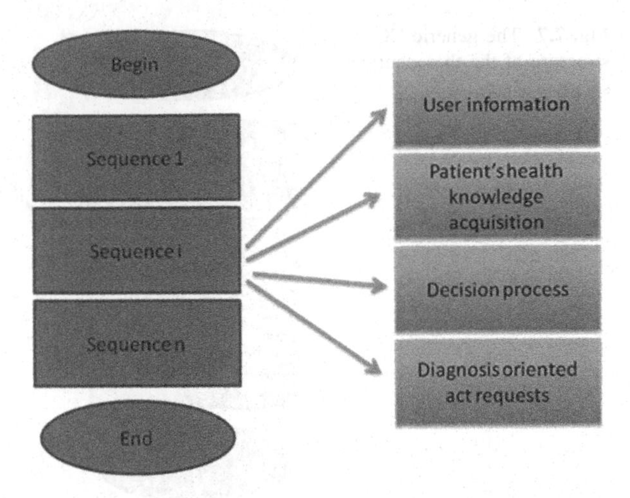

Fig. 7.8 The generic CG structure of the diagnosis section

Looking for a structure that could satisfy both diagnosis and therapeutic sections, we used the clinical information system component "instructions" that define the need for any act or drug request or prescription, or the setting up of nursing targets. We thereby obtained the final structure, a set of sequences, each of them including at least one of four phases, informing the user, acquiring knowledge about the patient's health, reasoning through a decision process in order to generate instructions of diagnosis oriented act requests recommendations or therapeutic prescriptions. This structure is summarized in Fig. 7.9.

We compared this structure with the other CG described in the material section, and found that the model could support the representation of any of their elements.

For example, the various elements from the ONCORIF's breast cancer CG could be mapped to the structure, as shown in Fig. 7.10.

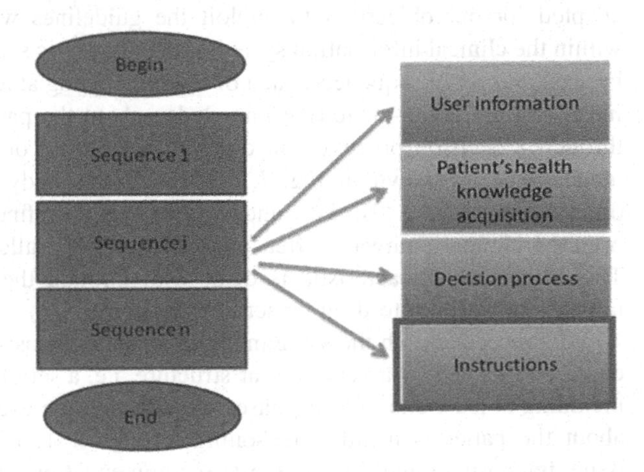

Fig. 7.9 A generic CG structure adapted for an immersive CDSS within the clinical information system

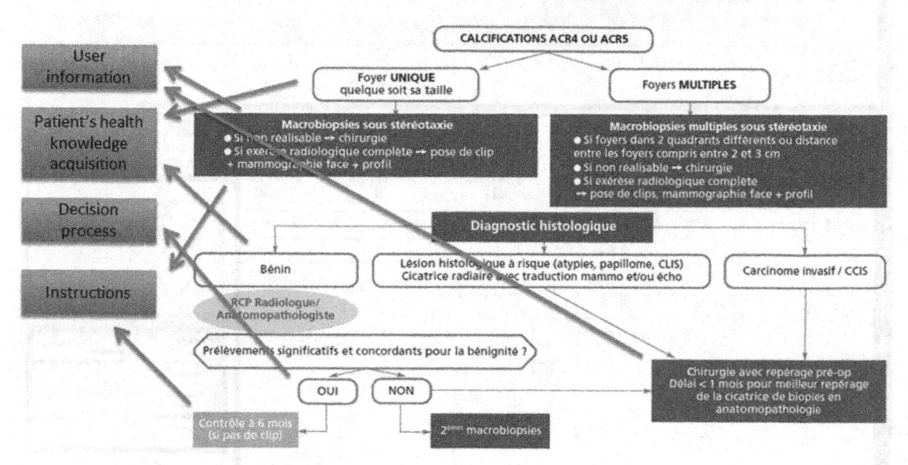

Fig. 7.10 Example of mapping of CG to the generic structure

7.2.3 Immersive CDSS Within the Clinical Information System

We knew at this point that we needed to define a CDSS able to run decision trees and to pilot the four processes of informing the user, acquiring knowledge about the patient's health through our semantic platform, processing decisions, and proposing instructions generating prescriptions and care plan, activity follow-up. We integrated the Microsoft Workflow Foundation (WF) ® with the Medasys DxCare® clinical information system elements piloting those four processes. Since users want to customise the CG according to their local needs and population settings, the CDSS has to be independent of any CG, supporting an authoring environment allowing "Referents" to author themselves their CG within the CDSS. Figure 7.11 shows the CDSS authoring environment, mixing WF® and DxCare® components.

The Microsoft WF® engine runs the decision trees, presents the informative elements included within the CG to the user. We developed.net methods to query the semantic interoperability platform from WF, and were able to find the occurrences of a concept within the patient's EHR, and others to manage hypotheses, conclusions and pathways. The usual user computer interaction components were exploited in order to allow the user to verify and complete inconsistent or incomplete information retrieved from the EHR. The decision process was implemented using the WF® native rule engine and we used the "Instruction" component from DxCare® to propose the generation of "drug order entries" "nurse targets" or "act requests" that are natively processed by the clinical information system.

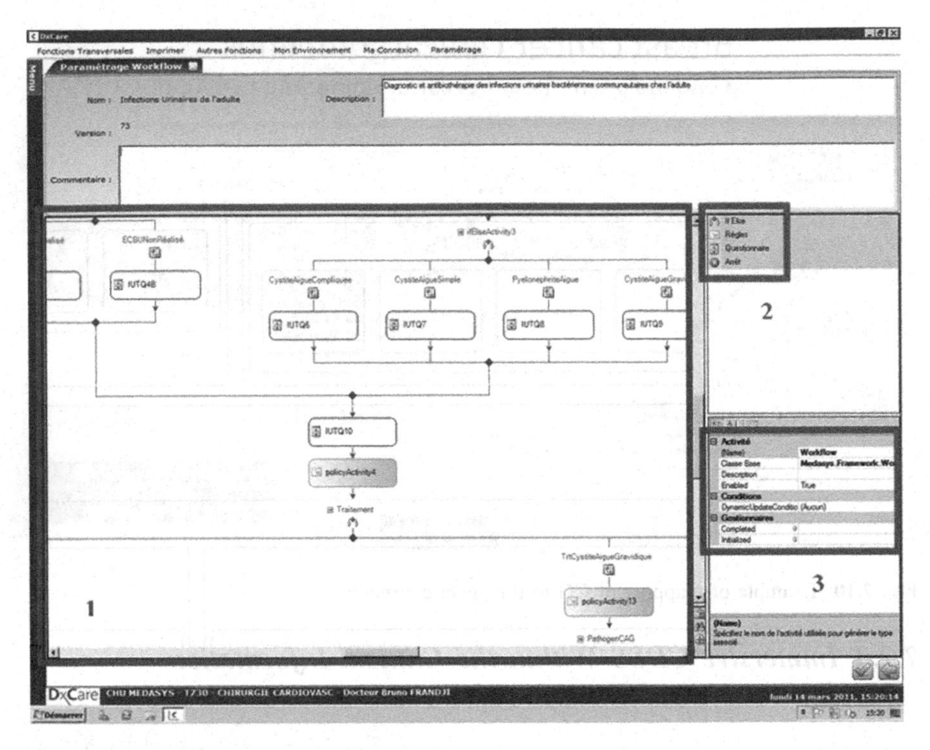

Fig. 7.11 Microsoft Worflow Foundation® piloting DxCare® clinical information system processes. User can author decision trees (1) by dragging CIS entry models, rule sets, or paths managers (2) and dropping them in Sect. 7.1, than entering the rules definition using a dedicated panel (3)

7.3 Results

Within the Debug-IT project, we formalized the French CG related to adults Urinary Tract Infections, first as Excel® based decision graphs composed of rules that partners could validate, then as sequences of user information, user machine interaction through questions-responses, user decisions and actions capture, rule sets, paths management, involving templates-based questioning and automatic generation of medical instructions. This model has been found adequate for the other CG that we have tested, like the breast cancer Oncorif's CG shown in Fig. 7.12:

The explanation of the comprehensive coverage of the CG expressiveness lies:

1. in the expressiveness of the Microsoft Workflow foundation® rule structures, able to process any.net® function. This is particularly useful when the CG describes rules involving temporal constraints, in addition to the filters already described in the semantic interoperability platform, since it makes it possible to compare date and time between information retrieved from the semantic interoperability platform and any referential time such as an event.

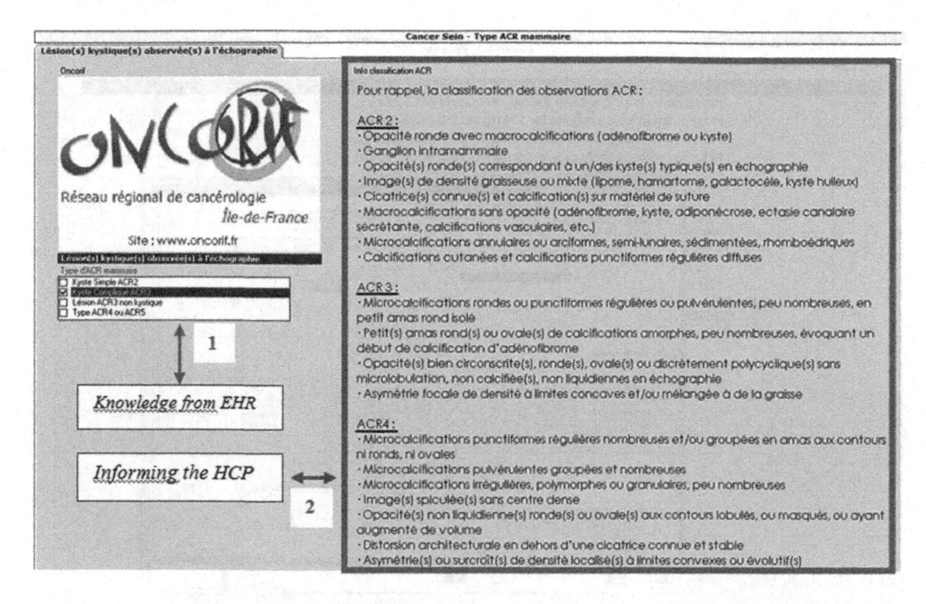

Fig. 7.12 Running the Breast cancer Oncorif's CG with the CDSS, a node within the decision tree performing "acquiring knowledge from the EHR" (1) and "informing the HCP" (2)

2. In the HCP controlled approach to present information obtained through the semantic interoperability platform, so that HCP are able to complete missing information or to correct inconsistent data (see Fig. 7.13),
3. in the "user Information" step that presents to the user any text from the CG that is appropriate in the user context,
4. in the ability of the CIS to represent any action through "instructions" of various types, each being taken charge of by one component of the CIS, able to express any semantic action, since it does this in daily routine. For example, an instruction for a drug prescription is taken charge of by the CPOE

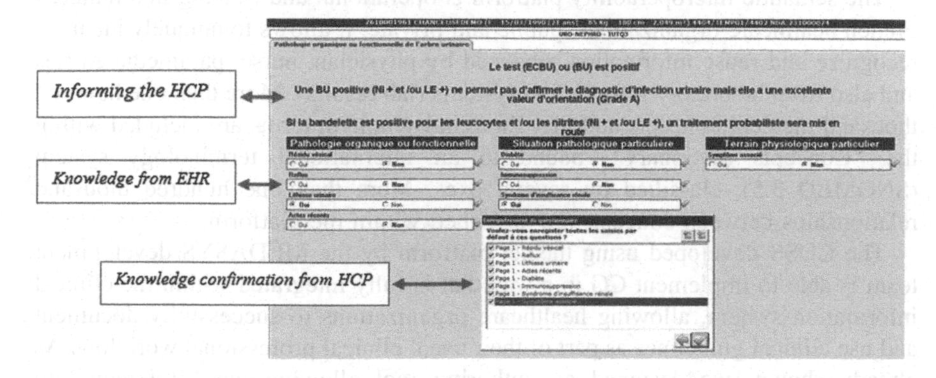

Fig. 7.13 Example of CDSS Human Computer Interaction allowing the HCP to confirm the knowledge retrieved from the patient's EHR

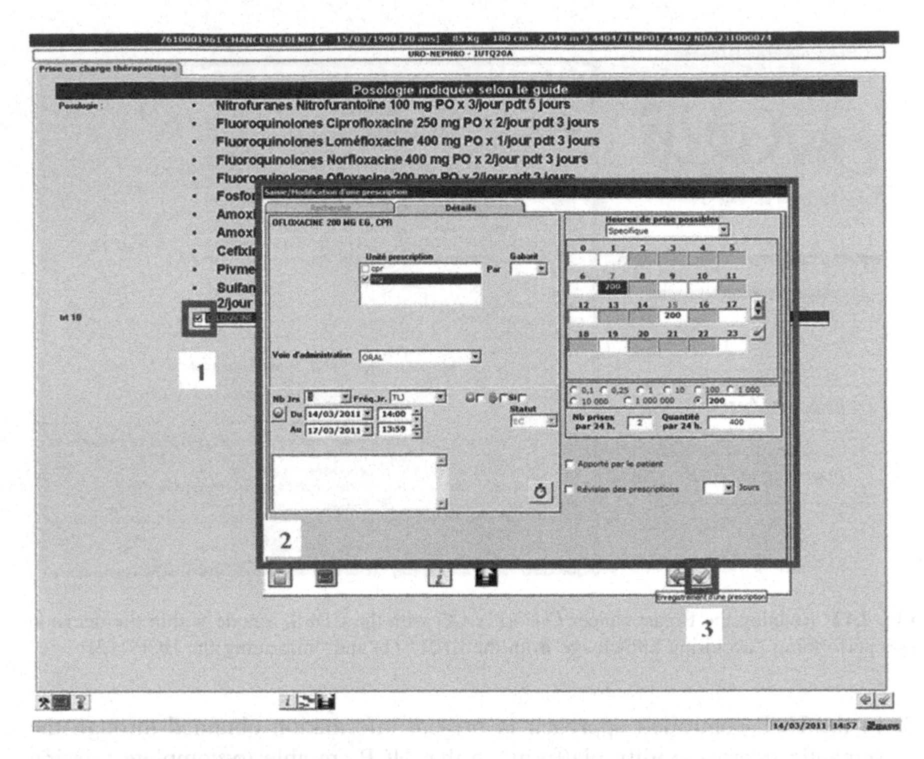

Fig. 7.14 Proposed pre-configured order entries using the clinical information system component. The click on the instruction (1) triggers the pre-configured order entry (2). The user just has to click on the validation button (3) to generate the prescription

(Clinical Prescription Order Entry) component from the CIS, that manages the expression of the temporal constraints related to each of the drug administrations, as shown in Fig. 7.14.

The semantic interoperability platform is operational and running in a hundred French healthcare organizations, public and private. It allows to uniquely identify, recognize and reuse information produced by physician, nurse, paramedic entries and also from laboratory information systems (lab results). More than one hundred thousand medical concepts and fifty thousand synonym terms are included within the "Concepts dictionary", bound to an international terminology system (SNOMED 3.5), classified by several axes. More than one hundred thousand relationships between concepts are described within the platform.

The CDSS developed using the.net platform by the MEDASYS development team is able to implement CG in a way that is fully integrated within the clinical information system, allowing healthcare organizations to successfully document and use clinical guidelines as part of their usual clinical professional workflow. As already shown, we designed an authoring tool allowing the "Referents" to implement themselves CG using the CDSS.

The CDSS performs the first processes: informing the user, acquiring the knowledge about the patient's health obtained through the semantic interoperability platform, allowing the user to complete or correct inconsistent or incomplete data, through the usual entry information components of the clinical information system as shown in Fig. 7.13.

The CDSS use WF® both as a workflow processor and as a rule inference engine to navigate within the decision graphs and draw hypotheses and conclusions. When it offers proposals to prescribe a drug or to carry out an act, the process manipulates the usual order entries component of the clinical information system. Orders are already pre-configured so that the user may only have to validate the form to generate the entry, as shown in Fig. 7.14.

The same mechanism applies for the positioning of nurse targets.

The information generated through the decision tree is recorded within the patient's EHR (Fig. 7.15).

As the orders entered by the CDSS are part of the clinical information system, they are immediately available to any HCP working on the patient's health (physician, nurses, paramedics, ancillary services, imagery, lab…). Figure 7.16 presents a graphical user interface component supporting the nurse activities, alimented from the CDSS with the drug administration to be done. The nurse just has to administrate the drug and mark the activity as done:

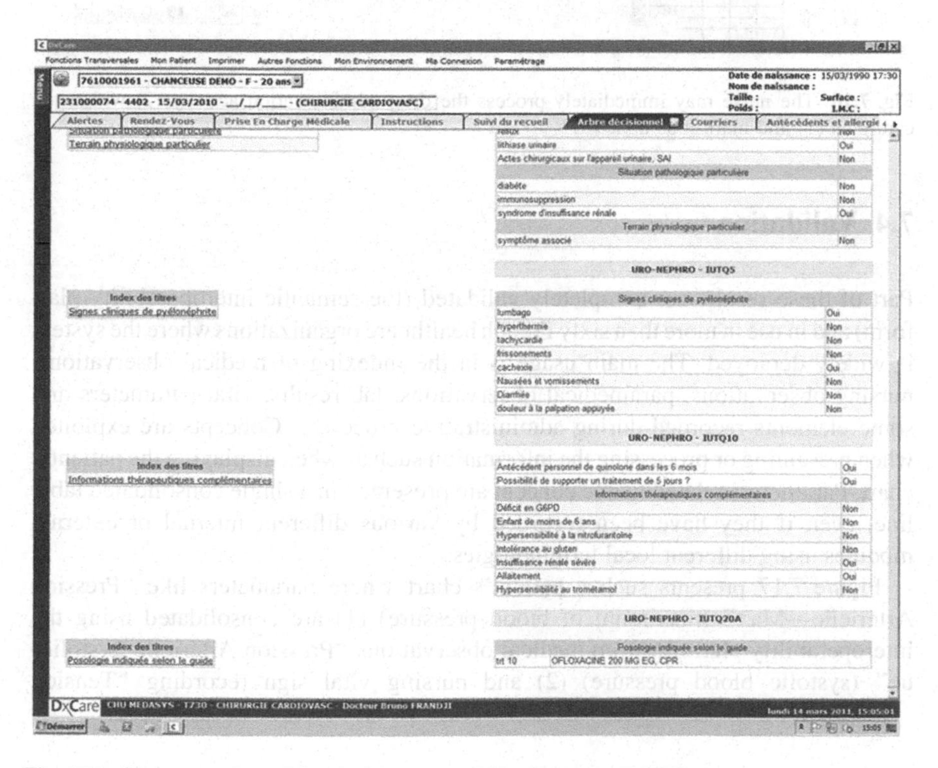

Fig. 7.15 Documentation of the decision tree within the patient's EHR

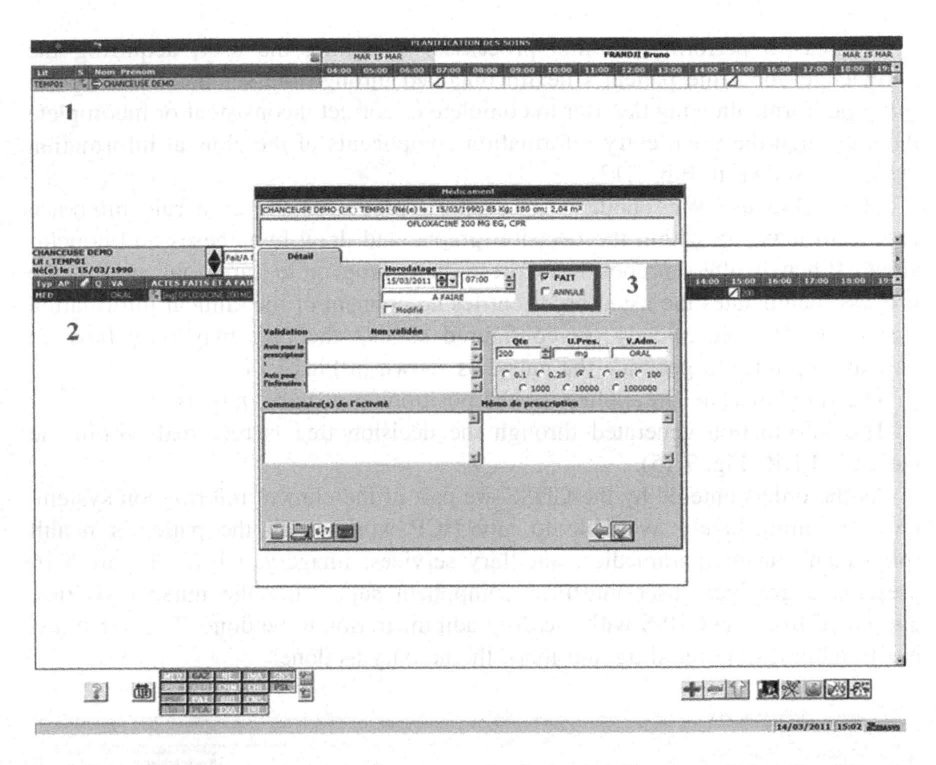

Fig. 7.16 The nurse may immediately process the drug administration activity (2) within the care plan (1) and mark it as done (3)

7.4 Validation

Part of these results are completely validated (the semantic interoperability platform) and in use in more than sixty French healthcare organizations where the system is widely deployed. The main usage is in the indexing of medical observations, nursing observations, paramedical observations, lab results, vital parameters and some elements recorded during administrative processes. Concepts are exploited when presenting or processing the information such as when displaying the patient's chart. Parameters related to one concept are presented in a single consolidated table line, even if they have been recorded by various different internal or external modules using different local terminologies.

Figure 7.17 presents such a patient's chart where parameters like "Pression Artérielle—Max" (maximum of blood pressure) (1) are consolidated using the interoperability platform from medical observations "Pression Artérielle Systolique" (systolic blood pressure) (2) and nursing vital sign recording "Tension

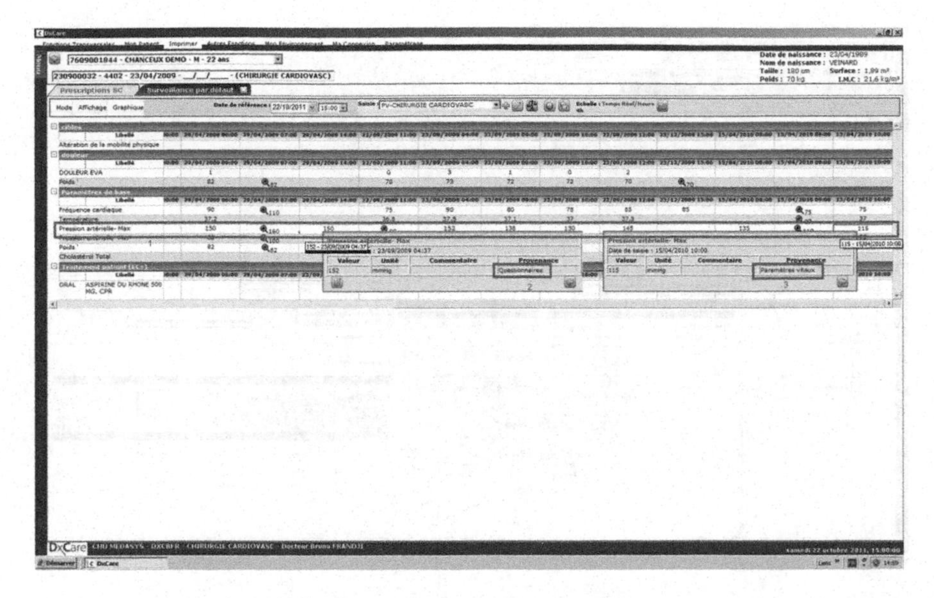

Fig. 7.17 Consolidated patient's chart using the interoperability platform

Artérielle Max" (maximum of arterial tension) (3). Within some healthcare organisations, parameters like "Fréquence cardiaque" (Cardiac frequency) are consolidated using the interoperability platform from data taken during medical observations, nursing vital sign recording and medical devices connected to the information system. The interoperability platform also provides the source of the information to the user in order to assess its reliability.

The interoperability platform is also of paramount importance in the generation of customised views over the EHR. Again, the information is consolidated by the interoperability platform and linked to concepts before being presented in tables. It is also possible to exploit the hierarchical relationships between SNOMED concepts to generate dynamically created views over the patient's EHR, as in Fig. 7.18 when asking the displays of "Acides gras" (fatty acides) sub-concepts values. The interoperability platform retrieves all results of the subsumption function (e.g. "triglycerides", "total cholesterol" "HDL-C" and "LDL-C") which are presented in a dynamically consolidated table (Fig. 7.18).

The semi-formal structure for CG adapted for CDSS and the CDSS supporting CG's authoring, integrated within the CIS, retrieving patient's health knowledge from the EHR have successfully passed all the experimentation tests with all the CG described in the material section. The CDSS is operational and delivered to the healthcare organizations as part of the DxCare® CIS. Several healthcare organizations have requested to exploit the CDSS and are currently setting up projects to deploy the system.

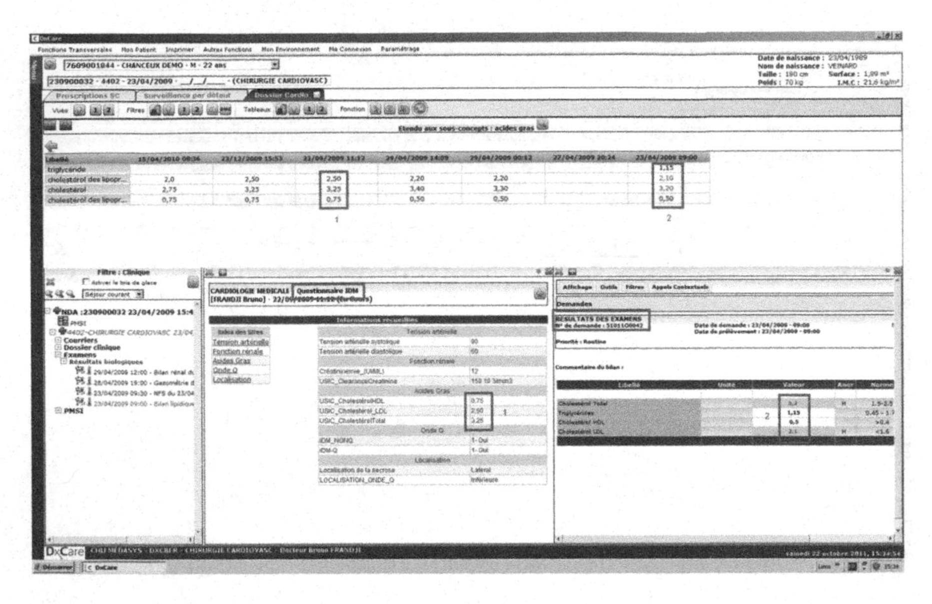

Fig. 7.18 Exploiting the interoperability platform on an EHR with the concept "Acides gras" (fatty acids). The interoperability platform knows enough of what a fatty acid is to be able to retrieve information about cholesterol in a questionnaire (1) and about triglyceride and cholesterol from a lab test (2) and presents it in a consolidated table

7.5 Discussion

The CDSS obtained should be able to provide real impact for CG usage. Thanks to the CG model designed, all the CG described in the material section could be implemented through an authoring process independent from system development.

In the context of CG's implementation, the semantic interoperability platform allows the CDSS to query the patient's EHR and to document the information generated through the decision tree navigation within the EHR. However, queries are currently only possible that do not require temporal queries involving automatic detection of repeated patterns (example of such unsupported queries in the CDSS: detecting consecutive episodes of simultaneous increase of a set of vital parameters. In such cases the value set would be presented to the HCP who would have to process the detection himself, even if other functions from the CIS might support such recognition). This restriction is also an axis of research within the RAVEL project. Another limitation stems by the nature of the information currently processed by the semantic interoperability platform. It is restricted to structured information, free text documents or images are not yet semantically indexed by the system.

From a wider perspective, the semantic interoperability platform permits the recognition and the reuse of the clinical information describing the patient's health. Moreover, its semantic content can be exploited not only by humans, but also by systems like CDSS. The RIM-based business abstract layer, the dictionary of concepts and the binding mechanism to an international standard terminology system, bring important components and technologies for meeting the objective of sharing information with other systems using different conceptual models, according to the principles defined by the ISO 13606,[7] the international standard governing EHR extracts communication between systems.

The CDSS uses native clinical information system components within a workflow and processes information with a rule inference engine allowing to navigate within the CG decision trees, so that the usage of the CDSS is fully integrated within the clinical information system. We have already identified several possibilities to increase the system abilities, through additional semantic indexing of information elements, improving the "Concepts dictionary" with more precise relationships from published concepts systems, binding mechanisms with other terminology systems, such as LOINC for biology and MedDRA for adverse effects. HCP do not have to process the semantic indexing. This is done by the system without requiring their involvement, so they do not have to learn about SNOMED or other terminology systems. This approach obviously reduces the risks of mistakes and erroneous inputs. Post coordinated concepts, e.g. concepts obtained as combinations of individual concepts, are added to the "concepts dictionary" when they are needed through an organizational procedure. They also can be recorded as hypotheses or conclusions within the decision process. For now, the semantic indexing is only possible with structured information. We plan further work, as part of the RAVEL project, to try to process also natural language input, texts and images. The rules defined when implementing CG manipulate codes issued from international standard terminology systems, so a certain level of portability between clinical information systems is ensured. However, it would require other CDSS to understand Microsoft.net® commands and to be able to implement decision trees in a similar way.

DxCare® being used by many health organizations, we hope that many CG will soon be implemented and deployed within the French healthcare system. The impact leverage is mainly constrained by the "Referents" control over the CG design process: although the authoring of guidelines within the CDSS is not complex, it requires.net® development capabilities that are not always available at the customer sites. But this issue can easily be overcome since it concerns system customization and not its usage by HCP. We expect that the CG coverage will be undertaken according to the user demand, the most urgently needed CG being authored first. The domain knowledge is evolving fast, and so CG maintenance requires a version management and deployment process. The CDSS' authoring tool should help the referents to maintain CG, through its user-friendly human computer interface and features like the management of various CG versions, with validity periods. An alternative that we also contemplate based on our experience

[7] ISO 13606 (2010) Health Informatics—Electronic health record communication (part 1 to 5), International Organisation for Standardization, www/iso/org, 2010.

in Prestige, is to design an interface adapted to our model from GLIF and/or from the ASTM E2210-12[8] standard specification for Guideline Elements Model, that would allow CG editing groups to deliver and maintain DCSS supported CG and to propose them to DxCare® users.

References

1. Abidi, S.R., Abidi, S.S., Hussain, S., Shepherd, M.: Ontology-based modeling of clinical practice guidelines: A clinical decision support system for breast cancer follow-up interventions at primary care settings. Stud. Health Technol. Inform. **129**, 845–849 (2007)
2. Åhlfeldt, H., Karlsson, D., Petersson, H., Chen, R., Nyström, M., Sundvall, E.: Advancement in the standardisation of the EHR 5th Scandinavian Conference on Health Informatics 2007
3. Beale, T.: Archetypes and the EHR. Stud. Health Technol. Inform. **96**, 238–244 (2003)
4. Damiani, G., Pinnarelli, L., Colosimo, S.C., Almiento, R., Sicuro, L., Galasso, R., Lorenzo Sommella, L., Ricciardi, W.: The effectiveness of computerized clinical guidelines in the process of care: A systematic review. BMC Health Serv. Res. **2010**(10), 2 (2010)
5. Degoulet, P., Marin, L., Lavril, M., Le Bozec, C., Delbecke, E., Meaux, J.J., Rose, L.: The HEGP component-based clinical information system. Int. J. Med. Inf. **69**(2–3), 115–126 (2003)
6. Deleger, L., Grouin, C., Zweigenbaum, P.: Extracting medical information from narrative patient records: the case of medication-related information. J. Am. Med. Inform. Assoc. **2010**(17), 555–558 (2010)
7. Isern, D., Moreno, A.: Computer-based execution of clinical guidelines: a review. Int. J. Med. Inform. **77**(12), 787–808 (2008)
8. Kirkpatrick, D., Burkman, R.T.: Does standardization of care through clinical guidelines improve outcomes and reduce medical liability? Obstet. Gynecol. **116**(5), 1022–1026 (2010)
9. Kulkarni, R.P., Ituarte, P.H., Gunderson, D., Yeh, M.W.: Clinical pathways improve hospital resource use in endocrine surgery. J Am Coll Surg. **212**(1), 35–41 (2011). Epub 2010 Nov 30
10. Latoszek-Berendsen, A., Tange, H., van den Herik, H.J., Hasman, A.: From clinical practice guidelines to computer-interpretable guidelines. A litterature overview. Methods Inf Med. **49**(6), 550–70 (2010)
11. Locatelli, F., Andrulli, S., Del Vecchio, L.: Difficulties of implementing clinical guidelines in medical practice. European Renal Association-European Dialysis and Transplant Association Editorial Comments. Nephrol. Dial. Transplant. **2000**(15), 1284–1287 (2000)
12. Lovis, C., Douglas, T., Pasche, E., Ruch, P., Colaert, D., Stroetmann, K.: DebugIT: Building a European distributed clinical data mining network to foster the fight against microbial diseases. Stud. Health Technol. Inform. **2009**(148), 50–59 (2009)
13. Papageorgiou, E., Stylios, C., Groumpos, P.: Novel architecture for supporting medical decision making of different data types based on fuzzy cognitive map framework. Conf. Proc. IEEE Eng. Med. Biol. Soc. **2007**, 1192–1195 (2007)
14. Peleg, M., Keren, S., Denekamp, Y.: Mapping computerized clinical guidelines to electronic medical records: Knowledge-data ontological mapper (KDOM). J. Biomed. Inform. **41**, 180–201 (2008)

[8] International A. ASTM Standard E2210 (2012)—12 Standard Specification for Guideline Elements Model version 3 (GEM III)-Document Model for Clinical Practice Guidelines. West Conshohocken, PA; 2012.

15. Prior, M., Guerin, M., Grimmer-Somers, K.: The effectiveness of clinical guideline implementation strategies–a synthesis of systematic review findings. J Eval Clin Pract. **14**(5), 888–97 (2008)
16. Rector, A.: Clinical terminology: Why is it so hard? Methods Inf. Med. **38**(4–5), 239–252 (1999). doi:10.1267/METH99040239. PMID 10805008
17. Rector, A.L., Qamar, R., Marley, T.: Binding ontologies and coding systems to electronic health records and messages. Appl. Ontol. (online) **4**(1), 51–69 (2009). Available at http://www.cs.manchester.ac.uk/~rector/papers/krmed2006-rector-binding-ontologies-to-ehrs.pdf
18. Schriger, D.L., Baraff, L.J., Buller, K., Shendrikar, M.A., Nagda, S., Lin, E.J., Mikulich, V.J., Cretin, S.: Implementation of clinical guidelines via a computer charting system: effect on the care of febrile children less than three years of age. J. Am. Med. Inform. Assoc. **7**(2), 186–95 (2000)
19. Sonnenberg, F.A., Hagerty, C.G.: Computer interpretable clinical practice guidelines—Where we are and where we are going? AMIA Yearbook **2006**(11), 45–158 (2006)
20. Stroetman,V., Kalra, D., Lewalle, P., Rector, A., Rodrigues, J., Stroetman, K., Surjan, G., Ustun, B., Virtanen, M., Zanstra, P.: Semantic interoperability for better health and safer healthcare. The European Commission. (2009). ISBN-13: 978-92-79-11139-6. DOI: 10.2759/38514. Available from: http://ec.europa.eu/information_society/activities/health/docs/publications/2009/2009semantic-health-report.pdf
21. Thiessard, F., Mougin, F., Diallo, G., Jouhet, V., Cossin, S., Garcelon, N., Campillo, B., Jouini, W., Grosjean, J., Massari, P., Griffon, N., Dupuch, M., Tayalati, F., Dugas, E., Balvet, A., Grabar, N., Pereira, S., Frandji, B., Darmoni, S., Cuggia, M.: RAVEL: Retrieval and visualization in electronic health records. Stud. Health Technol. Inform. **2012**(180), 194–198 (2012)
22. Wagholikar, K.B., MacLaughlin, K.L., Henry, M.R., Greenes, R.A., Hankey, R.A., Liu, H., Chaudhry, R.: Clinical decision support with automated text processing for cervical cancer screening. J. Am. Med. Inform. Assoc. **2012**(19), 833–839 (2012)

Chapter 8
Applying Semantics and Data Interoperability Principles for Cloud Management Systems

J. Martín Serrano-Orozco

Abstract In today's IT systems, the enormous amount of data and the increasing demand to produce knowledge (by aggregating data), generate the necessity of re-thinking if current information systems are optimally designed for managing such amount of information. If the answer is no, then what are the best practices to make more efficient the managing operations of data in those knowledge-based systems? As consequence of the generation of big quantities of data, it is becoming more difficult to design efficient management information systems. Likewise migrating to cloud-based systems the design of the systems and their respectively applications are more complex. This book chapter focuses on semantic interoperability, a particular ontology-engineering problem, which relates semantics and management systems. This chapter book aims at defining application design principles and methodological modelling procedures to create alternative solutions to the scientific and technological challenge of enabling information interoperability in cross-domain applications and management systems. Particular interest is focused on cloud management systems. This book chapter describes and explains the role semantics plays to provide solutions tackling the problem about information interoperability, discuss concepts about ontology engineering and introduce methodological approaches to formal representation of data and information models, in order to facilitate information interoperability between heterogeneous, complex and distributed information management systems.

J. M. Serrano-Orozco (✉)
National University of Ireland Galway—NUI Galway, Digital Enterprise Research Institute—DERI-CSET, Galway, Ireland
e-mail: martin.serrano@deri.org

C. Faucher and L. C. Jain (eds.), *Innovations in Intelligent Machines-4*, Studies in Computational Intelligence 514, DOI: 10.1007/978-3-319-01866-9_8,

8.1 Introduction

In the ICTs sector where metadata standards are proliferating at unprecedented levels, and automated information management systems need to collect and process information from, every day, sensors, devices, applications, systems, etc., the interoperability of the information and the knowledge exchange emerge as the major challenges. Even more, this continuously trend increases accordingly to the everyday new developing demands for smarter services and cloud applications.

A consequence of this diversity of metadata, efficient data collection and optimal processing mechanisims have to be implemented, thus service providers, operators and infrastructure managers have to operate with many different standards and vocabularies. The use of semantic techniques to face up this requirement seems as an optimum approach providing advantages in terms of facilitating data and information mapping. However those semantic techniques are becoming very costly, in terms of computing resources and time-response factors, resulting in lost of business opportunities because existing systems are not adaptable to those scalability demands.

To alleviate interoperability problems in ICT systems, the convergence of software solutions and managing systems controlling the networking infrastructures is seen as a viable alternative to generate solutions in short incubation periods and with high scalability demands. Current ICT research is focused on the integrated management of resources, networks, systems and cloud services and the way in how to exchange data standards facilitating this information interoperability and systems integration labor. These challenges in terms of realistic services and applications can be generalized as looking forward to providing seamless mobility services to, for example, personalize services automatically.

On the other hand, if this cross-domain and information sharing problem is translated into the communications domain, a common scenario where computer network infrastructures collect data with the objective of supporting diverse communication services and applications, is observed. The problem about interoperability of data and information exchange is evident, and it acquires more importance with the IT sector increasing demands about convergence and data exchange within heterogeneous communication systems. Interoperability problem is increased exponentially, at recent days, when most data applications are migrating into cloud systems, emphasizing the need of proper management techniques to facilitate information exchange.

This book chapter follow two main objectives (a) analyses the basis of semantic interoperability alike introduces concepts and illustrates the way to understand how to enable interoperability of the information using a methodological approach to formalize and represent data by using information models (knowledge engineering) and (b) offers guidance and good practices when semantics is applied in cloud-based services and information management systems.

This chapter focuses on linked data and information interoperability in cloud management systems. The extensible, reusable, common and manageable information linked data plane is critical for this deployment.

This linked data approach relies on the fact that high-level infrastructure representations do not use resources when they are not being required to support or deploy services. The optimisation of resources using this approach relies on classification and identification, by semantic descriptions in a knowledge-based fashion way, about what resources need to be used, thus enabling services to be executed and deployed by result of information processing.

Organization of this chapter is as follows: Sect. 8.2 presents the current state of the art and motivations for using linked data in terms of technology convergence and interoperability. Section 8.3 presents challenges for a linked data and interoperability in Internet systems where information exchange occurs to support composed Internet service creation and delivery. Section 8.4 introduces the data link approach in a form of meta-ontologies for facilitating information interoperability and acting as demonstrator supporting the information interoperability approach. Section 8.5 examines challenges and limitations on cloud computing systems in the form of system requirements. Section 8.6 describes scalable features about management of linked data and its benefits when used in cloud applications. Section 8.7 presents the summary and outlook and finally some relevant references used in this chapter are listed.

8.2 Technology Convergence and Interoperability

Convergence between services, applications and Internet technologies for communications networks has been a clear trend in the Information and Communications Technology (ICT) domain in the past few years. The semantic web has played a crucial role as enabler of many of the applications and services in this convergence, although widely discussed, in terms of implementation, the progress has not fully run in parallel, due to many complex issues involving non-interoperable aspects where social, economic and political dimensions take place.

Technology convergence in the Future Internet (FI) is currently seen as an opportunity to improve the network infrastructure addressing service-oriented, social trends and economic commitments. So challenges in the future communications systems mainly Internet-based systems, demand, in terms of end user requirements, personalized provisioning, service-oriented performance, and service-awareness networking. To support these demands, information interoperability and data model integration are crucial.

A visionary perspective for what the Future Internet looks like has been described in previous works [1–3]. The intention in this chapter is not to define what the Future Internet is, but rather to view the Future Internet in a service-oriented manner, coming through a revision about the role that linked data can play to satisfy part of the mentioned challenges. In Future Internet, services and networks follow a common guideline: providing solutions in form of implemented interoperable mechanisms.

Communications networks have undergone a radical shift from a traditional circuit-switched environment with heavy/complex signalling focused on applications-oriented perspective, towards a converged Service-Oriented Architecture space (SOA), mostly Internet interaction by customer as end-user and network operators as service providers. The business benefits of this shift significantly reflect cost reduction and increase systems flexibility to react to users' demands more efficiently and by replacing, in a best practice case, a plethora of proprietary hardware and software platforms with generic solutions supporting standardised development and deployment stacks. As an example of this shift, the emergence and wide-scale deployment of wireless access network technologies calls into question about the viability of basing the future Internet on IP and TCP—protocols that were never intended for use across highly unreliable and volatile wireless interfaces (information and data exchange).

Research initiatives addressing this SOA requirement argue that the future lies in layers of overlay networks that can meet various requirements whilst keeping a very simplistic, almost unmanaged, IP for the underlying Internet, GENI NSF-funded initiative to rebuild the Internet [4]. Others argue that the importance of wireless access networks requires a more fundamental redesign of the core Internet Protocols themselves [5, 6]. Whilst this debate races nothing, it is a clear outcome that in terms of information interoperability or data models, the information exchange/sharing is required.

Services agnostic designs are not anymore a way to achieve interactive solutions in terms of service composition and information sharing capabilities for heterogeneous infrastructure support. A narrow focus on designing optimal networking protocols in isolation is too limited. Instead, a more holistic and long-term view is required, in which networking issues are addressed in a manner various protocols delivering communications services can be supported, meeting rapidly changing communities of users' needs. This new holistic view increasingly stops to become a matter of critical infrastructure. Network operators are today coming to realise the lack in the promise of simpler all-IP networks, where new integrated Internet services are easier and quicker to design, develop, deploy and manage.

Figure 8.1 depicts the mentioned service-aware Future Internet holistic view and its implementation relies on the inference plane [7], or Data Link Plane where the exchange of information facilitates knowledge-driven support and generation of composed services with operations by enabling interoperable management information. The move towards converged IP-based communication networks increases the need for providing solutions to a number of significant technical issues by using more standard information exchange and thus promoting information interoperability. The objective of this switch towards interoperability allows the networks to be managed more effectively and most important offering open opportunities for a user knowledge-based service-oriented support. This final feature allows having a fundamental impact on future communications networks and services.

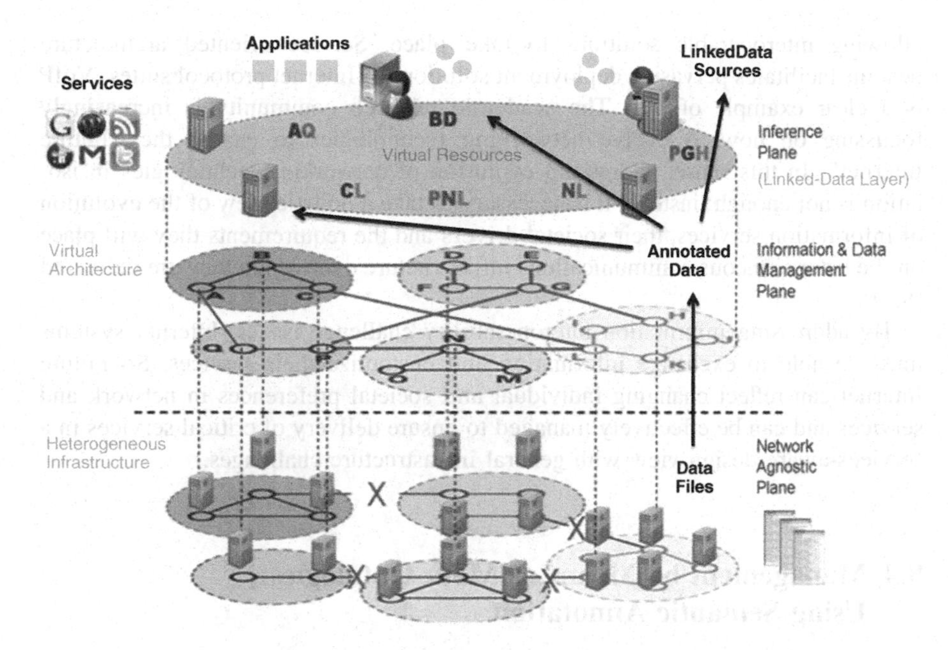

Fig. 8.1 Service composition by using linked data—Information interoperability

8.3 Challenges in Data Interoperability

Taking a broad view of the state of the art, current development of data link interactions and converging communications and information systems concentrates on offering access to data rather than exchange. Many of the problems present in current Data Link and Information management in the Internet are generated by interoperability problems where there are three persistent problems:

1. Users are offered relatively small numbers of Internet services, which they can *personalise* to meet their *evolving needs*; thus communities of users can not tailor *services* to help *create, improve and sustain* their *social interactions*;
2. The Information *services* in Internet that are offered are typically *technology-driven* and static, designed to maximise usage of capabilities of underlying network technologies and not to satisfy user requirements *per se*, and thus cannot be *readily adapted* to their changing *operational context;*
3. Network operators cannot *configure* their networks to *operate effectively* in the face of changing *service usage patterns* and rapid networking technology deployment; networks can only be *optimised*, on an *individual basis*, to meet specific low-level objectives, often resulting in sub-optimal operation in comparison to the more important *business and service user objectives.*

As the move towards convergence of communications networks and services continue; a more extended service-oriented architecture design gain momentum, allowing interoperable solutions to take place. Service oriented architecture designs facilitates pervasive deployment solutions of Internet protocol suites, VoIP is a clear example of this. The academic research community is increasingly focussing on how to evolve networking technologies to enable the "Future Internet". In this sense, addressing evolution of networking technologies in isolation is not enough; instead, it is necessary to take a holistic view of the evolution of information services, their societal drivers and the requirements they will place on the heterogeneous communications infrastructure over which they are delivered [8, 9].

By addressing information interoperability challenge issues, Internet systems must be able to exchange information and customize their services. So Future Internet can reflect changing individual and societal preferences in network and services and can be effectively managed to ensure delivery of critical services in a services-aware design view with general infrastructure challenges.

8.4 Management by Means of Meta-Ontologies: Using Semantic Annotation

A current activity in many research and development communities is the composition of data models for enabling information management control. It focuses on the semantic enrichment task of the management information described in both enterprise and networking data models with ontological data to provide an extensible, reusable, common and manageable data link plane, also named Inference Plane [7].

The proposal of using semantic annotation is for providing tools to integrate user data with the management service operations, and offers a more complete understanding of user's contents based on their social relationships and hence, a more inclusive governance of the management of resources, devices, networks, systems and services for promoting the data link of integrated management information within different management systems. This approach is to use ontologies as the mechanism to generate a formal description, which represents the collection and formal representation for network management data models and endows such models with the necessary semantic richness and formalisms to represent different types of information needed to be integrated in network management operations. Using a formal methodology, the user's contents represent values used in various service management operations, thus the knowledge-based approach over the inference plane [7] aims to be a solution that uses ontologies to support interoperability and extensibility required in the systems handling end-user contents for pervasive applications [10].

8.4.1 Service and Network

The meta-ontology approach, which is cited in this section, integrates concepts from the IETF policy standards [11, 12] as well as the TM Forum SID model [13, 14]. In this approach important classes that were originally defined in the IETF, SIM and DEN-ng models are re-used and implemented as such, some other extended or adapted for communication services adaptability (for more details see [15]). The meta-model proposed (Fig. 8.2), defines a set of interactions between the Context Data, Pervasive Management, and Communications System Domains in order to define relationships and interactions between the classes from the information models on these three different domains. The meta-Ontology construction process, which is a four-phased methodology, is the result of a formal study to build up ontologies contained and studied from [16, 17].

As cited before, the semantic web is the driver of service/end user oriented applications, for this reason the formal language used to build the set of ontologies is the web ontology language (OWL) [18, 19], which has been extended in order to apply to pervasive computing environments; these additional formal definitions act as complementary parts of the lexicon. The formal descriptions about the terminology related to management domain are included to build and enrich the proposal for integrating network and other management data with context information to more completely define the appropriate management operations using formal descriptions.

The proposed meta-ontology model integrates concepts from policy-based management systems [20, 21] to define a context-aware system that is managed by policies. Figure 8.2 shows the Ontology high-level representation. The image represents the integration of the entity information class (InfoEntity) related to the management operation class through the Event class. The Event class interacts with other classes from different domains in order to represent context information. Note that only the InfoEntity class from the context information model domain and the Event class from the service management domain are shown.

This representation simplifies the identification of interactions between the information models. These entity concepts, and their mutual relationships, are represented as lines in the figure. The InfoEntity class forms part of an Event class, and then the Event governs the policy functionality of a Managed Entity by taking into account context information contained in Events. This functionality too enables context to change the operation requested from a pervasive service or application, and is represented as interaction between Event and InfoEntity.

The meta-Ontology model is driven by a set of pervasive service management use cases that each requires service lifecycle policy-based management architecture as represented in Fig. 8.3. The service composition and its model representation contain the service lifecycle operations, as depicted in Fig. 8.3. In this figure, service management operations, as well as the relationships involved in the management service lifecycle process, are represented as classes. These classes will then be used, in conjunction with ontologies, to build a language that allows a

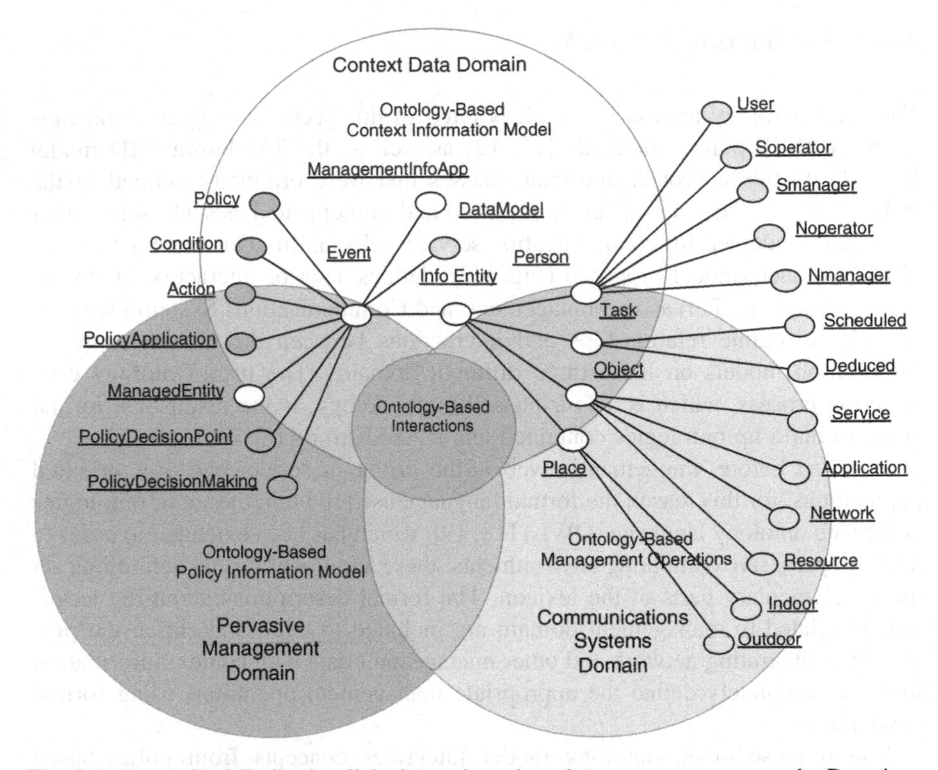

Fig. 8.2 View of InfoEntity data links in service-oriented management systems—by Domains

restricted form of English to be used to describe its policies. The annotated information about service management is underlayed in form of relationships, which has a correspondence with control activities called "events" that could be so related as context information.

The meta-model is founded on information model principles for context information and policy management promoting an integrated management, which is required by both pervasive as well as autonomic management applications. The combination of context-awareness, ontologies and policy-driven services motivates the definition of a new, extensible, and scalable semantic framework for the integration of these three diverse sources of knowledge to realize a scalable management platform.

From Fig. 8.3, the Service Editor Service Interface acts as the application that creates the new service. Assume that the service for deploying and updating the service code in certain network nodes has been created. This result in the creation of an event named "aServiceOn", which instantiates a relationship between the Application and Maintenance classes. This in turn causes the appropriate policies and service code to be distributed via the Distribution class as defined by the "aServiceAt" aggregation. The service distribution phase finds the nearest and/or most appropriate servers or nodes to store the service code and policies, and then

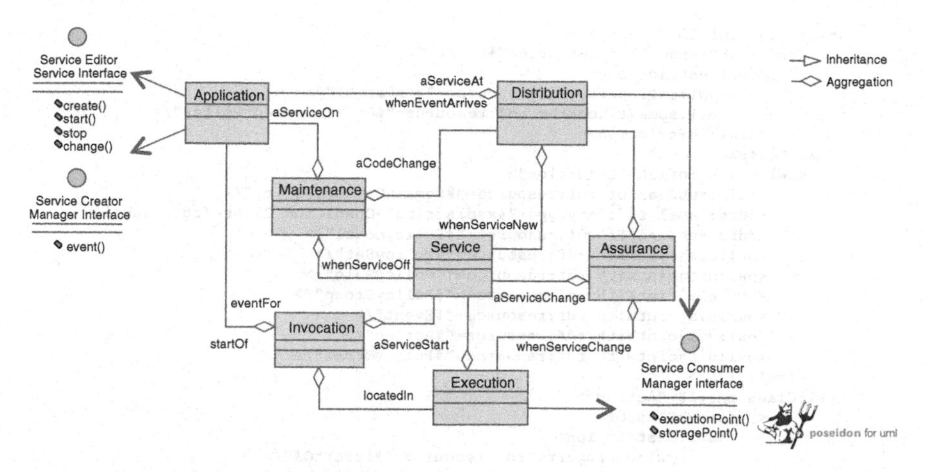

Fig. 8.3 Control and representation of service lifecycle inter actions—UML Graph representation

deploys them when the task associated with the "eventFor" aggregation is instantiated. When a service invocation arrives, as signalled in the form of one or more application events, the invocation phase detects these events as an indication of a context variation, and then instantiates the service by instantiating the association "aServiceStart". The next phase to be performed is the execution of the service. Any location-specific parameters are defined by the "locatedIn" aggregation. The execution phase implies the deployment of service code, as well as the possible evaluation of new policies to monitor and manage the newly instantiated service.

Monitoring is done using the service consumer manager interface, as it is the result of associations with execution. If maintenance operations are required, then these operations are performed using the appropriate applications, as defined by the"aServiceOn" aggregation, and completed when the set of events corresponding to the association "whenServiceOff" is received. Any changes required to the service code and/or polices for controlling the service lifecycle are defined by the events that are associated with the "whenServiceNew" and "aServiceChange" associations.

The service management operations are related to each other, and provide the necessary infrastructure to guarantee the monitoring and management of the services over time. The UML design shown in Fig. 8.3 captures these relationships, thus the pervasive service provisioning and deployment is on certain manner assured to provide service code and policies supporting such services to the service consumers. The coding process for formalizing the ontology follows the definitions and specifically the description contained in the field "relation" from the definitions provided in this chapter. It is very important to understand the real sense of the "relation" field, as this represents how the descriptions and concepts are used to build and enhance the proposed ontology.

```
<owl:Class rdf:ID="Event">
    <rdfs:subClassOf rdf:resource="#Policy"/>
        <owl:Restriction>
            <owl:onProperty rdf:resource="#isPartOf"/>
            <owl:someValuesFrom rdf:resource="#DomainConceptPolicy"/>
        </owl:Restriction>
</owl:Class>
    <owl:Class rdf:ID="Condition">
        <rdfs:subClassOf rdf:resource="#DomainConceptPolicy"/>
        <rdfs:label rdf:datatype="&xsd;string">Condition Class</rdfs:label>
        <rdfs:subClassOf rdf:resource="#PolicyModel"/>
        <owl:disjointWith rdf:resource="#PolicySet"/>
        <owl:disjointWith rdf:resource="#PolicyRule"/>
        <owl:disjointWith rdf:resource="#PolicyGroup"/>
        <owl:disjointWith rdf:resource="#Event"/>
        <owl:disjointWith rdf:resource="#Action"/>
        <owl:disjointWith rdf:resource="#PolicyModel"/>
    </owl:Class>
<owl:Class rdf:ID="Action">
        <rdfs:subClassOf>
            <owl:Restriction>
                <owl:onProperty rdf:resource="#isPartOf"/>
                <owl:someValuesFrom rdf:resource="#DomainConceptPolicy"/>
            </owl:Restriction>
        </rdfs:subClassOf>
        <rdfs:subClassOf rdf:resource="#PolicyModel"/>
        <owl:disjointWith rdf:resource="#PolicySet"/>
        <owl:disjointWith rdf:resource="#PolicyRule"/>
        <owl:disjointWith rdf:resource="#PolicyGroup"/>
        <owl:disjointWith rdf:resource="#Event"/>
        <owl:disjointWith rdf:resource="#Condition"/>
        <owl:disjointWith rdf:resource="#PolicyModel"/>
    </owl:Class>
<owl:Class rdf:ID="ManagedEntity">
    <rdfs:subClassOf rdf:resource="#Policy"/>
</owl:Class>
<owl:Class rdf:ID="Obligation">
    <rdfs:subClassOf rdf:resource="#Policy"/>
</owl:Class>
<owl:Class rdf:ID="Authorization">
    <rdfs:subClassOf rdf:resource="#Policy"/>
</owl:Class>
```

Fig. 8.4 Event InfoEntity description using OWL grammar representation

The base or this ontology code is fully described in [10]. This section refers to a partial description of some of the concepts more relevant when the ontology is created; descriptions, definitions and integration of concepts are also included and described.

Figure 8.4 shows the OWL grammar as an example of the ontology representation that describes the InfoEntity and its corresponding domain elements. It is a description of an object class for representing an InfoEntity, and represents the simplest definition and relationships in the sense of disjointness to the Application, Place, Person, DataBaseIM and Task classes. The disjoints represent a semantic tool for filtering the seek of information, thus the object classes including disjoints can be easily identified to be considered or not as part of the knowledge that is being seek.

The use of XML, and the resulting use of RDF extensions, aimed to improve the expressiveness of ontology languages. The RDF-Schema (RDFS) [22] language emerged as a set of extensions to provide increased semantics of RDF by

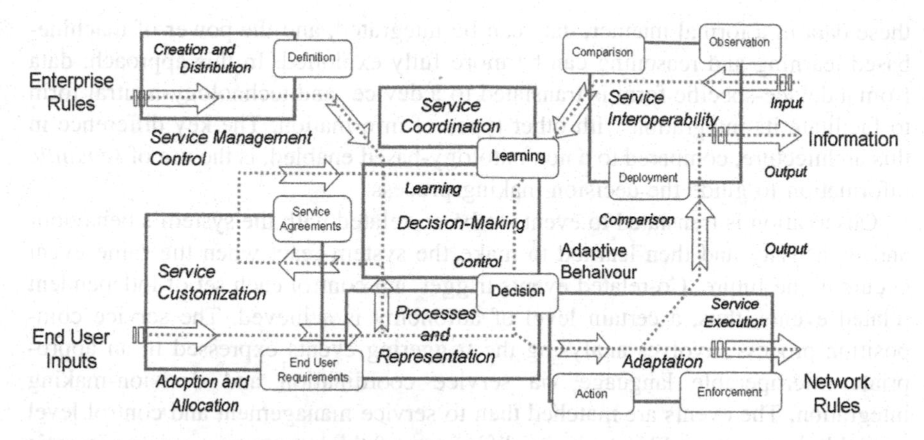

Fig. 8.5 Semantic-based service control engine—Functional architecture

providing basic ontological modelling primitives, like classes, properties, ranges and domains.

One of the advantages of using OWL to express the ontology is to provide a number of tools for parser and text editors. This enables new adopters to use a tool or set of tools that is best suited to their needs. OWL is used to define the set of concepts and constraints imposed by the information model over which it defines instances.

8.4.2 Functional Architecture Approach

A diagram for a functional architecture supporting the meta-Ontology with its functional components is depicted in Fig. 8.5. This diagram shows the interactions between the main components. The architecture controls with a certain high level domain-based view, how service composition is performed, thus the control of the behaviour is considered as added value functionalities using high level and formal representations expanding operations in other service application domains.

The architecture presented is using information to manage service operation and instructions with ontology-based information models using semantic mechanism. Based on previous implementation experience [23, 24], this architecture allows adaptability and dynamism to current and future Internet services with the advantages of incorporating context from users and applications in the form of events.

This functional approach offers more functionality than standard service-aware management approaches, as they cannot orchestrate system behaviour using knowledge from business, network, and other constituencies. By representing these data in a formal manner, data can be integrated, and the power of machine-based learning and reasoning can be more fully exploited. In this approach, data from a device-specific form is translated to a device- and technology-neutral form to facilitate its integration with other types of information. The key difference in this architecture, compared to a non ontology-based enabled, is the use of *semantic* information to guide the decision-making process.

Observation is translated to events to be co-related with the system's behaviour and its activity and then learned to make the system react when the same event occurs in the future. Co-related events trigger and control each set of independent related events, thus, a certain level of autonomy is achieved. The service composition process involves analysing the triggering events expressed in an appropriate interoperable language via service coordination and decision-making integration. The events are matched then to service management and control level available instructions [25] with the difference of this component using semantic descriptions to co-relate events with particular kind of conflicts that must be identified and evaluated. A detailed Semantic-Based Service Control Engine (2SCE) and its components as part of functional architecture is out of the scope of this chapter, however implementation results are being analyzed and interaction between different domain events tested [26].

8.5 Linked Data and Cloud Management Systems

The increasing popularity of "cloud", as a powerful distributed computing system with almost infinite computing processing capabilities, open the possibility to consider if cloud-based systems can enable information management systems to handle more efficiently the managing operations of data.

The design of cloud management systems (distributed management systems with extensive scalable capacities to run parallel control process) has attracted the interest of diverse academic and ITC's industry communities because it is seen as an opportunity to improve the network infrastructure and provide solutions for many different web-based era demands. In this race for designing and developing the Future Internet, many communities have realized that a common problem to tackle is the interoperability of the information. Unfortunately, current Internet architecture does not include means to achieve interoperability at a data level, so Information Interoperability has not fully run coordinated course in terms of implementation, due to many complex issues involving deployment approach; as a result, multiple non-interoperable solutions are in place. Following this problem, linked data is becoming an accepted best practice to exchange information in an interoperable and reusable way. Different communities on the Internet use linked data standards to provide and exchange interoperable information. It is a practice that is being adopted by ITC′s as well. This section describes the current state of the art in linked data and Information Interoperability best practices. Current common requirements of different communities active in the Future Internet and common practices by using standards for information Interoperability are presented. Results about linked data and management by using semantic annotation for management information data models contained in both enterprise and networking domains are described. Ontologies are used to support this approach. An introductory application scenario is depicted. This chapter supports the idea of looking at linked data as an independent layer in the Internet architecture, on top of the networking layer, but below the application layers, since it provides a common data model for all applications. It is interesting to identify what implications this imposes on the Future Internet Architecture, but also how future architectures and cloud system can benefit from this new layer.

Particular interest for cloud computing services and the use of virtual infrastructures supporting such services is also a result of the business model cloud computing offers, where bigger revenue and more efficient exploitation is envisaged [27]. Likewise, particular interest exists from the industry sector (where most of the implementations are taking place) for developing more management tools and solutions in the cloud. In the other hand academic communities point towards finding solutions for more powerful computing processing and at the same time more efficient. Thus generally problems on manageability, control of cloud and other research challenges are being investigated.

It is anticipated that cloud computing should reduce cost and time of computing and processing [28]. However while cost benefit is reflected mainly to the end-user, from a cloud service provider perspective, cloud computing is more than a simple arrangement of mostly virtual servers, offering the potential of tailored service offerings and theoretically infinite expansion. *Cloud computing system is a large number of tailored resources, which are interacting to facilitate the deployment, adaptation, and support of services*, and this situation represent significant infrastructure and information management challenges. In management terms there is a potential trend to adopt, refine, and test traditional management methods to exploit, optimize and automate [27] the management operations of cloud computing infrastructures, however this is proving difficult to implement, so designs for management by using new methodologies, techniques and paradigms are necessary.

Cloud management is a complex task [29] as clouds must support appropriate levels of tailored service performance to large groups of diverse users. A sector of services, named private clouds, coexists with and is provisioned through a bigger public cloud, where the services associated to those private clouds are accessed through (virtualized) wide area networks. In this section challenges for managing cloud service infrastructure are discussed in the form of scenarios. A special focus on management systems, which are essential for the provisioning and control access of virtual infrastructure resources [30], is taken in order to establish where such systems are able to address fundamental issues related to data processing scalability and reliability.

8.5.1 Cloud Monitoring

Monitoring in cloud is essential for automatic or autonomous adaptation to current data load, as well as for providing feedback on service logic. Scalability and security are essential for cloud monitoring. Without solving the problems of scalability and security, tools and technologies are almost impossible to be deployed in a cloud environment. Moreover, cloud monitoring will also require application-level information monitored in addition to the system usage data current tools can provide.

There are several related works in this area. Lattice is a distributed monitoring framework, which was experimented and validated in computing clouds [31] and in network clouds [30]. DSMon [32] introduces system monitoring for distributed environments and mainly focuses on fault tolerant aspects; NWS [33] also provides a distributed framework for monitoring and has the ability of forecasting performance changes. When compared to DSMon and NWS, another system resources monitoring tool called DR monitoring [34], requires less resources to run and supports multiple platforms (Linux and Windows). However, by the time of the publication of this document, the DR monitoring tool lacks scalability and fails to address security concerns. From the industrial viewpoint, HP Open View [35] and IBM Tivoli [36] have been developed to ease system monitoring and are primarily targeting the enterprise application environment. Although the commercial products are relatively portable across different operating systems, they are usually highly integrated with vendor specific applications. In the cloud environment, heterogeneity is one of the fundamental requirements for monitoring tools. Therefore, the industrial tools are unsuitable for more general purpose monitoring of the cloud. GoogleApp engine [37] and Hyperic [38] both provide monitoring tools for system status such as CPU, memory, and processes resource allocations. Such system usage data can be useful for general purpose cloud monitoring, but they may not be sufficient enough for an application level manager to make appropriate decisions.

8.5.2 Cloud Interconnection: Federation

Federation in the cloud [39, 40] would imply a requirement where user's applications or services shall still be able to execute across a federation of resources stemming from different cloud providers. It also refers to the ability for different cloud providers to scale their service offerings and to share capabilities to combine efforts and provide a better quality of service for their customers. While the technological aspects required supporting cloud federation is an ongoing research domain, there has been little work to support the holistic end-to-end monitoring and management of federated cloud services and resources. This approach requires users and multiple providers to both delegate, share and consume each others' resources in a peer-to-peer manner in a secure, managed, monitored and auditable way, with a particular focus on information interoperability between management and resource description approaches.

Federation presents as an approach for supporting the increasingly important requirement to orchestrate multiple vendors, operators and end user interactions [41, 42]. In this way can be seen the applicability of this concept in the cloud computing area. Cloud computing offers an end-user perspective where the use of one or any other infrastructure is transparent, in the best case the infrastructure is ignored by the cloud user [43]. However from the cloud operator perspective, there are heterogeneous shared network devices as part of diverse infrastructures that must be self-coordinated for offering distributed management or alternatively centrally managed in order to provide the services for which they have been configured. Furthermore, there must be support to facilitate composition of new services, which requires a total overview of available resources [44]. In such a federated system, the number of conflicts or problems that may arise when using diverse information referring to the same service or individuals with the objective of providing an end-to-end service across federated resources must be analysed by methodologies that can detect conflicts. In this sense, semantic annotation and semantic interoperability tools appear as a tentative approach solution. However this type of approach yet remains as a challenge for being investigated.

8.5.3 Cloud Management

The need to control multiple computers running applications and likewise, the interaction of multiple service providers supporting a common service, exacerbates the challenge of finding management alternatives for orchestrating between the different cloud-based systems and services [45]. Even though having full control of the management operations when a service is being executed is necessary, distributing this decision control is still an open issue. In cloud management systems supporting such complex management operations [27, 30] must address the challenging problem of coordinating multiple running applications' management operations, while prioritizing tasks for service interoperability between different cloud systems.

An emerging alternative to solve cloud computing decision control, from a management perspective, is the use of formal languages as a tool for information exchange between the diverse data and information systems participating in cloud service provisioning. These formal languages rely on an inference plane [46, 47]. By using semantic decision support, and enriched monitoring information, management decision support is enabled and facilitated. As a result of using semantics rephrase, a more complete control of service management operations can be offered, hence a more integrated management, which responds to business objectives. This semantically-enabled decision support gives better control in the management of resources, devices, networks, systems and services, thereby promoting the management of the cloud with formal information models [48].

There is a need to manage cloud systems by means of policies/rules as the mechanism to represent and contain Description Logic (DL) to operate operational rules. For example, the SWRL language [49, 50] can be used to formalize a policy language to build up a collection of model representations with the necessary semantic richness and formalisms to represent and integrate the heterogeneous information present in cloud management operations. This approach relies on the fact that high level infrastructure representations do not use resources when they are not being required to support or deploy services [51, 52]. Thus, with high-level instructions, the cloud infrastructure can be managed in a more dynamic and optimal way.

8.5.4 Cloud Configuration and Re-Configuration

There are several cloud usage patterns based on bandwidth, storage, and server instances over time [53]. Constant usage over time is typical for internal applications with small variations in usage. Cyclic internal loads are typical for batch and data processing of internal data. Highly predictable cyclic external loads are characteristic of web servers such as news, sports, whereas spiked external loads are seen on web pages with suddenly popular content. Spiked internal load are characteristic of internal one-time data processing and analysis, while steady growth over time is seen on start-up web pages.

The cloud paradigm enables applications to scale-up and scale-down on demand, and to more easily adapt to the usage patterns as outlined above. Depending on a number or type of requests, the application can change its configuration to satisfy given service criteria and, at the same time, optimize resource utilization and reduce the costs. Similarly, clients—which can run on a cloud as well—can reconfigure themselves based on application availability and service levels required.

8.6 Linked Data and Scalable Cloud Management

In this section, some features about management of cloud infrastructures pointing towards the advantages about what linked data can offer when it is used within the Cloud-based Service Management Control Loop (Fig. 8.3) are presented. In cloud system management, in order to apply deployment policies for a service request, the cloud manager has to check the logs of running services. Traditionally, the logs recorded from running services and computing nodes are processed in the data processing components usually called Data Correlation Engine under separated files that need to be loaded into relational tables to be accessed by a rule engine.

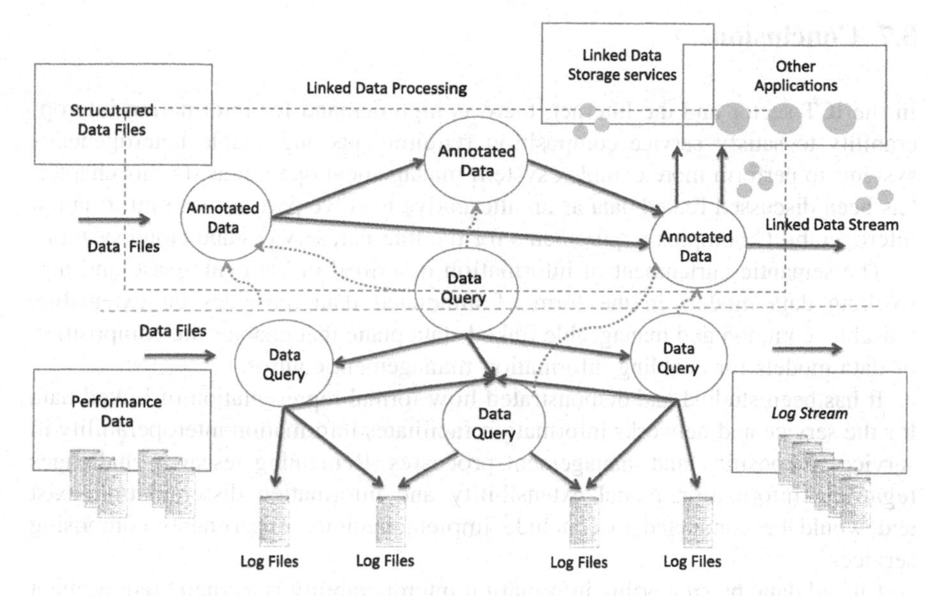

Fig. 8.6 Linked data stream processing layer

However, the logs and service requests are fed into Data Correlation Engine in the stream way. There are a wide variety of literatures in stream/event processing studied in [54, 55] showing that relational database engines do not perform well with continuous processing over log streams. Therefore, the stream/event processing engine should be integrated into the Data Correlation Engine.

To support the seamless integration of rule-based policy represented as SWRL on top of background knowledge represented in Description Logic with log streams, have been using linked data models for correlation of data. In this unified model, all the data are represented as a graph within the layered system representation as it is illustrated in Fig. 8.6. This graph is composed of RDF triples which can be used to represented data, knowledge base and rules. The upper layer of the layout is for static data such as ontologies, schema and business logics which is time-independent. The lower layer is used for linking data of log streams into vocabularies and logic rules in the upper layer. The linked data in the lower layer is called Linked Data Stream as depicted in Fig. 8.6.

The linked stream data model brings several advantages in data correlation operations. The first advantage comes from the data distribution. The graph-based layout gives the data processing operators the global view of the whole dataset. Therefore, the query processor can filter the irrelevant data to a query much earlier than the log-file approach does. Traditionally, the monitoring data recorded in separated log files are partitioned in individual services, processes, etc. thus, cross-correlating the relevant data items among them needs to load all the data into a relational storage before carrying out the correlation.

8.7 Conclusions

In the ICT sector and the Internet, there is high demand for information interoperability to satisfy service composition requirements and enable heterogeneous systems to perform more complex system management operations. In this chapter, has been discussed linked data as an alternative to solve part of those information interoperability complex requirements for the Internet, services and cloud systems.

The semantic enrichment of information described in both enterprise and networking data models in the form of annotated data, provides an extensible, reusable, common and manageable linked-data plane that enables the composition of data models for enabling information management control.

It has been studied and demonstrated how formal representation of linked data for the service and networks information facilitates information interoperability in service composition and management processes. Remaining research challenges regarding information model extensibility and information dissemination exist and would be conducted to conclude implementations, experiments composing services.

Linked data, by supporting information interoperability is a crucial requirement in Future Internet and cloud management systems. Their implications for networks and services is still an open issue for research and interoperability and information exchange must be validated via direct industrial investment, and roll out on real integrated test beds to trial Future Internet infrastructures (potentially overlay networks or new inference planes) for services.

Acknowledgments This work has been supported by the OpenIoT Project (Open source blueprint for large scale self-organizing cloud environments for IoT applications) [56], which is co-funded by the European Commission under FP7 framework program, contract number FP7-ICT-2011-7-287305-OpenIoT. It is also partially supported by Science Foundation Ireland under grant number SFI/08/CE/I1380 (Lion-II) CSET-DERI Centre for Science, Engineering and Technology—Digital Enterprise Research Institute.

References

1. Clark, D. et al.: New arch: future generation internet architecture. New arch final technical report, http://www.isi.edu/newarch/
2. Blumenthal, M., Clark, D: Rethinking the design of the Internet: the end to end arguments vs. the brave new world. ACM Transactions on Internet Technology, vol. 1, No. 1. Aug 2001
3. Feldmann, A.: Internet clean-slate design: what and why? ACM SIGCOM Comput. Comm. Rev. **37**(3), 59–64 (2007)
4. NSF-funded initiative to rebuild the Internet, http://www.geni.net/
5. Clean Slate program, Stanford University, http://cleanslate.stanford.edu
6. Architecture Design Project for New Generation Network, http://akari-project.nict.go.jp/eng/index2.htm
7. Strassner, J, Foghlú, M.Ó, Donnelly, W, Agoulmine, N.: Beyond the knowledge plane: an inference plane to support the next generation Internet. IEEE GIIS 2007, 2–6 July 2007

8. Irish Future Internet Forum, http://www.futureinternet.ie
9. SFI-SRC FAME: Scientific Research Cluster: Federated Autonomic Management of End to End Communications Services. http://www.fame.ie/
10. Serrano, M., Strassner, J., ÓFoghlú, M.: A formal approach for the inference plane supporting integrated management tasks in the future Internet. 1st IFIP/IEEE ManFI International Workshop, In conjunction with 11th IFIP/IEEE IM2009, 1–5 June 2009, at Long Island (2009)
11. Westerinen, A., Schnizlein, J., Strassner, J.: Terminology for Policy-Based Management. IETF Request for Comments (RFC 3198). November (2001)
12. Moore, E.: Policy core information model-extensions. IETF Request for comments (RFC 3460), January 2003. http://www.ietf.org/rfc/rfc3460.txt
13. TMF: The Shared Information and Data Model—Common Business Entity Definitions: Policy, GB922 Addendum 1-POL, July 2003
14. SID: Shared Information Data model. http://www.tmforum.org/InformationManagement/1684/home.html
15. Serrano, J.M.: Management and context integration based on ontologies for pervasive service operations in autonomic communication systems. PhD Thesis, UPC (2008)
16. Horridge, M., Knublauch, H., Rector, A., Stevens, R., Wroe, C.: A practical guide to building OWL ontologies using the Protégé-OWL Plugin and CO-ODE Tools Edition 1.0. Manchester University, August. (2004)
17. Gruber, T.: Towards "principles for the design of ontologies used for knowledge sharing". Int. J. Hum. Comput. Stud. **43**(5/6), 907–928 (1995)
18. OWL: Ontology Web Language. http://www.w3.org/2004/OWL
19. OWL-s http://www.daml.org/services/owl-s/
20. Sloman, M.: Policy driven management for distributed systems. J. Netw. Syst. Manag. **2**, 333–360 (1994)
21. Strassner, J.: Policy based network management. Morgan Kaufmann, ISBN 1-55860-859-1, (2004)
22. W3C Website. http://www.w3c.org/rdf
23. Serrano, J.M, Serrat, J, Strassner, J.: Ontology-based reasoning for supporting context-aware services on autonomic networks. 2007 IEEE/ICC International Conference on Communications, Glasgow, 24–28 June 2007
24. Serrano, J.M., Serrat, J., O'Sullivan, D.: Onto-context manager elements supporting autonomic systems: basis and approach. 1st IEEE MACE International Workshop as part of ManWeek 2006, Dublin, 23–27 Oct 2006
25. Schönwälder, J. Straub F.: Next generation structure of management information for the internet. In: Proceedings of the 10th IFIP/IEEE DSOM International Workshop, Zürich, (1999)
26. Keeney, J., Conlan, O., Holub, V., Wang, M., Chapel, L., Serrano, M.: A semantic monitoring and management framework for end-to-end services. In: Proceedings of 12th IFIP/IEEE international symposium on integrated management—IM2011. Dublin, IE, 23–27 May 2011
27. Serrano, J.M: Applied ontology engineering in cloud services, networks and management systems. to be released on March 2012, Springer Publishers, 2012. Hardcover, p. 222 pages, ISBN-10: 1461422353, ISBN-13: 978-1461422358
28. IBM Software Group, U.S.A.: Breaking through the haze: understanding and leveraging cloud computing. Route 100, Somers, NY. IBB0302-USEN-00. (2008)
29. Rochwerger, B., Caceres, J., Montero, R.S., Breitgand, D., Elmroth, E., Galis, A., Levy, E., Llorente, I.M., Nagin, K., Wolfsthal, Y., Elmroth, E., Caceres, J., Ben-Yehuda, M., Emmerich, W., Galan, F.: The RESERVOIR model and architecture for open federated cloud computing. IBM J. Res. Dev. **53**(4), 4 (2009)
30. Clayman, S., Galis, A., Mamatas, L., Monitoring Virtual Networks. 12th IEEE/IFIP network operations and management symposium (NOMS 2010)—International on management of the future internet, Osaka http://www.man.org/2010/, 19–23 Apr 2010
31. Clayman, S., Galis, A., Toffetti, G., Vaquero, L.M., Rochwerger, B., Massonet, P.: Towards a service-based internet. Lect. Notes Comput. Sci. **6481**, 215–217 (2010). doi:10.1007/978-3-642-17694-4_30

32. Shao, J, Wei, H., Wang, Q., Mei. H.: A runtime model based monitoring approach for cloud. In: Cloud Computing (CLOUD), 2010 IEEE 3rd International Conference on, pp. 313–320 July (2010)
33. Wolski, R., Spring, N.T., Hayes, J.: The network weather service: a distributed resource performance forecasting service for metacomputing. Future Gener. Comput. Syst. **15**, 757–768 (1999)
34. The 'Intercloud' and the Future of Computing, an interview: Vint Cerf at FORA.tv, the Churchill Club, Jan 7, 2010. SRI International Building, Menlo Park, CA, Online Jan 2011. http://www.youtube.com/user/ForaTv#p/search/1/r2G94ImcUuY
35. HP: Hp openview event correlation services, Nov. 2010. Available (online): http://www.managementsoftware.hp.com/products/ecs/ds/ecs ds.pdf
36. IBM: Tivoli support information center, Nov. 2010. Available (online): http://publib.boulder.ibm.com/tividd/td/IBMTivoliMonitoringforTransactionPerformance5.3.html
37. Google. Google app engine system status, Nov. 2010. Available (online): http://code.google.com/status/appengine
38. Hyperic. Cloudstatus ® powered by hyperic, Nov. 2010. Available (online): http://www.cloudstatus.com
39. Chapman, C., Emmerich, E., Marquez, F.G., Clayman, S., Galis, A.: Software architecture definition for on-demand cloud provisioning. Springer Journal on Cluster Computing—doi:10.1007/s10586-011-0152-0; May 2011
40. Reservoir Project: http://www.reservoir-fp7.eu/
41. Bakker, J.H.L., Pattenier, F.J.: The layer network federation reference point-definition and implementation. Bell Labs. Innovation, Lucent Technology, Huizen. In TINA Conferences Proceedings 1999. pp. 125–127, Oahu, ISBN: 0-7803-5785-X (1999)
42. Serrano, J.M., Van deer Meer, S., Holum, V., Murphy J., Strassner, J: Federation, a matter of autonomic management in the future Internet. 2010 IEEE/IFIP Network Operations and Management Symposium—NOMS 2010. Osaka International Convention Center, Osaka, 19–23 Apr 2010
43. Allee, V.: The Future of Knowledge: Increasing Prosperity through Value Networks. Butterworth-Heinemann, London (2003)
44. Kobielus, J.: New federation frontiers in ip network services. Publication: Business Communications Review. Date: Tuesday, 1 Aug 2006
45. Chapman, C., Emmerich, W., Galn, F., Clayman, S., Galis, A.: Elastic service management in computational clouds. 12th IEEE/IFIP NOMS2010/International Workshop on Cloud Management (CloudMan 2010), Osaka, 19–23 Apr 2010
46. Strassner, J., Ó Foghlú, M., Donnelly, W. Agoulmine, N. "Beyond the Knowledge Plane: An Inference Plane to Support the Next Generation Internet", IEEE GIIS 2007, 2-6 July, 2007
47. Serrano, J.M., Strassner, J., ÓFoghlú, M.: A formal approach for the inference plane supporting integrated management tasks in the future internet. 1st IFIP/IEEE ManFI International Workshop, In conjunction with 11th IFIP/IEEE IM2009, Long Island, 1–5 June 2009
48. Blumenthal, M., Clark, D.: Rethinking the design of the Internet: the end to end arguments vs. the brave new world. ACM Trans. Internet Technol. **1**(1), 70–109 (2001)
49. Bijan, P. et al.: Cautiously Approaching SWRL. http://www.mindswap.org/papers/CautiousSWRL.pdf. (2006)
50. Mei, J., Boley, H.: Interpreting SWRL rules in RDF graphs. Electron. Notes Theoret. Comput. Sci. **151**, 53–69 (2006). (Elsevier)
51. Neiger, G., Santoni, A., Leung, F., Rodgers D., Uhlig, R.: Intel virtualization technology: software-only virtualization with the IA-32 and Itanium architectures. Intel Technology Journal, vol. 10 Issue (03), August 2006
52. Cisco, VMWare.: DMZ Virtualization using VMware vSphere 4 and the Cisco Nexus. www.vmware.com/files/pdf/dmz-vsphere-nexus-wp.pdf. (2009)
53. Host your web site in the cloud, Jeff Barr, Sitepoint, 2010, ISBN 978-0-9805768-3-2

54. Le-Phuoc, D., Dao-Tran, M., Parreira, J.X., Hauswirth, M.: A native and adaptive approach for unified processing of linked streams and linked data. In: ISWC'11, 11 Oct 2011
55. Le-Phuoc, D., Nguyen, H., Quoc, M., Parreira, J.X., Hauswirth, M.: The linked sensor middleware: connecting the real world and the semantic Web. In: 9th Semantic Web Challenge co-located with 10th International Semantic Web Conference—ISWC 2011, Bonn, 23–27 Oct 2011
56. OpenIoT Project http://www.openiot.eu

Chapter 9
Paraconsistent Annotated Logic Programs and Application to Intelligent Verification Systems

Kazumi Nakamatsu and Jair M. Abe

Abstract Paraconsistent Logic is well known as a logical tool that can deal with inconsistency in a consistent logical system. In this chapter we introduce the development of a paraconsistent annotated logic program called Extended Vector Annotated Logic Program with Strong Negation (abbr. EVALPSN). First of all, we review paraconsistent annotated logics $P\mathcal{T}$ and their logic programs, and introduce the development from the paraconsistent logics $P\mathcal{T}$ to EVALPSN and their formal definitions. Furthermore, we introduce one version of EVALPSN named before-after (abbr. bf)-EVALPSN, which can deal with before-after relations between two time intervals, and its reasoning system consists of two kinds of inference rules with an example. We also provide a bf-EVALPSN based safety verification system for a pipeline process order control with simple examples.

9.1 Background and Introduction

Logical systems have been applied to various Intelligent Systems as their theoretical foundations or reasoning languages such as logic programs. However, some of them essentially include a lot of inconsistency such as contradictions, dilemma, conflicts, etc. It would be required to deal with inconsistency in the same intelligent systems logically if we try to formalize such contradictory systems aiming at

---

K. Nakamatsu (✉)
School of Human Science and Environment, University of Hyogo, Himeji 670-0092, Japan
e-mail: nakamatu@shse.u-hyogo.ac.jp

J. M. Abe
ICET-Paulista University, R. Dr. Bacelar CEP, Sao Paulo, SP 04026-022, Brazil

J. M. Abe
Institute of Advanced Studies, University of Sao Paulo, Rua Praa do Relogio, 109, Bloco K, 5 andar Cidade Universitaria, CEP, Sao Paulo, SP 05508-970, Brazil
e-mail: jairabe@uol.com.br

C. Faucher and L. C. Jain (eds.), *Innovations in Intelligent Machines-4*,
Studies in Computational Intelligence 514, DOI: 10.1007/978-3-319-01866-9_9,

constructing consistent intelligent systems. Ordinal Logical Systems cannot deal with inconsistency such as p and $\neg p$ consistently in the same system. A logical system that can deal with inconsistency is known as Paraconsistent Logic. The main purpose of paraconsistent logic is to deal with inconsistency in a framework of consistent logical systems.

It has been six decades since the first paraconsistent logical system was proposed by Jaskowski [10]. It was four decades later than the first paraconsistent logic by Jaskowski that a family of paraconsistent logic called Annotated Logics $P\mathcal{T}$[1] were proposed by da Costa et al. [6]. They can express inconsistency with many truth values called Annotations, which should be attached to every atomic formula explicitly, although the semantics of annotated logics is basically two-valued.

A paraconsistent annotated logic was developed from the viewpoint of logic programming, aiming at application to computer science such as the semantics for knowledge bases in [5, 11, 40, 41]. Furthermore, we have developed the paraconsistent annotated logic program in order to deal with inconsistency and non-monotonic reasoning in a framework of annotated logic programming by introducing ontological(strong) negation and the stable model semantics in [16], and named the developed annotated logic program Annotated Logic Program with Strong Negation(ALPSN for short). ALPSN has been applied to some non-monotonic reasoning systems, default logic [39], autoepistemic logic [15] and a non-monotonic Assumption Based Truth Maintenance System (ATMS) [7] as their computational models in [17, 27]. ALPSN can deal with not only inconsistency such as conflict but also non-monotonic reasoning such as default reasoning, however, it seems to be more important and useful from a practical viewpoint to deal with resolving conflicts in a logical way than just to express conflicts consistently. It is not so adequate for ALPSN to deal with conflict resolving or decision making in its logical framework.

Defeasible Logic is known as one of formalizations for non-monotonic reasoning called Defeasible Reasoning that can deal with conflict solving easily in a logical way [3, 34, 35]. Therefore, in order to deal with defeasible reasoning in a framework of paraconsistent annotated logic programming, we proposed a new version of ALPSN called Vector Annotated Logic Program with Strong Negation (VALPSN for short) and applied it to conflict solving in [18]. It also has been shown that VALPSN provides one computational model of a defeasible logic [19]. Furthermore, we have extended VALPSN to deal with deontic notions (obligation, forbiddance and permission) and shown that the extended version of VALPSN named Extended VALPSN(EVALPSN for short) can deal with deontic defeasible reasoning in [20, 21]. We also have shown EVALPSN can deal with safety verification based intelligent control. The basic ideas of EVALPSN safety verification are that each control system has norms such as guidelines or criteria for safe

[1] Generally the symbol $\mathcal{T}$ represents plural complete lattices of annotations, therefore, if there is no specification for the complete lattice $\mathcal{T}$, assume that $P\mathcal{T}$ represents plural annotated logics in this chapter.

control; such norms can be naturally formulated in EVALPSN deontic expressions; then the safety verification for control systems can be carried out by EVALPSN defeasible deontic reasoning. Compared to conventional safety verification methods, EVALPSN based safety verification has some advantages such that it can provide a formal safety verification method and a computational framework for control systems as logic programs.

EVALPSN has been applied to railway interlocking safety verification [23], robot action control [29], safety verification for air traffic control [22], traffic signal control [24], discrete event control [25, 26] and pipeline valve control [28].

Considering the safety verification for process control, there are many cases in which the safety verification for process order is significant. For example, suppose a pipeline network in which two kinds of liquids, nitric acid and caustic soda are used for cleaning the pipelines. If those liquids are processed continuously and mixed in the same pipeline by accident, explosion by neutralization would be caused. In order to avoid such a dangerous accident, the safety for process order should be strictly verified in a formal way. However, it seems to be a little difficult to utilize EVALPSN for the safety verification of process order control different from that of process control.

Moreover, we have developed EVALPSN toward treating before-after relations between time intervals and applied it to process order control [31], which has been named before-after (bf)-EVALPSN. The before-after relation reasoning system based on bf-EVALPSN consists of two groups of inference rules called the basic bf-inference rule and the transitive bf-inference rule.

The original ideas of treating such before-after relations in logic were proposed for developing practical planning and natural language understanding systems by Allen [1, 2]. In his logic, before-after relations between two time intervals are represented in some special predicates and treated in a framework of first order temporal logic. On the other hands, in bf-EVALPSN, before-after relations between two time intervals are regarded as paraconsistency between before and after degrees, and they can be represented more minutely in vector annotations of a special literal $R(p_i, p_j, t)$ representing the before-after relation between two processes(time intervals) at time t. Bf-EVALPSN based before-after relation reasoning system consists of two kinds of efficient inference rules that can be carried out in the framework of annotated logic programming.

This chapter is organized as follows: in Sect. 9.2, Paraconsistent Annotated Logics $P\mathcal{T}$ and their logic programs are reviewed; in Sect. 9.3, the development from ALPSN to VALPSN and EVALPSN and their formal definitions are introduced; in Sect. 9.4, bf-EVALPSN is formally defined and its simple reasoning example is introduced; in Sect. 9.5, the bf-EVALPSN reasoning system consisting of two kinds of inference rules is defined and explained in details with some examples; in Sect. 9.6, an example of a process order verification system based on bf-EVALPSN reasoning is provided; lastly, we conclude this chapter by describing the advantages and disadvantages of applying EVLPSAN to intelligent control systems.

We assume that the readers are familiar with the basic knowledge of logic and logic program.

9.2 Annotated Logic Program with Strong Negation

9.2.1 Paraconsistent Annotated Logics $P\mathcal{T}$

We recapitulate the syntax and semantics for propositional paraconsistent annotated logics $P\mathcal{T}$ in [5]. Generally, a truth value called an annotation is attached to each atomic formula explicitly in annotated logic, and the set of annotations constitutes a complete lattice. The paraconsistent annotated logics denoted by $P\mathcal{T}$ have the complete lattice $\mathcal{T}$ of annotations.

Definition 9.1 The primitive symbols of $P\mathcal{T}$ are defined as follows.

1. propositional symbols $p, q, \ldots, p_i, \ldots, q_j, \ldots$;
2. each member of $\mathcal{T}$ is an annotation constant (we may call it an annotation);
3. the connectives and parentheses $\wedge, \vee, \rightarrow, \neg, (,)$.

Formulas are defined recursively as follows.

4. if p is a propositional symbol and $\mu \in \mathcal{T}$ is an annotation constant, then $p{:}\mu$ is an annotated atomic formula (atom for short);
5. if F, F_1, and F_2 are formulas, then $\neg F, F_1 \wedge F_2, F_1 \vee F_2$ and $F_1 \rightarrow F_2$ are formulas.

We suppose the 4-valued lattice in Fig. 9.1 as the complete lattice $\mathcal{T}$, where the annotation **t** may be intuitively interpreted as the truth value true and the annotation **f** as the truth value false. It may be comprehensible that the annotations, $\perp$, **t**, **f** and $\top$. We use symbol $\leq$ to denote the ordering in terms of knowledge amount over the complete lattice $\mathcal{T}$, and symbols $\perp$ and $\top$ are used to denote the bottom and top elements in the complete lattice $\mathcal{T}$, respectively. In the paraconsistent annotated logic $P\mathcal{T}$, each annotated atomic formula can be interpreted epistemically, for example, p:**t** may be interpreted epistemically as "proposition p is known to be true".

There are two kinds of negation in the paraconsistent annotated logic $P\mathcal{T}$, one of them represented by symbol $\mathcal{T}$ in Definition 1 is called epistemic negation, and the epistemic negation followed by an annotated atomic formula is defined as a mapping between the elements of the complete lattice $\mathcal{T}$ as follows: $\neg(\perp) = \perp, \neg(\mathbf{f}) = \mathbf{t}, \neg(\mathbf{t}) = \mathbf{f}, \neg(\top) = \top$. This definition shows that the epistemic negation maps annotations to themselves without changing the knowledge

Fig. 9.1 Complete lattice $\mathcal{T}$

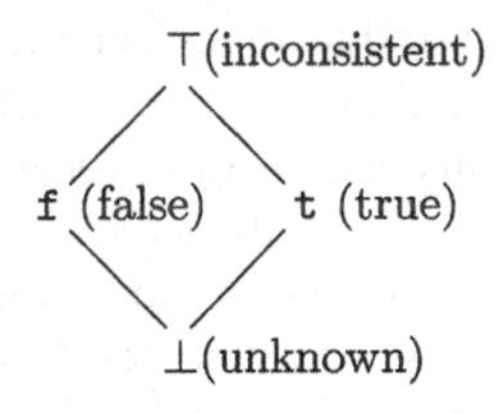

amounts of the annotations, therefore, the epistemic negation can be eliminated by the syntactical mapping. For example, the knowledge amount of annotation **t** is the same as that of annotation **f** as shown in the complete lattice $\mathcal{T}$, and we have the epistemic negation $\neg(p{:}\mathbf{t}) = p{:}\ \neg(\mathbf{t}) = p{:}\mathbf{f}$,[2] which shows that the knowledge amount in terms of proposition p cannot be changed by the epistemic negation. There is another negation called Ontological(Strong) Negation that is defined in Paraconsistent Annotated Logics $P\mathcal{T}$,

Definition 9.2 *(Ontological Negation* $\sim$*)* Let F be any formula,

$$\sim F =_{def} F \rightarrow ((F \rightarrow F) \wedge \neg(F \rightarrow F))$$

The epistemic negation in Definition 9.2 is not interpreted as a mapping between annotations since it is not followed by an annotated atomic formula. Therefore, the strongly negated formula F is intuitively interpreted so that if the formula F exists, the contradiction $((F \rightarrow F) \wedge \neg(F \rightarrow F)) \in$ is implied. Usually, strong negation is used for denying the existence of the formula following it.

The semantics for Paraconsistent Annotated Logics $P\mathcal{T}$ is defined as follows.

Definition 9.3 Let v be the set of all propositional symbols and F be the set of all formulas. An interpretation I is a function, $I : v \rightarrow \mathcal{T}$. To each interpretation I, we can associate the valuation function such that $v_I : \mathrm{F} \rightarrow \{0,1\}$, which is defined as follows.

1. let p be a propositional symbol and μ an annotation,
   $v_I(p{:}\ \mu) = 1$ iff $\mu \leq I(p)$ and $v_I(p{:}\ \mu) = 0$ iff otherwise,
2. let A and B be any formulas, and A not an annotated atom,
   $v_I(\neg A) = 1$ iff $v_I(A) = 0$ and $v_I(\sim B) = 1$ iff $v_I(B) = 0$,
3. other formulas $A \wedge B$, $A \vee B$, $A \rightarrow B$ are valuated as usual.

9.2.2 Generally Horn Program

We review a basic paraconsistent annotated logic program called Generally Horn Program (GHP for short) introduced in [5] as the basis of ALPSN. We assume the same complete lattice $\mathcal{T}$ in Fig. 9.1 and its ordering.

Definition 9.4 (*Annotated Literal*) If A is a literal and $\mu \in \mathcal{T}$ is an annotation, then $A{:}\mu$ is called an Annotated Literal. If μ is one of $\{\mathbf{t}, \mathbf{f}\}$, $A{:}\mu$ is called a Well-annotated literal, and μ is called a W-annotation.

[2] Hereafter expression $\neg p{:}\mu$ is conveniently used for expressing a negative annotated literal instead of $\neg(p{:}\mu)$ or $p{:}\neg(\mu)$.

Definition 9.5 Let A_0 be a positive literal and A_1, …, A_n literals.

$$A_1 \wedge \ldots \wedge A_n \rightarrow A_0$$

is called a Horn clause or a program clause.[3] The left part A_0 and right part $A_1 \wedge \ldots \wedge A_n$ of the symbol → are called the Head and Body of the program clause, respectively. Furthermore, the body $A_1 \wedge \ldots \wedge A_n$ is called the Definition of the literal A_0. A program claus $\rightarrow A_0$ that consists of only head part is called a Unit Program Clause, which is denoted by just A_0 without the right arrow symbol. A Logic Program is a set of program clauses.

Definition 9.6 (*Generally-Horn Program*) If A_0, …, A_n are literals and μ_0, …, μ_n are annotations,

$$A_1 : \mu_1 \wedge \ldots \wedge A_n : \mu_n \rightarrow A_0 : \mu_0$$

is called a Generalized Horn Clause (gh-clause for short). A Generally-Horn Program (GHP for short) is a set of gh-clauses.

Herbrand-like interpretation in [12] is considered for GHP as its semantics. The universe of individuals in the interpretation consists of all ground terms of the language being interpreted. Let the set of all individuals in models and interpretations be a Herbrand universe. A Herbrand interpretation I can be regarded as an interpretation such that $I : B_P \rightarrow \mathcal{T}$, where B_P is the Herbrand base (the set of all variable-free atoms) of the GHP P. The complete lattice of annotations and the epistemic negation in GHP are defined as well as the paraconsistent annotated logic $P\mathcal{T}$. Usually, an interpretation I for a GHP is denoted by the set, $\{(p : \sqcup\mu_i) | I \models p : \mu_1 \wedge \cdots \wedge p : \mu_n\}$, where $\sqcup\mu_i$ is the least upper bound of the set $\{\mu_1, \cdots, \mu_n\}$. Then the ordering $\leq$ over the complete lattice $\mathcal{T}$ is extended to an ordering over the set of interpretations.

Definition 9.7 Let I_1 and I_2 be any interpretations for a GHP P, and A an atom in the Herbrand base B_P.

$$I_1 \leq_I I_2 =_{def} (\forall A \in B_P)(I_1(A) \leq I_2(A)).$$

The fixpoint semantics [12] for GHP is defined as well as usual logic programs. Associate with each GHP P over the complete lattice $\mathcal{T}$, a monotonic operator T_P from Herbrand interpretations to themselves is defined.

Definition 9.8 Let P be a GHP, A, B_1,…, B_k literals, I an interpretation and μ, μ_1,…, μ_k, annotations. Then the monotonic operator T_P is defined as follows,

[3] We use the right arrow symbol → in program clauses as a logical connective as far as no confusion occurs.

$$T_P(I)(A) = \sqcup\left\{\mu \,\middle|\, \begin{array}{l} B_1 : \mu_1 \wedge \cdots \wedge B_k : \mu_k \rightarrow A : \mu \text{ ia a ground instance of} \\ \text{a gh clause in } P \text{ and } I \models B_1 : \mu_1 \wedge \cdots \wedge B_k : \mu_k \end{array}\right\},$$

where the symbol $\sqcup$ is used for denoting the least upper bound of a set.

Definition 9.9 Let Δ be a special interpretation for GHP that assigns the annotation $\perp$ to all members of B_P. Then the Upward Iteration of T_P is defined iteratively,

$$T_P \uparrow 0 = \Delta,\ T_P \uparrow \alpha = T_P(T_P \uparrow (\alpha - 1)),\ T_P \uparrow \lambda = \sqcup\{T_P \uparrow \eta | \eta < \lambda\},$$

where α is a successor ordinal and λ is a limit one.

Some well-known properties of the operator T_P are presented without proof, the proof for the following theorem can be found in [5].

Theorem 9.1

1. *T_P is a monotonic function.*
2. *Any GHP P has the least model M_P that is identical to the least fixpoint of the function* T_P.
3. $T_P \uparrow \omega = M_P$.

9.3 ALPSN to EVALPSN

In this section, we introduce the development of paraconsistent annotated logic programs called Annotated Logic Program with Strong Negation (ALPSN for short), Vector ALPSN (VALPSN for short) and Extended VALPSN (EVALPSN for short).

9.3.1 ALPSN and its Stable Model Semantics

Now we trace the development of paraconsistent annotated logic programs with strong negation. First of all, let us introduce Annotated Logic Program with Strong Negation (ALPSN for short), which is obtained by merging the strong negation into GHP, and its stable model semantics [16] is also introduced.

Definition 9.10 (*ALPSN*) Let L_0 ,..., L_n be well-annotated literals over the complete lattice $\mathcal{T}$ in Fig. 9.1, then,

$$L_1 \wedge \cdots \wedge L_i \wedge \sim L_{i+1} \wedge \cdots \wedge \sim L_n \rightarrow L_0$$

is called an Annotated Clause with Strong Negation (ALPSN clause or ASN-clause for short), where the symbol $\sim$ is the strong negation. ALPSN is defined as a finite set of ALPSN clauses.

We assume that ALPSN is a set of ground ALPSN clauses, then there is no loss of generality in making this assumption, since any logic program in the sense of Lloyd [12] may be viewed as such a set of ALPSN clauses by instanciating all variables occurring in the ALPSN-clauses.

The stable model semantics for general logic program was proposed by Gelfond and Lifschitz [9] and has been taken up as general semantics for various non-monotonic reasonings introduced in such as Nute [36]. Here we extend the stable model semantics for general logic program to ALPSN. First, in order to eliminate the strong negation in ALPSN the Gelfond-Lifschitz transformation for general logic program is modified.

Definition 9.11 (*Gelfond-Lifschitz Transformation for ALPSN*) Let I be a Herbrand interpretation for an ALPSN P. Then P^I, the Gelfond-Lifschitz transformation of the ALPSN P with respect to the Herbrand interpretation I, is a GHP obtained from the ALPSN P by deleting:

1. each ALPSN-clause that has a strongly negated annotated literal $\sim(C : \mu)$ in its body with $I \models C : \mu$, and
2. all strongly negated annotated literals in bodies of the remaining ALPSN clauses.

The Gelfond-Lifschitz transformation P^I has the unique least model given by $T_{P^I} \uparrow \omega$ as described in Theorem 1, since it includes neither epistemic nor strong negation. Therefore, the stable model for ALPSN is defined as follows.

Definition 9.12 (*Stable Model for ALPSN*) Let I be a Herbrand interpretation of an ALPSN P,

I is called the stable model of the ALPSN P iff $I = T_{P^I} \uparrow \omega$.

We have shown that ALPSN stable model can provide annotated semantics for some non-monotonic logics such as Reiter's default logic [39]. For example, we have proposed a translation from Reiter's default theory into ALPSN and proved that there is a one-to-one correspondence between the extension of the original default theory and the stable model for the ALPSN translation in [16, 17]. This result shows that default theory extension can be calculated by the corresponding ALPSN stable model computation. However, it is not so suitable for formalizing the semantics for other kinds of non-monotonic reasoning such as defeasible reasoning. In order to deal with defeasible reasoning, we have developed a new version of ALPSN.

9.3.2 VALPSN

A new version of ALPSN called Vector Annotated Logic Program with Strong Negation (VALPSN for short), which has a different kind of annotations from ALPSN and is able to deal with defeasible reasoning, has been introduced in [19]. An annotation in ALPSN represents simple truth values such as t(true) or f(false). On the other hand, an annotation in VALSPN called a Vector Annotation is a 2-dimensional non-negative integer vector (i, j) whose components i and j represent the amounts of positive and negative knowledge, respectively.

Definition 9.13 (*Vector Annotation*) A vector annotation is a 2-dimensional nonnegative integer vector, and the complete lattice $\mathcal{T}_v(n)$ of vector annotations is defined as follows,

$$\mathcal{T}_v(n) = \{(i,j) | 0 \leq i \leq n, 0 \leq j \leq n, i, j \text{ and } n \text{ are integers}\}$$

The ordering of the complete lattice $\mathcal{T}_v(n)$ of vector annotations is denoted by symbol $\preccurlyeq_v$ and defined as follows: let $\mathbf{v_1} = (x_1, y_1)$ and $\mathbf{v_2} = (x_2, y_2)$ be vector annotations,

$$\mathbf{v_1} \preccurlyeq_v \mathbf{v_2} \text{ iff } x_1 \leq x_2 \text{ and } y_1 \leq y_2.$$

The first component i of the vector annotation (i, j) denotes the amount of positive information to support the followed literal and the second one, j, does the amount of negative one as well. Vector annotated literals also can be interpreted epistemically as well as usual annotated literals in ALPSN. For example, vector annotated literal p:(2,0) can be interpreted positively that "the literal p is known to be true of strength 2", vector annotated literal p:(0,1) negatively that "the literal p is known to be false of strength 1", vector annotated literal p:(2,1) paraconsistently that "the literal p is known to be true of strength 2 and false of strength 1", and vector annotated literal p:(0, 0) that "the literal p is known to be neither true nor false" (there is no information in terms of the literal p). Therefore, the epistemic negation for vector annotated literals can be defined as a mapping for exchanging the first and second components of vector annotations.

Definition 9.14 (*Epistemic Negation in VALPSN*) Let (i, j) be a vector annotation. The epistemic negation, $\neg$ for vector annotated literals is defined as the following mapping over the complete lattice $\mathcal{T}_v(n)$, $\neg(i, j) = (j, i)$.

The epistemic negation followed by a vector annotated literal can be eliminated as well as ALPSN.

Definition 9.15 (*Well Vector Annotated Literal*) Let p be a literal. p:$(i,0)$ or p:$(0, j)$ are called Well Vector Annotated Literals (wva-literals for short), where i and j are non-negative integers.

Definition 9.16 (VALPSN) Let L_0 ,…, L_n be well vector annotated literals over the complete lattice $\mathcal{T}_v(n)$, then,

$$L_1 \wedge \cdots \wedge L_i \wedge \sim L_{i+1} \wedge \cdots \wedge \sim L_n \rightarrow L_0$$

is called a Vector Annotated Clause with Strong Negation (VALPSN clause for short). A Vector Annotated Logic Program with Strong Negation (VALPSN clause or VASN clause for short) is a finite set of VALPSN clauses. If a VALPSN or a VALPSN clause contain no strong negation, they may be called just a VALP or a VALP clause, respectively.

Defeasible logic is known as a non-monotonic formalism that can deal with defeasible reasoning [8]. A defeasible logic was originally introduced in [34]. Since then the defeasible logic had been developed in [3, 4, 35, 36]. In [19], we have proposed a translation from defeasible theory in [4] into VALPSN and shown that the VALPSN stable model for the defeasible theory can provide vector annotated semantics for defeasible reasoning. Furthermore, paraconsistent annotated logic program VALPSN has been developed to deal with deontic notions such as obligation.

9.3.3 EVALPSN

Deontic logic is known as one of the modal logics, which can reason normative behavior represented by modal operators such as obligation, permission and forbiddance. It has been applied to computer science in [13, 14] and modeling legal argument in [36]. Usually, symbol $\circ$ is used to represent obligation in deontic logic as follows: let p be a literal, then we have the following interpretations. p, "p is a fact". $\circ p$, "p is obligatory". $\circ\neg p$, "p is forbidden". $\neg\circ\neg p$, "p is permitted." Furthermore, a defeasible deontic logic to deal with deontic notions and defeasible deontic reasoning has been developed in [35] We have proposed EVALPSN by extending VALPSN to deal with deontic notions and shown that it can deal with defeasible deontic reasoning in [20, 21].

An extended annotation in EVALPSN has a form of $[(i, j), \mu]$ and is called an Extended Vector Annotation. The first component (i, j) of the extended vector annotation is a vector annotation in the complete lattice $\mathcal{T}_v(n)$ as well as VALPSN and the second component μ is an index of facts and deontic notions such as obligation, and the set of the second components constitutes the complete lattice,

$$\mathcal{T}_d = \{\bot, \alpha, \beta, \gamma, ^*1, ^*2, ^*3, \top\}.$$

The ordering ($\preccurlyeq_d$) of the complete lattice $\mathcal{T}_d$ is described by the Hasse's diagram in Fig. 9.2.

The intuitive meaning of each member of the complete lattice $\mathcal{T}_d$ is $\bot$(unknown), $\top$(inconsistency), α(fact), β(obligation), γ(non-obligation), $*_1$(fact and obligation), $*_2$(obligation and non-obligation), $*_3$(fact and non-obligation).

Definition 9.17 The complete lattice $\mathcal{T}_e(n)$ of extended vector annotations is defined as: $\mathcal{T}_e(n) = \mathcal{T}_v(n) \times \mathcal{T}_d$ and the ordering($\preccurlyeq_e$) of the complete lattice $\mathcal{T}_e(n)$ is defined as: let $[(i_1, j_1), \mu_1]$ and $[(i_2, j_2), \mu_2]$ be extended vector annotations,

$$[(i_1,j_1),\mu_1] \preccurlyeq_e [(i_2,j_2),\mu_2] \text{ iff } (i_1,j_1) \preccurlyeq_v (i_2,j_2) \text{ and } \mu_1 \preccurlyeq_d \mu_2.$$

EVALPSN has two kinds of epistemic negation, $\neg_1$ and $\neg_2$, which are defined as mappings over complete lattices $\mathcal{T}_v(n)$ and $\mathcal{T}_d$, respectively.

Definition 9.18 (*Epistemic Negations in EVALPSN, $\neg_1$ and $\neg_2$*)

$$\neg_1([(i,j),\mu]) = [(j,i),\mu], \qquad \text{where } \mu \in \mathcal{T}_d,$$
$$\neg_2([(i,j),\perp]) = [(i,j),\perp], \quad \neg_2([(i,j),\alpha]) = [(i,j),\alpha], \quad \neg_2([(i,j),\beta]) = [(i,j),\gamma],$$
$$\neg_2([(i,j),\gamma]) = [(i,j),\beta], \quad \neg_2([(i,j),*_1]) = [(i,j),\perp], \quad \neg_2([(i,j),*_2]) = [(i,j),*_2],$$
$$\neg_2([(i,j),*_3]) = [(i,j),*_1], \quad \neg_2([(i,j),\top]) = [(i,j),\top].$$

The epistemic negations, $\neg_1$ and $\neg_2$ followed by extended vector annotated literals can be eliminated by the syntactic operations in Definition 9.18. The strong negation $\sim$ in EVALPSN can be defined by one of the epistemic negations $\neg_1$ and $\neg_2$ as well as Definition 9.2.

Definition 9.19 (*well extended vector annotated literal*) Let p be a literal. p:$[(i,0), \mu]$ and p:$[(0, j), \mu]$ are called Well Extended Vector Annotated Literals (weva-literals for short), where $i, j \in \{1, 2,\ldots, n\}$, and $\mu \in \{\alpha, \beta, \gamma\}$.

Definition 9.20 (*EVALPSN*) If $L_0, \ldots, L_n$ are weva-literals,

$$L_1 \wedge \cdots \wedge L_i \wedge \sim L_{i+1} \wedge \cdots \wedge \sim L_n \rightarrow L_0$$

is called an Extended Vector Annotated Logic Program Clause with Strong Negation (EVALPSN clause for short). An Extended Vector Annotated Logic Program with Strong Negation (EVALPSN for short) is a finite set of EVALPSN clauses. If an EVALPSN or an EVALPSN clause contain no strong negation, they may be called just an EVALP or an EVALP clause, respectively.

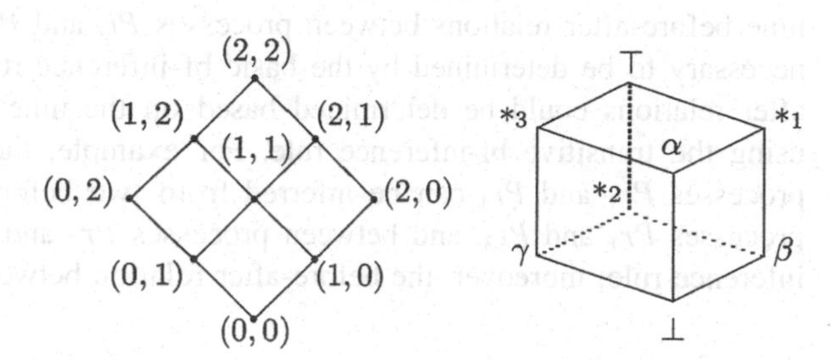

Fig. 9.2 Complete lattices $\mathcal{T}_v(2)$ and $\mathcal{T}_d$

9.4 Before-After EVALPSN

In this section, we introduce one version of EVALPSN called before-after (bf for hort)-EVALPSN, which can deal with before-after relations between two time intervals (processes) [30, 32, 33, 38, 39]. The reasoning system of bf-EVALPSN consists of two kinds of inference rules called the Basic Bf-inference Rule and the Transitive Bf-inference Rule, both of which can be translated into bf-EVALPSN and are useful implementation tools for bf-EVALPSN based real-time process order verification systems. Bf-EVALPSN has a special literal, $R(p_m, p_n, t)$ called Bf-literal whose vector annotation (i, j) represents the before-after relation between processes Pr_m and Pr_n at time t.[4] The integer components i and j of the vector annotation (i, j) represent the after and before degrees between processes Pr_m and Pr_n, respectively, and the before-after relation between processes Pr_m and Pr_n at time t can be represented in the vector annotation of the bf-literal para-consistently. For example, an annotated bf-literal, $R(p_m, p_n, t):[(i, j), \alpha]$ can be intuitively interpreted as: the before-after relation between processes $Pr_i(p_i)$ and $Pr_j(p_j)$ at time t is represented by the vector annotation (i, j).

In the bf-EVALPSN reasoning system, basic bf-inference rules are used for determining the vector annotation of a bf-literal in real-time according to start/finish time information of two processes. On the other hand, transitive bf-inference rules are used for determining the vector annotation of a bf-literal based on vector annotations of two related bf-literals in real-time. We describe the convenience of the transitive bf-inference rule. Suppose that there are three processes, Pr_0, Pr_1 and Pr_2 starting in sequence, then the before-after relation between processes Pr_0 and Pr_2 can be determined from two before-after relations between processes Pr_0 and Pr_1, and between processes Pr_1 and Pr_2. Such reasoning of process before-after relation is also formalized as transitive bf-inference rules in bf-EVALPSN. The transitive bf-inference reasoning system can contribute to reduce using times of the basic bf-inference rule and it is a unique remarkable feature of the bf-EVALPSN reasoning system. Suppose that there is a process order control system dealing with ten processes Pr_0, Pr_1, ... and Pr_9 starting in sequence. Without the transitive bf-inference rule, the process order control system has to deal with 45 before-after relations independently by the basic bf-inference rule. However, if we use the transitive bf-inference rule, just nine before-after relations between processes Pr_i and Pr_{i+1} $(i = 0,1, 2, \ldots 8)$ are necessary to be determined by the basic bf-inference rule, and the other before-after relations could be determined based on the nine before-after relations by using the transitive bf-inference rule. For example, the before-after relation of processes Pr_1 and Pr_4 can be inferred from two before-after relations between processes Pr_1 and Pr_3, and between processes Pr_3 and Pr_4 by the transitive bf-inference rule; moreover, the before-after relation between processes Pr_1 and Pr_3

[4] In this chapter, process Pr_{id} is represented as p_{id} in bf-EVALPSN literal.

is inferred from two before-after relations between processes Pr_1 and Pr_2, and between processes Pr_2 and Pr_3 by the transitive bf-inference rule.

Now we define bf-EVALPSN formally.

Definition 9.21 (*bf-EVALPSN*) An extended vector annotated bf-literal,

$$R(p_m, p_n, t) : [(i, j), \mu]$$

is called a Bf-EVALP literal. If an EVALPSN clause contains bf-EVALP literals it is called a Bf-EVALPSN Clause or a Bf-EVALP Clause if it contains no strong negation. A Bf-EVALPSN is a finite set of bf-EVALPSN clauses.

We provide a paraconsistent before-after interpretation for vector annotations to represent bf-relations in bf-EVALPSN, and such vector annotations are called Bf-annotations. Exactly speaking, bf-relations are classified into fifteen meaningful sorts according to bf-relations between each start/finish time of two processes. First of all, we define the most basic bf-relation in bf-EVALPSN.

Before(be)/After(af) Bf-relations Before and After are defined according to the bf-relation between each start time of two processes, which are represented by bf-annotations **be/af**, respectively. Suppose that there are process Pr_i with its start time x_s and finish time x_f, and process Pr_j with its start time y_s and finish time y_f. If one process has started before/after another one starts, then the bf-relations between them are defined as "before(**be**)/after(**af**)", respectively. They are described by the process time chart in Fig. 9.3 with the condition that process Pr_i has started before process Pr_j starts. The bf-relation between their start/finish times is denoted by inequality $x_s < y_s$.[5]

We define other kinds of bf-relations as follows.

Disjoint Before(db)/After(da) Bf-relations Disjoint Before/After between two processes are represented by bf-annotations **db/da**, respectively. The expression "disjoint before/after" implies that there is a time lag between the earlier process finish time and the later one's start time. They also are described by the process time chart in Fig. 9.4 with the condition that process Pr_i had finished before process Pr_j started. The bf-relation between their start/finish times is denoted by inequality $x_f < y_s$.

Immediate Before(mb)/After(ma) Bf-relations Immediate Before/After between two processes are represented by bf-annotations **mb/ma**, respectively. The expression "immediate before/after" implies that there is no time lag between the earlier process finish time and the later one's start time. The bf-relations are

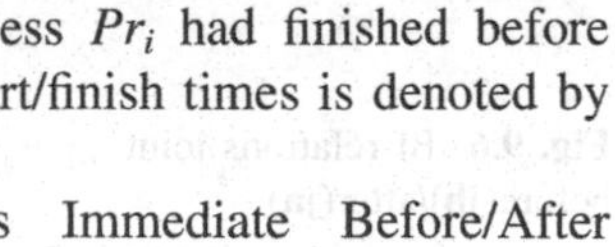

Fig. 9.3 Bf-relations before(**be**)/after(**af**)

[5] If time t_1 is earlier than t_2, we conveniently denote the relation by inequality.

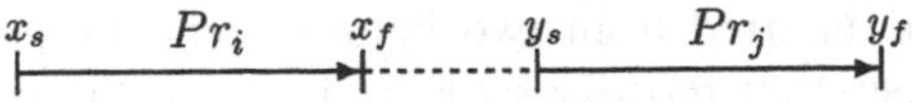

Fig. 9.4 Bf-relations disjoint before(**db**)/after(**af**)

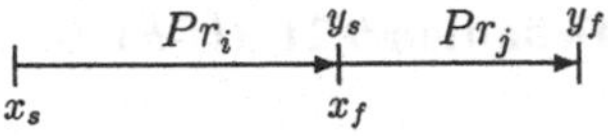

Fig. 9.5 Bf-relations immediate before(**mb**)/after(**ma**)

also described by the process time chart in Fig. 9.5 with the condition that process Pr_i had finished immediately before process Pr_j started. The bf-relation between their start/finish times is denoted by equality $x_f = y_s$.

Joint Before(jb)/After(ja) Bf-relations Joint Before/After between two processes are represented by bf-annotations **jb/ja**, respectively. The expression "joint before/after" implies that the two processes overlap and the earlier process had finished before the later one finished. The bf-relations are also described by the process time chart in Fig. 9.6 with the condition that process Pr_i had started before process Pr_j started and process Pr_i had finished before process Pr_j finished. The bf-relation between their start/finish times is denoted by inequalities

$$x_s < y_s < x_f < y_f.$$

S-included Before(sb)/After(sa) Bf-relations S-included Before/After between two processes are represented by bf-annotations **sb/sa**, respectively. The expression "s-included before/after" implies that one process had started before another one started and they have finished at the same time. The bf-relations are also described by the process time chart in Fig. 9.7 with the condition that process Pr_i had started before process Pr_j started and they have finished at the same time. The bf-relation between their start/finish times is denoted by equality $x_f < y_f$ and inequalities $x_s < y_s < x_f$.

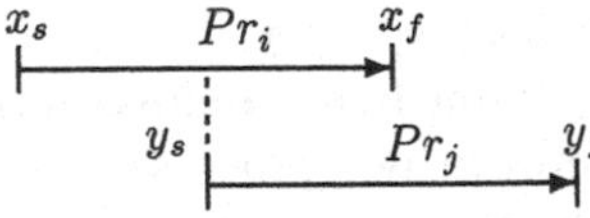

Fig. 9.6 Bf-relations joint before(**jb**)/after(**ja**)

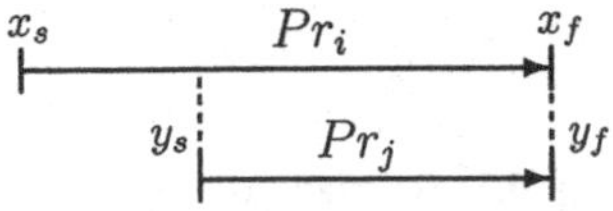

Fig. 9.7 Bf-relations S-included before(**sb**)/after(**sa**)

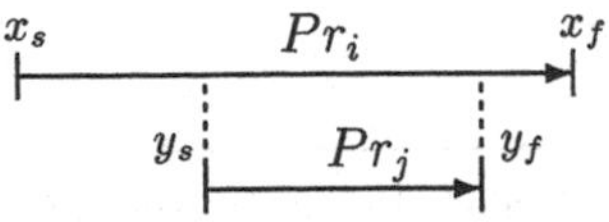

Fig. 9.8 Bf-relations included before(**ib**)/after(**ia**) Fig. 9.18: process time chart ex-3

Included Before(ib)/After(ia) Bf-relations Included Before/After between processes Pr_i and Pr_j are represented by bf-annotations **ib/ia**, respectively. The expression "included before/after" implies that one process had started/finished before/after another one started/finished, respectively. The bf-relations are also described by the process time chart in Fig. 9.8 with the condition that process Pr_i had started before process Pr_j started and process Pr_i has finished after process Pr_j finished. The bf-relation between their start/finish times is denoted by inequalities $x_s < y_s$ and $y_f < x_f$.

F-included Before(fb)/After(fa) Bf-relations F-included Before/After between two processes Pr_i and Pr_j are represented by bf-annotations **fb/fa**, respectively. The expression, "f-included before/after" implies that the two processes have started at the same time and one process had finished before another one finished.The bf-relations are also described by the process time chart in Fig. 9.9 with the condition that processes Pr_i and Pr_j had started at the same time and process Pr_i has finished after process Pr_j finishes. The bf-relation between their start/finish times is denoted by equality $x_s = y_s$ and inequality $y_f < x_f$.

Paraconsistent Before-after(pba) Bf-relation Paraconsistent Before-after between two processes Pr_i and Pr_j is represented by bf-annotation **pba**. The expression "paraconsistent before-after" implies that two processes had started at the same time and also finished at the same time. The bf-relation is described by the process time chart in Fig. 9.10 with the condition that processes Pr_i and Pr_j had started at the same time and also finished at the same time. The bf-relation between their start/finish times is denoted by equalities $x_s = y_s$ and $y_f = x_f$.

Now we consider the mapping of bf-relation annotations, {**be**, **af**, **db**, **da**, **mb**, **ma**, **jb**, **ja**, **ib**, **ia**, **sb**, **sa**, **fb**, **fa**, **pba**} into vector annotations. If we consider a before-after measure over the 15 bf annotations, obviously there exists a partial order($<_h$) in the before-after measure, where $\mu_1 < h\ \mu_2$ is intuitively interpreted that bf-annotation μ_1 is more "before" than bf-annotation μ_2, and μ_1, $\mu_2 \in${**be**, **af**, **db**, **da**, **mb**, **ma**, **jb**, **ja**, **ib**, **ia**, **sb**, **sa**, **fb**, **fa**, **pba**}.

If $\mu_1 <_h \mu_2$ and $\mu_2 >_h \mu_1$, then we denote it $\mu_1 \equiv_h \mu_2$. Then, we have the following ordering:

$$\mathbf{db} <_h \mathbf{mb} <_h \mathbf{jb} <_h \mathbf{sb} <_h \mathrm{ib} <_h \mathbf{fb} <_h \mathbf{pba} <_h \mathbf{ia} <_h \mathbf{ja} <_h \mathbf{ma} <_h \mathbf{da}$$

and

$$\mathbf{sb} \equiv_h \mathbf{be} <_h \mathbf{af} \equiv_h \mathbf{sa}$$

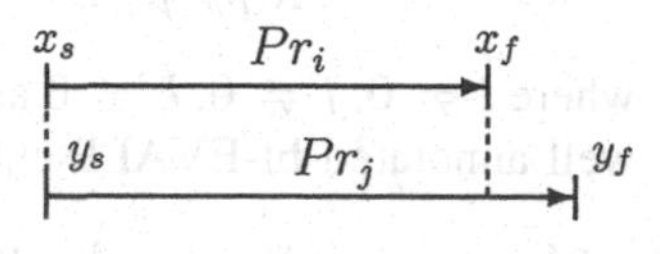

Fig. 9.9 Bf-relations F-included before(**fb**)/after(**fa**)

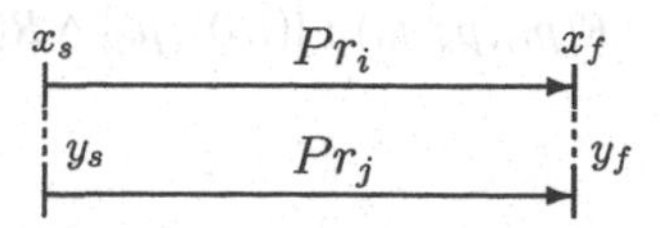

Fig. 9.10 Bf-relations paraconsistent before-after(**pba**)

On the other hand, if we take the knowledge(information) amount that each bf relation expresses into account as another measure, obviously there exists another partial order($<_v$) in terms of the knowledge amount, where a relation $\mu_1 <_v \mu_2$ means that the bf-annotation μ_1 has less knowledge amount in terms of before-after relation than the bf-annotation μ_2. If $\mu_1 <_v \mu_2$ and $\mu_2 <_v \mu_1$, we denote it $\mu_1 <_v \mu_2$. Then, we have the following ordering:

$$\mathbf{be} <_v \mu_1, \text{ where } \mu_1 \in \{\mathbf{db}, \mathbf{mb}, \mathbf{jb}, \mathbf{sb}, \mathbf{ib}, \mathbf{fb}, \mathbf{pba}\},$$
$$\mathbf{af} <_v \mu_2, \text{ where } \mu_2 \in \{\mathbf{da}, \mathbf{ma}, \mathbf{ja}, \mathbf{sa}, \mathbf{ia}, \mathbf{fa}, \mathbf{pba}\},$$
$$\mathbf{db} \equiv_v \mathbf{mb} \equiv_v \mathbf{jb} \equiv_v \mathbf{sb} \equiv_v \mathbf{ib} \equiv_v \mathbf{fb} \equiv_v \mathbf{pba} \equiv_v \mathbf{fa} \equiv_v \mathbf{ia} \equiv_v \mathbf{sa} \equiv_v \mathbf{ja} \equiv_v \mathbf{ma} \equiv_v \mathbf{da}$$
and
$$\mathbf{be} \equiv_v \mathbf{af}$$

If we regard the before-after degree as the horizontal measure and the before-after knowledge amount as the vertical one, we obtain the complete bi-lattice, $\mathcal{T}_v(12) = \{\perp_{12}(0,0),\ldots,$ **be**(0,8),..., **db**(0,12),..., **mb**(1,11),..., **jb**(2,10),..., **sb**(3,9),..., **ib**(4,8),..., **fb**(5,7),..., **pba**(6,6),..., **fa**(7,5),..., **ia**(8,4),..., **sa**(9,3),..., **ja**(10,2),..., **ma** (11,1),..., **da**(12,0),...,$\top_{12}(12,12)\}$ of bf-annotations, which is described by the Hasse's diagram in Fig. 9.11. Now we define the epistemic negation $\neg_1$ for bf-annotations, which maps bf-annotations to themselves.

Definition 9.22 (*Epistemic Negation* $\neg_1$ *for Bf-annotations*) The epistemic negation $\neg_1$ over the bf-annotations, {**be, af, db, da, mb, ma, jb, ja, ib, ia, sb, sa, fb, fa, pba**} is obviously defined as the following mapping:

$$\neg_1(\mathbf{af}) = \mathbf{be}, \neg_1(\mathbf{be}) = \mathbf{af}, \neg_1(\mathbf{da}) = \mathbf{db}, \neg_1(\mathbf{db}) = \mathbf{da}, \neg_1(\mathbf{ma}) = \mathbf{mb},$$
$$\neg_1(\mathbf{mb}) = \mathbf{ma}, \neg_1(\mathbf{ja}) = \mathbf{jb}, \neg_1(\mathbf{jb}) = \mathbf{ja}, \neg_1(\mathbf{jb}) = \mathbf{ja}, \neg_1(\mathbf{sa}) = \mathbf{sb},$$
$$\neg_1(\mathbf{sb}) = \mathbf{sa}, \neg_1(\mathbf{ia}) = \mathbf{ib}, \neg_1(\mathbf{ib}) = \mathbf{ib}, \neg_1(\mathbf{fa}) = \mathbf{fb}, \neg_1(\mathbf{fb}) = \mathbf{fa}, \neg_1(\mathbf{pba}) = \mathbf{pba}.$$

We note that a bf-EVALP literal $R(p_m, p_n, t)$:$[(i, j), \mu]$ would not be well annotated if neither $i \neq 0$ nor $j \neq 0$. However, since the bf-EVALP literal is equivalent to the conjunction of two well annotated bf-EVALP literals,

$$R(p_m, p_n, t) : [(i, 0), \mu] \wedge R(p_m, p_n, t) : [(0, j), \mu].$$

For example, suppose a non-well annotated bf-EVALP clause

$$R(p_m, p_n, t_0) : [(i, j), \mu_0] \rightarrow R(p_m, p_n, t_1) : [(k, l), \mu_1]$$

where $i \neq 0, j \neq 0, k \neq 0$ and $\neq l0$. It can be equivalently transformed into two well annotated bf-EVALP clauses,

$$R(p_m, p_n, t_0) : [(i, 0), \mu_0] \wedge R(p_m, p_n, t_0) : [(0, j), \mu_0] \rightarrow R(p_m, p_n, t_1) : [(k, 0), \mu_1],$$
$$R(p_m, p_n, t_0) : [(i, 0), \mu_0] \wedge R(p_m, p_n, t_0) : [(0, j), \mu_0] \rightarrow R(p_m, p_n, t_1) : [(0, l,), \mu_1].$$

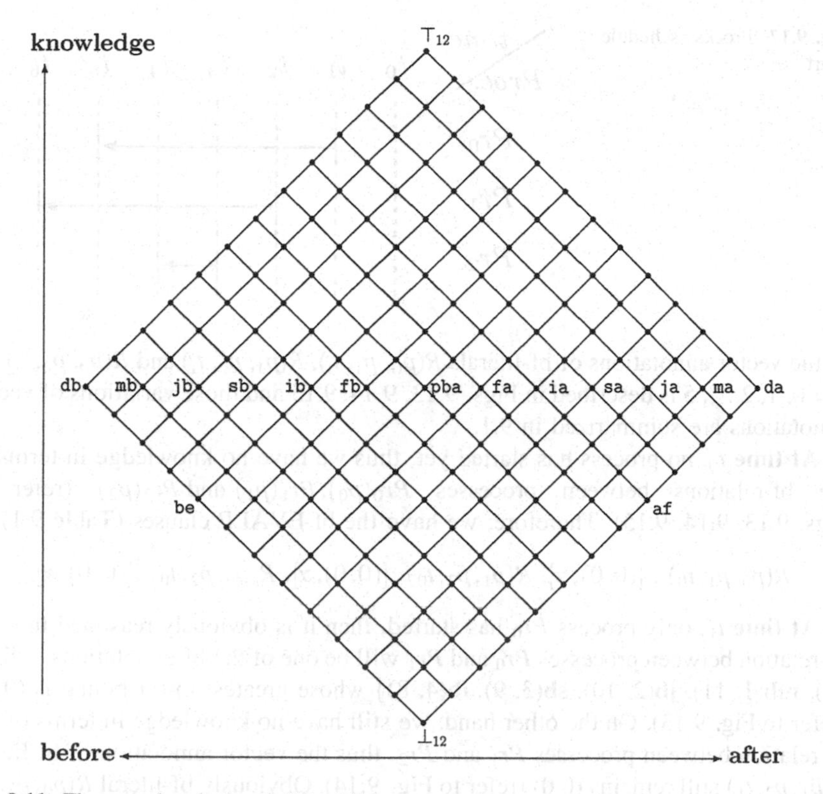

Fig. 9.11 The complete lattice $\mathcal{T}_v(12)$ of Bf-annotations

9.5 Reasoning System in Bf-EVALPSN

In this section, we provide the before-after relation reasoning system of bf-EVALPSN, which consists of the basic inference rule for reasoning a bf-relation according to the before-after relations of process start/finish times and the transitive inference rule for reasoning a bf-relation from two other bf-relations transitively.

9.5.1 Example of Bf-Relation Reasoning

First of all, we show a simple example of bf-relation reasoning by the bf-relation inference rule. Suppose that processes Pr_0, Pr_1 and Pr_2 are scheduled to be processed according to the time chart in Fig. 9.12, then we show how the bf-relations between those processes are reasoned at each time, $t_i (0 \leq i \leq 7)$. Each variation

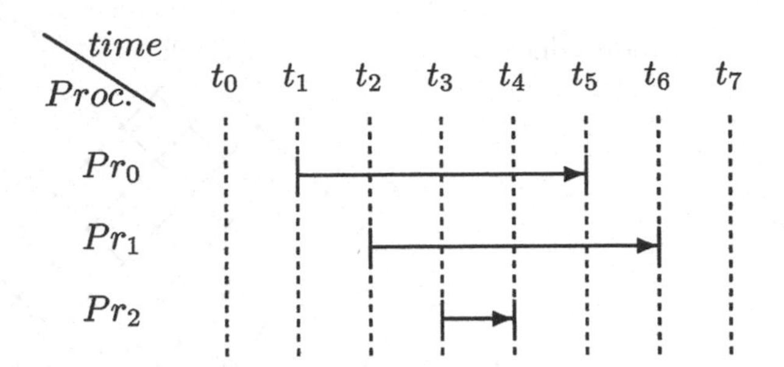

Fig. 9.12 Process schedule chart

of the vector annotations of bf-literals $R(p_0, p_1, t_i)$, $R(p_1, p_2, t_i)$ and $R(p_0, p_2, t_i)$ for $i = 0, 1, 2, \ldots, 5$ is described in Figs. 9.13, 9.14, 9.15 and those variations of vector annotations are summarized in 9.1.

At time t_0, no process has started yet, thus we have no knowledge in terms of the bf-relations between processes $Pr_0(p_0), Pr_1(p_1)$ and $Pr_2(p_2)$ (refer to Figs. 9.13, 9.14, 9.15). Therefore, we have the bf-EVALP clauses (Table 9.1),

$$R(p_0, p_1, t_0) : [(0,0), \alpha],\ R(p_1, p_2, t_0) : [(0,0), \alpha],\ R(p_0, p_2, t_0) : [(0,0), \alpha].$$

At time t_1, only process Pr_0 has started, then it is obviously reasoned that the bf-relation between processes Pr_0 and Pr_1 will be one of the bf-annotations, {**db**(0, 12), **mb**(1, 11), **jb**(2, 10), **sb**(3, 9), **ib**(4, 8)} whose greatest lower bound is (0, 8) (refer to Fig. 9.13). On the other hand, we still have no knowledge in terms of the bf-relation between processes Pr_1 and Pr_2, thus the vector annotation of bf-literal $R(p_1, p_2, t_1)$ still remains (0,0) (refer to Fig. 9.14). Obviously bf-literal $R(p_0, p_2, t_1)$ has the same vector annotation **be**(0,8) as the one of bf-literal $R(p_0, p_1, t_1)$ (refer to Fig. 9.15). Therefore, we have the bf-EVALP clauses,

$$R(p_0, p_1, t_1) : [(0,8), \alpha],\ R(p_1, p_2, t_1) : [(0,0), \alpha],\ R(p_0, p_2, t_1) : [(0,8), \alpha].$$

At time t_2, process Pr_1 has started before process Pr_0 finishes, then it is obviously reasoned that the bf-relation between processes Pr_0 and Pr_1 will be one of the bf-annotations, {**jb**(2, 10), **sb**(3, 9), **ib**(4, 8)} whose greatest lower bound is (2, 8) (refer to Fig. 9.13). As process Pr_2 has not started yet, the vector annotation of bf-literal $R(p_1, p_2, t_2)$ turns to (0, 8) from (0, 0) as well as the one of bf-literal $R(p_0, p_2, t_1)$ (refer to Fig. 9.14) and the vector annotation of bf-literal $R(p_0, p_2, t_2)$ still remains (0, 8) (refer to Fig. 9.15). Therefore, we have the bf-EVALP clauses,

$$R(p_0, p_1, t_2) : [(2,8), \alpha],\ R(p_1, p_2, t_2) : [(0,8), \alpha],\ R(p_0, p_2, t_2) : [(0,8), \alpha].$$

At time t_3, process Pr_2 has started before processes Pr_0 and Pr_1 finish, then the vector annotation of bf-literal $R(p_0, p_1, t_3)$ still remains (2, 8) (refer to Fig. 9.13), and the vector annotation (0, 8) of both bf-literals $R(p_0, p_1, t_3)$ and $R(p_0, p_2, t_3)$ turns to (2,8) as well as the one of bf-literal $R(p_0, p_1, t_2)$ (refer to Figs. 9.14 and 9.15). Therefore, we have the bf-EVALP clauses,

$$R(p_0, p_1, t_3) : [(2,8), \alpha],\ R(p_1, p_2, t_3) : [(2,8), \alpha],\ R(p_0, p_2, t_3) : [(2,8), \alpha].$$

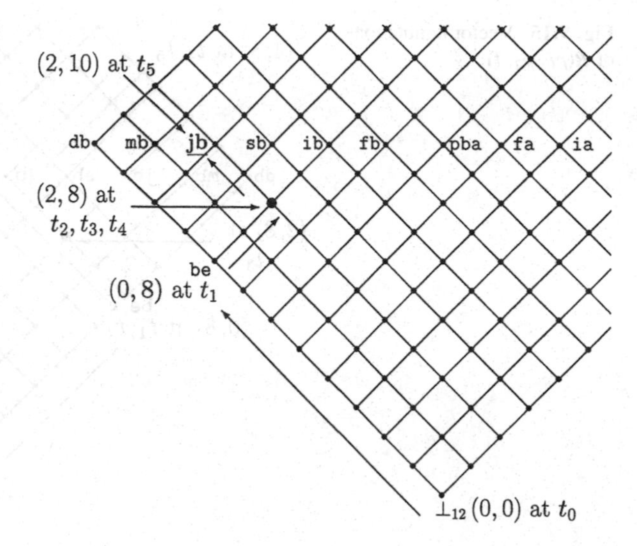

Fig. 9.13 Vector annotations of $R(p_0, p_1, t)$

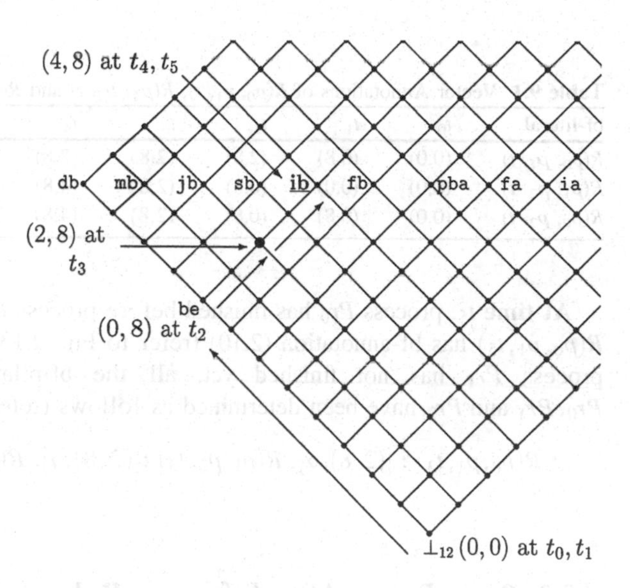

Fig. 9.14 Vector annotations of $R(p_1, p_2, t)$

At time t_4, only process Pr_2 has finished before processes Pr_0 or Pr_1 finish, then bf-literals $R(p_1, p_2, t_4)$ and $R(p_0, p_2, t_4)$ have the same bf-annotation (4,8) (refer to Figs. 9.14 and 9.15). On the other hand, the vector annotation of bf-literal $R(p_0, p_1, t_4)$ still remains (2,8) (refer to Fig. 9.13). Therefore, we have the bf-EVALP clauses,

$$R(p_0, p_1, t_4) : [(2,8), \alpha],\ R(p_1, p_2, t_4) : [\mathbf{ib}(4,8), \alpha],\ R(p_0, p_2, t_4) : [\mathbf{ib}(4,8), \alpha].$$

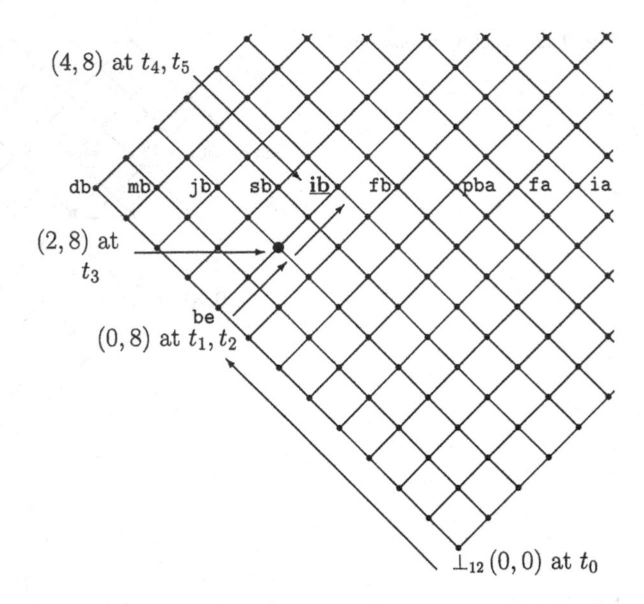

Fig. 9.15 Vector annotations of $R(p_0, p_2, t)$

Table 9.1 Vector Annotations of $R(p_0, p_1, t)$, $R(p_1, p_2, t)$ and $R(p_0, p_2, t)$

bf-literal	t_0	t_1	t_2	t_3	t_4	t_5	t_6	t_7
$R(p_0, p_1, t)$	(0,0)	(0,8)	(2,8)	(2,8)	(2,8)	(2,10)	(2,10)	(2,10)
$R(p_1, p_2, t)$	(0,0)	(0,0)	(0,8)	(2,8)	(4,8)	(4,8)	(4,8)	(4,8)
$R(p_0, p_2, t)$	(0,0)	(0,8)	(0,8)	(2,8)	(4,8)	(4,8)	(4,8)	(4,8)

At time t_5, process Pr_0 has finished before process Pr_1 finishes, then bf-literal $R(p_0, p_1, t_5)$ has bf-annotation (2,10) (refer to Fig. 9.13). Therefore, even though process Pr_1 has not finished yet, all the bf-relations between processes Pr_0, Pr_1 and Pr_2 have been determined as follows (refer to Figs. 9.14 and 9.15),

$$R(p_0, p_1, t_3) : [(2, 8), \alpha],\ R(p_1, p_2, t_3) : [(2, 8), \alpha],\ R(p_0, p_2, t_3) : [(2, 8), \alpha].$$

9.5.2 Basic Before-After Inference Rule

Now we construct the basic bf-inference rule by referring to the example in Sect. 5.1. In order to represent the basic bf-inference rule in bf-EVALPSN, we newly introduce two literals:

$st(p_i, t)$, which is intuitively interpreted as process Pr_i starts at time t, and
$fi(p_i, t)$, which is intuitively interpreted as process Pr_i finishes at time t.

Those literals are used for expressing process start/finish information and may have one of the vector annotations, $\{\perp(0,0), \mathbf{t}(1,0), \mathbf{f}(0,1), \top(1,1)\}$, where annotations t and f can be intuitively interpreted as "true" and "false", respectively.

Firstly, we show a group of basic bf-inference rules to be applied at the initial stage (time t_0) for bf-relation reasoning, which are named (0, 0)-*rules*.

(0, 0)-rules. Suppose that no process has started yet and the vector annotation of bf-literal $R(p_i, p_j, t)$ is (0, 0), which shows that there is no knowledge in terms of the bf-relation between processes Pr_i and Pr_j, then the following two basic bf-inference rules are applied at the initial stage.

(0, 0)-rule-1

If process Pr_i started before process Pr_j starts, then, the vector annotation (0, 0) of bf-literal $R(p_i, p_j, t)$ should turn to bf-annotation **be**(0, 8), which is the greatest lower bound of the bf-annotations, {**db**(0, 12), **mb**(1, 11), **jb**(2, 10), **sb**(3, 9), **ib**(4, 8)}.

(0, 0)-rule-2

If both processes Pr_i and Pr_j have started at the same time, then, it is reasonably anticipated that the bf-relation between processes Pr_i and Pr_j will be one of the bf-annotations, {**fb**(5, 7), **pba**(6, 6), **fa**(7, 5)} whose greatest lower bound is (5, 5) (refer to Fig. 9.11). Therefore, the vector annotation (0, 0) of bf-literal $R(p_i, p_j, t)$ should turn to (5, 5).

Basic bf-inference rules (0, 0)-rule-1 and (0, 0)-rule-2 may be translated into the bf-EVALPSN clauses,

$$R(p_i, p_j, t) : [(0,0), \alpha] \wedge st(p_i, t) : [\mathbf{t}, \alpha] \wedge \sim st(p_j, t) : [\mathbf{t}, \alpha] \\ \rightarrow R(p_i, p_j, t) : [(0,8), \alpha] \tag{9.1}$$

$$R(p_i, p_j, t) : [(0,0), \alpha] \wedge st(p_i, t) : [\mathbf{t}, \alpha] \wedge st(p_j, t) : [\mathbf{t}, \alpha] \rightarrow R(p_i, p_j, t) : [(5,5), \alpha] \tag{9.2}$$

Suppose that one of basic bf-inference rules (0,0)-rule-1 and 2 has been applied, then the vector annotation of bf-literal $R(p_i, p_j, t)$ should be either one of (0,8) or (5,5). Therefore, we need to consider two groups of basic bf-inference rules to be applied for following basic bf-inference rules (0,0)-rule-1 and (0,0)-rule-2, which are named (0,8)-*rules* and (5,5)-*rules*, respectively.

(0, 8)-rules. Suppose that process Pr_i has started before process Pr_j starts, then the vector annotation of bf-literal $R(p_i, p_j, t)$ should be (0, 8). We have the following inference rules to be applied for following basic bf-inference rule (0,0)-rule-1.

(0, 8)-rule-1

If process Pr_i has finished and process Pr_j has not started yet, then, it is reasonably anticipated that the bf-relation between processes Pr_i and Pr_j will be one of the bf-annotations, {**mb**(1, 11), **db**(0, 12)} whose greatest lower bound is (0, 11) (refer to Fig. 9.11). Therefore, the vector annotation (0, 8) of bf-literal $R(p_i, p_j, t)$ should turn to (0,11). Moreover, if process Pr_j starts immediately after process Pr_i, the vector annotation (0,11) of bf-literal $R(p_i, p_j, t)$ should turn to bf-annotation **mb**(1,11).

(0, 8)-rule-2

If process Pr_i has finished and process Pr_j has not started yet as well as (0,8)-rule-1, then, if process Pr_j does not start immediately after process Pr_i, the vector annotation (0,11) of bf-literal $R(p_i, p_j, t)$ should turn to bf-annotation **db**(0,12).

(0,8)-rule-3

If process Pr_j starts before process Pr_i finishes, then the vector annotation (0,8) of bf-literal $R(p_i, p_j, t)$ should turn to (2,8) that is the greatest lower bound of the bf-annotations, {**jb**(2,10), **sb**(3,9), **ib**(4,8)}.

Basic bf-inference rules (0,8)-rule-1, (0,8)-rule-2 and (0,8)-rule-3 may be translated into the bf-EVALPSN clauses,

$$R(p_i,p_j,t) : [(0,8),\ \alpha] \wedge fi(p_i,t) : [\mathbf{t},\alpha] \rightarrow R(p_i,p_j,t) : [(0,11),\ \alpha] \quad (9.3)$$

$$R(p_i,p_j,t) : [(0,11),\ \alpha] \wedge st(p_j,t) : [\mathbf{t},\alpha] \rightarrow R(p_i,p_j,t) : [(1,11),\ \alpha] \quad (9.4)$$

$$R(p_i,p_j,t) : [(0,11),\ \alpha] \wedge st(p_j,t) : [\mathbf{f},\alpha] \rightarrow R(p_i,p_j,t) : [(0,12),\ \alpha] \quad (9.5)$$

$$\begin{aligned} R(p_i,p_j,t) : [(0,8),\ \alpha] \wedge \sim fi(p_i,t) : [\mathbf{t},\alpha] \wedge st(p_j,t) : [\mathbf{t},\alpha] \\ \rightarrow R(p_i,p_j,t) : [(2,8),\ \alpha] \end{aligned} \quad (9.6)$$

(5, 5)-rules. Suppose that both processes Pr_i and Pr_j have already started at the same time, then the vector annotation of bf-literal $R(p_i, p_j, t)$ should be (5,5).We have the following inference rules to be applied for following basic bf-inference rule (0,0)-rule-2.

(5,5)-rule-1

If process Pr_i has finished before process Pr_j finishes, then the vector annotation (5,5) of bf-literal $R(p_i, p_j, t)$ should turn to bf-annotation **sb**(5,7).

(5,5)-rule-2

If both processes Pr_i and Pr_j have finished at the same time, then the vector annotation (5,5) of bf-literal $R(p_i, p_j, t)$ should turn to bf-annotation **pba**(6,6).

(5,5)-rule-3

If process Pr_j has finished before process Pr_i finishes, then the vector annotation (5,5) of bf-literal $R(p_i, p_j, t)$ should turn to bf-annotation **sa**(7,5).

Basic bf-inference rules (5,5)-rule-1, (5,5)-rule-2 and (5,5)-rule-3 may be translated into the bf-EVALPSN clauses,

$$R(p_i,p_j,t) : [(5,5),\ \alpha\wedge] fi(p_i,t) : [\mathbf{t},\alpha] \wedge \sim fi(p_j,t) : [\mathbf{t},\alpha] \rightarrow R(p_i,p_j,t) : [(5,7),\ \alpha] \quad (9.7)$$

$$R(p_i,p_j,t) : [(0,8),\ \alpha] \wedge fi(p_i,t) : [\mathbf{t},\alpha] \wedge fi(p_j,t) : [\mathbf{t},\alpha] \rightarrow R(p_i,p_j,t) : [(6,6),\ \alpha] \quad (9.8)$$

$$\begin{aligned} R(p_i,p_j,t) : [(0,8),\ \alpha] \wedge \sim fi(p_i,t) : [\mathbf{t},\alpha] \wedge fi(p_j,t) : [\mathbf{t},\alpha] \\ \rightarrow R(p_i,p_j,t) : [(7,5),\ \alpha] \end{aligned} \quad (9.9)$$

If basic bf-inference rules, (0, 8)-rule-1, (0, 8)-rule-2, (5, 5)-rule-1, (5, 5)-rule-2 and (5, 5)-rule-3, have been applied, bf-relations represented by bf-annotations such as **jb**(2, 10)/**ja**(10, 2) between two processes should be derived. On the other hand, even if basic bf-inference rule (0,8)-rule-3 has been applied, no bf-annotation could be derived. Therefore, a group of basic bf-inference rules called (2,8)-rules should be considered for following basic bf-inference rule (0,8)-rule-3.

(2,8)-rules. Suppose that process Pr_i has started before process Pr_j starts and process Pr_j has started before process Pr_i finishes, then the vector annotation of bf-literal $R(p_i, p_j, t)$ should be (2,8) and the following three rules should be considered.

(2,8)-rule-1

If process Pr_i finished before process Pr_j finishes, then the vector annotation (2,8) of bf-literal $R(p_i, p_j, t)$ should turn to bf-annotation **jb**(2,10).

(2,8)-rule-2

If both processes Pr_i and Pr_j have finished at the same time, then the vector annotation (2,8) of bf-literal $R(p_i, p_j, t)$ should turn to bf-annotation **fb**(3,9).

(2,8)-rule-3

If process Pr_j has finished before Pr_i finishes, then the vector annotation (2,8) of bf-literal $R(p_i, p_j, t)$ should turn to bf-annotation **ib**(4,8).

Basic bf-inference rules (2,8)-rule-1, (2,8)-rule-2 and (2,8)-rule-3 may be translated into the bf-EVALPSN clauses,

$$R(p_i,p_j,t) : [(2,8),\alpha] \wedge fi(p_i,t) : [\mathbf{t},\alpha] \wedge \sim fi(p_j,t) : [\mathbf{t},\alpha] \\ \rightarrow R(p_i,p_j,t) : [(2,10),\alpha] \tag{9.10}$$

$$R(p_i,p_j,t) : [(2,8),\alpha] \wedge fi(p_i,t) : [\mathbf{t},\alpha] \wedge fi(p_j,t) : [\mathbf{t},\alpha] \rightarrow R(p_i,p_j,t) : [(3,9),\alpha] \tag{9.11}$$

$$R(p_i,p_j,t) : [(2,8),\alpha] \wedge \sim fi(p_i,t) : [\mathbf{t},\alpha] \wedge fi(p_j,t) : [\mathbf{t},\alpha] \\ \rightarrow R(p_i,p_j,t) : [(4,8),\alpha] \tag{9.12}$$

The application orders of all basic bf-inference rules are summarized in Table 9.2.

9.5.3 Transitive Before-After Inference Rule

Here we construct the transitive bf-inference rule, which can reason a vector annotation of bf-literal transitively. Bf-relations could be derived efficiently by using the transitive bf-inference rule.

Suppose that there are three processes Pr_i, Pr_j and Pr_k starting sequentially, then, we consider to derive the vector annotation of bf-literal $R(p_i, p_k, t)$ from those of bf-literals $R(p_i, p_j, t)$ and $R(p_j, p_k, t)$ transitively. Firstly, we show some simple examples for understanding the basic idea of the transitive bf-inference rule.

Table 9.2 Application Order of Basic Bf-inference Rules

vector annotation	rule	vector annotation	rule	vector annotation	rule	vector annotation
(0,0)	rule-1	(0,8)	rule-1	(0,12)		
			rule-2	(1,11)		
			rule-3	(2,8)	rule-1	(2,10)
					rule-2	(3,9)
					rule-3	(4,8)
	rule-2	(5,5)	rule-1	(5,7)		
			rule-2	(6,6)		
			rule-3	(7,5)		

Example 9.1 Suppose that only process Pr_i has started at time t as shown in Fig. 9.16, then we obtain the vector annotation (0,8) of bf-literal $R(p_i, p_j, t)$ by basic bf- annotation (0,8). Thus, we have the following bf-EVALP clause as a transitive bf-inference rule, which is temporarily named trule-1,

trule-1

$$R(p_i, p_i, t) : [(0,8), \alpha] \wedge R(p_j, p_k, t) : [(0,0), \alpha] \rightarrow R(p_i, p_k, t) : [(0,8), \alpha].$$

Example 9.2 Suppose that processes Pr_i and Pr_j have started at the same time t as shown in Fig. 9.17, then, we obtain the vector annotation (5,5) of bf-literal $R(p_i, p_j, t)$ by basic bf-inference rule (0,0)-rule-2 (2) and the vector annotation (0,8) of bf-literal $R(p_j, p_k, t)$ by basic bf-inference rule (0,0)-rule-1 (1). Moreover, bf-annotation **be**(0,8) is reasoned as the vector annotation of bf-literal $R(p_i, p_k, t)$. Thus, we have the following bf-EVALP clause as another transitive bf-inference rule, which is temporarily named trule-2,

trule-2

$$R(p_i, p_i, t) : [(5,5), \alpha] \wedge R(p_j, p_k, t) : [(0,8), \alpha] \rightarrow R(p_i, p_k, t) : [(0,8), \alpha].$$

Example 9.3 Suppose that both processes Pr_i and Pr_j have already started at time t but process Pr_k has not started yet as shown in Fig. 9.18, then we have already obtained the vector annotation (2,8) of bf-literal $R(p_i, p_i, t)$ by basic bf-inference rule (0,8)-rule-3 (6) and the vector annotation (0,8) of bf-literal $R(p_j, p_k, t)$ by basic bf-inference rule (0,0)-rule-1 (1). Obviously bf-annotation **be**(0,8) is reasoned as the vector annotation of bf-literal $R(p_i, p_k, t)$. Thus, we also have the following bf-EVALP clause as a transitive bf-inference rule, which is temporarily named trule-3,

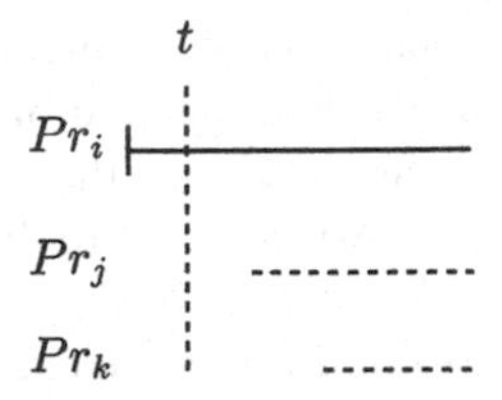

Fig. 9.16 Process time chart Ex-9.1

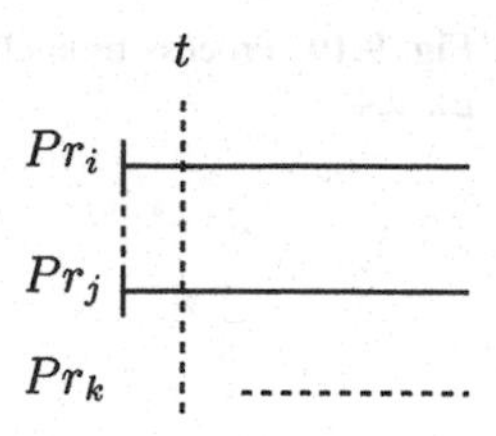

Fig. 9.17 Process time chart Ex-9.2

trule-3

$$R(p_i, p_i, t) : [(2, 8), \alpha] \wedge R(p_j, p_k, t) : [(0, 8), \alpha] \rightarrow R(p_i, p_k, t) : [(0, 8), \alpha].$$

We could efficiently derive bf-relations by using transitive bf-inference rules shown in those examples. However, there are exceptional cases in which bf-relations could not be determined uniquely by transitive bf-inference rules. Such a case is introduced in the following example.

Example 9.4 Here we introduce another example referring to three process time charts in Figs. 9.19, 9.20 and 9.21. The difference between those three process schedules is that process Pr_j starts at different times t_3, t_4 and t_5 in each time chart. The difference of the vector annotations of bf-literals $R(p_i, p_i, t_n)$, $R(p_j, p_k, t_n)$ and $R(p_i, p_k, t_n)$ $(1 \leq n \leq 7)$ in each process schedule is listed in Tables 9.3, 9.4 and 9.5.

If we focus on the variation of the vector annotations of bf-literals $R(p_i, p_i, t_3)$, $R(p_j, p_k, t_3)$ and $R(p_i, p_k, t_3)$ at time t_3 in Tables 9.3, 9.4 and 9.5, the following transitive bf-inference rule could be derived.

trule-4

$$R(p_i, p_i, t) : [(2, 8), \alpha] \wedge R(p_j, p_k, t) : [(2, 8), \alpha] \rightarrow R(p_i, p_k, t) : [(2, 8), \alpha].$$

Moreover, if we consider the difference of the vector annotations of bf-literals $R(p_i, p_i, t_4)$, $R(p_j, p_k, t_4)$ and $R(p_i, p_k, t_4)$ at time t_4 in Table 9.3, the following transitive bf-inference rule should be derived.

trule-5

$$R(p_i, p_i, t) : [(2, 10), \alpha] \wedge R(p_j, p_k, t) : [(2, 8), \alpha] \rightarrow R(p_i, p_k, t) : [(2, 10), \alpha].$$

As well as trule-5, the following two transitive bf-inference rules could be derived if we take the difference of the vector annotations of bf-literals $R(p_i, p_i, t_4)$,

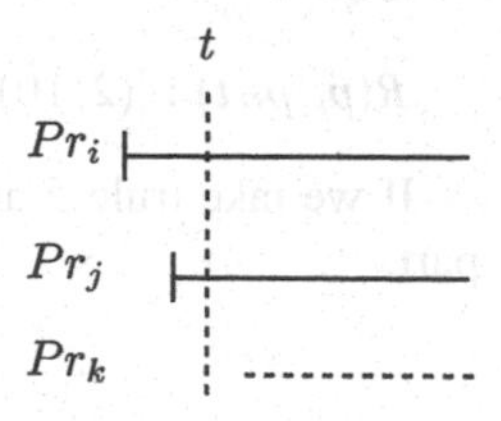

Fig. 9.18 Process time chart Ex-9.3

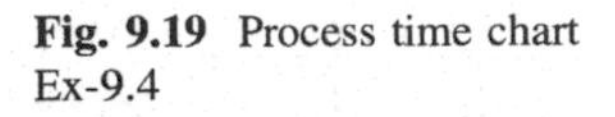

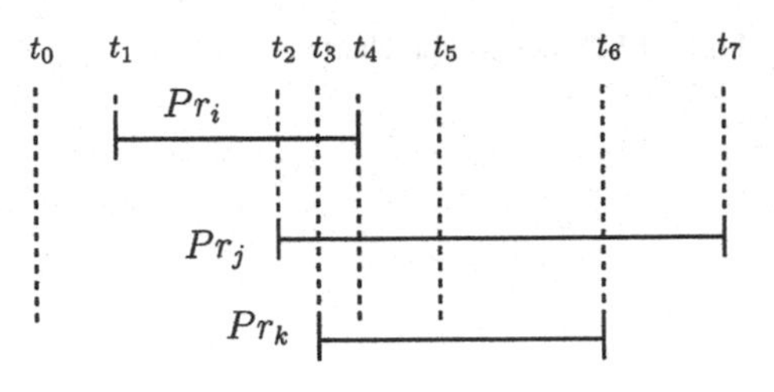

Fig. 9.19 Process time chart Ex-9.4

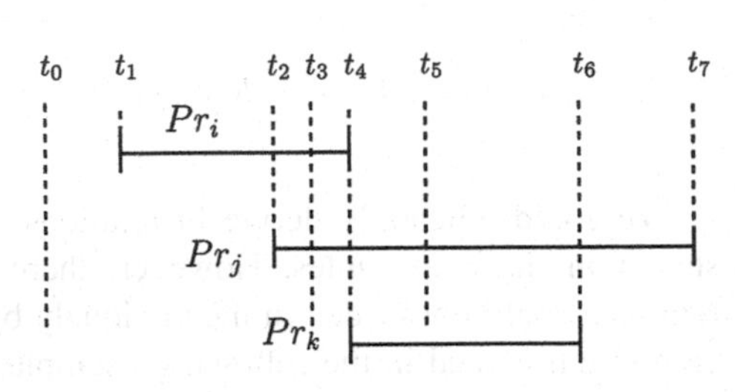

Fig. 9.20 Process time chart Ex-9.5

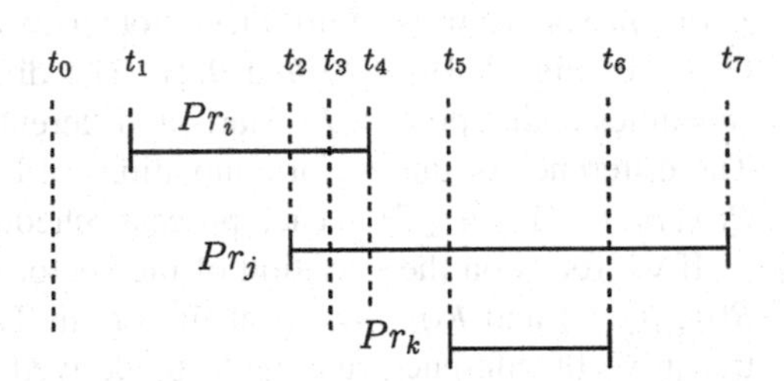

Fig. 9.21 Process time chart Ex-9.6

Table 9.3 Vector Annotations of Process Time Chart Ex-9.4

bf-literal	t_1	t_2	t_3	t_4	t_5	t_6	t_7
$R(p_i, p_i, t)$	(0, 8)	(2, 8)	(2, 8)	(2, 10)	(2, 10)	(2, 10)	(2, 10)
$R(p_j, p_k, t)$	(0, 0)	(0, 8)	(2, 8)	(2, 8)	(2, 8)	(4, 8)	(4, 8)
$R(p_i, p_k, t)$	(0, 8)	(0, 8)	(2, 8)	(2, 10)	(2, 10)	(2, 10)	(2, 10)

$R(p_j, p_k, t_4)$ and $R(p_i, p_k, t_4)$ at time t_4 in Tables 9.4 and 9.5 into account, respectively.

trule-6

$$R(p_i, p_i, t) : [(2, 10), \alpha] \wedge R(p_j, p_k, t) : [(2, 8), \alpha] \rightarrow R(p_i, p_k, t) : [(1, 11), \alpha].$$

trule-7

$$R(p_i, p_i, t) : [(2, 10), \alpha] \wedge R(p_j, p_k, t) : [(0, 8), \alpha] \rightarrow R(p_i, p_k, t) : [(0, 12), \alpha].$$

If we take trule-5 and trule-6 into account, obviously they have the same body part,

Table 9.4 Vector Annotations of Process Time Chart Ex-9.5

bf-literal	t_1	t_2	t_3	t_4	t_5	t_6	t_7
$R(p_i, p_i, t)$	(0, 8)	(2, 8)	(2, 8)	(2, 10)	(2, 10)	(2, 10)	(2, 10)
$R(p_j, p_k, t)$	(0, 0)	(0, 8)	(0, 8)	(2, 8)	(2, 8)	(4, 8)	(4, 8)
$R(p_i, p_k, t)$	(0, 8)	(0, 8)	(0, 8)	(1, 11)	(1, 11)	(1, 11)	(1, 11)

Table 9.5 Vector Annotations of Process Time Chart Ex-9.6

bf-literal	t_1	t_2	t_3	t_4	t_5	t_6	t_7
$R(p_i, p_i, t)$	(0, 8)	(2, 8)	(2, 8)	(2, 10)	(2, 10)	(2, 10)	(2, 10)
$R(p_j, p_k, t)$	(0, 0)	(0, 8)	(0, 8)	(0, 8)	(2, 8)	(4, 8)	(4, 8)
$R(p_i, p_k, t)$	(0, 8)	(0, 8)	(0, 8)	(0, 12)	(0, 12)	(0, 12)	(0, 12)

$$R(p_i, p_i, t) : [(2, 10), \alpha] \wedge R(p_j, p_k, t) : [(2, 8), \alpha] \tag{9.13}$$

and different head parts, $R(p_i, p_k, t)$:[(2,10), α] and $R(p_i, p_k, t)$:[(1,11), α]. Therefore, it implies that such transitive bf-inference rules cannot be deterministically applied. We analyze the reason why there are more than two applicable transitive bf-inference rules at time t_4. If we focus on the application order of transitive bf-inference rules during time t_2 to time t_4, obviously there are two different transitive bf-inference rule application orders in which trule-5 and trule-6 are applied after trule-4 and trule-3, respectively as follows.

Order-1: trule-3 → trule-4 → trule-5
Order-2: trule-3 → trule-6

Moreover, whenever the same body, bf-EVALP clause (13) of trule-5 and trule-6 is satisfied, the body part, $R(p_i, p_i, t)$:[(2,10), α] ∧ $R(p_j, p_k, t)$:[(0,8), α] of trule-7 is also satisfied at time t_4, then, its head, bf-EVALP clause $R(p_i, p_k, t)$:[(0,12), α] could be derivable. Thus, we have another application order of transitive bf-inference rules trule-3 and trule-7.

Order-3: trule-3 → trule-7

Therefore, without considering the application order of transitive bf-inference rules we may not derive bf-relations correctly by the transitive bf-inference rule. We list all transitive bf-inference rules with their application orders. For simplicity, we represent a transitive bf-inference rule,

$$R(p_i, p_i, t) : [(n_1, n_2), \alpha] \wedge R(p_j, p_k, t) : [(n_3, n_4), \alpha] \rightarrow R(p_i, p_k, t) : [(n_5, n_6), \alpha].$$

by omitting bf-literals as follows: $(n_1, n_2) \wedge (n_3, n_4) \rightarrow (n_5, n_6)$ in the list of transitive bf-inference rules.[6]

---

[6] The bottom vector annotation (0,0) in transitive bf-inference rules implies that any vector annotation (i,j) satisfies it.

Transitive Bf-inference Rules

$$\mathbf{TR0}(0,0) \wedge (0,0) \rightarrow (0,0)$$
$$\mathbf{TR1}(0,8) \wedge (0,0) \rightarrow (0,8)$$
$$\mathbf{TR1-1}(0,12) \wedge (0,0) \rightarrow (0,12)$$
$$\mathbf{TR1-2}(1,11) \wedge (0,8) \rightarrow (0,12)$$
$$\mathbf{TR1-3}(1,11) \wedge (5,5) \rightarrow (1,11) \quad (9.14)$$
$$\mathbf{TR1-4}(2,8) \wedge (0,8) \rightarrow (0,8)$$
$$\mathbf{TR1-4-1}(2,10) \wedge (0,8) \rightarrow (0,12)$$
$$\mathbf{TR1-4-2}(4,8) \wedge (0,12) \rightarrow (0,8)$$
$$\mathbf{TR1-4-3}(2,8) \wedge (2,8) \rightarrow (2,8)$$
$$\mathbf{TR1-4-3-1}(2,10) \wedge (2,8) \rightarrow (2,10) \quad (9.15)$$
$$\mathbf{TR1-4-3-2}(4,8) \wedge (2,10) \rightarrow (2,8)$$
$$\mathbf{TR1-4-3-3}(2,8) \wedge (4,8) \rightarrow (4,8)$$
$$\mathbf{TR1-4-3-4}(3,9) \wedge (2,10) \rightarrow (2,10)$$
$$\mathbf{TR1-4-3-5}(2,10) \wedge (4,8) \rightarrow (3,9)$$
$$\mathbf{TR1-4-3-6}(4,8) \wedge (3,9) \rightarrow (4,8)$$
$$\mathbf{TR1-4-3-7}(3,9) \wedge (3,9) \rightarrow (3,9) \quad (9.16)$$
$$\mathbf{TR1-4-4}(3,9) \wedge (0,12) \rightarrow (0,12)$$
$$\mathbf{TR1-4-5}(2,10) \wedge (2,8) \rightarrow (1,11)$$
$$\mathbf{TR1-4-6}(4,8) \wedge (1,11) \rightarrow (2,8)$$
$$\mathbf{TR1-4-7}(3,9) \wedge (1,11) \rightarrow (1,11)$$
$$\mathbf{TR1-5}(2,8) \wedge (5,5) \rightarrow (2,8) \quad (9.17)$$
$$\mathbf{TR1-5-1}(4,8) \wedge (5,7) \rightarrow (2,8)$$
$$\mathbf{TR1-5-2}(2,8) \wedge (7,5) \rightarrow (4,8)$$
$$\mathbf{TR1-5-3}(3,9) \wedge (5,7) \rightarrow (2,10)$$
$$\mathbf{TR1-5-4}(2,10) \wedge (7,5) \rightarrow (3,9) \quad (9.18)$$
$$\mathbf{TR2}(5,5) \wedge (0,8) \rightarrow (0,8)$$
$$\mathbf{TR2-1}(5,7) \wedge (0,8) \rightarrow (0,12)$$
$$\mathbf{TR2-2}(7,5) \wedge (0,12) \rightarrow (0,8)$$

$$\mathbf{TR2-3}(5,5) \wedge (2,8) \rightarrow (2,8)$$
$$\mathbf{TR2-3-1}(5,7) \wedge (2,8) \rightarrow (2,10) \tag{9.19}$$
$$\mathbf{TR2-3-2}(7,5) \wedge (2,10) \rightarrow (2,8)$$
$$\mathbf{TR2-3-3}(5,5) \wedge (4,8) \rightarrow (4,8)$$
$$\mathbf{TR2-3-4}(7,5) \wedge (3,9) \rightarrow (4,8)$$
$$\mathbf{TR2-4}(5,7) \wedge (2,8) \rightarrow (1,11) \tag{9.20}$$
$$\mathbf{TR2-5}(7,5) \wedge (1,11) \rightarrow (2,8)$$
$$\mathbf{TR3}(5,5) \wedge (5,5) \rightarrow (5,5)$$
$$\mathbf{TR3-1}(7,5) \wedge (5,7) \rightarrow (5,5) \tag{9.21}$$
$$\mathbf{TR3-2}(5,7) \wedge (7,5) \rightarrow (6,6)$$

Here we comment two important points in terms of the above list of transitive bf-inference rules.

(1) The name of a transitive bf-inference rule such as TR1-4-3 indicates the application sequence of transitive bf-inference rules until the transitive bf-inference rule has been applied. For example, if transitive bf-inference rule TR1 has been applied, one of transitive bf-inference rules TR1-1,TR1-2, and TR1-5 should be applied at the following stage; and if transitive bf-inference rule TR1-4 has been applied after transitive bf-inference rule TR1, one of transitive bf-inference rules TR1-4-1,TR1-4-2, ···, or TR1-4-7 should be applied at the following stage; on the other hand, if one of transitive bf-inference rules TR1-1, TR1-2 and TR1-3 has been applied after transitive bf-inference rule TR1, there should be no transitive bf-inference rule to be applied at the following stage because bf-annotations **db**(0,12) or **mb**(1,11) _ have been derived as the vector annotations of bf-literal $R(p_i, p_k, t)$.

(2) Transitive bf-inference rules, TR1-4-2(14), TR1-4-3-2(15), TR1-4-6(16), TR1-5-1(17), TR2-2(18), TR2-3-2(19), TR2-5(20), TR3-1(21) have no following rule to be applied, even though they cannot derive the final bf-relations between processes represented by bf-annotations such as **jb**(2,10). For example, suppose that transitive bf-inference rule TR1-4-3-2 has been applied, then the vector annotation (2,8) of bf-literal $R(p_i, p_k, t)$ just indicates that the final bf-relation between processes *Pri* and *Prk* is represented by one of three bf-annotations, **jb**(2,10), **sb**(3,9) and **ib**(4,8) because vector annotation (2,8) is the greatest lower bound of the bf-annotations. Therefore, if one of transitive bf-inference rules (14), (15), (16), (17), (18), (19), (20) and (21) has been applied, one of basic bf-inference rules, (0,8)-rules, (2,8)-rules or (5,5)-rules should be applied for deriving the final bf-annotation at the

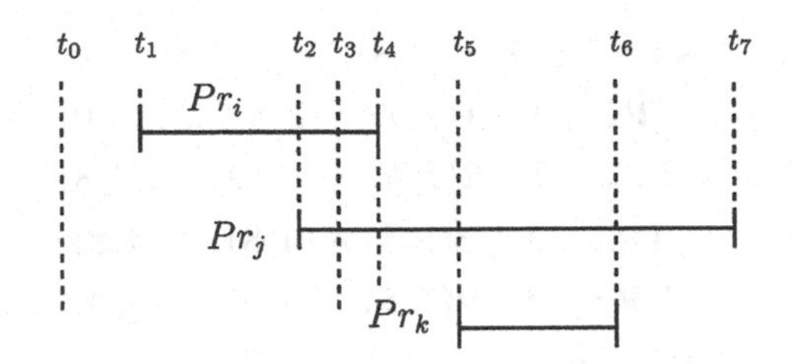

Fig. 9.22 Process time chart for Pr_i, Pr_j, and Pr_k

following stage. For example, if transitive bf-inference rule TR1-4-3-2 has been applied, basic bf-inference rules, (2,8)-rules should be applied at the following stage.

9.5.4 Example of Transitive Bf-Relation Reasoning

Now we show bf-relation transitive reasoning by the transitive bf-inference rule with an example of the process time chart in Fig. 9.22.

At time t_1, transitive bf-inference rule TR1 is applied and we have the bf-EVALP clause, $R(p_i, p_k, t_1)$:[(0,8), α].

At time t_2, transitive bf-inference rule TR1-2 is applied, however bf-literal $R(p_i, p_k, t_2)$ has the same vector annotation (0,8) as time t_2. Therefore, we have the bf-EVALP clause, $R(p_i, p_k, t_2)$:[(0,8), α].

At time t_3, no transitive bf-inference rule can be applied, since the vector annotations of bf-literals $R(p_i, p_j, t_3)$ and $R(p_j, p_k, t_3)$ are the same as time t_2. Therefore, we still have the bf-EVALP clause having the same vector annotation, $R(p_i, p_k, t_3)$:[(0,8), α].

At time t_4, transitive bf-inference rule TR1-2-1 is applied and we obtain the bf-EVALP clause having bf-annotation **db**(0,12), $R(p_i, p_k, t_4)$:[(0,12), α].

9.6 Bf-EVALPSN Safety Verification for Process Order

In this section, we present a simple example for application of the bf-relation reasoning system based on bf-EVALPSN to process order verification.

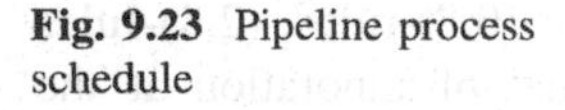

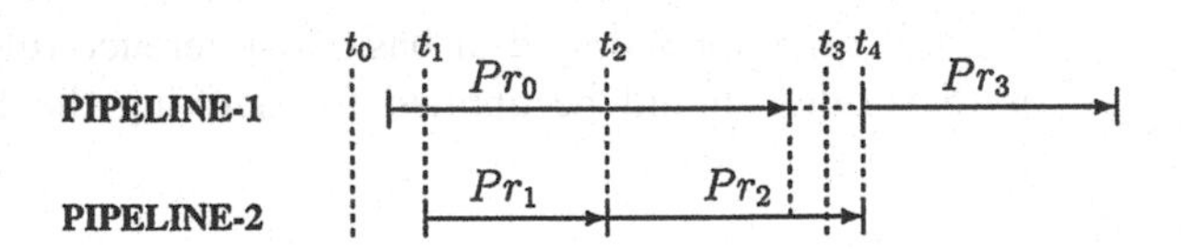

Fig. 9.23 Pipeline process schedule

9.6.1 Safety Property for Process Order

We assume a pipeline processing system consists of two pipelines, PIPELINE-1 and PIPELINE-2, which deal with pipeline processes Pr_0 and Pr_3, and Pr_1 and Pr_2, respectively. The schedule of those processes is shown in Fig. 9.23. Moreover, we assume that the pipeline processing system has four safety properties *SPR-i*($i = 0$, 1, 2, 3) to be strictly assured for process order safety.

SPR-0 process Pr_0 must start before any other processes, and process Pr_0 must finish before process Pr_2 finishes,
SPR-1 process Pr_1 must start after process Pr_0 starts,
SPR-2 process Pr_2 must start immediately after process Pr_1 finishes,
SPR-3 process Pr_3 must start immediately after both processes Pr_0 and Pr_2 finish.

All safety properties *SPR-i*($i = 0$, 1, 2, 3) can be translated into the following bf-EVALPSN clauses.

SPR-0

$$\sim R(p_0, p_1, t) : [(0,8), \alpha] \rightarrow st(p_1, t) : [\mathbf{f}, \beta] \tag{9.22}$$

$$\sim R(p_0, p_2, t) : [(0,8), \alpha] \rightarrow st(p_2, t) : [\mathbf{f}, \beta] \tag{9.23}$$

$$\sim R(p_0, p_3, t) : [(0,8), \alpha] \rightarrow st(p_3, t) : [\mathbf{f}, \beta] \tag{9.24}$$

$$st(p_1, t) : [\mathbf{f}, \beta] \wedge st(p_2, t) : [\mathbf{f}, \beta] \wedge st(p_3, t) : [\mathbf{f}, \beta] \rightarrow st(p_0, t) : [\mathbf{f}, \gamma] \tag{9.25}$$

$$\sim fi(p_0, t) : [\mathbf{f}, \beta] \rightarrow fi(p_0, t) : [\mathbf{f}, \gamma] \tag{9.26}$$

where bf-EVALPSN clauses (9.22), (9.23) and (9.24) declare that if process Pr_0 has not started before all other processes Pr_i($i = 1$, 2, 3) start, it should be forbidden to start each process Pr_i($i = 1$, 2, 3); bf-EVALPSN clause (9.25) declares that if each process Pr_i($i = 1$, 2, 3) is forbidden from starting, it should be permitted to start process Pr_0; and bf-EVALPSN clause (9.26) declares that if there is no forbiddance from finishing process Pr_0, it should be permitted to finish process Pr_0.

SPR-1

$$\sim st(p_1, t) : [\mathbf{f}, \beta] \rightarrow st(p_1, t) : [\mathbf{f}, \gamma] \tag{9.27}$$

$$\sim fi(p_1, t) : [\mathbf{f}, \beta] \rightarrow fi(p_1, t) : [\mathbf{f}, \gamma] \tag{9.28}$$

where bf-EVALPSN clause (9.27)/(9.28) declares that if there is no forbiddance fromstarting/finishing process Pr_1, it should be permitted to start/finish process Pr_1, respectively.

SPR-2

$$\sim R(p_2, p_1, t) : [(11,0), \alpha] \rightarrow st(p_2, t) : [\mathbf{f}, \beta] \tag{9.29}$$

$$\sim st(p_2, t) : [\mathbf{f}, \beta] \rightarrow st(p_2, t) : [\mathbf{f}, \gamma] \tag{9.30}$$

$$\sim R(p_2, p_0, t) : [(10, 2), \alpha] \rightarrow fi(p_2, t) : [\mathbf{f}, \beta] \tag{9.31}$$

$$\sim fi(p_2, t) : [\mathbf{f}, \beta] \rightarrow fi(p_2, t) : [\mathbf{f}, \gamma] \tag{9.32}$$

where bf-EVALPSN clause (9.29) declares that if process Pr_1 has not finished before process Pr_2 starts, it should be forbidden to start process Pr_2; the vector annotation (11,0) of bf-literal $R(p_2, p_1, t)$ is the greatest lower bound of $\{$**da**(12, 0), **ma**(11, 1)$\}$, which implies that process Pr_1 has finished before process Pr_2 starts; bf-EVALPSN clauses (9.30)/(9.32) declares that if there is no forbiddance from starting/finishing process Pr_2, it should be permitted to start/finish process Pr_2, respectively; and bf-EVALPSN clauses (9.31) declares that if process Pr_0 has not finished before process Pr_2 finishes, it should be forbidden to finish process Pr_2.

SPR-3

$$\sim R(p_3, p_0, t) : [(11, 0), \alpha] \rightarrow st(p_3, t) : [\mathbf{f}, \beta] \tag{9.33}$$

$$\sim R(p_3, p_1, t) : [(11, 0), \alpha] \rightarrow st(p_3, t) : [\mathbf{f}, \beta] \tag{9.34}$$

$$\sim R(p_3, p_2, t) : [(11, 0), \alpha] \rightarrow st(p_3, t) : [\mathbf{f}, \beta] \tag{9.35}$$

$$\sim st(p_3, t) : [\mathbf{f}, \beta] \rightarrow st(p_3, t) : [\mathbf{f}, \gamma] \tag{9.36}$$

$$\sim fi(p_3, t) : [\mathbf{f}, \beta] \rightarrow fi(p_3, t) : [\mathbf{f}, \gamma] \tag{9.37}$$

where bf-EVALPSN clauses (9.33), (9.34) and (9.35) declare that if one of processes $Pr_i (i = 0, 1, 2)$ has not finished yet, it should be forbidden to start process Pr_3; and bf-EVALPSN clauses (9.36)/(9.37) declares that if there is no forbiddance from starting/finishing process Pr_3, it should be permitted to start/finish process Pr_3, respectively.

9.6.2 Bf-EVALPSN Process Order Safety Verification

Here, we show how the bf-EVALPSN process order safety verification is carried out at five time points, t_0, t_1, t_2, t_3 and t_4 in the process schedule (Fig. 9.19). We consider five bf-relations between processes Pr_0, Pr_1, Pr_2 and Pr_3 represented by the vector annotations of bf-literals, $R(p_0, p_1, t)$, $R(p_0, p_2, t)$, $R(p_0, p_3, t)$, $R(p_1, p_2, t)$, $R(p_2, p_3, t)$, which should be verified based on safety properties *SPR*-0, *SPR*-1, *SPR*-2 and *SPR*-3 in real-time.

Initial Stage (at time t_0), no process has started at time t_0, thus, the bf-EVALP clauses,

$$R(p_0, p_1, t_0) : [(0, 0), \alpha] \tag{9.38}$$

$$R(p_1, p_2, t_0) : [(0, 0), \alpha] \tag{9.39}$$

$$R(p_2, p_3, t_0) : [(0,0), \alpha] \tag{9.40}$$

are obtained; also, the bf-EVALP clauses,

$$R(p_0, p_2, t_0) : [(0,0), \alpha] \tag{9.41}$$

$$R(p_0, p_3, t_0) : [(0,0), \alpha] \tag{9.42}$$

are obtained by transitive bf-inference rule TR0; then, bf-EVALP clauses (9.38), (9.41) and (9.42) satisfy each body of bf-EVALPSN clauses (9.22), (9.23) and (9.24), respectively, therefore, the forbiddance from starting each process $Pr_i (i = 1, 2, 3)$,

$$st(p_1, t_0) : [\mathbf{f}, \beta] \tag{9.43}$$

$$st(p_2, t_0) : [\mathbf{f}, \beta] \tag{9.44}$$

$$st(p_3, t_0) : [\mathbf{f}, \beta] \tag{9.45}$$

are derived; moreover, since bf-EVALP clauses (9.43), (9.44) and (9.45) satisfy the body of bf-EVALPSN clause (9.25), the permission for starting process Pr_0, $st(p_0, t_0) : [\mathbf{f}, \gamma]$ is derived; therefore, process Pr_0 is permitted to start at time t_0.

2nd Stage (at time t_1), process Pr_0 has already started but all other processes $Pr_i (i = 1, 2, 3)$ have not started yet; then, the bf-EVALP clauses,

$$R(p_0, p_1, t_1) : [(0,8), \alpha] \tag{9.46}$$

$$R(p_1, p_2, t_1) : [(0,0), \alpha] \tag{9.47}$$

$$R(p_2, p_3, t_1) : [(0,0), \alpha] \tag{9.48}$$

are obtained, where the bf-EVALP clause (9.46) is derived by basic bf-inference rule (0,0)-rule-1; moreover, the bf-EVALP clauses,

$$R(p_0, p_2, t_1) : [(0,8), \alpha] \tag{9.49}$$

$$R(p_0, p_3, t_1) : [(0,8), \alpha] \tag{9.50}$$

are obtained by transitive bf-inference rule TR1; as bf-EVALP clause (9.46) does not satisfy the body of bf-EVALPSN clause (9.22), the forbiddance from starting process Pr_1,

$$st(p_1, t_1) : [\mathbf{f}, \beta] \tag{9.51}$$

cannot be derived; then, since there is not forbiddance (9.50), the body of bf-EVALPSN clause (9.27) is satisfied, and the permission for starting process Pr_1, $st(p_1, t_1) : [\mathbf{f}, \gamma]$ is derived; on the other hand, since bf-EVALP clauses (9.49) and (9.50) satisfy the body of bf-EVALPSN clauses (9.29) and (9.33) respectively, the forbiddance from starting both processes Pr_2 and Pr_3, $st(p_2, t_1) : [\mathbf{f}, \beta]$ and $st(p_3, t_1) : [\mathbf{f}, \beta]$ are derived; therefore, process Pr_1 is permitted to start at time t_1.

3rd Stage (at time t_2), process Pr_1 has just finished and process Pr_0 has not finished yet; then, the bf-EVALP clauses,

$$\begin{aligned} R(p_0, p_1, t_2) &: [(4, 8), \alpha] \\ R(p_1, p_2, t_2) &: [(1, 11), \alpha] \\ R(p_2, p_3, t_2) &: [(0, 8), \alpha] \end{aligned} \tag{9.52}$$

are derived by basic bf-inference rules (2,8)-rule-3, (0,8)-rule-2 and (0,0)-rule-1, respectively; moreover, the bf-EVALP clauses,

$$R(p_0, p_2, t_2) : [(2, 8), \alpha] \text{ and} R(p_0, p_3, t_2) : [(0, 12), \alpha]$$

are obtained by transitive bf-inference rules TR1-4-6 and TR1-2, respectively; then, since bf-EVALP clause (9.52) does not satisfy the body of bf-EVALPSN clause (9.29), the forbiddance from starting process Pr_2,

$$st(p_2, t_2) : [\mathbf{f}, \beta] \tag{9.53}$$

cannot be derived; since there is not the forbiddance (9.53), it satisfies the body of bf-EVALPSN clause (9.30), and the permission for starting process Pr_2, $st(p_2, t_2) : [\mathbf{f}, \gamma]$, is derived; on the other hand, since bf-EVALP clause (9.53) satisfies the body of bf-EVALPSN clause (9.33), the forbiddance from starting process Pr_3, $st(p_3, t_2) : [\mathbf{f}, \beta]$ is derived; therefore, process Pr_2 is permitted to start, however process Pr_3 is still forbidden to start at time t_2.

4th Stage (at time t_3), process Pr_0 has finished, process Pr_2 has not finished yet, and process Pr_3 has not started yet; then, the bf-EVALP clauses,

$$\begin{aligned} R(p_0, p_1, t_3) &: [(4, 8), \alpha] \\ R(p_1, p_2, t_3) &: [(1, 11), \alpha] \\ R(p_2, p_3, t_3) &: [(0, 8), \alpha] \end{aligned} \tag{9.54}$$

in which the vector annotations are the same as the previous stage are obtained because bf-annotations of bf-EVALP clauses (9.54) have been already reasoned, and the before-after relation between processes Pr_2 and Pr_3 is the same as the previous stage; moreover, the bf-EVALP clauses,

$$\begin{aligned} R(p_1, p_2, t_4) &: [(1, 11), \alpha] \\ R(p_0, p_3, t_3) &: [(0, 12), \alpha] \end{aligned} \tag{9.55}$$

are obtained, where bf-EVALP clause (9.55) is derived by basic bf-inference rule (2,8)-rule-1; then, bf-EVALP clause (9.54) satisfies the body of bf-EVALP clause (35), and the forbiddance from starting process Pr_3, $st(p_3, t_3) : [\mathbf{f}, \beta]$ is derived; therefore, process Pr_3 is still forbidden to start because process Pr_2 has not finished yet at time t_3.

5th Stage (at time t_4), process Pr_2 has just finished and process Pr_3 has not started yet; then, the bf-EVALP clauses,

$$R(p_0, p_1, t_4) : [(4, 8), \alpha]$$
$$R(p_1, p_2, t_4) : [(1, 11), \alpha] \quad (9.56)$$
$$R(p_2, p_3, t_4) : [(0, 11), \alpha]$$

$$R(p_0, p_2, t_4) : [(2, 10), \alpha]$$
$$R(p_0, p_3, t_4) : [(0, 12), \alpha] \quad (9.57)$$

are obtained; bf-EVALP clause (9.56) is derived by basic bf-inference rule (0,8)-rule-1 (3); moreover, the bf-EVALP clause,

$$R(p_1, p_3, t_4) : [(0, 12), \alpha] \quad (9.58)$$

is reasoned by transitive bf-inference rule TR1-2; then, since bf-EVALP clauses (9.56), (9.57) and (9.58) do not satisfy the bodies of bf-EVALP clauses (9.33), (9.34) and (9.35), the forbiddance from starting process Pr_3, $st(p_3,t_4)$:[**f**, β] cannot be derived; therefore, the body of bf-EVALPSN clause (9.36) is satisfied, and the permission for starting process Pr_3, $st(p_3,t_4)$:[**f**, γ] is derived; therefore, process Pr_3 is permitted to start because all processes Pr_0, Pr_1 and Pr_2 have finished at time t_4.

9.7 Concluding Remarks

In this chapter, we have introduced the development of paraconsistent annotated logic program EVALPSN and its new version bf-EVALPSN that can deal with before-after relations between processes. Moreover, the reasoning system in bf-EVALPSN and its application to process order safety verification have been provided.

We would like to conclude this chapter by describing the advantages and disadvantages of intelligent control based on EVALPSN/bf-EVALPSN safety verification.

Advantages

- If the safety verification EVALPSN is locally stratified, it can be easily implemented in Prolog, C language, PLC (Programmable Logic Controller) ladder program etc.
- It has been proved that the method can be implemented as electronic circuits on micro chips [39]. Therefore, if real-time processing is required in the system, the method might be very useful.
- Our bf-EVALP safety verification method for process order control is a quite essential application of paraconsistent annotated logic in a sense of treating before-after relations between processes in paraconsistent vector annotations.
- The safety verification methods for both process control and process order control can be implemented under the same environment.

Disadvantages

- Since EVALPSN itself is basically not a tool of formal safety verification, it includes complicated and redundant expressions to construct safety verification systems. Therefore, it should be better to develop safety verification oriented tool or programming language based on EVALPSN if EVALPSN is applied to formal safety verification.
- It is difficult to understand how to utilize EVALPSN fully to do practical implementation due to paraconsistent annotated logic.

References

1. Allen, J.F.: Towards a general theory of action and time. Artif. Intell. **23**, 123–154 (1984)
2. Allen, J.F., Ferguson, G.: Actions and events in interval temporal logic. J. Logic Comput. **4**, 531–579 (1994)
3. Billington, D.: Defeasible logic is stable. J. Logic Comput. **3**, 379–400 (1997)
4. Billington, D.: Conflicting literals and defeasible logic. Proceedings 2nd Australian Workshop Commonsense Reasoning, pp. 1–15. Australian Artificial Intelligence Institute, Perth, Australia, (1997)
5. Blair, H.A., Subrahmanian, V.S.: Paraconsistent logic programming. Theoret. Comput. Sci. **68**, 135–154 (1989)
6. da Costa, N. C. A. et al.: The paraconsistent logics P*T*. Zeitschrift für Mathematische Logic und Grundlangen der Mathematik **37**, 139–148 (1989)
7. Dressler, O.: An extended basic ATMS. Proceedings of 2nd International Workshop on Non-monotonic Reasoning. LNCS, vol. 346, pp. 143–163 (1988)
8. Geerts, P. et al.: Defeasible logics. Handbook of Defeasible Reasoning and Uncertainty Management Systems, vol. 2, pp. 175–210. Springer (1998)
9. Gelfond, M., Lifschitz, V.: The stable model semantics for logic programming. Proceedings of 5th International Conference and Symposium Logic Programming (ICLP/SLP88), pp. 1070–1080. The MIT Press (1989)
10. Jaskowski, S.: Propositional calculus for contradictory deductive system. Stud. Logica. **24**, 143–157 (1948). (English translation of the original Polish paper)
11. Kifer, M., Subrahmanian, V.S.: Theory of generalized annotated logic programming and its applications. J. Logic Program. **12**, 335–368 (1992)
12. Lloyd, J.W.: Foundations of Logic Programming. 2nd edn. Springer (1987)
13. McNamara, P., Prakken, H.: (eds.) Norms, Logics and Information Systems. Frontiers in Artificial Intelligence and Applications, vol. 49, IOS Press (1999)
14. Meyer, J-J.C., Wiering, R.J.: (eds.) Logic in Computer Science. Wiley (1993)
15. Moor, R.: Semantical considerations on non-monotonic logic. Artif. Intell. **25**, 75–94 (1985)
16. Nakamatsu, K., Suzuki, A.: Annotated semantics for default reasoning. Proceedings of 3rd Pacific Rim International Conference Artificial Intelligence (PRICAI94), pp. 180–186. International Academic Publishers (1994)
17. Nakamatsu, K., Suzuki, A.: A nonmonotonic ATMS based on annotated logic programs. LNAI *1441*, 79–93, Springer (1998)
18. Nakamatsu, K., et al.: Defeasible reasoning between conflicting agents based on VALPSN. Proceedings of AAAI Workshop Agents' Conflicts, pp. 20–27. AAAI Press (1999)
19. Nakamatsu, K.: On the relation between vector annotated logic programs and defeasible theories. Logic Logical Philosophy **8**, 181–205 (2001)

20. Nakamatsu, K., et al.: A defeasible deontic reasoning system based on annotated logic programming. Proceedings of 4th International Conference Computing Anticipatory Systems (CASYS2000), AIP Conference Proceedings, vol. 573, pp. 609–620. American Institute of Physics, (2001)
21. Nakamatsu, K., et al.: Annotated semantics for defeasible deontic reasoning. LNAI **2005**, 432–440 (2001)
22. Nakamatsu, K., et al.: Paraconsistent logic program based safety verification for air traffic control. Proceedings of IEEE International Conference System, Man and Cybernetics 02(SMC02), CD-ROM, IEEE SMC (2002)
23. Nakamatsu, K., et al.: A railway interlocking safety verification system based on abductive paraconsistent logic programming. Proceedings of Soft Computing Systems (HIS02), Frontiers in Artificial Intelligence and Applications, vol. 7, pp. 775–784. IOS Press (2002)
24. Nakamatsu, K., Seno, T., Abe, J. M., Suzuki, A.: Intelligent real-time traffic signal control based on a paraconsistent logic program EVALPSN. In: Rough Sets, Fuzzy Sets, Data Mining and Granular Computing (RSFDGrC2003), LNAI vol. 2639, pp. 719–723. (2003)
25. Nakamatsu, K., KOmaba, H., Suzuki, A.: Defeasible deontic control for discrete events based on EVALPSN. Proceedings of 4th International Conference Rough Sets and Current Trends in Computing (RSCTC2004), LNAI, vol. 3066, pp. 310–315 (2004)
26. Nakamatsu, K., Ishikawa, R., Suzuki, A.: A paraconsistent based control for a discrete event cat and mouse. Proceedings of 8th International Conference Knowledge-Based Intelligent Information and Engineering Systems (KES2004), LNAI, vol. 3214, pp. 954–960 (2004)
27. Nakamatsu, K., Suzuki, A.: Autoepistemic theory and paraconsistent logic program. Advances in Logic Based Intelligent Systems, Frontiers in Artificial Intelligence and Applications, vol. 132, pp. 177–184. IOS Press (2005)
28. Nakamatsu, K.: Pipeline valve control based on EVALPSN safety verification. J Adv. Comput. Intell. Intell. Inf. **10**, 647–656 (2006)
29. Nakamatsu, K., Mita, Y., Shibata, T.: An intelligent action control system based on extended vector annotated logic program and its hardware implementation. J. Intell. Autom. Soft Comput. **13**, 222–237 (2007)
30. Nakamatsu, K., Akama, S., Abe, J.M.: Transitive reasoning of before-after relation based on Bf-EVALPSN. Knowl. Based Intell. Inf. Eng. Syst. LNCS **5178**, 474–482 (2008)
31. Nakamatsu, K.: The paraconsistent annotated logic program EVALPSN and its application. In: Computational Intelligence: A Compendium, Studies in Computational Intelligence, vol. 115, pp. 233–306. Springer (2008)
32. Nute, D.: Defeasible reasoning. Proceedings of 20th Hawaii International Conference System Science (HICSS87) vol. 1, pp. 470–477 (1987)
33. Nute, D.: Basic defeasible logics. Intensional Logics for Programming, pp. 125–154. Oxford University Press (1992)
34. Nute, D.: Defeasible logic. In: Gabbay D.M., et al. (eds) Handbook of Logic in Artificial Intelligence and Logic Programming, vol. 3, pp. 353–396. Oxford University Press (1994)
35. Nute, D.: Apparent obligatory. Defeasible Deontic Logic, Synthese Library, vol. 263, pp. 287–316. Kluwer Academic Publishers (1997)
36. Prakken, H.: Logical Tools for Modelling Legal Argument. Law and Philosophy Library, vol. 32, Kluwer Academic Publishers (1997)
37. Reiter, R.: A logic for default reasoning. Artif. Intell. **13**, 81–123 (1980)
38. Subrahmanian, V.S.: On the semantics of qualitative logic programs. Proceedings the 1987 Symposium Logic Programming (SLP87), pp. 173–182. IEEE CS Press (1987)
39. Subrahmanian, V.S.: Amalgamating knowledge bases. ACM Trans. Database Syst. **19**, 291–331 (1994)

Part III
Applications

Chapter 10
Fuzzy and Neuro-Symbolic Approaches in Personal Credit Scoring: Assessment of Bank Loan Applicants

Ioannis Hatzilygeroudis and Jim Prentzas

Abstract Credit scoring is a vital task in the financial domain. An important aspect in credit scoring involves the assessment of bank loan applications. Loan applications are frequently assessed by banking personnel regarding the ability/possibility of satisfactorily dealing with loan demands. Intelligent methods may be employed to assist in the required tasks. In this chapter, we present the design, implementation and evaluation of two separate intelligent systems that assess bank loan applications. The systems employ different knowledge representation formalisms. More specifically, the corresponding intelligent systems are a fuzzy expert system and a neuro-symbolic expert system. The former employs fuzzy rules based on knowledge elicited from experts. The latter is based on neurules, a type of neuro-symbolic rules that combine a symbolic (production rules) and a connectionist (adaline unit) representation. A characteristic of neurules is that they retain the naturalness and modularity of symbolic rules. Neurules were produced from available patterns. Evaluation showed that the performance of both systems is close although their knowledge bases were derived from different types of source knowledge.

I. Hatzilygeroudis (✉)
Department of Computer Engineering and Informatics, School of Engineering,
University of Patras, 26500 Patras, Greece
e-mail: ihatz@ceid.upatras.gr

J. Prentzas
Department of Education Sciences in Pre-School Age, Laboratory of Informatics,
School of Education Sciences, Democritus University of Thrace, 68100 Nea Chili,
Alexandroupolis, Greece
e-mail: dprentza@psed.duth.gr

C. Faucher and L. C. Jain (eds.), *Innovations in Intelligent Machines-4*,
Studies in Computational Intelligence 514, DOI: 10.1007/978-3-319-01866-9_10,

10.1 Introduction

Credit scoring is a process rating the creditworthiness of persons, corporations, banks, financial institutions and countries [18], [21]. Without doubt it is an important task in the financial sector that may assist lenders and investors in evaluating potential risks. An important aspect in credit scoring concerns evaluation of loan applications. Lenders use credit scoring methods to evaluate loan applications and discriminate between risky ones that may result to financial losses and promising ones that are likely to bring in revenue. Therefore, credit scoring assists lenders in deciding whether a loan application should be approved or not. The scope of credit scoring is beyond evaluation of loan applications and involves other tasks such as setting credit card limits and approval of additional credit to existing clients.

Credit scoring methods are employed in many private and public institutions such as banks, banking authorities, credit rating agencies, government departments and regulators. Nevertheless, a very frequent task involves the assessment of applications for bank loans. Such an assessment is important due to the involved risks for both clients (individuals or corporations) and banks. In the recent financial crisis, banks suffered losses from a steady increase of customers' defaults on loans [31]. So, banks should avoid approving loans for applicants that eventually may not comply with the involved terms. It is significant to approve loans that satisfy all critical requirements and will be able to afford corresponding demands. To this end, various parameters related to applicant needs should be considered. Furthermore, different loan programs from different banks should also be taken into account, to be able to propose a loan program best tailored to the specific applicant's status. Computer-based systems for evaluating loan applicants and returning the most appropriate loan program would be useful since valuable assessment time would be spared and potential risks could be reduced. Banking authorities encourage developing models to better quantify financial risks [20].

Three main types of credit scoring methods exist [18] that may be used to assess loan applications: expert scoring methods, statistical and Artificial Intelligence (AI) techniques. Expert scoring methods are time-consuming and depend on the expert's expertise. It is difficult to perform the task manually due to the large number of loan applications [29] and the enormous quantity of stored data involving past applications. For these reasons, statistical and AI techniques are usually applied [18, 21]. Statistical techniques usually employed in credit scoring are Linear Discriminant Analysis, Logistic Regression Analysis and Multivariate Adaptive Regression Splines [18, 21, 29]. A drawback of the statistical methods is that they make certain assumptions concerning the problem that may not be valid in reality [29]. In this aspect, AI techniques are superior. Furthermore, previous studies concerning credit scoring have demonstrated that certain AI techniques outperform statistical methods in terms of classification accuracy [29].

Due to the complexity/significance of the assessment process, intelligent assessment systems have been used to reduce the cost of the process and the risks of bad loans, to save time and effort and generally to enhance credit decisions [1]. Various AI techniques have been applied to credit scoring. Usual AI techniques employed are mainly based on neural networks [3] and support vector machines [16, 31] but approaches based on genetic algorithms [1], case-based reasoning [30], decision trees [29], and other methods [2, 14] have been applied too. One may discern three main types of AI techniques applied to credit scoring: (a) approaches involving a single (or individual) technique, (b) hybrid approaches combining two or more different techniques and (c) ensemble methods that generate multiple hypotheses using the same base learner [17, 19, 29, 31]. Hybrid and ensemble methods have recently been applied to credit scoring and seem to offer advantages. Hybrid methods have been applied to several domains since combination of different techniques is useful in handling complex problems [6, 10, 11, 13, 22, 25].

There are some requirements in designing an intelligent system for the assessment of loan applications. First, the system needs to include loan programs offered by different banks, to be able to return the most suitable one(s) for the applicant. Second, experience of banking staff specialized in loans is useful in order to outline loan attributes, applicant attributes, assessment criteria and the stages of the assessment process. Third, available cases from past loan applications are required to design and test the system. Fourth, explanation concerning the reached decision is necessary.

In this chapter, we present the design, implementation and evaluation of two intelligent systems that assess bank loan applications: a fuzzy expert system and a neuro-symbolic expert system. To construct those systems different types of source knowledge were exploited. The fuzzy expert system is based on rules that represent expert knowledge regarding the assessment process. The neuro-symbolic expert system is based on neurules, a type of hybrid rules integrating a symbolic (production rules) and a connectionist representation (adaline unit) [26]. Neurules exhibit advantages compared to pure symbolic rules such as, improved inference performance [7], ability to reach conclusions from unknown inputs and construct knowledge bases from alternative sources (i.e. symbolic rules or empirical data) [7, 8, 23, 24]. Neurules were produced from available cases (past loan applications). Evaluation of the fuzzy expert system encompassing rule-based expert knowledge and the neurule-based expert system encompassing empirical knowledge showed that their performances were close. This chapter is a revised and extended version of a paper presented at AIAI 2011 [14].

The rest of the chapter is organized as follows. Section 10.2 introduces the domain knowledge involved and the stages of the assessment process. Section 10.3 discusses development issues of the fuzzy expert system. Section 10.4 briefly presents neurules and discusses development issues of the neurule-based expert system. Section 10.5 presents evaluation results for both systems. Finally, Sect. 10.6 concludes.

10.2 Domain Knowledge Modeling

In this section, we discuss issues involving the primary loan and applicant attributes modeled in the systems as well as the basic stages of the inference process. The corresponding knowledge was derived from experts.

10.2.1 Modeled Loan Attributes

Each loan involves a number of attributes that need to be taken into account during inference. A basic attribute is the type of loan. Various types of loans exist. The two main types involve loans addressed to individuals and loans addressed to corporations.

Each main type of loan is further discerned to different categories. Loans addressed to individuals are discerned to personal, consumer and housing loans. Loans addressed to corporations are discerned to capital, fixed installations and other types of loans. The type of loan affects other attributes such as the amount of money that can be approved for loaning, the amount of installments and the interest.

Moreover, according to the type of loan, different requirements and applicant characteristics are taken into account. For instance, in the case of personal and consumption loans, a main applicant attribute taken into account is the net annual income. The maximum amount of money that can be approved for loaning is up to a certain percent of the annual income subtracting the amounts of existing loans in any bank. For example, if the maximum amount of money that can be approved for loaning is up to a 70 % of the annual income, then an applicant with annual income €20,000 may borrow up to €13,000. In case there is a pending loan of €5,000, then he/she may borrow up to €8,000.

Applications for housing loans are thoroughly examined since the involved amount of money is usually large and the risks for banks are high. Various applicant attributes need to be considered such as property status and net annual income. For instance, the net annual income should be at least €15,000 and annual installments should not exceed roughly 40 % of the net annual income.

So, loan attributes such as the following are considered:

- Type of loan.
- *The reason for applying* (e.g., purchase of items, purchase of a car, university studies, transfer of loan residues, etc.).
- *Supporting documents* (e.g. for consumer loans, proofs of item purchasing are required).
- Name of bank.
- *Type of interest.* Two basic types of loan interests involve fixed and floating interests. Floating interests may fluctuate (increase/decrease) during time according to general interest tendency in the market.

- Commencement of loan payment.
- Termination of loan payment.
- Amount of money loaned.
- *Loan expenses.* Loan expenses concern an amount of money paid when the loan is initially approved, with the first installment, with each installment or in specific time periods.
- *Way of payment.* The way of loan payment usually involves monthly installments. Installments include interest, capital and potential expenses. The amount of money deriving from interest gradually decreases.

10.2.2 Modeled Applicant Attributes

In order to approve a loan, an overall assessment of the applicant is required. The assessment process examines an applicant's ability to deal satisfactorily with loan demands. A thorough assessment assists in avoiding potential high risks involved in loans. Various applicant parameters are considered. The most significant attributes are the following:

- *Net annual income.* Expenses, obligations (e.g. installments) are excluded.
- *Financial and property status.* Possession of property is considered important (or even obligatory) for certain types of loans. This involves available bank accounts, bonds, stocks, real estate property, etc.
- *Personal attributes.* Personal attributes such as age, number of depending children, trade are considered important. Trade is considered a parameter of the applicant's social status. The number of depending children corresponds to obligations. Banks usually do not loan money to persons younger than twenty and older than seventy years old due to high risks (bank perspective).
- *Warrantor.* A primary parameter for the overall assessment of an applicant is also the warrantor, who accepts and signs the bank's terms. In case the client cannot comply with obligations concerning the loan, the warrantor will undertake all corresponding responsibilities.

10.2.3 Inference Process Stages

The inference process involves four main stages outlined in the following. Table 10.1 summarizes the outputs produced in each stage.

In the first stage, basic inputs are given to the inference process concerning the requested loan. More specifically, the type of loan and the reason for loan application are given by responding to relevant system queries. Both inputs are stored as facts in the systems and are used to retrieve relevant loan programs from the Loan Programs Base. The approach involves loan programs from four different banks. The retrieved loan programs are used in subsequent stages.

Table 10.1 Summary of outputs for the four stages of the inference process

Stage	Outputs	
Stage 1	*Retrieves relevant loan programs* from the Loan Programs Base according to basic inputs (e.g. type of loan and reason for loan application)	
Stage 2	*Applicant assessment* *Warrantor assessment*	*Overall assessment of the applicant*
Stage 3	*Restrictions* involving the funding of a loan are taken into account	
Stage 4	*Relevant loan programs are returned.* For each loan program return: interest, approved amount of money, installment and loan payment period	

The second stage involves an overall assessment of the applicant. This stage consists of three tasks: (a) applicant assessment, (b) warrantor assessment and (c) overall applicant assessment. The third task takes as input the results of the other two tasks. Such a process is applied since an overall applicant assessment is based on assessment of both applicant and warrantor. Rule-based inference performs all tasks. The variables involved in applicant and warrantor assessment are summarized in Tables 10.2 and 10.3 respectively.

Applicant assessment is considered an intermediate variable. Its evaluation is based on the values of input variables such as net annual income, overall financial status, number of depending children and age. Variable 'net annual income' can take three values: bad, fair and good. Variable 'overall financial status' can take three values: bad, fair and good. Variable 'number of depending children' can take three values: few children (corresponding to 0–2 children), fair number of children (corresponding to 3-5 children) and many children (corresponding to at least six children). Finally, variable 'age' can take three values: young, normal and old. Based on the values of these variables, the value of the intermediate variable 'applicant assessment' is evaluated. This variable takes three values: bad, fair and good. The design of systems has also taken into consideration that variables 'net annual income' and 'overall financial status' are more important in performing

Table 10.2 Summary of variables involved in applicant assessment (second stage)

Variables
Net annual income (bad, fair, good)
Overall financial status (bad, fair, good)
Number of depending children (few, fair, many)
Age (young, normal, old)

Table 10.3 Summary of variables involved in warrantor assessment (second stage)

Variables
Net annual income (bad, fair, good)
Overall financial status (bad, fair, good)
Age (young, normal, old)
Social status (bad, fair, good)

applicant assessment compared to the other two input variables (i.e. 'number of depending children' and 'age').

Warrantor assessment is also considered an intermediate variable. Such an assessment is significant in approving an application for a loan. The role of a warrantor is considered important in case of loans involving high risks. A warrantor is acceptable in case certain criteria (mainly financial) are satisfied. Evaluation of intermediate variable 'warrantor assessment' is based on warrantor attributes (input variables) such as net annual income, overall financial status, social status and age. Variables 'net annual income', 'overall financial status' and 'age' have similar representations as in the case of the applicant. Variable 'social status' depends on two parameters: monthly income and trade. This variable can take three values: bad, fair and good. Based on the values of the aforementioned variables, the value of the intermediate variable 'warrantor assessment' is evaluated. This variable takes three values: bad, fair and good.

The values of the intermediate variables ('applicant assessment' and 'warrantor assessment') are used to evaluate the value of the variable 'overall applicant assessment'. This variable takes five values: very bad, bad, fair, good and very good. Figure 10.1 depicts the interconnection among the involved parameters in applicant, warrantor and overall applicant assessment.

In the third stage, restrictions involving the funding of a loan are taken into account. As mentioned in a previous section, in case of personal and consumption loans, the maximum amount of money that can be approved for loaning may be up to a certain percentage of the annual income subtracting the amounts of existing loans in any bank. In case of home loans, annual installments should not exceed 40 % of the annual income. Furthermore, applicant obligations (e.g. existing loans) are given as input.

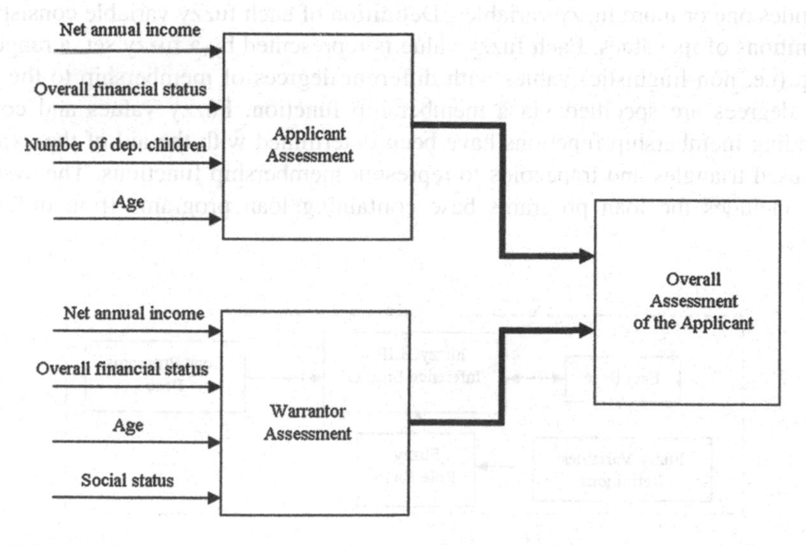

Fig. 10.1 Overview of the second stage of the inference process

In the fourth stage, results of all previous stages are taken into account. The maximum amount of money that can be approved for loaning is calculated. In case it is assessed that not all of the desired amount of money can be loaned to the applicant, the maximum possible amount of money that can be loaned is calculated. Relevant loan programs are produced as output. For each relevant loan program, all corresponding attributes are returned: interest, approved amount of money, installment and loan payment period.

10.3 The Fuzzy Expert System

Domain knowledge is characterized by inaccuracy since several terms do not have a clear-cut interpretation. Fuzzy logic makes it possible to define inexact domain entities via fuzzy sets. One of the reasons is that fuzzy logic provides capabilities for approximate reasoning, which is reasoning with inaccurate (or fuzzy) variables and values, expressed as linguistic terms [28]. All variables involved in the second stage of the inference process are represented as fuzzy variables. Inference is based on fuzzy rules acquired from domain experts. Rules provide a practical way of representing domain knowledge. Rules exhibit advantages as knowledge representation formalism such as naturalness and modularity. Naturalness refers to the fact that it is easy to comprehend the knowledge encompassed in rules. Modularity refers to the fact that each rule is autonomous and therefore it is easy to insert new knowledge into or remove existing knowledge from the rule base.

The developed fuzzy expert system has the typical structure of such systems as shown in Fig. 10.2. The fact base contains facts given as inputs or produced during inference. The rule base of the expert system contains fuzzy rules. A fuzzy rule includes one or more fuzzy variables. Definition of each fuzzy variable consists of definitions of its values. Each fuzzy value is represented by a fuzzy set, a range of crisp (i.e. non-linguistic) values with different degrees of membership to the set. The degrees are specified via a membership function. Fuzzy values and corresponding membership functions have been determined with the aid of the expert. We used triangles and trapezoids to represent membership functions. The system also includes the loan programs base containing loan programs from different banks.

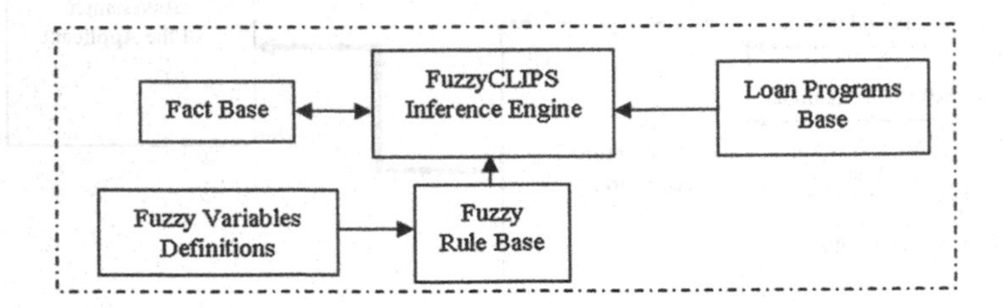

Fig. 10.2 Architecture of the fuzzy expert system

Reasoning in such a system includes three stages: fuzzification, inference, defuzzification. In fuzzification, the crisp input values (from the working memory) are converted to membership degrees (fuzzy values). In the inference stage, the MIN method is used for the combination of a rule's conditions, to produce the membership value of the conclusion, and the MAX method is used to combine the conclusions of the rules. In defuzzification, the centroid method is used to convert a fuzzy output to a crisp value, where applicable. In defuzzification, we also tested the mean of maxima method but results were worse compared to the centroid method.

Figures 10.3, 10.4, 10.5, 10.6, 10.7 provide examples of membership function definitions for various fuzzy sets corresponding to system fuzzy variables.

The system has been implemented in the FuzzyCLIPS expert system shell [32]. FuzzyCLIPS extends the CLIPS expert system shell by enabling representation and handling of fuzzy concepts. The knowledge encompassed in rules was acquired with the aid of experts from four different banks specialized in loan programs. An important task during knowledge acquisition was to cover all or most of the possible loan parameter combinations. Tables were used to record the possible situations for applicant, warrantor and overall assessment. Each table cell corresponds to a rule. Such tables are depicted in the following.

Table 10.5 depicts rules acquired in the case of applicant assessment. Table rows correspond to variables 'overall financial status' and 'age' whereas columns correspond to variables 'net annual income' and 'number of depending children'. Each variable corresponds to three sets and thus the table consists of nine (3^2) rows and nine columns. Each table cell corresponds to a rule (based on the corresponding combination of the four involved input variables) and each cell value corresponds to a value of the intermediate variable 'applicant assessment' (i.e. rule conclusion). Table 10.4 depicts the encoding of variable names and fuzzy sets used in Table 10.5.

Based on the data recorded in Table 10.5, 81 rules were initially extracted. For instance, the rules extracted from the first six cells in the first row are depicted in Table 10.6.

An examination of the rules showed that certain of them having the same conclusion could be merged resulting into a reduction of the total number of rules. For instance, the first three rules depicted in Table 10.6 could be merged into a single one as follows:

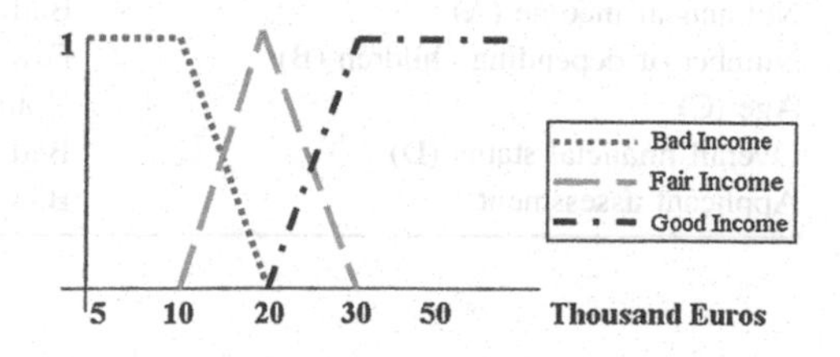

Fig. 10.3 Membership functions for applicant's net annual income

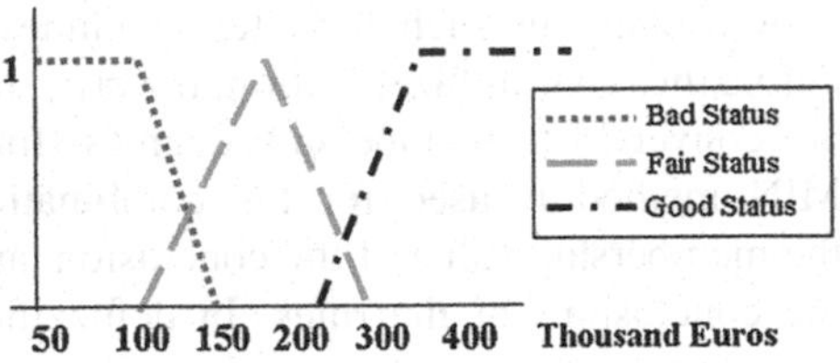

Fig. 10.4 Membership functions for applicant's overall financial status

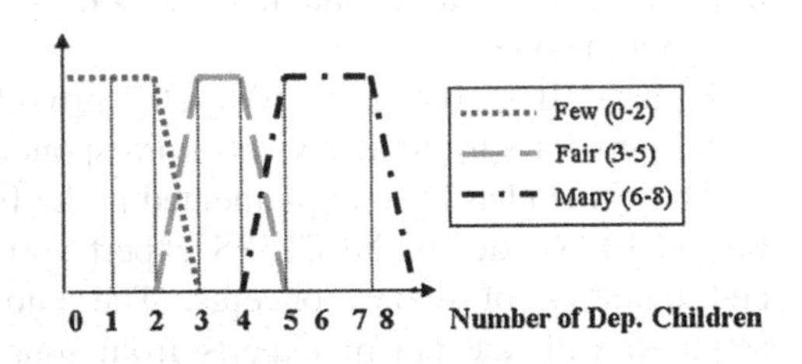

Fig. 10.5 Membership functions for applicant's number of depending children

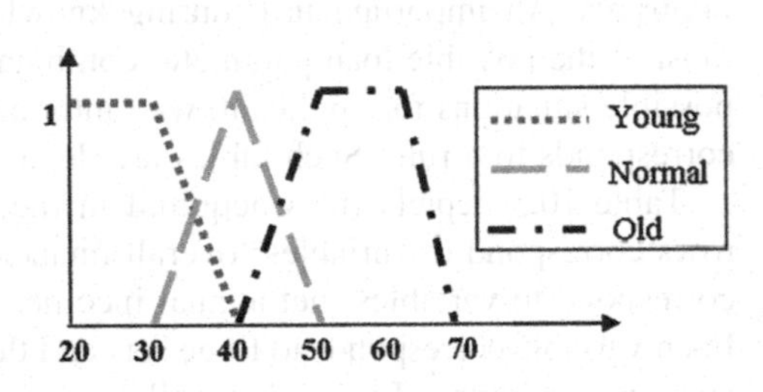

Fig. 10.6 Membership functions for applicant's age

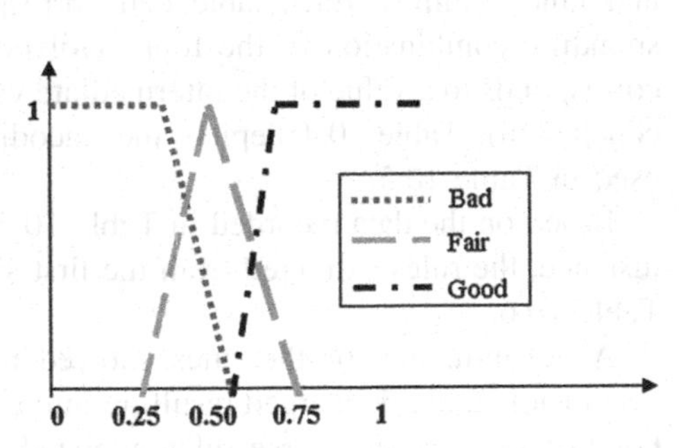

Fig. 10.7 Membership functions for applicant assessment

Table 10.4 Encoding of variable parameters involved in applicant assessment

Variables	Fuzzy sets		
	1	2	3
Net annual income (A)	Bad	Fair	Good
Number of depending children (B)	Few	Fair	Many
Age (C)	Young	Normal	Old
Overall financial status (D)	Bad	Fair	Good
Applicant assessment	Bad	Fair	Good

Table 10.5 Tabular depiction of initially recorded rules for applicant assessment

CD	AB								
	11	12	13	21	22	23	31	32	33
11	1	1	1	2	1	1	2	2	2
12	2	2	1	3	2	1	3	2	2
13	2	2	2	3	2	2	3	3	2
21	1	1	1	2	2	1	2	2	1
22	2	2	1	2	2	2	3	2	2
23	3	2	2	3	3	3	3	3	3
31	1	1	1	2	2	1	2	1	1
32	1	1	1	2	2	2	3	2	2
33	1	2	1	3	2	2	3	3	3

Table 10.6 Rules extracted from the first six cells in the first row in Table 10.5

Rule 1: If net_annual_income is bad and number_of_depending_children is few and age is young and overall_financial_status is bad then applicant_assessment is bad

Rule 2: If net_annual_income is bad and number_of_depending_children is fair and age is young and overall_financial_status is bad then applicant_assessment is bad

Rule 3: If net_annual_income is bad and number_of_depending_children is many and age is young and overall_financial_status is bad then applicant_assessment is bad

Rule 4: If net_annual_income is fair and number_of_depending_children is few and age is young and overall_financial_status is bad then applicant_assessment is fair

Rule 5: If net_annual_income is fair and number_of_depending_children is fair and age is young and overall_financial_status is bad then applicant_assessment is bad

Rule 6: If net_annual_income is fair and number_of_depending_children is many and age is young and overall_financial_status is bad then applicant_assessment is bad

"**If** net_annual_income is bad **and** number_of_depending_children is few or fair or many **and** age is young **and** overall_financial_status is bad **then** applicant_assessment is bad"

Generally speaking, rules having the same conclusion and corresponding to the same table row and column were grouped into sets. In the aforementioned example, rules corresponding to the same row (i.e. first row) were grouped. So finally, the initial 81 rules for applicant assessment were reduced to approximately 28. The reduction in the number of rules means reduction in computational time during reasoning.

We worked similarly in the case of warrantor and overall assessment. Table 10.8 depicts rules acquired in the case of warrantor assessment. Table 10.7 depicts the encoding of variable names and fuzzy sets used in Table 10.8. 81 rules for warrantor assessment were initially extracted based on the data contained in Table 10.8. After processing the initial rules by using the same approach as in the case of applicant assessment rules, approximately 27 rules were created.

Table 10.7 Encoding of variable parameters involved in warrantor assessment

Variables	Fuzzy sets		
	1	2	3
Net annual income (A)	Bad	Fair	Good
Overall financial status (B)	Bad	Fair	Good
Social status (C)	Bad	Fair	Good
Age (D)	Young	Normal	Old
Warrantor assessment	Bad	Fair	Good

Table 10.8 Tabular depiction of initially recorded rules for warrantor assessment

CD	AB								
	11	12	13	21	22	23	31	32	33
11	1	1	2	2	2	2	2	2	3
12	1	1	2	1	2	2	2	3	3
13	1	1	1	1	2	1	2	2	2
21	1	1	2	1	2	2	2	2	3
22	1	1	2	2	2	2	2	3	3
23	1	1	2	1	2	3	2	3	3
31	1	1	2	1	2	3	2	3	3
32	2	2	3	3	3	3	2	3	3
33	1	2	3	2	2	3	2	3	3

10.4 The Neurule-Based Expert System

In this section, we briefly present background knowledge concerning the syntax and semantics of neurules. We also outline the mechanisms of the neurule-based expert system.

10.4.1 Syntax and Semantics of Neurules

Symbolic rules and neural networks have been used in many intelligent systems [15] [4],. Their complementary advantages and disadvantages are a main reason for their integration [10]. Neuro-symbolic approaches are a usual type of hybrid or integrated intelligent systems [6, 10]. An important portion of neuro-symbolic approaches combine symbolic rules and neural networks. Neurules are a type of hybrid rules integrating symbolic rules with neurocomputing giving pre-eminence to the symbolic component [26]. Neurocomputing is used within the symbolic framework to improve the inference performance of symbolic rules [7]. In contrast to other hybrid approaches (e.g. [4], [5]), the constructed knowledge base retains the modularity of production rules, since it consists of autonomous units (neurules), and also retains their naturalness in a great degree, since neurules look much like symbolic rules. The inference mechanism is a tightly integrated process resulting in more efficient inferences than those of symbolic rules [7] and other hybrid approaches [12]. Explanations in the form of if–then rules can be produced [27].

The form of a rule is depicted in Fig. 10.8a. Each condition C_i is assigned a number sf_i, called its *significance factor*. Moreover, each rule itself is assigned a number sf_0, called its *bias factor*. Internally, each rule is considered as an adaline unit (Fig. 10.8b). The *inputs* C_i $(i = 1, \ldots, n)$ of the unit are the *conditions* of the rule. The weights of the unit are the significance factors of the rule and its bias is the bias factor of the neurule. Each input takes a value from the following set of discrete values: [1 (true), −1 (false), 0 (unknown)]. This gives the opportunity to easily distinguish between the falsity and the absence of a condition in contrast to symbolic rules. The *output D*, which represents the *conclusion* (decision) of the rule, is calculated via the formulas:

$$D = f(\mathbf{a}), \quad \mathbf{a} = sf_0 + \sum_{i=1}^{n} sf_i C_i$$

$$f(\mathbf{a}) = \begin{cases} 1 & \text{if a} \geq 0 \\ -1 & \text{otherwise} \end{cases}$$

where a is the activation value and f(x) the activation function, a threshold function. Hence, the output can take one of two values (‘−1’, ‘1’) representing failure and success of the rule respectively. The general syntax of a condition Ci and the conclusion D is:

<condition>::= <variable> <l-predicate> <value>
<conclusion>::= <variable> <r-predicate> <value>
Where <variable> denotes a variable, that is a symbol representing a concept in the domain, e.g. ‘net annual income’, ‘age’ etc., in a banking domain. <l-predicate> denotes a symbolic or a numeric predicate. The symbolic predicates are {is, isnot} whereas the numeric predicates are {<,>, =}. <r-predicate> can only be a symbolic predicate. <value> denotes a value. It can be a symbol or a number. The significance factor of a condition represents the significance (weight) of the condition in drawing the conclusion(s). Table 10.9 presents an example neurule, from a banking domain.

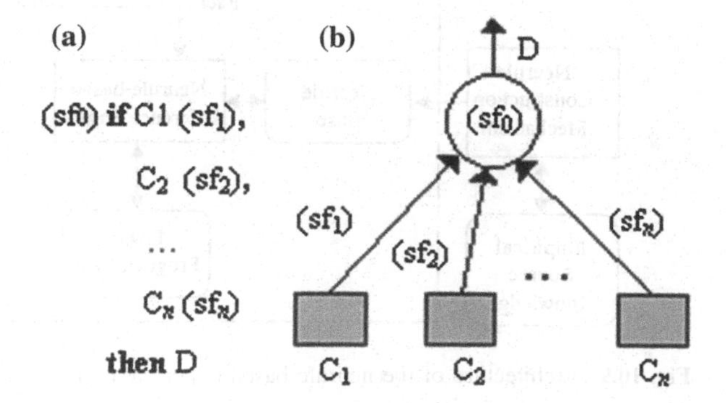

Fig. 10.8 **a** Form of a neurule **b** a neurule as an adaline unit

Table 10.9 An example neurule

N1
(−7.7) if social_status is good (6.6),
net_annual_income is bad (3.4),
age is normal (3.4),
property_status is bad (3.3),
property_status is fair (2.7)
then warrantor_assessment is fair

Variables are discerned to input, intermediate or output ones. An input variable takes values from the user (input data), whereas intermediate or output variables take values through inference since they represent intermediate and final conclusions respectively. We distinguish between intermediate and output neurules. An intermediate neurule is a neurule having at least one intermediate variable in its conditions and intermediate variables in its conclusions. An output neurule is one having an output variable in its conclusions.

Neurules can be constructed either from symbolic rules thus exploiting existing symbolic rule bases [7, 23] or from empirical data (i.e. training patterns) [8, 12, 24]. In each process, an adaline unit is initially assigned to each intermediate and final conclusion and the corresponding training set is determined. Each unit is individually trained via the Least Mean Square (LMS) algorithm (see e.g. [4]). When the training set is inseparable, more than one neurule having the same conclusion are produced.

In Fig. 10.9, the architecture of the neurule-based expert system is presented. For the implementation of the system, the HYMES tool [9] was used. The run-time system (in the dashed shape) consists of the following modules: the working memory, the neurule-based inference engine, the explanation mechanism, the neurule base and the loan programs base. The neurule base contains neurules.

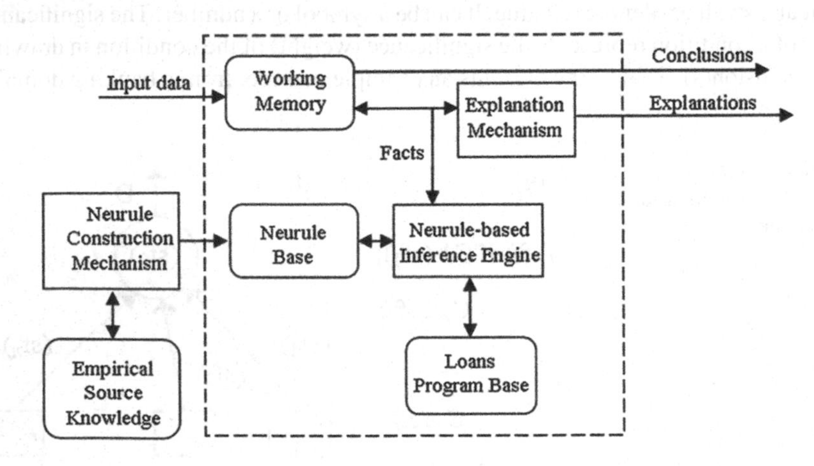

Fig. 10.9 Architecture of the neurule-based expert system

These neurules are produced off-line from available empirical knowledge concerning past loan cases (see Sect. 10.5). The construction process is performed by the neurule construction mechanism [8, 12]. The neurule-based inference engine reaches conclusions based on input data given by the user and facts derived during the inference process [7, 12]. Input data and derived facts are stored in the working memory. The explanation mechanism provides explanations for the reached conclusions in the form of if–then rules [27].

10.4.2 Construction of the Neurule Base from Empirical Source Knowledge

The neurule construction mechanism is based on the LMS training algorithm. It tries to produce one neurule for each intermediate or output conclusion. However, due to possible non-linearity of the data, this is not usually feasible. So, more than one neurule having the same conclusion may be produced for each intermediate or output conclusion. The construction process from empirical data is described in detail in [8].

The mechanism takes as input the following three types of data:

- A set of domain variables V, with their possible values
- Dependency information f_{DI}^{V}
- A set S of empirical data

Variables and values represent concepts in the problem domain. Let us consider a finite set of *domain variables* $V = \{V_i\}$, $1 \leq i \leq m$, which represent the concepts of the problem domain involved in making inferences. Each variable V_i can take values from a (corresponding) set of discrete values $S_{Vi} = \{v_{ij}\}$, $1 \leq j \leq k$.

Dependency information is given by the expert and indicates how variables depend on each other in making decisions. Dependency information is represented as a set of ordered pairs:

$$f_{DI}^{V} = \{(V_i, V_j) : V_i, V_j \in V,\ i \neq j\}$$

Each (V_i, V_j) is interpreted as "V_i depends on V_j". Dependency information can be illustrated by means of a matrix (see Table 10.10).

The set of empirical data contains a number of training examples. Each training example is a tuple of values corresponding to the domain variables. The training set of each neurule consists of positive and negative training examples i.e., examples with goal value equal to true or false respectively.

The main steps of the neurule construction process are the following:

1. Construct initial neurules, based on dependency information.
2. Extract an initial training set for each initial neurule from S.
3. For each initial neurule R_k do

Table 10.10 Dependency information for the bank loan domain

	Net annual income	Overall financial status	Num of dep. children	Age	Social status	Applicant assessment	Warrantor assessment
Applicant assessment	×	×	×	×			
Warrantor assessment	×	×		×	×		
Overall applicant assessment						×	×

3.1 Train R_k using the LMS algorithm and the corresponding training set.
3.2 If training is successful, produce the corresponding neurule.
3.3 Otherwise, split the training set into two subsets and apply steps 3.1–3.3 for each subset separately.

We may use different strategies to split a training set into two subsets. All strategies satisfy the following requirements:

- Each training subset contains all the negative examples of the initial training set to protect corresponding neurule from misactivations.
- Each training subset contains at least one positive example to guarantee the activation of the corresponding neurule.
- The two subsets created by splitting a (sub)set do not have common positive examples to avoid having different neurules activated by the same positive example(s).

In our bank loan domain, the set of domain variables involved in the second stage of the inference process and the corresponding dependency information are the following:

V={net_annual_income, overall_financial_status, children_num, age, social_status, applicant_assessment, warrantor assessment, overall_applicant_assessment}

f_{DI}^{V} ={(net_annual_income, applicant_assessment),
(overall_financial_status, applicant_assessment),
(num_of_dep_children, applicant_assessment),
(age, applicant_assessment),
(net_annual_income, warrantor_assessment)
(overall_financial_status, warrantor_assessment),
(age, warrantor_assessment),
(social_status, warrantor_assessment),
(applicant_assessment, overall_applicant_assessment),
(applicant_assessment, overall_applicant_assessment)}

The dependency information is depicted in Table 10.10 (where × means 'depends on').

The training examples were constructed from past loan cases mainly deriving from two banks (see Sect. 10.5).

10.4.3 Neurule-Based Inference Engine

The neurule-based inference engine performs a task of classification: based on the values of the condition variables and the weighted sums of the conditions, conclusions are reached. It gives pre-eminence to symbolic reasoning, based on a backward chaining strategy [7, 12]. As soon as the initial input data is given and put in the working memory, the output neurules are considered for evaluation. One of them is selected for evaluation. A neurule fires if the output of the corresponding adaline unit is computed to be '1' after evaluation of its conditions. A neurule is said to be 'blocked' if the output of the corresponding adaline unit is computed to be '−1' after evaluation of its conditions.

A condition evaluates to 'true', if it matches a fact in the working memory, that is there is a fact with the same variable, predicate and value. A condition evaluates to 'unknown', if there is a fact with the same variable, predicate and 'unknown' as its value. A condition cannot be evaluated if there is no fact in the working memory with the same variable. In this case, either a question is made to the user to provide data for the variable, in case of an input variable, or an intermediate neurule with a conclusion containing the variable is examined, in case of an intermediate variable. A condition with an input variable evaluates to 'false', if there is a fact in the working memory with the same variable, predicate and different value. A condition with an intermediate variable evaluates to 'false' if additionally to the latter there is no unevaluated intermediate neurule that has a conclusion with the same variable. Inference stops either when one or more output neurules are fired (success) or there is no further action (failure).

Conditions of neurules are organized according to the descending order of their significance factors. This facilitates inference. When a neurule is examined in the inference process, not all of its conditions need to be evaluated. Evaluation of a neurule's conditions proceeds until their weighted sum exceeds the remaining sum (i.e., sum of the absolute values of the unevaluated conditions' significance factors).

An advantage of neurule-based reasoning compared to symbolic rule-based reasoning is the ability to reach conclusions from neurules even if some of the conditions are unknown. This is not possible in symbolic rule-based reasoning. A symbolic rule needs all its conditions to be known in order to produce a conclusion.

10.5 Evaluation of the Systems

To evaluate the fuzzy expert system (FES) and the neurule-based expert system (NBES), we used 100 past loan cases mainly deriving from the National Bank of Greece and the Bank of Cyprus. 30 % of those cases were randomly chosen and used to test both systems. Random choice of test cases was performed in a way that an equal number of test cases corresponded to each of the five values of overall applicant assessment (i.e. very bad, bad, fair, good, very good). The rest 70 % of the available cases were used to construct the neurules contained in NBES. Thus, the knowledge base contents of the systems were derived from different source types (i.e. expert rules and cases).

Table 10.11 Evaluation results for applicant assessment

Applicant assessment	Accuracy		Sensitivity		Specificity	
	FES	*NBES*	*FES*	*NBES*	*FES*	*NBES*
Bad	0.90	0.83	0.80	0.87	0.95	0.81
Fair	0.73	0.76	0.60	0.60	0.80	0.85
Good	0.86	0.86	0.80	0.83	0.85	0.88
Average	*0.83*	*0.82*	*0.73*	*0.77*	*0.87*	*0.85*

Table 10.12 Evaluation results for warrantor assessment

Warrantor assessment	Accuracy		Sensitivity		Specificity	
	FES	*NBES*	*FES*	*NBES*	*FES*	*NBES*
Bad	0.93	0.93	0.80	0.80	1.00	1.00
Fair	0.76	0.76	0.70	0.80	0.80	0.75
Good	0.83	0.83	0.80	0.70	0.85	0.90
Average	*0.84*	*0.84*	*0.77*	*0.77*	*0.88*	*0.88*

Table 10.13 Evaluation results for overall applicant assessment

Overall Applicant assessment	Accuracy		Sensitivity		Specificity	
	FES	*NBES*	*FES*	*NBES*	*FES*	*NBES*
Very bad	0.93	0.90	0.66	0.83	1.00	0.90
Bad	0.96	0.93	0.83	0.83	1.00	0.95
Fair	0.86	0.90	0.50	0.83	0.95	0.91
Good	0.80	0.90	0.83	0.50	0.79	1.00
Very good	0.96	0.96	0.50	1.00	0.95	0.95
Average	*0.90*	*0.92*	*0.66*	*0.80*	*0.94*	*0.94*

Evaluation results for applicant assessment, warrantor assessment and overall applicant assessment are presented in Tables 10.11, 10.12 and 10.13 respectively. In each table, we present separate results for each one of the involved classes as well as average results (shown in italics) for all classes. As mentioned in Sect. 10.2, applicant and warrantor assessment involve three classes whereas overall applicant assessment involves five classes.

We use 'accuracy' accompanied by 'specificity' and 'sensitivity' as evaluation metrics:

$$\begin{aligned} accuracy &= (a + d)/(a + b + c + d), \\ sensitivity &= a/(a + b), \\ specificity &= d/(c + d) \end{aligned}$$

where a is the number of positive cases correctly classified, b is the number of positive cases that are misclassified, d is the number of negative cases correctly classified and c is the number of negative cases that are misclassified. By 'positive' we mean that a case belongs to the corresponding assessment class and by 'negative' that it doesn't. Results show that performances of both systems are comparable considering all metrics. In overall applicant assessment, NBES performed slightly better in terms of accuracy and much better as far as sensitivity is concerned.

10.6 Conclusions

In this chapter, we present the design, implementation and evaluation of two intelligent systems for performing a credit scoring task in banks i.e. assessment of bank loan applicants. Loan programs from different banks are taken into consideration. The intelligent systems involve a fuzzy expert system and a neuro-symbolic expert system constructed from expert rules and available cases respectively. The knowledge base of the neuro-symbolic expert system contains neurules, a type of hybrid rules integrating symbolic rules with neurocomputing.

Evaluation results for both systems are comparable, despite the different types of knowledge sources. In certain aspects, one system performs better than the other. This means that both types of knowledge sources can be exploited in producing outputs. Based on the results, an integrated (or hybrid) approach could be developed. More specifically, a hybrid system could be developed involving both systems as separate cooperating modules. A future direction of our work involves the acquisition of additional past loan cases in order to perform further experiments.

References

1. Abdou, H.A.: Genetic programming for credit scoring: the case of Egyptian public sector banks. Expert Syst. Appl. **36**, 11402–11417 (2009). http://dx.doi.org/10.1016/j.eswa.2009.01.076
2. Chen, Y.-S., Cheng, C.-H.: Hybrid models based on rough set classifiers for setting credit rating decision rules in the global banking industry. Knowl.-Based Syst. **39**, 224–239 (2013). http://dx.doi.org/10.1016/j.knosys.2012.11.004
3. Eletter, S.F., Yaseen, S.G., Elrefae, G.A.: Neuro-based artificial intelligence model for loan decisions. Am. J. Econ. Bus. Adm. **2**, 27–34 (2010)
4. Gallant, S.I.: Neural Network Learning and Expert Systems. The MIT Press, Cambridge, MA (1993)
5. Ghalwash, A.Z.: A recency inference engine for connectionist knowledge bases. Appl. Intell. **9**, 201–215 (1998). doi:10.1023/A:1008311702940
6. Hammer, B., Hitzler, P. (eds.): Perspectives of neural-symbolic integration, vol. 77. Springer, Berlin, Heidelberg (2010). (Studies in Computational Intelligence)
7. Hatzilygeroudis, I., Prentzas, J.: Neurules: improving the performance of symbolic rules. Int. J. AI Tools **9**, 113–130 (2000). doi:10.1142/S0218213000000094
8. Hatzilygeroudis, I., Prentzas, J.: Constructing modular hybrid rule bases for expert systems. Int. J. AI Tools **10**, 87–105 (2001). doi:10.1142/S021821300100043X
9. Hatzilygeroudis, I., Prentzas, J.: HYMES: a HYbrid modular expert system with efficient inference and explanation. In: Manolopoulos, Y., Evripidou, S. (eds.) Proceedings of the 8th Panhellenic Conference on Informatics, vol.1. Livanis Publishing Organization, Athens (2001)
10. Hatzilygeroudis, I., Prentzas, J.: Neuro-symbolic approaches for knowledge representation in expert systems. Int. J. Hybrid Intell. Syst. **1**, 111–126 (2004)
11. Hatzilygeroudis, I., Prentzas, J.: Knowledge representation in intelligent educational systems. In: Ma, Z. (ed.) Web-based intelligent e-learning systems: technologies and applications. Information Science Publishing, Hershey, PA (2006)
12. Hatzilygeroudis, I., Prentzas, J.: Neurules: integrated rule-based learning and inference. IEEE Trans. Knowl. Data Eng. **22**, 1549–1562 (2010). doi:10.1109/TKDE.2010.79
13. Hatzilygeroudis, I., Prentzas, J. (eds.) (2011) Combinations of intelligent methods and applications. In: Proceedings of the 2nd International Workshop, CIMA 2010, Smart Innovation, Systems and Technologies, vol. 8. Springer, Berlin, Heidelberg
14. Hatzilygeroudis, I., Prentzas, J.: Fuzzy and neuro-symbolic approaches to assessment of bank loan applicants. In: Iliadis, L., Maglogiannis, I., Papadopoulos, H. (eds.) IFIP Advances in Information and Communication Technology (AICT), vol. 36. Springer-Verlag, Berlin Heidelberg (2011)
15. Hatzilygeroudis, I., Koutsojannis, C., Papavlasopoulos, C., Prentzas, J.: Knowledge-based adaptive assessment in a Web-based intelligent educational system. In: Proceedings of the Sixth International Conference on Advanced Learning Technologies, IEEE Computer Society. Los Alamitos, CA (2006)
16. Hens, A.B., Tiwari, M.K.: Computational time reduction for credit scoring: an integrated approach based on support vector machine and stratified sampling method. Expert Syst. Appl. **39**, 6774–6781 (2012). doi:10.1016/j.eswa.2011.12.057
17. Kamos, E., Matthaiou, F., Kotsiantis, S.: Credit rating using a hybrid voting ensemble. In: Maglogiannis, I., Plagianakos, V., Vlahavas, I. (eds.) Proceedings of the Hellenic Conference on Artificial Intelligence, Lecture Notes in Artificial Intelligence, vol. 7297. Springer, Berlin Heidelberg (2012)
18. Li, X.-L., Zhong, Y.: An overview of personal credit scoring: techniques and future work. Int. J. Intell. Sci. **2**, 181–189 (2012). doi:10.4236/ijis.2012.224024

19. Marques, A.I., Garcia, V., Sanchez, J.S.: Exploring the behavior of base classifiers in credit scoring ensembles. Expert Syst. Appl. **39**, 10244–10250 (2012). doi:10.1016/j.eswa.2012.02.092
20. Min, J.H., Lee, Y.-C.: A practical approach to credit scoring. Expert Syst. Appl. **35**, 1762–1770 (2008). doi:10.1016/j.eswa.2007.08.070
21. Nurlybayeva, K.: Algorithmic scoring methods. Appl. Math. Sci. **7**, 571–586 (2013)
22. Prentzas. J., Hatzilygeroudis, I., Koutsojannis, C.: A Web-based ITS controlled by a hybrid expert system. In: Proceedings of the IEEE International Conference on Advanced Learning Technologies, IEEE Computer Society, Los Alamitos, CA (2001)
23. Prentzas, J., Hatzilygeroudis, I.: Rule-based update methods for a hybrid rule base. Data Knowl. Eng. **55**, 103–128 (2005). doi:10.1016/j.datak.2005.02.001
24. Prentzas, J., Hatzilygeroudis, I.: Incrementally updating a hybrid rule base based on empirical data. Expert Syst. **24**, 212–231 (2007). doi:10.1111/j.1468-0394.2007.00430.x
25. Prentzas, J., Hatzilygeroudis, I.: Combinations of case-based reasoning with other intelligent methods. Int. J. Hybrid Intell. Syst. **6**, 189–209 (2009). doi:10.3233/HIS-2009-0096
26. Prentzas, J., Hatzilygeroudis, I.: Neurules—a type of neuro-symbolic rules: an overview. In: Hatzilygeroudis, I., Prentzas, J. (eds.) Combinations of Intelligent Methods and Applications: Proceedings of the 2nd International Workshop, CIMA 2010, Smart Innovation, Systems and Technologies, vol. 8. Springer, Berlin, Heidelberg (2011)
27. Prentzas, J., Hatzilygeroudis, I.: An explanation mechanism for integrated rules. In: Hatzilygeroudis, I., Palade, V. (eds.) Proceedings of the 3rd International Workshop on Combinations of Intelligent Methods and Applications. Montpellier, France (2012)
28. Ross, T.J.: Fuzzy logic with engineering applications, 3rd edn. Wiley, Chichester, West Sussex (2010)
29. Wang, G., Ma, J., Huang, L., Xu, K.: Two credit scoring models based on dual strategy ensemble trees. Knowl.-Based Syst. **26**, 61–68 (2012). doi:10.1016/j.knosys.2011.06.020
30. Vukovic, S., Delibasic, B., Uzelac, A., Suknovic, M.: A case-based reasoning model that uses preference theory functions. Expert Syst. Appl. **39**, 8389–8395 (2012). doi:10.1016/j.eswa.2012.01.181
31. Zhou, L., Lai, K.K., Yu, L.: Least squares support vector machines ensemble models for credit scoring. Expert Syst. Appl. **37**, 127–133 (2010). doi:10.1016/j.eswa.2009.05.024
32. http://awesom.eu/~cygal/archives/2010/04/22/fuzzyclips_downloads/index.html

Chapter 11
Prediction of Long-Term Government Bond Yields Using Statistical and Artificial Intelligence Methods

Marco Castellani and Emanuel A. dos Santos

Abstract This chapter investigates the use of different artificial intelligence and classical techniques for forecasting the monthly yield of the US 10-year Treasury bonds from a set of four economic indicators. The task is particularly challenging due to the sparseness of the data samples and the complex interactions amongst the variables. At the same time, it is of high significance because of the important and paradigmatic role played by the US market in the world economy. Four data-driven artificial intelligence approaches are considered: a manually built fuzzy logic model, a machine learned fuzzy logic model, a self-organising map model, and a multi-layer perceptron model. Their prediction accuracy is compared with that of two classical approaches: a statistical ARIMA model and an econometric error correction model. The algorithms are evaluated on a complete series of end-month US 10-year Treasury bonds yields and economic indicators from 1986:1 to 2004:12. In terms of prediction accuracy and reliability, the best results are obtained by the three parametric regression algorithms, namely the econometric, the statistical, and the multi-layer perceptron model. Due to the sparseness of the learning data samples, the manual and the automatic fuzzy logic approaches fail to follow with adequate precision the range of variations of the US 10-year Treasury bonds. For similar reasons, the self-organising map model performs unsatisfactorily. Analysis of the results indicates that the econometric model has a slight edge over the statistical and the multi-layer perceptron models. This suggests that pure data-driven induction may not fully capture the complicated mechanisms

---

This work was carried out while Dr. dos Santos was member of the Board of IGCP—Portuguese Public Debt Agency.

M. Castellani (✉)
Department of Biology, University of Bergen, 5020 Bergen, Norway
e-mail: Marco.Castellani@bio.uib.no

E. A. dos Santos
Banco de Portugal, 1150-012 Lisbon, Portugal
e-mail: emanuelagsa@sapo.pt

C. Faucher and L. C. Jain (eds.), *Innovations in Intelligent Machines-4*,
Studies in Computational Intelligence 514, DOI: 10.1007/978-3-319-01866-9_11,

ruling the changes in interest rates. Overall, the prediction accuracy of the best models is only marginally better than the prediction accuracy of a basic one-step lag predictor. This result highlights the difficulty of the modelling task and, in general, the difficulty of building reliable predictors for financial markets.

11.1 Introduction

Changes in interest rates are important to macroeconomic analysis and economic growth. However, it is in the financial markets that they have a more valuable impact and are most closely monitored. This is because interest rates represent the price for borrowing money, and as such they determine the value of financial assets.

Starting from the business world, a large amount of information on future and forward contracts on bonds can be collected and used for building a model for the so-called term-structure of interest rates.[1] In a framework of certainty equilibrium, forward rates[2] must coincide with future spot rates.[3] Assuming that the economic agents' rational expectations equal true statistical expected values, it is possible to create specific models which can be calibrated, empirically tested, and used for predicting interest rates.[4] However, such deterministic environment does not exist. For example, future interest rates reflect human expectations on many factors not under control. Moreover, given the increasing internationalisation of economies and financial markets, the prediction of interest rates has become more complex, since changes in one country influence other countries as well.

Modelling financial time-series represents a complex task in terms of knowledge acquisition and representation.

Classical financial modelling theory is based on accurate mathematical identification of the observed system behaviour, modelling and forecasting economic variables using classic econometrics [26] or time series theory [9, 25]. The econometric approach to time series modelling makes use of economic theory to define the structure of the relationship between a specific economic variable (endogenous) and a set of explanatory variables. In most cases, the postulated functional form can be a linear or non-linear function. The unknown parameters of the model are then estimated using statistical techniques such as least squares. The estimated model is used to make forecasts that can be statistically evaluated.

---

[1] The term-structure of interest rates measures the relationship among the yields of risk-free securities that differ only in their term to maturity.

[2] Forward rates apply to contracts for delivery at some future date.

[3] Spot rates, on the contrary of forward rates, apply to contracts for immediate delivery.

[4] The Cox, Ingersoll, Ross [10] model is an example of an equilibrium asset pricing model for the term structure of interest rates.

Univariate ARIMA models [5] employ pure statistical methods to estimate and forecast the future values of a variable. In this case, current and past values are the only data used for the estimation.

Unfortunately, the complexity of financial markets and the intrinsic uncertainties regarding their dynamics make the expression of precise analytical relationships often impossible, impractical or just unmanageably complex. Moreover, due to the non-linear, non-parametric nature of economic systems, standard linear econometric modelling has often turned out to be unsatisfactory.

The study of biological nervous systems has shown that highly accurate and robust mappings can be achieved by learning appropriate sets of condition-response pairs. In the field of artificial intelligence (AI) two main approaches have emerged, each modelling cognitive processes at different levels of abstraction. The first method focuses on high-level symbolic associations and expresses complex stimulus–response relationships through sets of if–then rules. Fuzzy logic (FL) is a symbolic AI paradigm that extends Aristotle's classical logic to account for uncertainty about real world knowledge [22].

The second approach postulates that the computational capabilities of nervous systems are based on the parallel distributed processing of massively connected networks of simple computing units. Artificial neural networks (ANNs) represent the connectionist AI effort to model the architecture of biological information processing systems [27].

ANNs and FL share common features and complementary limitations. Both paradigms provide a choice of mapping algorithms capable to perform model-free identification of any arbitrarily complex non-linear function [24, 46]. The approximate nature of their pattern matching and association processes makes them particularly suitable to deal with ill-defined and uncertain problem domains.

These two AI approaches appear as alternatives to modelling and forecasting economic variables using classic theory. This chapter compares the ability of AI and classical econometric and ARIMA models of forecasting the US 10-year Treasury bonds yields. The task is chosen because of its complexity as a modelling problem and because of the role played by the US economy in the world market. A complete series of end-month US 10-year Treasury bonds yields and economic indicators covering 19 years between 1986:1 and 2004:12 are available [6]. The models are fitted using data regarding the first 18 years and evaluated on their capability of forecasting the US 10-year Treasury bonds yields for the remaining 12-months. The root mean square error (RMSE) of the 12-month out of sample forecasts is used to measure the modelling accuracy.

The remainder of the chapter is organized as follows. Section 11.2 presents the problem domain. Section 11.3 introduces the FL and ANN models. Section 11.3 describes the econometric and ARIMA models. Section 11.4 presents the experimental results and compares the performance of the models. Section 11.5 discusses the results. Section 11.6 concludes the chapter and proposes areas for further investigation.

11.2 Problem Domain and Experimental Data

The proposed case study concerns the forecasting of the US Treasury bonds yields from the measures of four economic indicators. A complete set of 228 monthly data covering the 19 years between 1986:1 and 2004:12 are available. There are no missing attributes.

The AI and classical modelling approaches are evaluated on their accuracy of predicting the correct monthly figure for the US Treasury bonds yields. This figure must be estimated based on the corresponding monthly figure of the four economic indicators. The forecasting task requires the identification of the input–output relationship between the dependent variable (the US bonds yields) and the four independent variables (the indicators). 216 data samples relative to the first 18 years are used to fit the models, and the remaining 12 data samples (2004:1 to 2004:12) are used to evaluate the modelling accuracy.

The choice of forecasting US Treasury bonds yields is motivated by the fact that the US economy is a paradigmatic market playing an important role in the world economy. In particular, developments in the US economy have impact on the other two main economic areas—Europe and Japan. The full extent of this influence became clear during the recent world great depression in the wake of the 2007 United States subprime financial crisis

In the case of Europe, it is recognised the existence of a significant correlation between the yields of US treasuries and the yields of German bunds given a stable exchange market. Figure 11.1 visualises this correlation during the period 2000:12–2004:12.

Since the German bund is nowadays the benchmark for the bonds issued by the other countries in the euro area, the forecast of US long-term interest rates could help to foresee the future evolution of interest rates on sovereign debt in any other European country. Baele et al. [1] pointed out that the government bonds yields in countries belonging to the euro area are sensitive to regional and global shocks but not to idiosyncratic shocks, supporting the assumption of an increasing interrelationship of the financial markets at world level.

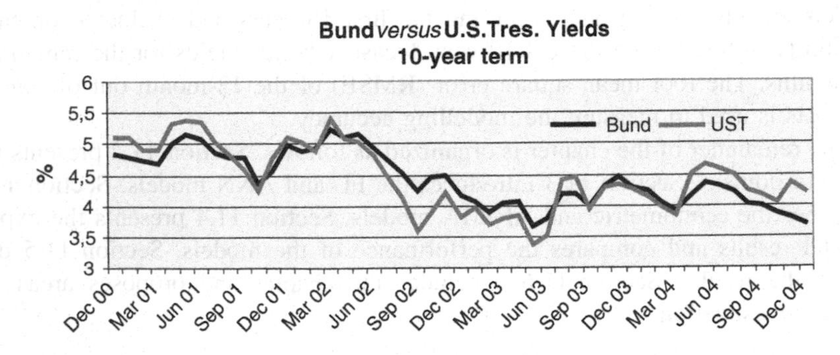

Fig. 11.1 US treasuries yields versus German bunds yields

11.2.1 Dependent Variable

The 10-year U.S. Treasury bond is one of the fixed maturity securities for which the U.S. Treasury calculates a daily yield. However, monthly data are used in order to have the same periodicity as that of the explanatory variables. The other maturities are, currently, 1, 3 and 6 months and 1, 2, 3, 5, 7 and 20 years. The 10-year maturity is selected because it is a widespread benchmark used in financial markets.

11.2.2 Independent Variables

Four economic indicators are chosen as explanatory variables to predict the US Treasury bonds yields, namely, the Purchasing Managers' Index (PMI), the Consumer Price Index (CPI), the London Interbank Offering Rate (Libor) and the Volatility Index (VIX).

The economic situation is important to interest rates. When the economy is flourishing and there is a high demand for funds, the price of borrowing money grows, leading to increasing interest rates. Conversely, in economic recessions, everything else being equal, there is downward pressure on interest rates. The most important economic indicator for the output of goods and services produced in a country is the gross domestic product (GDP). However, this indicator is published only on a quarterly and annual basis. The PMI published monthly by the Institute for Supply Management (ISM) appears to be a good proxy for the GDP, as it generally shows a high correlation with the overall economy. For example, according to ISM analysis, a PMI in excess of 42.7 % over a period of time indicates an expansion of the economy. This month-to-month indicator is a composite index based on the following five indicators for the manufacturing sector of the U.S. economy: new orders, production, employment, supplier deliveries and inventories.

Inflation is important to interest rates as well. Higher-than-expected inflation can cause yields and interest rates to rise, as investors want to preserve the purchasing power of their money. The most important measure of inflation is the average change over time in prices included in the CPI. A more accurate measure of the underlying rate of inflation is obtained when the volatile food and energy prices are excluded from the CPI. The latter measure, sometimes referred as the "core" CPI, is selected for this study as one of the four explanatory variables. The year-on-year (yoy) rate of change is used in place of the raw core CPI index. The source of the data is the Bureau of Labour Statistics.

Another major factor affecting the interest rates is the monetary policy of central banks. For example, the Federal Reserve (Fed) increases or decreases the Fed Funds rate—the key-rate for lending money to the other banks—according to the economic condition. When the economy is growing above its potential and unemployment is low, a central bank will increase rates to curb inflationary pressures. In a recession, a central bank will cut rates to stimulate economic growth and reduce unemployment. In this study, the Libor is used instead of the Fed Funds rate for the three-month term. The Libor is an average of the interest rate on dollar-denominated deposits traded between banks in London. The Libor reflects every change in the Fed Funds rate and has the advantage of having a daily market-driven fixing. As the source of data the British Bankers Association is used.

A further important factor setting the course of bonds yields is the stock exchange state. When the demand in the capital market shifts from government bonds to equities, bonds prices tend to decrease and bonds yields to increase since these variables are negatively correlated. To capture this relationship an indicator for the stock market volatility is used in this study: the VIX index compiled by the Chicago Board Options Exchange (CBOE). The VIX is calculated using options on the S&P 500 index, the widely recognised benchmark for U.S. equities. The VIX index measures market expectations of near-term volatility and has been considered by many to be the world's premier barometer of investor sentiment. To obtain a long series starting in 1986:1, two indices have to be reconciled: the VOX (1986:1 to 1989:12) and VIX (1990:1 to 2004:12). For the whole period, the most recent indicator VIX (1990:1 to 2004:12) is kept as released by the CBOE and its value for the period 1986:1 to 1989:12 is calculated by using the implicit rates of change in the old series.

Figure 11.2a–e show the evolution of the 10-year U.S. Treasury bonds yields and the four explanatory variables over the 19 years period. For each plot, the vertical dashed line marks the division between the 18-years modelling samples and the one-year evaluation samples. The two horizontal lines show the range of variation of the variable over the evaluation period. Table 11.1 summarises the main statistical measures of the time series.

11.3 AI Modelling Approaches

ANNs are composed of a certain number of elementary processing units called neurons, organised in layers. Each unit receives inputs from other neurons or the external world via adjustable connections called weights, and maps these inputs to the output space via a transformation function. The transformation function can vary widely according to the ANN architecture but is usually common within a layer. A wide survey of ANN architectures can be found in [27, 32].

ANNs require no prior knowledge about the task to be performed. Typically, the network undergoes a learning phase where the weights of the connections between neurons are adjusted. This procedure modifies the system response by modifying the way the incoming signals to the units are scaled.

ANNs can be divided into supervised and unsupervised, according to the learning paradigm they use [32]. Supervised ANNs are trained under the guide of an omniscient teacher that gives examples of input and output data pairs. This is

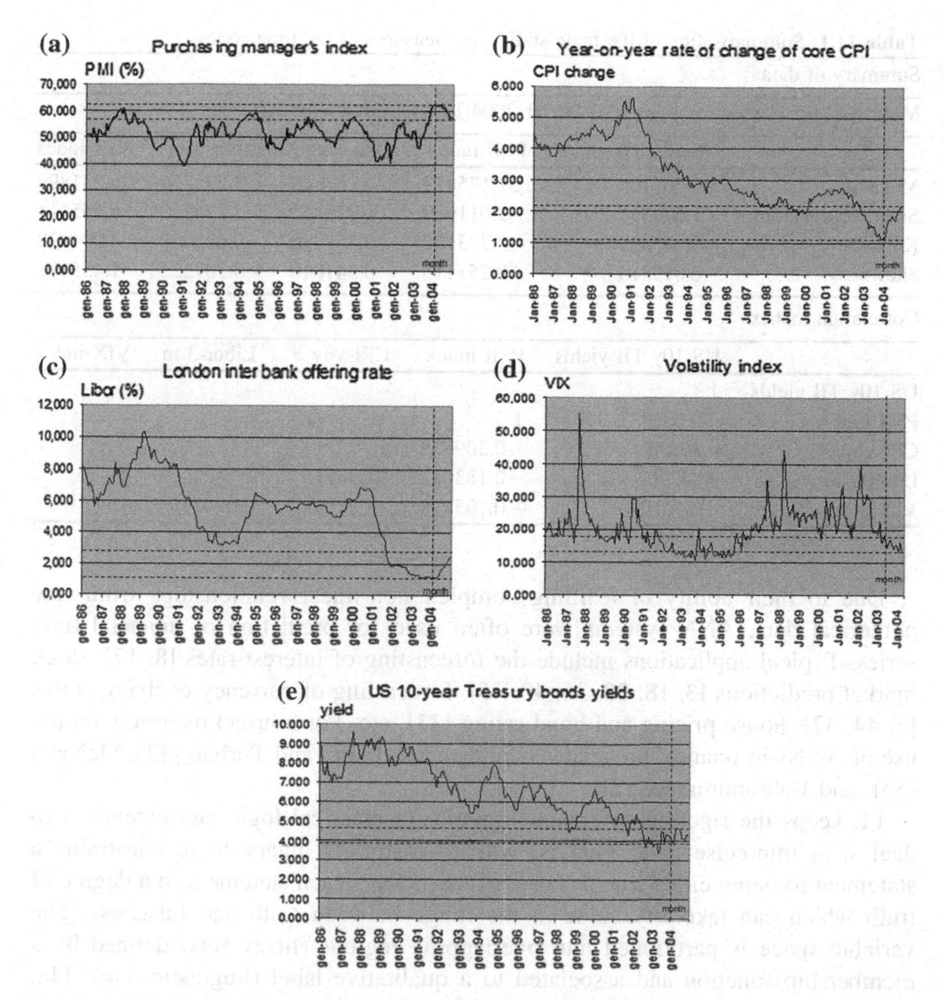

Fig. 11.2 Input and output variables. **a** PMI index (input variable). **b** CPI index (input variable). **c** LIBOR index (input variable). **d** VIX index (input variable). **e** US 10-year Treasury bonds yields (output variable)

the most commonly used ANN model for financial prediction tasks. Unsupervised ANNs are left free to organise themselves to find the best partition of the input space. In this study, a supervised and an unsupervised ANN are evaluated.

In an ANN, input–output mapping relationship is distributed among several neurons, and data are processed in parallel layer by layer. Thanks to the non-linear mapping of the individual units, ANNs are capable of modelling any arbitrarily complex function [32]. Moreover, their learning and generalisation capabilities remove the need for time-consuming system identification. However, because of its distributed nature, the decision-making policy is not retrievable after the training has ended. ANNs act similarly to a black box of which only the input and the output can be observed.

Table 11.1 Summarisation of the main statistical measures of the time series

Summary of data

Main statistical measures: sample: 1986:01 2004:12—Observations: 228

	US 10y TB yields	PMI index	CPI-yoy	Libor-3 m	VIX index
Mean	6.507105	52.175439	3.153809	5.291756	20.231899
Standard deviation	1.540538	5.011609	1.075363	2.270260	6.555111
Kurtosis	−0.955555	−0.233122	−0.855340	−0.625088	4.094172
Skewness	0.019810	−0.251113	0.341810	−0.207222	1.455191

Correlation matrix

	US 10y TB yields	PMI index	CPI-yoy	Libor-3 m	VIX index
US 10y TB yields	1	–	–	–	–
PMI index	−0.002433	1	–	–	–
CPI-yoy	0.86356	−0.309878	1	–	–
Libor-3 m	0.842462	−0.183082	0.696916	1	–
VIX index	−0.101080	−0.105170	−0.061067	0.070075	1

Due to their ability of learning complex non-linear relationships from raw numerical data, ANN systems were often used for prediction of financial time series. Typical applications include the forecasting of interest rates [8, 17], stock market predictions [3, 18, 20, 36, 40, 45], forecasting of currency exchange rates [7, 44, 47], house pricing and bond rating [11], etc. For a broad overview on the use of ANNs in finance, the reader is referred to Trippi and Turban [42], McNelis [35], and Bahrammirzaee [2].

FL keeps the rigorous inference structure of classical logic and extends it to deal with imprecise data. That is, whilst Aristotle's binary logic constrains a statement to being either true or false, FL associates each statement to a degree of truth which can take any value in the range between truth and falseness. The variable space is partitioned into overlapping regions (fuzzy sets) defined by a membership function and associated to a qualitative label (linguistic term) [48, 49]. The membership function defines the degree of truth that a numeric datum belongs to the associated label.

FL systems are usually composed of four blocks, namely the fuzzifier, the rule base (RB), the inference engine and the defuzzifier [31]. The RB and input and output fuzzy sets are often referred to as the knowledge base (KB) of the system. The fuzzifier translates crisp data into fuzzy sets and is the interface between the quantitative sensory inputs and the qualitative fuzzy knowledge. The inference engine processes the rules and generates an overall response in the form of a fuzzy set. The defuzzifier converts the fuzzy output into a crisp number.

FL modelling is akin to non-parametric basis function regression, where each rule can be thought of as a basis function. Fuzzy mappings can be used to model qualitatively complex systems, when analytically modelling would be too complex and time-consuming. If a set of qualitative 'rules of thumb' is available, the human-like representation of fuzzy rules makes it easy for experts to express their knowledge about the problem domain. The approximate nature of fuzzy relationships also makes the system more robust to noise and data corruption since the matching process is not bounded by perfect correspondence.

When expertise is not available, fuzzy rules can be obtained from data via machine learning techniques. Different strategies can be used to create or modify fuzzy mappings: new rules can be added or deleted, the input and the output space partitions (i.e. the membership functions) can be modified, or both the operations can be performed simultaneously.

The lack of a standard data induction algorithm makes the automatic acquisition of fuzzy rules less straightforward. Nonetheless, several studies addressed the application of FL to financial modelling and decision making [4, 23, 30, 34, 43]. In this study, a manually designed FL model and an automatically generated FL system are tested. A wide survey and an analysis on fuzzy identification systems is presented in [28].

For all the machine learned AI models, accuracy results are estimated on the average of 10 independent learning trials. The full details of the AI methods used are described by Castellani and dos Santos [6].

11.3.1 Supervised Artificial Neural Network Model

The multi-layer perceptron (MLP) [32] is perhaps the best known and most successful type of ANN. It is characterised by a fully connected feedforward structure made of three or four layers of processing elements. Figure 11.3 shows a typical MLP architecture.

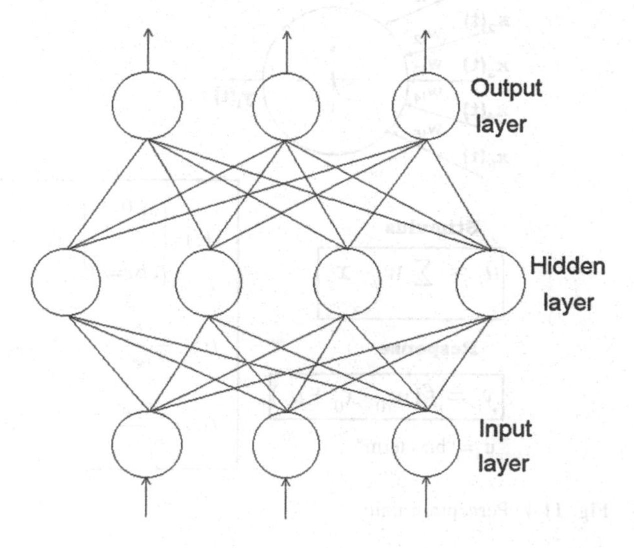

Fig. 11.3 Multi-layer perceptron

The basic unit of this kind of ANN is the perceptron. The perceptron performs a weighted summation of the input signals and transforms it via a non-linear step-wise transfer function. Figure 11.4 shows a perceptron unit.

The input layer fans the incoming signals out to the neurons of the next layer. Since the monthly forecasts for the US Treasury bonds are based on four economic indicators, four input neurons are used.

One or more hidden layers of perceptron units split the input space into several decision regions, each neuron building onto the partition of the previous layer. The complexity of the overall mapping depends on the number and the size of hidden layers. However, it was proved that no more than two hidden layers are required to model any arbitrarily complex relationship [32].

In this study, the configuration is chosen by experimentally, that is, by training different MLP structures and assessing their merit on the learning accuracy. The best prediction results for the US Treasury bonds yields are obtained using one hidden layer of 50 units.

The output layer collects the signals from the last hidden layer and further processes them to give the final ANN response. Since only one output is required (the monthly forecast for the US treasury bonds yields), this layer is composed of a single perceptron unit.

The mapping capabilities of the MLP stem from the nonlinearities used within the nodes. The proposed ANN model uses the hyperbolic tangent function for the hidden units and the sigmoidal function for the output node. Since the mapping range of the sigmoidal function is within the interval [0,1], the output of the MLP model is multiplied by a factor 10 to obtain a [0,10] mapping range.

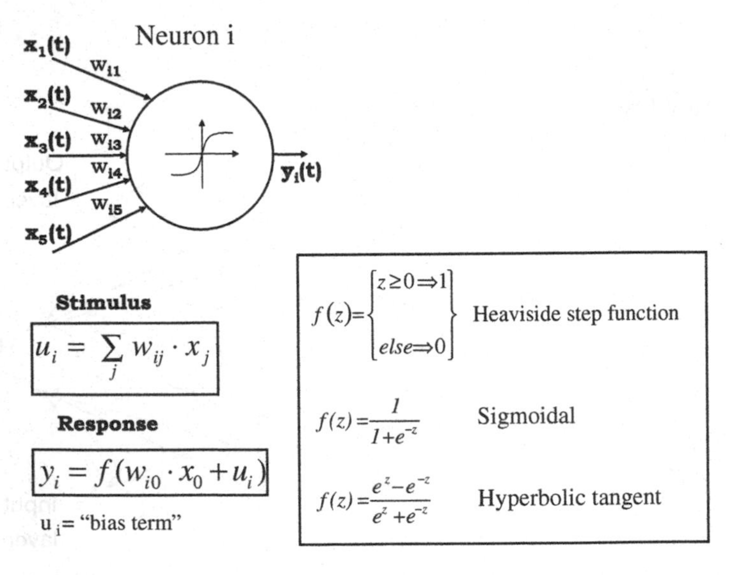

Fig. 11.4 Perceptron unit

The network is trained using the standard error backpropagation (BP) rule with momentum term [41]. According to this algorithm, the MLP uses the set of training patterns to learn the desired behaviour via least squares minimisation of the output error. The algorithm is run for a fixed number of iterations which is manually set to optimise the learning accuracy. Learning via backpropagation is akin to stochastic approximation of the input–output relationship. The learning parameters of the BP algorithm are optimised according to experimental trial and error.

Once the architecture is optimised and the ANN is trained, the system is ready to operate. Table 11.2 summarises the final MLP structure and BP rule settings.

11.3.2 Unsupervised Artificial Neural Network Model

Kohonen's self-organising feature map (SOM) [29] was originally created to reproduce the organisation of biological sensory maps of the brain. This ANN model implements a clustering algorithm that is akin to K-means. Figure 11.5 illustrates a typical SOM architecture. Due to its simple architecture, versatility, and ease of implementation, the SOM is the most popular kind of unsupervised ANN system. SOMs found application in several financial domains [12].

The SOM is made of two layers of nodes, namely the feature layer and the output layer. The feature layer collects the ANN input and forwards it to the neurons of the next layer. The output layer is composed of a two-dimensional grid of processing units. Each neuron measures the similarity between the input pattern and a reference vector stored in the values of the incoming weights. Similarity is measured as the Euclidean distance between the reference vector and the input vector.

Table 11.2 Summarisation of the final MLP structure and BP rule settings

Multi-Layer perceptron settings	
Input nodes	4
Output nodes	1
Hidden nodes	50
Activation function of hidden layer nodes	Hyper-tangent
Activation function of output layer nodes	Sigmoidal
Initialisation range for MLP weights	[−0.3, 0.3]
Backpropagation rule settings	
Learning coefficient	0.06
Momentum term	0.1
Learning trials	10
Learning iterations	100

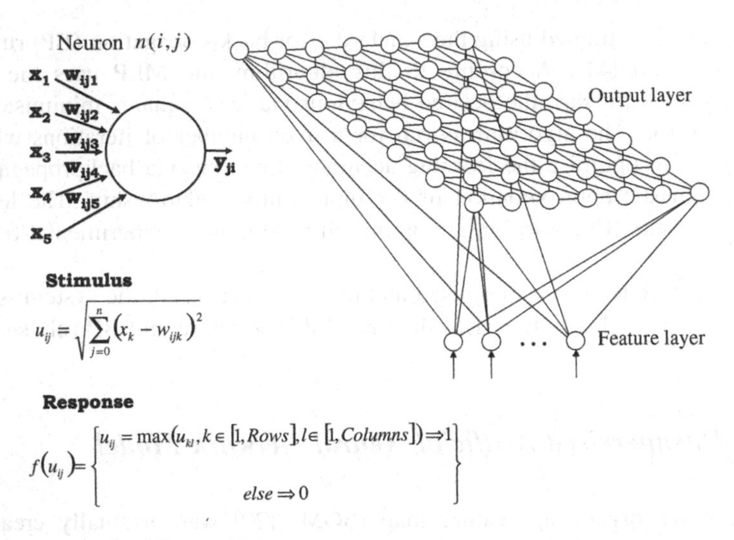

Fig. 11.5 Kohonen's self-organising map

Neurons of the output layer operate in a competitive fashion, that is, only the unit having the best matching reference vector is allowed to respond to the input pattern (winner-take-all rule).

Learning generates a vector quantiser by adjusting the incoming weights of the winner neuron to resemble more closely the input pattern. Other neurons in the neighbourhood have their weights modified of an amount that is increasingly scaled down as their distance from the winner unit widens. The magnitude of the weight correction factor is controlled via a neighbourhood function. In biological systems, competitive activation and neighbourhood learning are obtained respectively via inhibitory and excitatory synapses.

Upon iterative presentation of the input patterns, the ANN self-organises to respond with topologically close nodes to physically similar patterns. That is, the reference vectors of the output nodes move toward the centres of the clusters of training data samples. As learning proceeds, the amount of weight adaptation is decreased to allow finer adjustments of the SOM behaviour. Changes are also made more local by increasing the dampening of the weight correction factor with the distance from the winner neuron. At the end of the process, only the weights of the winner node are adjusted. The final setting of the reference vectors tends to approximate the maximal points of the probability density function of the training data [29].

SOMs can be used in model approximation by presenting the network with input vectors composed by input–output pairs of training patterns. The ANN adjusts its behaviour to cluster similar sets of condition-response pairs. During the validation phase, only the input pattern is fed and matched with the corresponding elements of the reference vector (i.e. the condition). The remaining weights of the winner neuron (i.e. the response) define the ANN model response.

Because of the topological organisation of the output layer, neighbouring conditions elicit similar responses, thus ensuring a smooth mapping of the desired relationship. Accordingly, previously unseen data are mapped according to the most similar training examples.

The SOM architecture and the learning parameters are set according to experimental trial and error. For the proposed study, a SOM having an input layer of 5 units (4 monthly economic indicators plus the corresponding US bonds yield) and an output layer of 15 × 15 units is built. The number of nodes was set taking into account the low sampling of the training space, since a large number of neurons ensures a smoother coverage of the input space. Table 11.3 summarises the main SOM settings and training parameters.

11.3.3 Manually Designed Fuzzy Logic Model

A standard Mamdani-type [33] FL system is used. The block diagram of this system is shown in Fig. 11.6.

Fuzzy sets are defined via trapezoidal membership functions (MFs), while output defuzzification is performed via the height method [37]. Since no expert knowledge is available in the form of fuzzy if–then rules, the FL model is built solely on the basis of the available data samples.

The partition of the input and the output spaces is determined according to experimental trial and error. The space of each of the four input variables is divided into seven evenly spaced linguistic terms. The output space is divided into nine linguistic terms spanning the interval [0, 10].

The rule base is built by creating a fuzzy rule out of each of the 216 training examples. Rules are generated by associating the fuzzy terms that better match the values of the input variables to the term that better matches the desired output. Duplicate rules are removed. Rules having the same input but different output are resolved by choosing the case that best fits the training examples. At the end of the process, the span of each input MF is slightly enlarged to reflect the uncertainty about the space partition.

Table 11.3 Summarisation of the main SOM settings and training parameters

Self-organising map settings	
Input nodes	5
Output nodes	15 × 15 grid
Initialisation range for weights	[−0.3, 0.3]
Learning parameters	
Learning trials	10
Learning iterations (τ)	10000
Learning coefficient (at iteration t)	$1 - t/\tau$
Neighbourhood function	Gaussian
Spread of Gaussian neighbourhood	$15*(1 - t/\tau)$

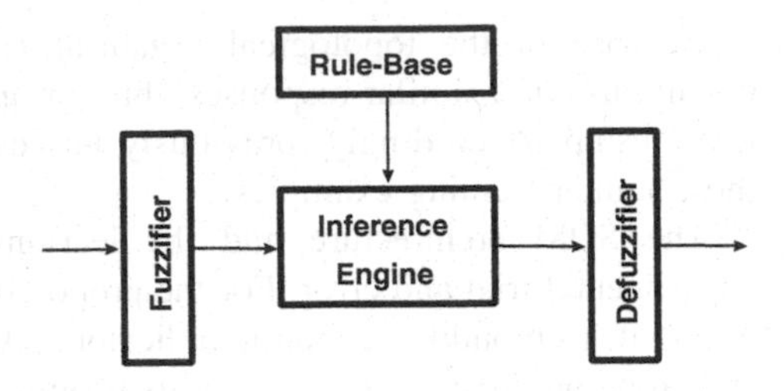

Fig. 11.6 Mamdani-type fuzzy logic system

This procedure generates 110 rules that are then used to forecast the values of the remaining 12-month out of sample 10-year Treasury bonds yields.

11.3.4 Automatically Designed Fuzzy Logic Model

Although the design of the fuzzy model is conceptually straightforward, much effort is required to generate the fuzzy mapping. Keeping the same Mamdani-type FL system used in the previous test, an alternative inductive machine learning approach is investigated for automatic identification of the 10-year Treasury bonds time series.

The generation of FL systems is essentially a search problem, where the solution space is represented by the large number of possible system configurations. Evolutionary algorithms (EAs) [13, 21] are a class of global search techniques that provide an ideal framework for the task. As well as allowing the optimisation of both the RB and the MFs, EAs only need a small amount of problem domain expertise for implementation.

EAs are modelled on Darwin's theory of natural evolution. This stipulates that a species improves its adaptation to the environment by means of a selection mechanism that favours the reproduction of those individuals of highest fitness. In EAs, the adaptation of an individual to the environment is defined by its ability to perform the required task. A problem-specific fitness function is used for the quality assessment of a candidate solution.

In EAs, a population of candidate solutions (i.e., FL systems) is iteratively made to evolve until a stopping criterion is met. The population is driven toward the optimal point(s) of the search space by means of stochastic search operators inspired by the biological mechanisms of genetic selection, mutation and recombination. At the end of the process, the best exemplar is chosen as the solution to the problem.

The EA used in this study [38] generates Mamdani-type FL systems through simultaneous evolution of the RB and MFs. The algorithm uses the generational replacement reproduction scheme [21] and an adaptive selection operator [39] that aims at maintaining the selection pressure constant throughout the whole evolution process. A set of crossover and mutation procedures each concerned with a different level of KB optimisation is used, namely, RB optimisation, MFs optimisation, and optimisation of both RB and MFs simultaneously.

Each member of the starting population is initialised with a blank RB and a random partition of the input and output spaces. During the fitness evaluation phase, candidate solutions are tested on the series of training data points. At each step, a solution forecasts the value of the US bonds yield by searching its KB for the rules best matching the set of input conditions. The algorithm creates a new rule if the set of input conditions having the highest matching degree does not lead to an existing rule action. The consequent of the newly generated rule is randomly determined. The aim of the procedure is to limit the RB growth only to the most relevant instances.

The fitness of the candidate solutions is evaluated as the measure of their root mean square modelling error over the set of training patterns. The lower the error is, the higher the chances are that the solution is selected for reproduction. Each learning cycle, a new population is created through crossover and mutation of the individuals that are selected for reproduction (i.e., the best performing ones). This procedure is repeated until a pre-defined number of iterations have elapsed and the fittest solution of the last generation is picked.

The learning parameters are set according to experimental trial and error. Table 11.4 summarises the main EA settings.

11.4 Classical Modelling Approaches

In this study two classes of traditional models are tested. The first model is a univariate model, in which future values of the variable are predicted only using current and past values of the own variable. For this reason, it belongs to the class of statistical models. The second model uses a set of variables chosen according to economic theories about the nature of the relationship with the variable to be forecast. Since the second model combines economics, mathematics and statistics, it is an example of econometric model. The full details of the two classical methods used are described by Castellani and dos Santos [6].

Table 11.4 Summarisation of the main EA settings

EA parameters	
Population size	80
Learning trials	10
Learning iterations	500
Crossover rate	1
Mutation rate	0.1
Max number of terms per variable	6
Initialisation parameters	
Number of terms per variable	4
Rule base	Empty
Fitness function settings	
Evaluation steps	216
Error measure	Root MSE

11.4.1 ARIMA Model

One of the most popular univariate time series model is the general ARIMA(p,d,q) model, where p is the number of autoregressive terms, d is the number of differences (order of integration), and q is the number of lagged disturbance terms. Its representation form is

$$\theta(L) \cdot (1 - L)^d \cdot y_t = c + \varphi(L) \cdot \varepsilon_t \tag{11.1}$$

where y_t is the time series, ε_t is the random error, c is a constant term and L is the backshift operator: $L \cdot y_t = y_t - 1$.

$\theta(L)$ is the autoregressive operator that is represented as a polynomial in the back shift operator, that is,

$$\theta(L) = 1 - \theta_1 \cdot L - \theta_2 \cdot L^2 - \cdots - \theta_p \cdot L^p \tag{11.2}$$

finally, $\varphi(L)$ is the moving-average operator, represented as a polynomial in the back shift operator, that is,

$$\varphi(L) = 1 - \varphi_1 \cdot L - \varphi_2 \cdot L^2 - \cdots - \varphi_p \cdot L^p \tag{11.3}$$

where t indexes time.

The first step to build the ARIMA model is the data identification process. For this purpose, visual inspection of the correlograms of the autocorrelation and partial autocorrelation functions is often recommended. The order of differentiation is related to the need to work with stationary time series. In many economic variables, first-difference is enough to achieve this objective.

Since the Dickey-Fuller test [14, 15] indicates that the U.S. 10-year Treasury bonds yield is an integrated variable of first order, the ARIMA model is estimated in first-difference. Following extensive experimental estimations, it is concluded that the ARIMA(2,1,2) is the best model for the available data sample in terms of forecasting performance and parsimony of parameters. The augmented Dickey-Fuller unit root test for all the variables is presented in Table 11.5.

The output from the ARIMA estimation is shown in Table 11.6. *AR*(*p*) is the component containing just the *p* lagged dependent variable terms that are statistically meaningful in the past history of the process. *MA*(*q*) is the disturbance component of the model. All *AR* and *MA* terms have high levels of statistical significance. Moreover, the inverted roots of the polynomials have absolute value smaller than one. Before using the estimated equation to forecast the 12-month values ahead of the variable, the performance of an augmented Dickey-Fuller test is used to confirm that the residuals of the equation are white noise disturbances.

Table 11.5 The augmented Dickey-Fuller unit root test for all the variables

Augmented Dickey-Fuller unit root test		
Variable	ADF test statistic	
	Level	First diff.
US 10y TB yields	−1.300460	−7.138813
PMI index	−3.312909	−6.633664
CPI-yoy	−0.849709	−5.350012
Libor-3 m	−1.410705	−5.008327
VIX index	−2.906936	−9.735054
1 % Critical Value −3.4612		
5 % Critical Value −2.5737		
10 % Critical Value −2.5737		

Table 11.6 The output from the ARIMA estimation

ARIMA(2, 1, 2) model output		
Dependent variable: Δ(U.S. 10-year Treasury bond yield		
Method: least squares		
Number of observations: 213 after adjusting endpoints		
Variable	*Coefficient*	*t-statistic*
Constant	−0.016974	−0.844646
AR(1)	−0.090818	−5.377226
AR(2)	−0.944908	−52.59363
MA(1)	0.092203	15.93348
MA(2)	0.99182	183.6825
S.E. of regression	0.287393	
Durbin-Watson statistic	1.823712	
F-statistic	4.150752	
Prob(F-statistic)	0.002953	
Inverted AR roots	−0.05 + 0.97i	
Inverted MA roots	−0.05 + 0.99i	

11.4.2 Econometric Model

The error correction model (ECM) [19] is the econometric method chosen to provide an alternative forecast for the time series of the bonds yields. This method differs from the standard regression model as it includes an error correction term to account for co-integration of the independent variables' time series. In the specialised literature, the ECM is widely reputed to possess a high predictive accuracy, and be appropriate to capture both the long and short-term dynamics of a time series.

Despite the variables being first-order integrated, it was chosen to estimate them in levels according to the ECM framework suggested by Engle and Granger [19]. This method preserves the long-run relationship between the variables, and takes into account the short-run dynamics caused by the deviations of the variables from their long-term trend.

The following equation is estimated

$$y_t = \beta_0 + \beta_1 \cdot x_{1t} + \cdots + \beta_4 \cdot x_{4t} + \lambda(y_{t-1} - \hat{y}_{t-1}) + \varepsilon_t \tag{11.4}$$

where y_t is the dependent variable, $x_{1t}, \ldots, x_{4t}$ are the independent variables, $y_{t-1} - \hat{y}_{t-1}$ is the modelling error of the previous time step, β_0 is a constant term, $\beta_1, \ldots, \beta_4$ are multiplicative coefficients of the independent variables, t represents the time step, λ is the speed at which y adjusts to the error in the previous time step, and ε_t is the random residual.

An alternative statistical model would have been the vector auto-regression model (VAR). In a context where some variables are weakly exogenous, a VAR model has the virtue of obviating a decision as to what variables are exogenous and what are not. In the proposed case, some causal effects from the dependent variable to the independent variables cannot be ruled out. However, the forecasting performance of an ad-hoc VAR model estimated using all the five variables over the same period compared poorly with the forecasting performance of the other models.

The output from the econometric estimation is shown in Table 11.7. Given the output of the regression, it is concluded that all coefficients are statistically significant within the given standard levels of confidence. The residuals from the regression are a white noise series. R-squared is the coefficient of determination: when multiplied by 100, it represents the percentage of variability in the dependent variable that is explained by the estimated regression equation. For this reason, it is a measure of the strength of the regression relationship.

All the coefficients have the expected signals predicted by economic theory as summarised earlier in Sect. 11.2. The coefficients of the variables related to economic growth, inflation and reference interest rates are positive indicating a direct relationship with the yields on long-term Treasury bonds. The negative coefficient of the volatility index suggests a negative correlation between the bonds market and the stock exchange condition as it is very often observed.

11.5 Experimental Results

This section compares the accuracy results obtained using econometric, statistical, and AI models for forecasting the US 10-year Treasury bonds yields. In all the cases, the models are fitted using the 216 data samples covering the 18-years span between 1986:1 and 2003:12. The evaluation of the models is based on the RMSE of the 12-month out of sample forecasts (2004:1 to 2004:12). For the machine

Table 11.7 The output from the econometric estimation

Econometric model output		
Dependent variable: Δ(U.S. 10-year Treasury bond yield		
Method: least squares		
Number of observations: 213 after adjusting endpoints		
Variable	*Coefficient*	*t-statistic*
Constant	1.511884	2.328155
X_1 purchasing managers' index	0.031313	3.542232
X_2 Core CPI, yoy rate of previous period	0.361839	3.186955
X_3 3-month LIBOR on US dollar	0.454422	8.690526
X_4 volatility index of CBOE	−0.009144	−2.505165
Error of previous period	0.907958	29.40349
R-squared	0.972524	
Adjusted R-squared	0.971863	
S.E. of regression	0.248929	
Durbin-Watson statistic	1.883824	
F-statistic	1472.428	
Prob(F-statistic)	0.000000	

learned AI models, accuracy results are estimated on the average of 10 independent learning trials. Table 11.8 gives the accuracy results of the six modelling approaches. For the sake of comparison, Table 11.8 includes also the RMSE of a one-step lag predictor, that is, a basic algorithm that predicts the yields of the US 10-year Treasury bonds from the figure of the previous month.

For each method, Figs. 11.7a–f show the evolution of the actual Treasury bonds yields and the forecasts of the models over the 12-month evaluation period. For the machine learning approaches Figs. 11.7a–f show a sample result.

The two FL modelling approaches give the worst accuracy results. Considering the high standard deviation of the RMSE of the EA-generated models, the difference in accuracy between the two approaches is statistically not significant. However, the automatic method creates more compact solutions. These solutions are characterised by a RB that is on average half the size of the RB of the manually designed FL system.

Table 11.8 The accuracy results of the six modelling approaches

RMSE modelling error

Model	Accuracy	Std. deviation
Fuzzy (manual)	0.3523	–
Fuzzy (learned)	0.3325	0.0872
SOM	0.2693	0.0248
One-step lag	0.2574	–
MLP	0.2480	0.0016
ARIMA(2, 1, 2)	0.2464	–
Econometric	0.2376	–

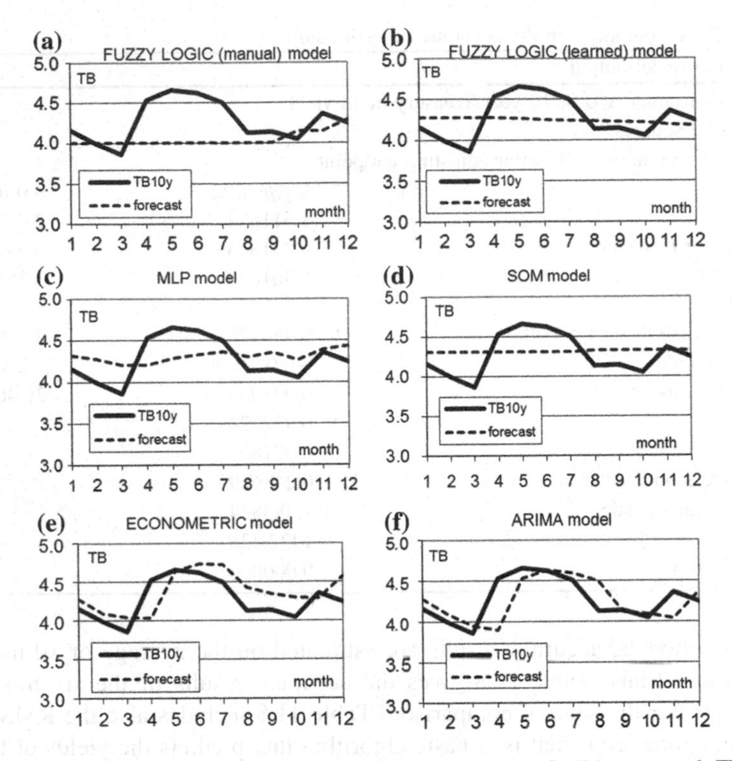

Fig. 11.7 Modelling results. **a** Manually designed FL model. **b** EA-generated FL model (sample). **c** MLP model (sample). **d** SOM model (sample). **e** ARIMA model. **f** ECM model

Figure 11.7a, b show the behaviour of two sample FL models. The forecast of the manually fitted FL system for the US 10-year Treasury bonds yields is constant throughout most of the year (9 months). During the period that the response is flat, the output of the model is decided exclusively by one rule. Since the overall system behaviour is built by assembling condition-response associations taken from the history of the past years, the predominantly flat output of the manually built FL system suggests that insufficient data may have prevented a finer modelling of the desired relationship. Indeed, inspection of Fig. 11.2a–e shows that the combination of values taken by the input variables during the 12-month evaluation period has little history in the past. Namely, the range of values taken by the CPI and the Libor indexes mostly reflects the history of the last years, while the range of values taken by the PMI and VIX indexes finds correspondence in more remote times.

Since the mapping of the EA-generated FL system is also learned from the same data samples, the results achieved using the automatic FL model do not improve those obtained using the manual FL system. Similarly to the case of the manually fitted FL model, the output of the sample EA-generated model appears to be dominated by very few rules, roughly modelling the main trend of the US 10-year Treasury bonds yields over the evaluation span.

The SOM system outperforms the two fuzzy models in terms of learning accuracy and robustness. The superior performance of the SOM model stems from the generalisation capability of the ANN. Although the learning algorithm is equally based on the mapping of input–output training pairs, the action of the neighbourhood function (see Sect. 11.3.2) partially fills the gaps between the centres of the training data clusters with the response of neurons that are not clearly committed to any cluster. The large number of neurons utilised in this experiment is likely to have helped the SOM system to provide a better response to previously unseen input data. Figure 11.7d shows the response of a sample SOM model. Analysis of Fig. 11.7d shows that the ANN output resembles a slowly varying interpolation of the desired curve. Also in this case, due to the lack of similar training examples only the overall trend of the evaluation period is modelled.

The MLP is the AI system that gives the best modelling results. In terms of average accuracy, the MLP solutions improve of about 10 % the results obtained by the SOM. The very small standard deviation indicates the high consistency of the learning procedure. This result is due to the 'global' nature of the training procedure used for this type of ANN. That is, the FL and SOM learning rules employed in this study use each training pattern to adjust the input–output mapping only in a neighbourhood of the pattern. This neighbourhood is defined by the width of the MFs in the FL models, and the neighbourhood function in the SOM. In MLP systems, the learning of each training pattern affects the setting of all the ANN weights. That is, each training example is used to adjust the whole input–output relationship. This "global" fitting of the data improves the MLP generalisation capability, particularly when the training set does not cover the whole input space.

Figure 11.7c shows the response of a sample MLP system throughout the evaluation year. The curve follows more closely the evolution of the US 10-year Treasury bonds yields, even though the quality of the mapping is still quite coarse.

The accuracy of the statistical model is within one standard deviation from the average accuracy of the MLP model. Given that the MLP learning algorithm performs a stochastic approximation of the desired input–output relationship, the equivalence of the modelling results reflects the similar statistical nature of the two modelling approaches. The similarity between the MLP and the statistical model extends also to their global approach to curve fitting. In the case of the ARIMA model, the model is fit by adjusting the global system response through the ARIMA parameters. Figure 11.7e shows the output of the ARIMA(2,1,2) model. The curve resembles a one-step lag prediction model.

Finally, the econometric ECM model obtains the best forecasting accuracy. Given the lack of historical data that affected the performance of the AI and pure statistical models, it is likely that the superior performance of the ECM model is due to the embedded problem domain knowledge. Figure 11.7f shows the output of the econometric model. Also in this case, the gross system response seems to resemble a one-step lag prediction model.

11.6 Discussion

The main difficulty of the modelling task results from the sparseness of the data that are used to build the models. Indeed, only 216 data points are available to identify the highly complex mapping from the 4-dimensional vector of economic indicators to the US Treasury bonds yields. This lack of historical data puts severely to the test the generalisation capability and the reliability of the modelling algorithms under evaluation. Unfortunately, such situation is not uncommon in the field of financial market prediction, where the range of the data samples is restricted within the boundaries of past market fluctuations. Given that some economic indicators are published only on periodical basis (e.g., the monthly PMI), the availability of historical figures is further restricted.

The sparseness of the data samples affects the accuracy of the six models. In particular, the two FL systems and the unsupervised ANN give unsatisfactory results in terms of precision of the forecasts and reliability of the learning procedure. The main reason for the failure of these three methods is in the modelling algorithm, which is based on the composition of several local input–output associations inferred from the distribution of the training data. Such approach is liable to produce poor prediction results when the input conditions are dissimilar from the cases covered by the training data. It is important to note that, in the case of the two FL systems, the poor prediction results are related to the chosen data-driven induction algorithms. A different modelling approach, such as the encoding of expert knowledge (if available), could produce a FL system capable of entirely acceptable performances.

The MLP and the two classical algorithms share the same global approach to modelling, based on parametric regression of the functional relationship. Their prediction results clearly improve the results obtained by the FL and the SOM systems. However, due to the lack of data samples, the prediction results are only marginally better than the forecasts made using a simple one-step lag algorithm (see Table 11.8).

The MLP, the ARIMA and the ECM models give similar RMSE results. To assess the statistical significance of the differences between their predictions, the accuracy of these three models is compared using the Diebold-Mariano [16] test.

Given a series and two competing predictions, the Diebold-Mariano test applies a loss criterion (such as squared error or absolute error) and then calculates a number of measures of predictive accuracy that allow the null hypothesis of equal accuracy to be tested. The procedure tests whether the mean difference between the loss criteria for the two predictions is zero using a long-run estimate of the variance of the difference series. The most common formula used to perform the Diebold-Mariano test is the following:

$$DM(A,B) = \frac{AVERAGE\left(E_A^2 - E_B^2\right)}{STDVA\left(E_A^2 - E_B^2\right)} \tag{11.5}$$

Table 11.9 The statistics of the Diebold-Mariano test for the comparisons of the three parametric regression algorithms

Comparison of prediction accuracies (Diebold-Mariano test)

Variable		Forecasts			Squared residuals			Difference of squared residuals		
Month	10Y TB	ECM	ARIMA	MLP	E^2_{ECM}	E^2_{ARIMA}	E^2_{MLP}	$E^2_{MLP}-E^2_{ARIMA}$	$E^2_{MLP}-E^2_{ECM}$	$E^2_{ECM}-E^2_{ARIMA}$
2004:01	4.16	4.28	4.28	4.32	0.0154	0.0152	0.0263	0.0111	0.0109	0.0002
2004:02	3.99	4.09	4.09	4.28	0.0105	0.0103	0.0830	0.0727	0.0726	0.0002
2004:03	3.86	4.04	3.94	4.21	0.0329	0.0071	0.1216	0.1145	0.0887	0.0258
2004:04	4.53	4.03	3.89	4.21	0.2457	0.4092	0.1030	−0.3062	−0.1427	−0.1635
2004:05	4.66	4.64	4.53	4.28	0.0003	0.0175	0.1418	0.1242	0.1414	−0.0172
2004:06	4.62	4.73	4.63	4.33	0.0118	0.0001	0.0846	0.0846	0.0728	0.0118
2004:07	4.50	4.72	4.60	4.36	0.0485	0.0095	0.0189	0.0094	−0.0295	0.0389
2004:08	4.13	4.43	4.50	4.30	0.0891	0.1357	0.0305	−0.1052	−0.0586	−0.0466
2004:09	4.14	4.35	4.11	4.35	0.0440	0.0007	0.0452	0.0446	0.0013	0.0433
2004:10	4.05	4.30	4.09	4.27	0.0633	0.0016	0.0477	0.0460	−0.0157	0.0617
2004:11	4.36	4.29	4.04	4.39	0.0052	0.1046	0.0008	−0.1038	−0.0044	−0.0994
2004:12	4.24	4.57	4.37	4.44	0.1110	0.0169	0.0390	0.0222	−0.0720	0.0942
							Std dva	0.1211	0.0788	0.0712
							Mean	0.0012	0.0054	−0.0042
							DM	0.0098	0.0685	0.0592

where DM is the Diebold-Mariano statistic, A and B are two models, E_A and E_B are their prediction errors, and the average and the standard deviation are calculated over the entire validation span.

Table 11.9 gives the statistics of the Diebold-Mariano test for the comparisons of the three parametric regression algorithms. Using a one-tailed test at a level of significance of 0.05, the critical value for rejecting the null hypothesis can be inferred from the Standard Normal Distribution to be equal to 1.645. Since the Diebold-Mariano test for the three parametric regression algorithms gives results that all are clearly lower than the critical value, the hypothesis that the three models have a similar forecasting accuracy cannot be rejected.

As a conclusion, although the three parametric regression algorithms can not be considered statistically different according to the Diebold-Mariano test, the relative size of the MSE points to a better performance of the econometric model as against the ARIMA and the AI models.

11.7 Conclusions and Further Work

Six AI and classical algorithms are evaluated on their ability of forecasting the monthly yield of the US 10-year Treasury bonds from a set of four economic indicators. The study compares the MSE of four AI models (a manually built and a machine learned FL model, a SOM model and a MLP model) with that of two classical models (a statistical ARIMA model and an econometric ECM model). 216 monthly data samples from 1986:1 to 2003:12 are used to fit the models, and 12 monthly data samples from 2004:1 to 2004:12 are used to validate the results.

In spite of the long observation period, the 216 data samples cover only sparsely the range of possible market fluctuations, representing thus a challenging test for the reliability and the accuracy of the algorithms under evaluation. Experimental evidence indicates the ECM model has a slight edge over the other algorithms, closely followed by the MLP and the ARIMA model. The better performance of the ECM model is likely due to the problem-specific knowledge that is embedded in the algorithm.

The two FL models failed to provide reliable and accurate forecasts for the US Treasury bonds yields. The main reason for their failure is probably due to the data-driven nature of their modelling algorithms, which in combination with the local mapping of the individual fuzzy rules gave poor results in the presence of conditions far from the training examples. For similar reasons, also the SOM model produced an unsatisfactory performance.

Examination of the prediction results of the six models showed that the AI systems tend to approximate the main trend of the modelling variable. However, the lack of an exhaustive training set of examples prevented the AI systems from capturing more detailed oscillations of the US Treasury bond yields. Conversely, the two classical systems showed a behaviour more resembling a one-step lagged system.

The MSE obtained by the best models is only marginally better that the MSE produced by a basic one-step lag predictor, that is, by predicting the yields of the US 10-year Treasury bonds from the figure of the previous month. This result underlines the difficulty of the modelling test and, in general, the difficulty of building reliable predictors for financial markets. The conclusions of this study suggest that pure data-driven induction can not fully capture the behaviour of the desired variable. A combination of statistical or machine learning techniques with expert knowledge of the financial markets is likely to provide the best predictive accuracy.

Further work should aim at building more powerful hybrid systems, combining machine learned and statistical information with economic theory and expert knowledge. A viable approach would be to incorporate expert knowledge into a FL framework, either as a complement or a complete replacement of the data-driven model fitting algorithms tested in this study. The main difficulty in this approach is represented by the often problematic process of knowledge elicitation from human experts. An alternative hybrid approach to FL modelling would be to combine the output of different predictors into a final forecast. The main difficulty of this approach concerns the definition of the weighing criterion for the output of the different models.

Acknowledgments The authors are grateful to António Afonso, professor of the Department of Economics of ISEG/UTL—Technical University of Lisbon, for his very helpful comments.

References

1. Baele, L., Ferrando, A., Hordal, P., Krilova, E., Monnet, C.: Measuring financial integration in Euro Area. ECB Occasional Paper Series 14, http://ideas.repec.org/p/ecb/ecbops/20040014.html (2004). Accessed April 2013
2. Bahrammirzaee, A.: A comparative survey of artificial intelligence applications in finance: artificial neural networks, expert system and hybrid intelligent systems. Neural Comput. Appl. **19**(8), 1165–1195 (2010)
3. Bartlmae, K., Gutjahr, S., Nakhaeizadeh, G.: Incorporating prior knowledge about financial markets through neural multitask learning. In: Weigend et al. (ed) Proceedings Fourth International Conference on Neural Networks in the Capital Markets (NNCM'96), Pasadena, Nov 1996. World Scientific Publishing Company, Singapore (1997)
4. Bojadziev, G.: Fuzzy Logic for Business, Finance, and Management, 2nd edn. World Scientific Publishing Company, Singapore (2007)
5. Box, G.E.P., Jenkins, G.M.: Time Series Analysis: Forecasting and Control. Holden-Day, San Francisco, CA (1976). Revised edition
6. Castellani, M., dos Santos, E.A.: Forecasting Long-Term Government Bond Yields: an Application of Statistical and AI Models. WP series, Economics and Management Institute, Technical University Lisbon, https://www.repository.utl.pt/handle/10400.5/2613 (2006). Accessed April 2013
7. Chen, A.S., Leung, M.T.: Performance evaluation of neural network architectures: the case of predicting foreign exchange correlations. J. Forecast. **24**(6), 403–420 (2005)
8. Cheng, W., Wagner, L., Lin, C.H.: Forecasting the 30-year U.S. Treasury bond with a system of neural networks. J. Comput. Intell. Finance **4**, 10–16 (1996)

9. Clements, M.P., Hendry, D.F.: Forecasting Economic Time Series. Cambridge University Press, Cambridge, UK (1998)
10. Cox, J.C., Ingersoll, J.E., Ross, S.A.: A theory of the term structure of interest rates. Econometrica **53**(2), 385–407 (1985)
11. Daniels, H., Kamp, B., Verkooijen, W.: Application of neural networks to house pricing and bond rating. Neural Comput. Appl. **8**, 226–234 (1999)
12. Deboeck, G.J.: Financial applications of self-organising maps. Neural Netw. World **8**(2), 213–241 (1998)
13. De Jong, K.: Evolutionary Computation, A Unified Approach. MIT Press, Cambridge, MA (2006)
14. Dickey, D., Fuller, W.: Distribution of the estimators for autoregressive time series with a unit root. J. Am. Stat. Assoc. **74**, 427–431 (1979)
15. Dickey, D., Fuller, W.: Likelihood ratio tests for autoregressive time series with a unit root. Econometrica **49**, 1057–1072 (1981)
16. Diebold, F.X., Mariano, R.S.: Comparing predictive accuracy. J. Bus. Econ. Stat. **13**, 253–263 (1995)
17. Din, A.: Optimization and forecasting with financial time series. Geneva Research Collaboration, Technical Report, Notes from seminar at CERN, 25 June 2002. http://grc.web.cern.ch/grc/documents/GRC-June-seminar-notes.PDF (2002). Accessed April 2013
18. Dunis, C.L., Jalilov, J.: Neural network regression and alternative forecasting techniques for predicting financial variables. Neural Netw. World **12**(2), 113–140 (2002)
19. Engle, R., Granger, C.: Co-integration and error correction representation, estimation and testing. Econometrica **35**, 251–276 (1987)
20. Feng, H.M., Chou, H.C.: Evolutional RBFNs prediction systems generation in the applications of financial time series data. Expert Syst. Appl. **38**(7), 8285–8292 (2011)
21. Fogel, D.B.: Evolutionary Computation: Toward a New Philosophy of Machine Intelligence, 2nd edn. IEEE Press, New York (2000)
22. Ganesh, M.: Introduction to Fuzzy Sets and Fuzzy Logic. Prentice-Hall of India Private Limited, New Delhi (2006)
23. Goonatilake, S., Campbell, J.A., Ahmad, N.: Genetic-Fuzzy systems for financial decision making. In: Furuhashi, T. (ed.) Advances in Fuzzy Logic, Neural Networks and Genetic Algorithms, pp. 202–223. Lecture Notes in Artificial Intelligence 1011. Springer, USA (1995)
24. Goonatilake, S., Khebbal, S.: Intelligent Hybrid Systems. Wiley, New York (1995)
25. Granger, C.W.J., Newbold, P.: Forecasting Time Series, 2nd edn. Academic Press, New York (1986)
26. Greene, W.: Econometric Analysis, 7th edn. Prentice Hall (2008)
27. Haykin, S.O.: Neural Networks and Learning Machines: International Version: A Comprehensive Foundation, 3rd edn. Pearson Education Inc (2008)
28. Hellendoorn, H., Driankov, D. (ed.): Fuzzy Model Identification: Selected Approaches. Springer, Berlin (1997)
29. Kohonen, T.: Self-organisation and Associative Memory. Springer, Heidelberg (1984)
30. Lai, R.K., Fan, C.Y., Huang, W.H., Chang, P.C.: Evolving and clustering fuzzy decision tree for financial time series data forecasting. Expert Syst. Appl. **36**(2–2), 3761–3773 (2009)
31. Lee, C.C.: Fuzzy logic in control systems: fuzzy logic controller, Part I & Part II. IEEE Trans. Syst. Man Cybern. **20**(2):404–418, 419–435 (1990)
32. Lippmann, R.P.: An introduction to computing with neural nets. IEEE ASSP Mag. 4–22 (1987)
33. Mamdani, E.H.: Application of fuzzy algorithms for control of simple dynamic plant. Proc. Inst. Electr. Eng. **121**(12), 1585–1588 (1974)
34. Mohammadian, M., Kingham, M.: Hierarchical fuzzy logic for financial modelling and prediction. J. Comput. Intell. Finance 16–18
35. McNelis, P.D.: Neural Networks in Finance–Gaining Predictive Edge in the Market. Academic Press, USA (2005)

36. Nair, B.B., Sai, S.G., Naveen, A.N., Lakshmi, A., Venkatesh, G.S., Mohandas, V.P.: A GA-artificial neural network hybrid system for financial time series forecasting. In: Proceedings of International Conference on Information Technology and Mobile Communication (AIM 2011), Nagpur, Maharashtra, India, 21–22 April, pp. 499–506. Springer, Berlin-Heidelberg, (2011)
37. Pham, D.T., Castellani, M.: Action aggregation and defuzzification in Mamdani-type fuzzy systems. Proc. ImechE, Part C **216**(7), 747–759 (2002)
38. Pham, D.T., Castellani, M.: Outline of a new evolutionary algorithm for fuzzy systems learning. Proc. ImechE, Part C **216**(5), 557–570 (2002)
39. Pham, D.T., Castellani, M.: Adaptive selection routine for evolutionary algorithms. J. Syst. Control Eng. **224**(16), 623–633 (2010)
40. Refenes, A.N., Bentz, Y., Bunn, D.W., Burgess, A.N., Zapranis, A.D.: Financial time series modelling with discounted least squares backpropagation. Neurocomputing **14**, 123–138 (1997)
41. Rumelhart, D.E., McClelland, J.L.: Parallel Distributed Processing: Exploration in the Micro-structure of Cognition, vol. 1–2. MIT Press, Cambridge, Mass (1986)
42. Trippi, R., Turban, E.: Neural Networks in Finance and Investing, 2nd edn. Irwin Professional, Chicago (1996)
43. van den Berg, J., Kaymak, U., van den Bergh, W.M.: Financial markets analysis by using a probabilistic fuzzy modelling approach. Int. J. Approximate Reasoning **35**(3), 291–305 (2004)
44. Walczak, S.: An empirical analysis of data requirements for financial forecasting with neural networks. J. Manage. Inf. Syst. **17**(4), 203–222 (2001)
45. Wang, D., Chang, P.C., Wu, J.L., Zhou, C.: A partially connected neural evolutionary network for stock price index forecasting. In: Huang, D.S. et al. (ed.) Bio-Inspired Computing and Applications, pp. 14–19. Lecture Notes in Computer Science 6840. Springer, Berlin Heidelberg (2012)
46. White, D.A., Sofge, A.: Handbook of Intelligent Control: Neural, Fuzzy and Adaptive Approaches. Van Nostrand Reinhold, New York (1992)
47. Yu, L., Lai, K.K., Wang, S.: Multistage RBF neural network ensemble learning for exchange rates forecasting. Neurocomputing **71**(16–18), 3295–3302 (2008)
48. Zadeh, L.A.: Fuzzy algorithms. Inf. Control **12**, 94–102 (1968)
49. Zadeh, L.A.: Outline of a new approach to the analysis of complex systems and decision processes. IEEE Trans. Syst. Man Cybern. **3**(1), 28–44 (1973)

Chapter 12
Optimization of Decision Rules Based on Dynamic Programming Approach

Beata Zielosko, Igor Chikalov, Mikhail Moshkov and Talha Amin

Abstract This chapter is devoted to the study of an extension of dynamic programming approach which allows optimization of approximate decision rules relative to the length and coverage. We introduce an uncertainty measure that is the difference between number of rows in a given decision table and the number of rows labeled with the most common decision for this table divided by the number of rows in the decision table. We fix a threshold γ, such that $0 \leq \gamma < 1$, and study so-called γ-decision rules (approximate decision rules) that localize rows in subtables which uncertainty is at most γ. Presented algorithm constructs a directed acyclic graph $\Delta_\gamma T$ which nodes are subtables of the decision table T given by pairs "attribute = value". The algorithm finishes the partitioning of a subtable when its uncertainty is at most γ. The chapter contains also results of experiments with decision tables from UCI Machine Learning Repository.

B. Zielosko (✉) · I. Chikalov · M. Moshkov · T. Amin
Computer, Electrical and Mathematical Sciences and Engineering Division, King Abdullah University of Science and Technology, Thuwal 23955-6900, Saudi Arabia
e-mail: beata.zielosko@us.edu.pl

I. Chikalov
e-mail: igor.chikalov@kaust.edu.sa

M. Moshkov
e-mail: mikhail.moshkov@kaust.edu.sa

T. Amin
e-mail: talha.amin@kaust.edu.sa

B. Zielosko
Institute of Computer Science, University of Silesia, 39, Będzińska St
41-200 Sosnowiec, Poland

C. Faucher and L. C. Jain (eds.), *Innovations in Intelligent Machines-4*, Studies in Computational Intelligence 514, DOI: 10.1007/978-3-319-01866-9_12,

12.1 Introduction

Decision rules are used widely for knowledge representation in data mining and knowledge discovery. Significant advantage of decision rules is their simplicity and easy way for understanding and interpreting by humans. Exact decision rules can be overfitted, i.e., dependent essentially on the noise or adjusted too much to the existing examples. If decision rules are considered as a way of knowledge representation then instead of exact decision rules with many attributes, it is more appropriate to work with approximate decision rules which contain smaller number of attributes and have relatively good accuracy.

There are many approaches to the construction of decision rules: Apriori algorithm [1], brute-force approach which is applicable to tables with relatively small number of attributes, Boolean reasoning [21, 22, 25], genetic algorithms [8, 26], separate-and-conquer approach (algorithms based on a sequential covering procedure) [10, 12–15], ant colony optimization [16], algorithms based on decision tree construction [17, 19, 23], different kinds of greedy algorithms [20, 21]. Each method has different modifications. For example, in the case of decision trees we can use greedy algorithms based on different uncertainty measures, i.e., Gini index, number of unordered pairs of rows with different decisions and other, to construct decision rules.

In this chapter, we present one more approach based on an extension of dynamic programming. We introduce an uncertainty measure that is the difference between number of rows in a given decision table and the number of rows labeled with the most common decision for this table divided by the number of rows in the decision table. We fix a threshold γ, $0 \leq \gamma < 1$, and study so-called γ-decision rules (approximate decision rules) that localize rows in subtables that uncertainty is at most γ.

We are interested in the construction of short rules which cover many objects. The choice of short rules is connected with the Minimum Description Length principle [24]. The rule coverage is important to discover major patterns in the data. Unfortunately, the problems of minimization of length and maximization of coverage of decision rules are NP-hard. We try to avoid this "restriction" for relatively small decision tables by the use of an extension of dynamic programming approach. The algorithm constructs a directed acyclic graph $\Delta_\gamma(T)$ that nodes are subtables of the decision table T given by descriptors (pairs "attribute = value"). The algorithm finishes the partitioning of a subtable when its uncertainty is at most γ. The threshold γ helps to control computational complexity and makes the algorithm applicable to solving more complex problems. The constructed graph allows to describe the whole set of so-called irredundant γ-decision rules. Then we can make optimization of such rules. We can describe all irredundant γ-decision rules with the minimum length, and after that among these rules find all rules with the maximum coverage. We can change the order of optimization: find all irredundant γ-decision rules with the maximum coverage, and after that find among such rules all irredundant γ-decision rules with the minimum length.

We prove that by removal of some descriptors from the left-hand side of each γ-decision rule that is not irredundant and by changing the decision on the right-hand side of this rule we can obtain an irredundant γ-decision rule which length is at most the length of initial rule and the coverage is at least the coverage of initial rule. It means that we work not only with optimal rules among irredundant γ-decision rules but also with optimal ones among all γ-decision rules.

Similar approach to the decision tree optimization was considered in [2, 3, 11, 18]. First results for decision rules based on dynamic programming approach were obtained in [27]. The aim of this study was to find one decision rule with the minimum length for each row. In [5], we studied dynamic programming approach for exact decision rule optimization. In [7], we studied dynamic programming approach for so-called β-decision rule optimization and we used uncertainty measure that is the number of unordered pairs of rows with different decisions in decision table T. In [6], we considered length and coverage of partial decision rules and we used uncertainty measure that is the difference between number of rows in a given decision table and the number of rows labeled with the most common decision in this table. In [28], we presented results of sequential optimization of approximate decision rules relative to the length, coverage and number of misclassifications.

To make some comparative study according to the length and coverage of γ-decision rules we also present a greedy algorithm for construction of approximate decision rules.

In the paper, we present also results of experiments with some decision tables from UCI Machine Learning Repository [9] based on Dagger [4] software system created in King Abdullah University of Science and Technology (KAUST).

The chapter consists of ten sections. Section 12.2 contains main notions connected with decision tables and decision rules. Section 12.3 is devoted to the consideration of irredundant γ-decision rules. In Sect. 12.4, we study a directed acyclic graph which allows to describe the whole set of irredundant γ-decision rules. In Sect. 12.5, we consider a procedure of optimization of this graph (really, corresponding rules) relative to the length, and in Sect. 12.6—relative to the coverage. In Sect. 12.7, we discuss possibilities of sequential optimization of rules relative to the length and coverage. In Sect. 12.8, we present a greedy algorithm for γ-decision rule construction. Section 12.9 contains results of experiments with decision tables from UCI Machine Learning Repository, and Sect. 12.10—conclusions.

12.2 Main Notions

In this section, we consider definitions of notions corresponding to decision tables and decision rules.

A *decision table* T is a rectangular table with n columns labeled with conditional attributes $f_1, \ldots, f_n$. Rows of this table are filled by nonnegative integers which are interpreted as values of conditional attributes. Rows of T are pairwise different and each row is labeled with a nonnegative integer (decision) that is interpreted as a value of the decision attribute. It is possible that T is empty, i.e., has no rows. An example of decision table is presented in Table 12.1.

A minimum decision value that is attached to the maximum number of rows in T is called *the most common decision for T*. The most common decision for empty table is equal to 0.

We denote by $N(T)$ the number of rows in the table T and by $Nmcd(T)$, we denote the number of rows in the table T labeled with the most common decision for T. We will interpret the value $G(T) = (N(T) - Nmcd(T))/N(T)$ as an *ucertainty* of nonempty table T.

The table T is called *degenerate* if T is empty or all rows of T are labeled with the same decision.

A table obtained from T by the removal of some rows is called a *subtable* of the table T. Let T be nonempty, $f_{i(1)},\ldots,f_{i(s)} \in \{f_1,\ldots,f_n\}$ and $a_1,\ldots,a_s$ be nonnegative integers. We denote by $T(f_{i(1)},a_1)\ldots(f_{i(s)},a_s)$ the subtable of the table T that contains only rows that have numbers $a_1,\ldots,a_s$ at the intersection with columns $f_{i(1)},\ldots f_{i(s)}$. Such nonempty subtables (including the table T) are called *separable subtables* of T.

We denote by $E(T)$ the set of attributes from $\{f_1,\ldots,f_n\}$ which are not constant on T. For any $f_i \in E(T)$, we denote by $E(T,f_i)$ the set of values of the attribute f_i in T.

The expression:

$$f_{i(1)} = a_1 \wedge \ldots \wedge f_{i(s)} = a_s \rightarrow d \tag{12.1}$$

is called a *decision rule over T* if $f_{i(1)},\ldots,f_{i(s)} \in \{f_1,\ldots,f_n\}$, and $a_1,\ldots,a_s,d$ are non-negative integers. It is possible that $s = 0$. In this case (12.1) is equal to the rule:

$$\rightarrow d. \tag{12.2}$$

Let $r = (b_1,\ldots,b_n)$ be a row of T. We will say that the rule (12.1) is *realizable for r*, if $a_1 = b_{i(1)},\ldots, a_s = b_{i(s)}$. If $s = 0$ then the rule (12.2) is realizable for any row from T.

Let γ be a nonnegative real number and $0 \leq \gamma < 1$. We will say that the rule (12.1) is *γ-true for T* if d is the most common decision for $T' = T\left(f_{i(1)}, a_1\right)\ldots\left(f_{i(s)}, a_s\right)$ and $G(T') \leq \gamma$. If $s = 0$ then the rule (12.2) is γ-true for T if d is the most common decision for T and $G(T) \leq \gamma$.

Table 12.1 Decision table T_0

$T_0 =$

	f_1	f_2	f_3	
r_1	0	0	0	2
r_2	0	1	1	2
r_3	0	0	1	3
r_4	1	0	1	3
r_5	1	1	1	1

If the rule (12.1) is γ-true for T and realizable for r, we will say that (12.1) is a *γ-decision rule for T and r*. Note that if $\gamma = 0$ we have an exact decision rule for T and r.

12.3 Irredundant γ-Decision Rules

We will say that the rule (12.1) with $s > 0$ is an *irredundant* γ-decision rule for T and r if (12.1) is a γ-decision rule for T and r and the following conditions hold:

(i) $f_{i(1)} \in E(T)$, and if $s > 1$ then $f_{i(j)} \in E(T(f_{i(1)},a_1)\ldots(f_{i(j-1)},a_{j-1}))$ for $j = 2,\ldots,s$;
(ii) $G(T) > \gamma$, and if $s > 1$ then $G(T(f_{i(1)},a_1)\ldots(f_{i(j)},a_j)) > \gamma$ for $j = 1,\ldots,s\text{-}1$.

If $s = 0$ then the rule (12.2) is an *irredundant* γ-decision rule for T and r if (12.2) is a γ-decision rule for T and r, i.e., if d is the most common decision for T and $G(T) \le \gamma$.

Lemma 12.1: *Let T be a decision table with $G(T) > \gamma$, $f_{i(1)} \in E(T)$, $a_1 \in E(T,f_{i(1)})$, and r be a row of the table $T' = T(f_{i(1)},a_1)$. Then the rule* (12.1) *with $s \ge 1$ is an irredundant γ-decision rule for T and r if and only if the rule:*

$$f_{i(2)} = a_2 \wedge \ldots \wedge f_{i(s)} = a_s \rightarrow d \quad (12.3)$$

is an irredundant γ-decision rule for T' and r (if $s = 1$ then (12.3) *is equal to $\rightarrow d$).*

Proof It is clear that (12.1) is a γ-decision rule for T and r if and only if (12.3) is a γ-decision rule for T' and r.

It is easy to show that the statement of lemma holds if $s = 1$. Let now $s > 1$.
Let (12.1) be an irredundant γ-decision rule for T and r. Then from (i) it follows that $f_2 \in E(T')$ and if s > 2 then, for $j = 3,\ldots,s$, $f_{i(j)} \in E(T'(f_{i(2)},a_2)\ldots(f_{i(j-1)}, a_{j-1}))$. From (ii) it follows that $G(T') > \gamma$ and if $s > 2$ then $G(T'(f_{i(2)},a_2)\ldots(f_{i(j)},a_j)) > \gamma$ for $j = 2,\ldots,s\text{-}1$. Therefore, (12.3) is an irredundant γ-decision rule for T' and r.
Let (12.3) be an irredundant γ-decision rule for T' and r. Then, for $j = 2,\ldots, s$,

$$f_{i(j)} \in E(T(f_{i(1)},a_1)\ldots(f_{i(j-1)},a_{j-1})).$$

Also, we know that $f_{i(1)} \in E(T)$. Therefore the condition (i) holds. Since (12.3) is an irredundant γ-decision rule for T' and r, we have $G(T(f_{i(1)},a_1)) > \gamma$, and if $s > 2$ then $G(T(f_{i(1)},a_1)\ldots(f_{i(j)},a_j)) > \gamma$ for $j = 2,\ldots,s - 1$. We know also that $G(T) > \gamma$. Therefore the condition (ii) holds, and (12.1) is an irredundant γ-decision rule for T and r.
Let τ be a decision rule over T and τ be equal to (12.1).
The number s of conditions on the left-hand side of τ is called the *length* of this rule and is denoted by $l(\tau)$. The length of decision rule (12.2) is equal to 0.

The *coverage* of τ is the number of rows in T for which τ is realizable and which are labeled with the decision d. We denote it by $c(\tau)$. The coverage of decision rule (12.2) is equal to the number of rows in T which are labeled with the decision d.

Proposition 12.1: *Let T be a nonempty decision table, r be a row of T and τ be a γ-decision rule for T and r which is not irredundant. Then by removal of some descriptors from the left-hand side of τ and by changing the decision on the right-hand side of τ we can obtain an irredundant γ-decision rule $irr(\tau)$ for T and r such that $l(irr(\tau)) \leq l(\tau)$ and $c(irr(\tau)) \geq c(\tau)$.*

Proof Let τ be equal to (12.1). Let T be a table for which $G(T) \leq \gamma$ and k be the most common decision for T. One can show that the rule $\rightarrow k$ is an irredundant γ-decision rule for T and r. We denote this rule by $irr(\tau)$. It is clear that $l(irr(\tau)) \leq l(\tau)$ and $c(irr(\tau)) \geq c(\tau)$.

Let T be a table for which $G(T) > \gamma$. Let $t \in \{1,\ldots,m\}$ be the minimum number such that $T' = T(f_{i(1)},a_1)\ldots(f_{i(t)},a_t)$, and $G(T') \leq \gamma$. If $t < s$ then we denote by τ' the decision rule:

$$f_{i(1)} = a_1 \wedge \ldots \wedge f_{i(t)} = a_t \rightarrow q,$$

where q is the most common decision for T'. If $t = s$ then $\tau' = \tau$. It is clear that τ' is a γ-decision rule for T and r. If $f_{i(1)} \notin E(T)$ then we remove the condition $f_{i(1)} = a_1$ from τ'. For any $j \in \{2,\ldots,t\}$, if $f_{i(j)} \notin E(T(f_{i(1)},a_1)\ldots(f_{i(j-1)},a_{j-1}))$ then we remove the condition $f_{i(j)} = a_j$ from the left-hand side of the rule τ'.

One can show that the obtained rule is an irredundant γ-decision rule for T and r. We denote this rule by $irr(\tau)$. It is clear that $l(\tau) \geq l(irr(\tau))$. One can show that $c(\tau) \leq c(\tau') = c(irr(\tau))$.

12.4 Directed Acyclic Graph $\Delta_\gamma T$

Now, we consider an algorithm that constructs a directed acyclic graph $\Delta_\gamma(T)$ which will be used to describe the set of irredundant γ-decision rules for T and for each row r of T. Nodes of the graph are separable subtables of the table T. During each step, the algorithm processes one node and marks it with the symbol *. At the first step, the algorithm constructs a graph containing a single node T which is not marked with the symbol *.

Let the algorithm have already performed p steps. Let us describe the step $(p + 1)$. If all nodes are marked with the symbol * as processed, the algorithm finishes work and presents the resulting graph as $\Delta_\gamma(T)$. Otherwise, choose a node (table) Θ, which has not been processed yet. Let d be the most common decision for Θ. If $G(\Theta) \leq \gamma$ label the considered node with the decision d, mark it with symbol * and proceed to the step $(p + 2)$. If $G(\Theta) > \gamma$, for each $f_i \in E(\Theta)$, draw a bundle of edges from the node Θ. Let $E(\Theta,f_i) = \{b_1,\ldots, b_t\}$. Then draw t edges from Θ and label these edges with pairs $(f_i,b_1),\ldots,(f_i,b_t)$ respectively. These edges enter to nodes $\Theta(f_i,b_1),\ldots,\Theta(f_i,b_t)$. If some of nodes $\Theta(f_i,b_1),\ldots,\Theta(f_i,b_t)$ are absent in the graph then add these nodes to the graph. We label each row r of Θ with the set of attributes $E_{\Delta\gamma(T)}(\Theta,r) = E(\Theta)$, some attributes from this set can be removed later during a procedure of optimization. Mark the node Θ with the symbol * and proceed to the step $(p + 2)$.

The graph $\Delta_\gamma(T)$ is a directed acyclic graph. A node of such graph will be called *terminal* if there are no edges leaving this node. Note that a node Θ of $\Delta_\gamma(T)$ is terminal if and only if $G(\Theta) \le \gamma$.

Later, we will describe the procedures of optimization of the graph $\Delta_\gamma(T)$. As a result we will obtain a graph G with the same sets of nodes and edges as in $\Delta_\gamma(T)$. The only difference is that any row r of each nonterminal node Θ of G is labeled with a nonempty set of attributes $E_G(\Theta,r) \subseteq E(\Theta)$. It is possible also that $G = \Delta_\gamma(T)$.

Now, for each node Θ of G and for each row r of Θ, we describe the set of γ-decision rules $Rul_G(\Theta,r)$. We will move from terminal nodes of G to the node T.

Let Θ be a terminal node of G labeled with the most common decision d for Θ. Then

$$Rul_G(\Theta, r) = \{\leftarrow d\}.$$

Let now Θ be a nonterminal node of G such that for each child Θ′ of Θ and for each row r' of Θ′, the set of rules $Rul_G(\Theta',r')$ is already defined. Let $r = (b_1,\ldots, b_n)$ be a row of Θ. For any $f_i \in E_G(\Theta,r)$, we define the set of rules $Rul_G(\Theta,r,f_i)$ as follows:

$$Rul_G(\Theta, r, f_i) = \{f_i = b_i \wedge \sigma \rightarrow k : \sigma \rightarrow k \in Rul_G(\Theta(f_i, b_i), r)\}.$$

Then

$$Rul_G(\Theta, r) = \bigcup_{f_i \in E_G(\Theta,r)} Rul_G(\Theta, r, f_i).$$

Theorem 12.1: *For any node Θ of $\Delta_\gamma(T)$ and for any row r of Θ, the set $Rul_{\Delta\gamma(T)}(\Theta,r)$ is equal to the set of all irredundant γ-decision rules for Θ and r.*

Proof We will prove this statement by induction on nodes in $\Delta_\gamma(T)$. Let Θ be a terminal node of $\Delta_\gamma(T)$ and d be the most common decision for Θ. One can show that the rule $\rightarrow d$ is the only irredundant γ-decision rule for Θ and r. Therefore the set $Rul_{\Delta\gamma(T)}(\Theta,r)$ is equal to the set of all irredundant γ-decision rules for Θ and r.

Let Θ be a nonterminal node of $\Delta_\gamma(T)$ and for each child of Θ, the statement of theorem hold. Let $r = (b_1,\ldots,b_n)$ be a row of Θ. It is clear that $G(\Theta) > \gamma$. Using Lemma 12.1 we obtain that the set $Rul_{\Delta\gamma(T)}(\Theta,r)$ contains only irredundant γ-decision rules for Θ and r.

Let τ be an irredundant γ-decision rule for Θ and r. Since $G(\Theta) > \gamma$, the left-hand side of τ is nonempty. Therefore τ can be represented in the form $f_i = b_i$ & $\sigma \rightarrow k$, where $f_i \in E(\Theta)$. By Lemma 12.1, $\sigma \rightarrow k$ is an irredundant

γ-decision rule for $\Theta(f_i,b_i)$ and r. From the inductive hypothesis we get that the rule $\sigma \rightarrow k$ belongs to the set $Rul_{\Delta\gamma(T)}(\Theta(f_i,b_i),r)$. Therefore, $\tau \in Rul_{\Delta\gamma(T)}(\Theta,r)$.

Example 12.1: To illustrate the algorithm presented in this section, we consider decision table T_0 depicted in Table 12.1. In the example we set $\gamma = 0.5$, so during the construction of the graph $\Delta_{0.5}(T_0)$ we stop the partitioning of a subtable Θ of T_0 when $G(\Theta) \leq 0.5$. We denote $G = \Delta_{0.5}(T_0)$ (see Fig. 12.1).

For each node Θ of the graph G and for each row r of Θ, we describe the set $Rul_G(\Theta,r)$. We will move from terminal nodes of G to the node T_0. Terminal nodes of the graph G are Θ_1, Θ_2, Θ_3, Θ_4, Θ_5 and Θ_6. For these nodes,

$$Rul_G(\Theta_1,r_1) = Rul_G(\Theta_1,r_2) = Rul_G(\Theta_1,r_3) = \{\rightarrow 2\};$$
$$Rul_G(\Theta_2,r_4) = Rul_G(\Theta_2,r_5) = \{\rightarrow 1\};$$
$$Rul_G(\theta_3,r_1) = Rul_G(\Theta_3,r_3) = Rul_G(\Theta_3,r_4) = \{\rightarrow 3\};$$
$$Rul_G(\Theta_4,r_2) = Rul_G(\Theta_4,r_5) = \{\rightarrow 1\};$$
$$Rul_G(\Theta_5,r_1) = \{\rightarrow 2\};$$
$$Rul_G(\Theta_6,r_2) = Rul_G(\Theta_6,r_3) = Rul_G(\Theta_6,r_4) = Rul_G(\Theta_6,r_5) = \{\rightarrow 3\}.$$

All children of the table T_0 are already treated so we can describe the sets of rules corresponding to rows of T_0:

$$Rul_G(T_0,r_1) = \{f_1 = 0 \rightarrow 2, f_2 = 0 \rightarrow 3, f_3 = 0 \rightarrow 2\},$$
$$Rul_G(T_0,r_2) = \{f_1 = 0 \rightarrow 2, f_2 = 1 \rightarrow 1, f_3 = 1 \rightarrow 3\},$$
$$Rul_G(T_0,r_3) = \{f_1 = 0 \rightarrow 2, f_2 = 0 \rightarrow 3, f_3 = 1 \rightarrow 3\},$$
$$Rul_G(T_0,r_4) = \{f_1 = 1 \rightarrow 1, f_2 = 0 \rightarrow 3, f_3 = 1 \rightarrow 3\},$$
$$Rul_G(T_0,r_5) = \{f_1 = 1 \rightarrow 1, f_2 = 1 \rightarrow 1, f_3 = 1 \rightarrow 3\}.$$

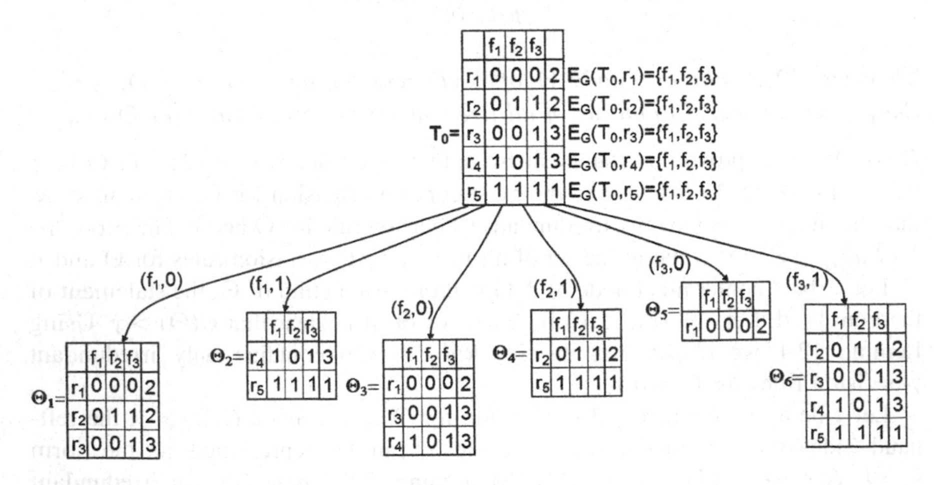

Fig. 12.1 Directed acyclic graph $G = \Delta_{0.5}(T_0)$

Presented sets of rules contain only rules with the length equal to 1, i.e., they contain only optimal rules relative to the length.

12.5 Procedure of Optimization Relative to Length

We consider the procedure of optimization of the graph G relative to the length l. For each node Θ in the graph G, this procedure corresponds to each row r of Θ the set $Rul^l_G(\Theta,r)$ of γ-decision rules with the minimum length from $Rul_G(\Theta,r)$ and the number $Opt^l_G(\Theta,r)$—the minimum length of a γ-decision rule from $Rul_G(\Theta,r)$.

The idea of the procedure is simple. It is clear that, for each terminal node Θ of G and for each row r of Θ, the following equalities hold:

$$Rul^l_G(\Theta,r) = Rul_G(\Theta,r) = \{\rightarrow d\},$$

where d is the most common decision for Θ, and

$$Opt^l_G(\Theta,r) = 0.$$

Let Θ be a nonterminal node of G, and $r = (b_1,\ldots,b_n)$ be a row of Θ. We know that

$$Rul_G(\Theta,r) = \bigcup_{f_i \in E_G(\Theta,r)} Rul_G(\Theta,r,f_i)$$

and, for $f_i \in E_G(\Theta,r)$,

$$Rul_G(\Theta,r,f_i) = \{f_i = b_i \wedge \sigma \rightarrow k : \sigma \rightarrow k \in Rul_G(\Theta(f_i,b_i),r)\}.$$

For $f_i \in E_G(\Theta,r)$, we denote by $Rul^l_G(\Theta,r,f_i)$ the set of all γ-decision rules with the minimum length from $Rul_G(\Theta,r,f_i)$ and by $Opt^l_G(\Theta,r,f_i)$—the minimum length of a γ-decision rule from $Rul_G(\Theta,r,f_i)$.

One can show that

$$Rul^l_G(\Theta,r,f_i) = \{f_i = b_i \wedge \sigma \rightarrow k : \sigma \rightarrow k \in Rul^l_G(\Theta(f_i,b_i),r)\},$$

$$Opt^l_G(\Theta,r,f_i) = Opt^l_G(\Theta(f_i,b_i),r) + 1,$$

and

$$\begin{aligned} Opt^l_G(\Theta,r) &= \min\{Opt^l_G(\Theta,r,f_i) : f_i \in E_G(\Theta,r)\} \\ &= \min\{Opt^l_G(\Theta(f_i,b_i),r) + 1 : f_i \in E_G(\Theta,r)\}. \end{aligned}$$

It is easy to see also that

$$Rul^l_G(\Theta,r) = \bigcup_{f_i \in E_G(\Theta,r),\, Opt^l_G(\Theta(f_i,b_i),r)+1 = Opt^l_G(\Theta,r)} Rul^l_G(\Theta,r,f_i).$$

We now describe the procedure of optimization of the graph G relative to the length l.

We will move from terminal nodes of the graph G to the node T. We will correspond to each row r of each table Θ the number $Opt^l_G(\Theta, r)$ which is the minimum length of a γ-decision rule from $Rul_G(\Theta,r)$ and we will change the set $E_G(\Theta,r)$ attached to the row r in Θ if Θ is a nonterminal node of G. We denote the obtained graph by G^l.

Let Θ be a terminal node of G. Then we correspond the number

$$Opt^l_G(\Theta, r) = 0$$

to each row r of Θ.

Let Θ be a nonterminal node of G and all children of Θ have already been treated. Let $r = (b_1,\ldots,b_n)$ be a row of Θ. We match the number

$$Opt^l_G(\Theta, r) = \min\{Opt^l_G(\Theta(f_i, b_i), r) + 1 : f_i \in E_G(\Theta, r)\}$$

with the row r in the table Θ and we set

$$E_{G^l}(\Theta, r) = \{f_i : f_i \in E_G(\Theta, r), Opt^l_G(\Theta(f_i, b_i), r) + 1 = Opt^l_G(\Theta, r)\}.$$

From the reasoning before the description of the procedure of optimization relative to the length the next statement follows.

Theorem 12.2: *For each node Θ of the graph G^l and for each row r of Θ, the set $Rul_{G^l}(\Theta,r)$ is equal to the set $Rul^l_G(\Theta, r)$ of all γ-decision rules with the minimum length from the set $Rul_G(\Theta,r)$.*

The directed acyclic graph G^l obtained from the graph G (see Fig. 12.1) by the procedure of optimization relative to the length is the same as depicted in Fig. 12.1. Sets of rules corresponding to rows of T_0 are the same as presented in Example 12.1. Each rule has the length equal to 1.

12.6 Procedure of Optimization Relative to Coverage

We consider the procedure of optimization of the graph G relative to the coverage c. For each node Θ in the graph G, this procedure corresponds to each row r of Θ the set $Rul^c_G(\Theta, r)$ of γ-decision rules with the maximum coverage from $Rul_G(\Theta,r)$ and the number $Opt^c_G(\Theta, r)$—the maximum coverage of a γ-decision rule from $Rul_G(\Theta,r)$.

The idea of the procedure is simple. It is clear that for each terminal node Θ of G and for each row r of Θ, the following equalities hold:

$$Rul^c_G(\Theta, r) = Rul_G(\Theta, r) = \{\rightarrow d\},$$

where d is the most common decision for Θ, and $Opt^c_G(\Theta, r)$ is equal to the number of rows in Θ which are labeled with the decision d.

Let Θ be a nonterminal node of G, and $r = (b_1,\ldots, b_n)$ be a row of Θ. We know that

$$Rul_G(\Theta, r) = \bigcup_{f_i \in E_G(\Theta,r)} Rul_G(\Theta, r, f_i),$$

and, for $f_i \in E_G(\Theta,r)$,

$$Rul_G(\Theta, r, f_i) = \{f_i = b_i \wedge \sigma \rightarrow k : \sigma \rightarrow k \in Rul_G(\Theta(f_i, b_i), r)\}.$$

For $f_i \in E_G(\Theta, r)$, we denote by $Rul^c_G(\Theta, r, f_i)$ the set of all γ-decision rules with the maximum coverage from $Rul_G(\Theta,r,f_i)$ and by $Opt^c_G(\Theta, r, f_i)$—the maximum coverage of a γ-decision rule from $Rul_G(\Theta,r,f_i)$.

One can show that

$$Rul^c_G(\Theta, r, f_i) = \{f_i = b_i \wedge \sigma \rightarrow k : \sigma \rightarrow k \in Rul^c_G(\Theta(f_i, b_i), r)\},$$

$$Opt^c_G(\Theta, r, f_i) = Opt^c_G(\Theta(f_i, b_i), r),$$

and

$$\begin{aligned} Opt^c_G(\Theta, r) &= \max\{Opt^c_G(\Theta, r, f_i) : f_i \in E_G(\Theta, r)\} \\ &= \max\{Opt^c_G(\Theta(f_i, b_i), r) : f_i \in E_G(\Theta, r)\}. \end{aligned}$$

It is easy to see also that

$$Rul^c_G(\Theta, r) = \bigcup_{f_i \in E_G(\Theta,r), Opt^c_G(\Theta(f_i,b_i),r)=Opt^c_G(\Theta,r)} Rul^c_G(\Theta, r, f_i).$$

We now describe the procedure of optimization of the graph G relative to the coverage c.

We will move from terminal nodes of the graph G to the node T. We will correspond to each row r of each table Θ the number $Opt^c_G(\Theta, r)$ which is the maximum coverage of a γ-decision rule from $Rul_G(\Theta,r)$ and we will change the set $E_G(\Theta,r)$ attached to the row r in Θ if Θ is a nonterminal node of G. We denote the obtained graph by G^c.

Let Θ be a terminal node of G and d be the most common decision for Θ. Then we correspond to each row r of Θ the number $Opt^c_G(\Theta, r)$ that is equal to the number of rows in Θ which are labeled with the decision d.

Let Θ be a nonterminal node of G and all children of Θ have already been treated. Let $r = (b_1,\ldots,b_n)$ be a row of Θ. We match the number

$$Opt^c_G(\Theta, r) = \max\{Opt^c_G(\Theta(f_i, b_i), r) : f_i \in E_G(\Theta, r)\}$$

with the row r in the table Θ and we set

$$E_{G^c}(\Theta, r) = \{f_i : f_i \in E_G(\Theta, r), Opt^c_G(\Theta(f_i, b_i), r) = Opt^c_G(\Theta, r)\}.$$

From the reasoning before the description of the procedure of optimization relative to the coverage the next statement follows.

Theorem 12.3: *For each node* Θ *of the graph* G^c *and for each row* r *of* Θ, *the set* $Rul_G^c(\Theta,r)$ *is equal to the set* $Rul_G^c(\Theta,r)$ *of all* γ*-decision rules with the maximum coverage from the set* $Rul_G(\Theta,r)$.

Figure 12.2 presents the directed acyclic graph G^c obtained from the graph G (see Fig. 12.1) by the procedure of optimization relative to the coverage.

Using the graph G^c we can describe for each row r_i, $i = 1,\ldots,5$, of the table T_0 the set $Rul_G^c(T_0,r_i)$ of all irredundant 0.5-decision rules for T_0 and r_i with the maximum coverage. We will give also the value $Opt_G^c(T_0,r_i)$ which is equal to the maximum coverage of an irredundant 0.5-decision rule for T_0 and r_i. This value was obtained during the procedure of optimization of the graph G relative to the coverage. We have:

$$Rul_G^c(T_0,r_1) = \{f_1 = 0 \rightarrow 2, f_2 = 0 \rightarrow 3\}, Opt_G^c(T_0,r_1) = 2,$$
$$Rul_G^c(T_0,r_2) = \{f_1 = 0 \rightarrow 2, f_3 = 1 \rightarrow 3\}, Opt_G^c(T_0,r_2) = 2,$$
$$Rul_G^c(T_0,r_3) = \{f_1 = 0 \rightarrow 2, f_2 = 0 \rightarrow 3, f_3 = 1 \rightarrow 3\}, Opt_G^c(T_0,r_3) = 2,$$
$$Rul_G^c(T_0,r_4) = \{f_2 = 0 \rightarrow 3, f_3 = 1 \rightarrow 3\}, Opt_G^c(T_0,r_4) = 2,$$
$$Rul_G^c(T_0,r_5) = \{f_3 = 1 \rightarrow 3\}, Opt_G^c(T_0,r_5) = 2.$$

12.7 Sequential Optimization

Theorems 12.2 and 12.3 show that we can make sequential optimization relative to the length and coverage. We can find all irredundant γ-decision rules with the maximum coverage and after that among these rules find all rules with the

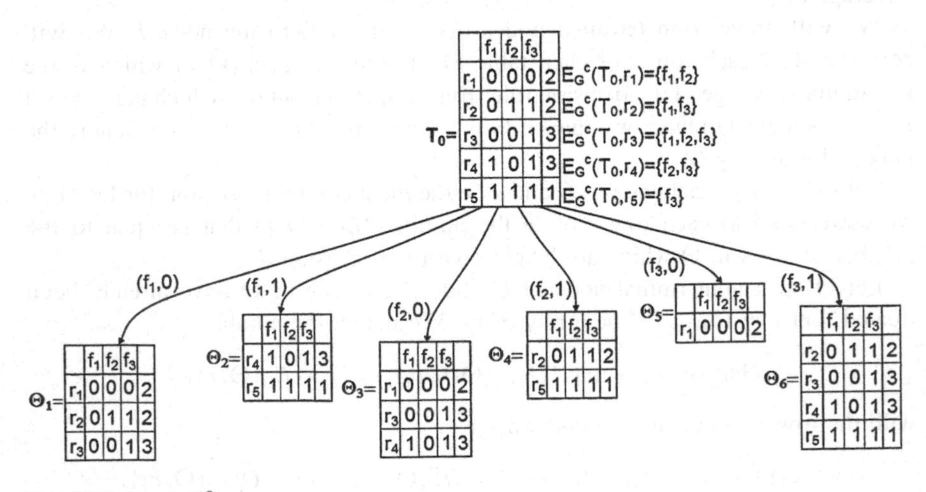

Fig. 12.2 Graph G^c

minimum length. We can also change the order of optimization: find all irredundant γ-decision rules with the minimum length, and after that find among such rules all rules with the maximum coverage.

We will say that a γ-decision rule for T and r is *totally optimal relative to the length and coverage* if it has the minimum length and the maximum coverage among all irredundant γ-decision rules for T and r. We can describe all totally optimal γ-decision rules using procedures of optimization relative to the length and coverage.

We set $G = \Delta_\gamma(T)$ and apply the procedure of optimization relative to the coverage to the graph G. As a result we obtain the graph G^c and, for each row r of T—the value $Opt_G^c(T,r)$ that is equal to the maximum coverage of γ-decision rule for T and r.

Now, we apply the procedure of optimization relative to the length to the graph G. As a result we obtain the graph G^l. After that, we apply the procedure of optimization relative to the coverage to the graph G^l. As a result we obtain the graph G^{lc} and, for each row r of T, the value $Opt^c{}_{G^l}(T,r)$ which is equal to the maximum coverage of irredundant γ-decision rule for T and r among all irredundant γ-decision rules for T and r with the minimum length.

One can show that a totally optimal relative to the length and coverage irredundant γ-decision rule for T and r exists if and only if $Opt_G^c(T,r) = Opt^c{}_{G^l}(T,r)$. If the last equality holds then the set $Rul_G{}^{lc}(T,r)$ is equal to the set of all totally optimal relative to the length and coverage irredundant γ-decision rules for T and r.

It is clear that the results of sequential optimization of irredundant γ-decision rules for T and r depend on the order of optimization (length + coverage or coverage + length) if and only if there are no totally optimal irredundant γ-decision rules for T and r.

For the decision table T_0 presented in Example 12.1, the directed acyclic graph G^l is the same as the graph G depicted in Fig. 12.1. A graph G^{lc} obtained from the graph G^l by the procedure of optimization relative to the coverage, is the same as the directed acyclic graph G^c depicted in Fig. 12.2. Sets of rules corresponding to rows of T_0 are the same as presented at the end of Sect. 12.6. So, using the graph G^{lc} we can describe for each row r_i, $i = 1,\ldots,5$, of the table T_0 the set $Rul_G{}^{lc}(T_0,r_i)$ of all irredundant 0.5-decision rules for T_0 and r_i that have the maximum coverage among all irredundant 0.5-decision rules for T_0 and r_i with the minimum length. For $i = 1,\ldots,5$, values $Opt^c{}_{G^l}T_0,r_i$ (obtained during the procedure of optimization of the graph G^l relative to the coverage) are equal to values $Opt_G^c(T_0,r_i)$ presented at the end of Sect. 6. Therefore, for $i = 1,\ldots,5$, $Rul_G{}^{lc}(T_0,r_i) = Rul_G{}^c(T_0,r_i)$ is the set of all totally optimal relative to the length and coverage 0.5-decision rules for T_0 and r_i.

It is possible to show (see analysis of similar algorithms in [19], page 64) that the time complexities of algorithms that construct the graph $\Delta_\gamma(T)$ and make sequential optimization of γ-decision rules relative to the length and coverage are bounded from above by polynomials on the number of separable subtables of T, and the number of attributes in T. In [18], it was shown that the number of separable subtables for each decision table with attributes from a restricted infinite information system is bounded from above by a polynomial on the number of attributes in the table. Examples of restricted infinite information system were considered, in particular, in [19].

12.8 Greedy Algorithm

We would like to make some comparative study of the length and the coverage of decision rules constructed using greedy approach and approach based on dynamic programming. So, in this section, we present a greedy algorithm for approximate decision rule construction.

Algorithm 12.1: Greedy algorithm for γ-decision rule construction

Require: Decision table T with conditional attributes $f_1,\ldots,f_n$, row $r = (b_1,\ldots,b_n)$ of T, and real number γ such that $0 \leq \gamma < 1$.

Ensure: γ-decision rule for T and r.

$Q \leftarrow \emptyset$;

$$T' \leftarrow T;$$

while $G(T') > \gamma$ **do**

select $f_i \in \{f_1,\ldots,f_n\}$ with the minimum index such that $G(T'(f_i, b_i))$ is the minimum;

$$T' \leftarrow T'(f_i, b_i);$$

$$Q \leftarrow Q \cup \{f_i\};$$

end while

$\wedge_{f_i \in Q}(f_i = b_i) \rightarrow d$ where d is the most common decision for T'.

Presented algorithm at each iteration chooses an attribute $f_i \in \{f_1,\ldots,f_n\}$ with the minimum index such that uncertainty of the corresponding subtable is the minimum. We apply the greedy algorithm sequentially to the table T and each row r of T. As a result, for each row of the decision table T we obtain one γ-decision rule. Such rules form a vector of rules $vec_{rule} = (rule_1,\ldots,rule_{N(T)})$. For decision table

T_0 depicted in Table 12.1, and $\gamma = 0.5$, the vector of γ-decision rules is the following: $vec_{rule} = (f_3 = 0 \rightarrow 2,\ f_1 = 0 \rightarrow 2,\ f_1 = 0 \rightarrow 2,\ f_2 = 0 \rightarrow 3,\ f_1 = 1 \rightarrow 1)$.

12.9 Experimental Results

We studied a number of decision tables from UCI Machine Learning Repository [9]. Some decision tables contain conditional attributes that take unique value for each row. Such attributes were removed. In some tables there were equal rows with, possibly, different decisions. In this case each group of identical rows was replaced with a single row from the group with the most common decision for this group. In some tables there were missing values. Each such value was replaced with the most common value of the corresponding attribute.

Let T be one of these decision tables. We consider for this table values of γ from the set $\Gamma(T) = \{G(T) \times 0.001, G(T) \times 0.01, G(T) \times 0.1, G(T) \times 0.2\}$.

We studied the minimum length of irredundant γ-decision rules. Results can be found in Table 12.2. For each row r of T, we find the minimum length of an irredundant γ-decision rule for T and r. After that, we find for rows of T the minimum length of a decision rule with the minimum length (column "Min"), the maximum length of such rule (column "Max"), and the average length of rules with the minimum length—one for each row (column "Avg"). Presented results show that the length of irredundant γ-decision rules is nonincreasing when the value of γ is increasing.

We studied the maximum coverage of irredundant γ-decision rules. Results can be found in Table 12.3. For each row r of T, we find the maximum coverage of an irredundant γ-decision rule for T and r. After that, we find for rows of T the minimum coverage of a decision rule with maximum coverage (column "Min"), the maximum coverage of such rule (column "Max"), and the average coverage of rules with maximum coverage—one for each row (column "Avg"). We can observe that the coverage of irredundant γ-decision rules is nondecreasing when the value of γ is increasing.

Table 12.2 Minimum length of γ-decision rules for $\gamma \in \Gamma(T)$

Name of decision table	$\gamma = G(T) \times 0.001$			$\gamma = G(T) \times 0.01$			$\gamma = G(T) \times 0.1$			$\gamma = G(T) \times 0.2$		
	Min	Avg	Max	Min	Avg	Max	Min	Avg	Max	Min	Avg	Max
Adult-stretch	1	1.25	2	1	1.25	2	1	1.25	2	1	1.25	2
Agaricus-lepiota	1	1.18	2	1	1.18	2	1	1.00	1	1	1.00	1
Balance-scale	3	3.20	4	3	3.20	4	2	2.84	4	2	2.67	4
Breast-cancer	1	2.67	6	1	2.67	6	1	2.67	6	1	2.54	6
Cars	1	2.43	6	1	2.43	6	1	2.43	6	1	2.43	6
Hayes-roth-data	1	2.15	4	1	2.15	4	1	2.15	4	1	2.15	4
Lymphography	1	1.99	4	1	1.99	4	1	1.97	4	1	1.80	3
Nursery	1	3.12	8	1	3.12	8	1	2.79	7	1	2.52	7
Shuttle-landing	1	1.40	4	1	1.40	4	1	1.40	4	1	1.40	4
Soybean-small	1	1.00	1	1	1.00	1	1	1.00	1	1	1.00	1
Teeth	1	2.26	4	1	2.26	4	1	2.26	4	1	2.26	4
Zoo-data	1	1.56	4	1	1.56	4	1	1.56	4	1	1.52	4

Table 12.3 Maximum coverage of γ-decision rules for $\gamma \in \Gamma(T)$

Name of decision table	$\gamma = G(T) \times 0.001$			$\gamma = G(T) \times 0.01$			$\gamma = G(T) \times 0.1$			$\gamma = G(T) \times 0.2$		
	Min	Avg	Max	min	Avg	Max	Min	Avg	Max	Min	Avg	Max
Adult-stretch	4	7.00	8	4	7.00	8	4	7.00	8	4	7.00	8
Agaricus-lepiota	72	2135.46	2688	72	2149.80	2688	1008	3200.72	3408	2048	3270.39	3408
Balance-scale	1	4.21	5	1	4.21	5	1	10.94	24	1	13.85	24
Breast-cancer	1	9.53	25	1	9.53	25	1	9.53	25	1	12.27	26
Cars	1	332.76	576	1	332.76	576	1	332.76	576	1	332.82	576
Hayes-roth-data	1	6.52	12	1	6.52	12	1	6.52	12	1	6.52	12
Lymphography	2	21.54	32	2	21.54	32	2	24.38	32	3	36.84	53
Nursery	1	1531.04	4320	1	1531.04	4320	2	1602.99	4320	2	1663.75	4320
Shuttle-landing	1	2.13	3	1	2.13	3	1	2.13	3	1	2.13	3
Soybean-small	10	12.53	17	10	12.53	17	10	12.83	17	10	12.83	17
Teeth	1	1.00	1	1	1.00	1	1	1.00	1	1	1.00	1
Zoo-data	3	11.07	19	3	11.07	19	3	11.71	19	4	12.78	19

Table 12.4 presents some results about totally optimal rules relative to the length and coverage, i.e., irredundant γ-decision rules with the minimum length and the maximum coverage among all irredundant γ-decision rules for T and r. Column "Rows" contains number of rows in T. For the values of γ from the set $\Gamma(T)$, for each of the considered decision tables T, we count the number of rows r such that there exists a totally optimal irredundant γ-decision rule for T and r (column "T-opt"). We also find for rows of T, the minimum (column "Min"), the average (column "Avg") and the maximum (column "Max") number of totally optimal irredundant γ-decision rules for T and r among all rows r of T. Note that two irredundant rules with different order of the same descriptors (pairs attribute = value) are considered as two different rules.

Presented results show that for each value of γ, for five data sets ("Adult-stretch", "Balance-scale", "Cars", "Hayes-roth-data", "Teeth"), at least one totally optimal irredundant γ-decision rule exists for each row. For data set "Nursery", for each row there exists at least one irredundant γ-decision rule with the minimum length and the maximum coverage with the exception of $\gamma = G(T) \times 0.2$.

In Table 12.5, we present results of sequential optimization of irredundant γ-decision rules. For the values of γ from the set $\Gamma(T)$, for tables with rows that have no totally optimal rules relative to the length and coverage, we make two steps of optimization—relative to the length and then relative to the coverage (column "l + c"). After that, we find the average length (column "l-avg") and the average coverage (column "c-avg") of rules after two steps of optimization. We consider also the reversed order of optimization—relative to the coverage and then relative to the length (column "c + l").

Table 12.6 presents the average length and the average coverage of γ-decision rules constructed by the greedy algorithm. Column "Decision table" contains the name of a decision table T, column "Rows" contains the number of rows in T, column "Attr" contains the number of conditional attributes in T, column "Length" contains the average length of γ-decision rules, and column "Coverage"—the average coverage of γ-decision rules.

Table 12.7 presents, for $\gamma \in \Gamma(T)$, comparison of the average length of γ-decision rules constructed by the dynamic programming algorithm and the greedy algorithm. Each cell of this table contains a relative difference that is equal to:

$$\frac{GreedyLength - OptimumLength}{OptimumLength},$$

where *GreedyLength* denotes the average length of γ-decision rules constructed by the greedy algorithm, *OptimumLength* denotes the average length of γ-decision rules constructed by the dynamic programming algorithm. The last line contains the average relative difference for all data sets presented in Table 12.7.

Based on the results, we can see that the average length of γ-decision rules constructed by the dynamic programming and greedy algorithms is the same for 15 cases (such values are in bold).

Table 12.4 Totally optimal irredundant γ-decision rules for $\gamma \in \Gamma(T)$

Decision table	Rows	$\gamma = G(T) \times 0.001$				$\gamma = G(T) \times 0.01$				$\gamma = G(T) \times 0.1$				$\gamma = G(T) \times 0.2$			
		T-opt	Min	Avg	Max	T-opt	Min	Avg	Max	T-opt	Min	Avg	Max	T-opt	Min	Avg	Max
Adult-stretch	16	16	1	1.5	2	16	1	1.5	2	16	1	1.5	2	16	1	1.5	2
Agaricus-lepiota	8124	612	0	0.1	2	612	0	0.1	2	3672	0	0.5	1	3600	0	0.4	1
Balance-scale	625	625	6	13.4	24	625	6	13.4	24	625	2	9.4	24	625	2	8.1	24
Breast-cancer	266	133	0	12.3	720	133	0	12.3	720	133	0	12.3	720	136	0	12.0	720
Cars	1728	1728	1	76.2	720	1728	1	76.2	720	1728	1	76.2	720	1728	1	76.2	720
Hayes-roth-data	69	69	1	4.5	24	69	1	4.5	24	69	1	4.5	24	69	1	4.5	24
Lymphography	148	52	0	1.0	24	52	0	1.0	24	63	0	1.1	24	80	0	1.1	6
Nursery	12960	12960	1	151.6	40320	12960	1	151.6	40320	1290	1	102.8	5040	12918	0	64.8	5040
Shuttle-landing	15	13	0	2.6	24	13	0	2.6	24	13	0	2.6	24	13	0	2.6	24
Soybean-small	47	37	0	1.9	5	37	0	1.9	5	38	0	2.2	5	38	0	2.2	5
Teeth	23	23	2	13.5	72	23	2	13.5	72	23	2	13.5	72	23	2	13.5	72
Zoo-data	59	44	0	8.4	408	44	0	8.4	408	45	0	8.5	408	52	0	8.7	408

Table 12.5 Sequential optimization of irredundant γ-decision rules for $\gamma \in \Gamma(T)$

Decision table	$\gamma = G(T) \times 0.001$				$\gamma = G(T) \times 0.01$				$\gamma = G(T) \times 0.1$				$\gamma = G(T) \times 0.2$			
	l + c		c + l		l + c		c + l		l + c		c + l		l + c		c + l	
	l-avg	c-avg	l-avg	c-avg	l-avg	c-avg	l-avg	c-avg	l-avg	c-avg	l-avg	c-avg	l-avg	c-avg	l-avg	c-avg
Agaricus-lepiota	1.2	1370.1	2.5	2135.5	1.2	1370.1	2.5	2149.8	1.0	2351.5	2.5	3200.7	1.0	2453.6	2.0	3270.4
Breast-cancer	2.7	7.0	3.4	9.5	2.7	7.0	3.4	9.5	2.7	7.0	3.4	9.5	2.5	8.7	3.3	12.3
Lymphography	2.0	15.2	2.9	21.5	2.0	15.2	2.9	21.5	2.0	18.4	2.9	24.4	1.8	30.8	2.5	36.8
Shuttle-landing	1.4	1.9	1.7	2.1	1.4	1.9	1.7	2.1	1.4	1.9	1.7	2.1	1.4	1.9	1.7	2.1
Soybean-small	1.0	12.2	1.2	12.5	1.0	12.2	1.2	12.5	1.0	12.4	1.2	12.8	1.0	12.4	1.2	12.8
Zoo-data	1.6	10.5	1.9	11.1	1.6	10.5	1.9	11.1	1.6	11.0	1.8	11.7	1.5	12.2	1.7	12.8

Table 12.6 Average length and average coverage of γ-decision rules, $\gamma \in \Gamma(T)$, constructed by the greedy algorithm

Decision table	Rows	Attr	$\gamma = G(T) \times 0.001$		$\gamma = G(T) \times 0.01$		$\gamma = G(T) \times 0.1$		$\gamma = G(T) \times 0.2$	
			Length	Coverage	Length	Coverage	Length	Coverage	Length	Coverage
Adult-stretch	16	4	1.75	6.25	1.75	6.25	1.75	6.25	1.75	6.25
Agaricus-lepiota	1824	22	1.18	949.81	1.18	949.81	1.18	949.81	1.00	1427.11
Balance-scale	625	4	3.32	3.71	3.32	3.71	3.32	3.71	2.84	12.47
Breast-cancer	266	9	3.71	4.26	3.71	4.26	3.71	4.26	3.58	6.14
Cars	1728	6	2.73	331.41	2.73	331.41	2.73	331.41	2.72	331.54
Hayes-roth-data	69	5	2.32	6.07	2.32	6.07	2.32	6.07	2.32	6.07
Lymphography	148	18	2.76	9.55	2.76	9.55	2.76	9.55	2.02	17.86
Nursery	12960	8	3.51	1508.20	3.51	1508.20	3.51	1508.20	2.81	1641.29
Shuttle-landing	15	6	1.40	1.87	1.40	1.87	1.40	1.87	1.40	1.87
Soybean-small	47	35	1.00	8.89	1.00	8.89	1.00	8.89	1.00	8.89
Teeth	23	8	2.26	1.00	2.26	1.00	2.26	1.00	2.26	1.00
Zoo-data	59	16	1.63	10.05	1.63	10.05	1.63	10.05	1.59	10.44

Table 12.7 Comparison of the average length of γ-decision rules for $\gamma \in \Gamma(T)$

Decision table	$\gamma = G(T) \times 0.001$	$\gamma = G(T) \times 0.01$	$\gamma = G(T) \times 0.1$	$\gamma = G(T) \times 0.2$
Adult-stretch	0.40	0.40	0.40	0.40
Agaricus-lepiota	**0.00**	**0.00**	0.18	**0.00**
Balance-scale	0.04	0.04	0.17	0.06
Breast-cancer	0.39	0.39	0.39	0.41
Cars	0.12	0.12	0.12	0.12
Hayes-roth-data	0.08	0.08	0.08	0.08
Lymphography	0.39	0.39	0.40	0.12
Nursery	0.13	0.13	0.26	0.12
Shuttle-landing	**0.00**	**0.00**	**0.00**	**0.00**
Soybean-small	**0.00**	**0.00**	**0.00**	**0.00**
Teeth	**0.00**	**0.00**	**0.00**	**0.00**
Zoo-data	0.04	0.04	0.04	0.05
Average	0.13	0.13	0.17	0.11

Note: Bold number in table are important. They mean no difference between greedy and dynamic programming approaches

Table 12.8 presents, for $\gamma \in \Gamma(T)$, comparison of the average coverage of γ-decision rules constructed by the dynamic programming algorithm and the greedy algorithm. Each cell of this table contains a relative difference that is equal to:

$$\frac{OptimumCoverage - GreedyCoverage}{OptimumCoverage},$$

where *OptimumCoverage* denotes the average coverage of γ-decision rules constructed by the the dynamic programming, *GreedyCoverage* denotes the average coverage of γ-decision rules constructed by the greedy algorithm. The last line contains the average relative difference for all data sets presented in Table 12.8.

Table 12.8 Comparison of the average coverage of γ-decision rules for $\gamma \in \Gamma(T)$

Decision table	$\gamma = G(T) \times 0.001$	$\gamma = G(T) \times 0.01$	$\gamma = G(T) \times 0.1$	$\gamma = G(T) \times 0.2$
Adult-stretch	0.11	0.11	0.11	0.11
Agaricus-lepiota	0.56	0.56	0.70	0.56
Balance-scale	0.12	0.12	0.66	0.10
Breast-cancer	0.55	0.55	0.55	0.50
Cars	**0.00**	**0.00**	**0.00**	**0.00**
Hayes-roth-data	0.07	0.07	0.07	0.07
Lymphography	0.56	0.56	0.61	0.52
Nursery	0.01	0.01	0.06	0.01
Shuttle-landing	0.12	0.12	0.12	0.12
Soybean-small	0.29	0.29	0.31	0.31
Teeth	**0.00**	**0.00**	**0.00**	**0.00**
Zoo-data	0.09	0.09	0.14	0.18
Average	0.21	0.21	0.28	0.21

Note: Bold number in table are important. They mean no difference between greedy and dynamic programming approaches

Presented results show that the average coverage of γ-decision rules constructed by the dynamic programming and greedy algorithms is the same for 8 cases (such values are in bold). We can also observe that the relative difference in the case of coverage is greater often than in the case of length.

12.10 Conclusions

We studied an extension of dynamic programming approach to the optimization of irredundant γ-decision rules relative to the length and coverage. The considered approach allows us to describe the whole set of irredundant γ-decision rules and optimize these rules sequentially relative to the length and coverage or relative to the coverage and length. We discussed the notion of totally optimal γ-decision rules, i.e., rules with the minimum length and the maximum coverage in the same time. We also presented results of experiments with decision tables from UCI Machine Learning Repository [9] and make some comparison with the average length and average coverage of γ-decision rules constructed by the greedy algorithm. Presented results show that the relative difference connected with the average coverage of γ-decision rules constructed by the dynamic programming algorithm and the greedy algorithm is greater than the relative difference connected with the average length.

Optimizations of irredundant γ-decision rules relative to the length and coverage and construction of totally optimal rules can be considered as a tool that supports design of classifiers. To predict a value of the decision attribute for a new object we can use in a classifier only rules after sequential optimization relative to the coverage, and then relative to the length. Rules that cover many objects and have small number of descriptors can be useful also in knowledge discovery to represent knowledge extracted from decision tables. In this case, rules with smaller number of descriptors are more understandable. Sequential optimization of γ-decision rules can be considered as a problem of multi-criteria optimization of decision rules with hierarchically dependent criteria. In the future, we will consider as a cost function the number of misclassifications also. Then we can make a sequential optimization of irredundant γ-decision rules relative to arbitrary subset and order of cost functions length, coverage, number of misclassifications.

References

1. Agrawal, R., Srikant, R.: Fast algorithms for mining association rules in large databases. In: Bocca, J.B., Jarke, M., Zaniolo C. (eds.) Proceedings of the 20th International Conference on Very Large Data Bases, VLDB'94, pp. 487–499. Morgan Kaufmann (1994)
2. Alkhalid, A., Chikalov, I., Husain, S., Moshkov, M.: Extensions of dynamic programming as a new tool for decision tree optimization. In: Ramanna, S., Jain, L.C., Howlett, R.J. (eds.) Emerging Paradigms in Machine Learning, *Smart Innovation, Systems and Technologies*, vol. 13, pp. 11–29. Springer, Heidelberg (2013)

3. Alkhalid, A., Chikalov, I., Moshkov, M.: On algorithm for building of optimal α-decision trees. In: Szczuka, M.S., Kryszkiewicz, M., Ramanna, S., Jensen, R., Hu, Q. (eds.) RSCTC 2010, *LNCS*, vol. 6086, pp. 438–445. Springer, Heidelberg (2010)
4. Alkhalid, A., Amin, T., Chikalov, I., Hussain, S., Moshkov, M., Zielosko, B.: Dagger: A tool for analysis and optimization of decision trees and rules. In: Ficarra, F.V.C. (ed.) Computational Informatics, Social Factors and New Information Technologies: Hypermedia Perspectives and Avant-Garde Experiences in the Era of Communicability Expansion, pp. 29–39. Blue Herons, Bergamo, Italy (2011)
5. Amin, T., Chikalov, I., Moshkov, M., Zielosko, B.: Dynamic programming approach for exact decision rule optimization. In: Skowron, A., Suraj, Z. (eds.) Rough Sets and Intelligent Systems—Professor Zdzisław Pawlak in Memoriam, *Intelligent Systems Reference Library*, vol. 42, pp. 211–228. Springer, Heidelberg (2013)
6. Amin, T., Chikalov, I., Moshkov, M., Zielosko, B.: Dynamic programming approach for partial decision rule optimization. Fundam. Inform. **119**(3–4), 233–248 (2012)
7. Amin, T., Chikalov, I., Moshkov, M., Zielosko, B.: Dynamic programming approach to optimization of approximate decision rules. Inf. Sci. **221**, 403–418 (2013)
8. Ang, J., Tan, K., Mamun, A.: An evolutionary memetic algorithm for rule extraction. Export Syst. Appl. **37**(2), 1302–1315 (2010)
9. Asuncion, A., Newman, D.J.: UCI Machine Learning Repository (2007). http://www.ics.uci.edu/~mlearn/
10. Błaszczyński, J., Słowiński, R., Szeląg, M.: Sequential covering rule induction algorithm for variable consistency rough set approaches. Inf. Sci. **181**(5), 987–1002 (2011)
11. Chikalov, I.: On algorithm for constructing of decision trees with minimal number of nodes. In: Ziarko, W., Yao, Y.Y. (eds.) RSCTC 2000, *LNCS*, vol. 2005, pp. 139–143. Springer, Heidelberg (2001)
12. Clark, P., Niblett, T.: The cn2 induction algorithm. Mach. Learn. **3**(4), 261–283 (1989)
13. Dembczyński, K., Kotłowski, W., Słowiński, R.: Ender: a statistical framework for boosting decision rules. Data Min. Knowl. Discov. **21**(1), 52–90 (2010)
14. Fürnkranz, J.: Separate-and-conquer rule learning. Artif. Intell. Rev. **13**(1), 3–54 (1999)
15. Grzymała-Busse, J.W.: Lers—a system for learning from examples based on rough sets. In: Słowiński, R. (ed.) Intelligent Decision Support. Handbook of Applications and Advances of the Rough Sets Theory, pp. 3–18. Kluwer Academic Publishers (1992)
16. Liu, B., Abbass, H.A., McKay, B.: Classification rule discovery with ant colony optimization. In: IAT 2003, pp. 83–88. IEEE Computer Society (2003)
17. Michalski, S., Pietrzykowski, J.: iAQ: A Program That Discovers Rules. AAAI-07 AI Video Competition (2007). URL http://videolectures.net/aaai07_michalski_iaq/
18. Moshkov, M., Chikalov, I.: On algorithm for constructing of decision trees with minimal depth. Fundam. Inform. **41**(3), 295–299 (2000)
19. Moshkov, M., Zielosko, B.: Combinatorial Machine Learning—A Rough Set Approach, *Studies in Computational Intelligence*, vol. 360. Springer, Heidelberg (2011)
20. Moshkov, M., Piliszczuk, M., Zielosko, B.: Partial Covers, Reducts and Decision Rules in Rough Sets—Theory and Applications, *Studies in Computational Intelligence*, vol. 145. Springer, Heidelberg (2008)
21. Nguyen, H.S.: Approximate boolean reasoning: foundations and applications in data mining. In: Peters, J.F., Skowron, A. (eds.) T. Rough Sets, *LNCS*, vol. 4100, pp. 334–506. Springer (2006)
22. Pawlak, Z., Skowron, A.: Rough sets and boolean reasoning. Inf. Sci. **177**(1), 41–73 (2007)
23. Quinlan, J.R.: C4.5: Programs for Machine Learning. Morgan Kaufmann Publishers Inc. (1993)
24. Rissanen, J.: Modeling by shortest data description. Automatica **14**(5), 465–471 (1978)
25. Skowron, A., Rauszer, C.: The discernibility matrices and functions in information systems. In: Słowinski, R. (ed.) Intelligent Decision Support. Handbook of Applications and Advances of the Rough Set Theory, pp. 331–362. Kluwer Academic Publishers, Dordrecht (1992)
26. Ślęzak, D., Wróblewski, J.: Order based genetic algorithms for the search of approximate entropy reducts. In: Wang, G., Liu, Q., Yao, Y., Skowron, A. (eds.) RSFDGrC 2003, *LNCS*, vol. 2639, pp. 308–311. Springer (2003)

27. Zielosko, B., Moshkov, M., Chikalov, I.: Optimization of decision rules based on methods of dynamic programming. Vestnik of Lobachevsky State University of Nizhny Novgorod **6**, 195–200 (2010). (in Russian)
28. Zielosko, B.: Sequential optimization of γ-decision rules. In: Ganzha, M., Maciaszek, L.A., Paprzycki, M. (eds.) Proceedings of FedCSIS 2012, Wrocław, Poland, 9–12 Sept 2012, pp. 339–346 (2012)

Chapter 13
Relationships Among Various Parameters for Decision Tree Optimization

Shahid Hussain

Abstract In this chapter, we study, in detail, the relationships between various pairs of cost functions and between uncertainty measure and cost functions, for decision tree optimization. We provide new tools (algorithms) to compute relationship functions, as well as provide experimental results on decision tables acquired from UCI ML Repository. The algorithms presented in this paper have already been implemented and are now a part of Dagger, which is a software system for construction/optimization of decision trees and decision rules. The main results presented in this chapter deal with two types of algorithms for computing relationships; first, we discuss the case where we construct approximate decision trees and are interested in relationships between certain cost function, such as depth or number of nodes of a decision trees, and an uncertainty measure, such as misclassification error (accuracy) of decision tree. Secondly, relationships between two different cost functions are discussed, for example, the number of misclassification of a decision tree versus number of nodes in a decision trees. The results of experiments, presented in the chapter, provide further insight.

13.1 Introduction

A *decision tree* is a finite directed tree with the root in which terminal nodes are labeled with *decisions*, nonterminal nodes with *attributes*, and edges are labeled with *values of attributes*. Decision trees are widely used as *predictors* [1], as a way of *representing knowledge* [2], and as *algorithms* for problem solving [3]. Each such use has a different optimization objective. That is, we need to minimize the

S. Hussain (✉)
Computer, Electrical and Mathematical Science and Engineering Division,
King Abdullah University of Science and Technology,
Thuwal 23955-6900, Saudi Arabia
e-mail: shahid.hussain@kaust.edu.sa

C. Faucher and L. C. Jain (eds.), *Innovations in Intelligent Machines-4*,
Studies in Computational Intelligence 514, DOI: 10.1007/978-3-319-01866-9_13,

number of misclassifications in order to achieve more accurate decision trees (from the perspective of *prediction*). To have more understandable decision trees we need to minimize the number of nodes in a decision tree (*knowledge representation*). Decision trees, when used as *algorithms*, need to be shallow i.e., we need to minimize either the depth or average depth (or in some cases both) of a decision tree in order to reduce algorithm complexity. Unfortunately, almost all problems connected with decision trees optimization are NP-hard [3, 4].

Several exact algorithms for decision tree optimization are known including brute-force algorithms [5], algorithms based on dynamic programming [6–8], and algorithms using branch-and-bound technique [9]. Similarly, different algorithms and techniques for construction and optimization of approximate decision trees have been extensively studied by researchers in the field, for example, using genetic algorithms [10], simulated annealing [11], and ant colony [12]. Most approximation algorithms for decision trees are greedy, in nature. Generally, these algorithms employ a top-down approach and at each step minimize some impurity. Several different impurity criteria are known in literature, for example information-theoretic [13], statistical [2], and combinatorial [3, 14]. See [13, 15–21] for comparison of different impurity criteria.

We have created a software system for decision trees (as well as decision rules) called dagger—a tool based on dynamic programming which allows us to optimize decision trees (and decision rules) relative to various cost functions such as depth (length), average depth (average length), total number of nodes, and number of misclassifications sequentially [22–25].

Decision tree optimization and sequential optimization naturally lead to questions such as what is the *relationship* between two *cost functions* for construction of decision trees, or how *uncertainty* effect the overall structure of trees (i.e., how *number of nodes* or average depth of decision trees is associated with *entropy*, or *misclassification error*?) These questions are very important from the practical point of view, related to building optimal yet cost effective decision trees. To this end, we have created algorithms to answer such questions. In this chapter, we present algorithms for computing relationship between a pair of cost functions as well as computing relationships for effect of uncertainty measure on construction of decision trees for different cost functions. We also give details about experimental results for several datasets acquired from UCI ML Repository [26] as well as demonstrate working of algorithms on simple examples.

The presented algorithms and their implementation in the software tool Dagger together with similar algorithms devised by the authors (see for example [27]) can be useful for investigations in Rough Sets [28, 29] where decision trees are used as classifiers [30].

This chapter is divided into five sections including the Introduction. Section 13.2 presents some basic notions and preliminaries related to decision trees, cost functions, and uncertainty measures. Section 13.3 describes, in detail, the main algorithms and examples. Section 13.4 considers the experimental results related to relationship algorithms and presents plots for the two kinds of relationship studies for the decision trees. Section 13.5 concludes the chapter.

13.2 Preliminaries and Basic Notions

In this section, we introduce notions related to decision tables, decision trees, and other relevant definitions and results.

13.2.1 Decision Tables and Decision Trees

We consider only decision tables with discrete attributes. These tables do not contain missing values and equal rows. Consider a *decision table* T depicted in Fig. 13.1.

Here $f_1, \ldots, f_m$ are the conditional attributes; $c_1, \ldots, c_N$ are nonnegative integers which can be interpreted as the decisions (values of the decision attribute d); b_{ij} are nonnegative integers which are interpreted as values of conditional attributes (we assume that the rows $(b_{11}, \ldots, b_{1m}), \ldots, (b_{N1}, \ldots, b_{Nm})$ are pairwise different). We denote by $E(T)$ the set of attributes (columns of table T), each of which contains different values. For $f_i \in E(T)$, let $E(T, f_i)$ be the set of values from the column f_i. We denote by $N(T)$ the number of rows in the decision table T.

Let $f_{i_1}, \ldots, f_{i_t} \in \{f_1, \ldots, f_m\}$ and $a_1, \ldots, a_t$ be nonnegative integers. We denote by $T(f_{i_1}, a_1) \ldots (f_{i_t}, a_t)$ the subtable of table T, which consists of such and only such rows of T that at the intersection with columns $f_{i_1}, \ldots, f_{i_t}$ have numbers $a_1, \ldots, a_t$, respectively. Such nonempty tables (including table T) will be called *separable subtables* of table T. For a decision table T, we denote by $\mathrm{SEP}(T)$, the set of separable subtables of T. Figure 13.2 shows a separable subtable Θ of table T in general case.

A *decision tree* Γ *over table* T is a finite directed tree with a root in which each terminal node is labeled with a decision. Each nonterminal node is labeled with a conditional attribute, and for each nonterminal node, the outgoing edges are labeled with pairwise different nonnegative integers. Let v be an arbitrary node of Γ. We now define a subtable $T(v)$ of table T. If v is the root then $T(v) = T$. Let v be a node of Γ that is not the root, nodes in the path from the root to v be labeled with attributes $f_{i_1}, \ldots, f_{i_t}$, and edges in this path be labeled with values $a_1, \ldots, a_t$, respectively. Then $T(v) = T(f_{i_1}, a_1) \ldots (f_{i_t}, a_t)$. Figure 13.3 gives an example of a decision tree showing the separable subtables for a simple decision table given in the figure.

Fig. 13.1 Decision table

f_1	⋯	f_m	d
b_{11}	⋯	b_{1m}	c_1
⋮		⋮	⋮
b_{N1}	⋯	b_{Nm}	c_N

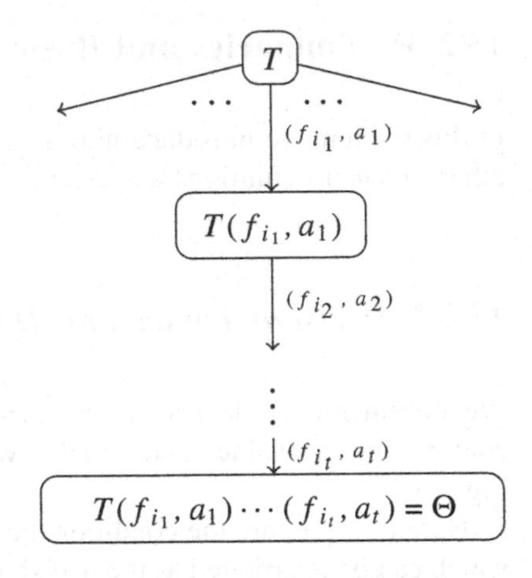

Fig. 13.2 A separable subtable Θ of table T

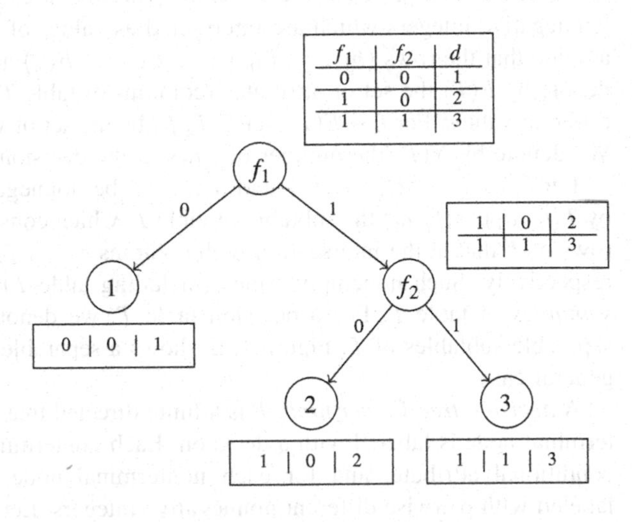

Fig. 13.3 An example of separable subtable for the given table

We say that a decision table T is a *degenerate* table if T is empty or all the rows in T are labeled with the same decision. Such a decision is called the *common decision* for T. For a *non-degenerate* table T, a decision b is called the *most common decision*, if most rows in T are labeled with the decision b.

13.2.1.1 Exact Decision Trees

We say that Γ is an *exact decision tree for T* if any node v of Γ satisfies the following conditions:

- If all rows of $T(v)$ are labeled with the same decision then, v is a terminal node labeled with this decision.

- Otherwise, v is labeled with an attribute $f_i \in E(T(v))$ and, if $E(T(v), f_i) = \{a_1, \ldots, a_t\}$ then, t edges leave node v, and these edges are labeled with $a_1, \ldots, a_t$, respectively.

13.2.1.2 α-Decision Trees

Let $\mathcal{U}$ be an uncertainty measure defined on the set of decision tables (the notion of uncertainty measures for decision tables with examples of different uncertainty measures is discussed in detail in Sect. 13.2.4). We say that Γ is an *α-decision tree for T* for some nonnegative real α, if any node v of Γ satisfies the following conditions:

- If $\mathcal{U}(T(v)) \leq \alpha$ then v is a terminal node labeled with the most common decision for $T(v)$.
- Otherwise, v is labeled with an attribute $f_i \in E(T(v))$ and, if $E(T(v), f_i) = \{a_1, \ldots, a_t\}$ then t edges leave node v, and these edges are labeled with $a_1, \ldots, a_t$, respectively.

For any α, an α-decision tree for T is called a decision tree for T. Note that the parameter α characterizes the accuracy of decision trees. In particular, 0-decision tree for a table T is an exact decision tree for T.

13.2.2 The Graph $\Delta(T)$

Consider an algorithm for construction of a graph $\Delta(T)$, which will be used to study relationships. The nodes of this graph are some separable subtables of table T. During each step we process one node and mark it with the symbol *. We start with the graph that consists of one node T and finish when all the nodes of the graph are processed.

Assume the algorithm has already performed p steps. We now describe step number $(p + 1)$. If all nodes are processed then the work of the algorithm is finished, and the resulting graph is $\Delta(T)$. Otherwise, choose a node (table) Θ that has not been processed yet. Let b be the most common decision for Θ. If Θ is a degenerate table then, label the considered node with b, mark it with symbol * and proceed to step number $(p + 2)$. If Θ is non-degenerate table, then for each $f_i \in E(\Theta)$ draw a bundle of edges from the node Θ (this bundle of edges will be called *f_i-bundle*). Let $E(\Theta, f_i) = \{a_1, \ldots, a_t\}$. Then draw t edges from Θ and label these edges with pairs $(f_i, a_1), \ldots, (f_i, a_t)$ respectively. These edges enter into nodes $\Theta(f_i, a_1), \ldots, \Theta(f_i, a_t)$. If some of the nodes $\Theta(f_i, a_1), \ldots, \Theta(f_i, a_t)$ are not present in the graph then add these nodes to the graph. Mark the node Θ with the symbol * and proceed to step number $(p + 2)$.

Now for each node Θ of the graph $\Delta(T)$, we describe the set $D(\Theta)$ of decision trees corresponding to the node Θ. We will move from terminal nodes, which are labeled with numbers, to the node T. Let Θ be a node, which is labeled with a number b. Then the only trivial decision tree depicted in Fig. 13.4 corresponds to the node Θ.

Let Θ be a nonterminal node (table) then there is a number of bundles of edges starting in Θ. We consider an arbitrary bundle and describe the set of decision trees corresponding to this bundle. Let the considered bundle be an f_i-bundle where $f_i \in (\Theta)$ and $E(\Theta, f_i) = \{a_1, \ldots, a_t\}$. Let $\Gamma_1, \ldots, \Gamma_t$ be decision trees from sets corresponding to the nodes $\Theta(f_i, a_1), \ldots, \Theta(f_i, a_t)$. Then the decision tree depicted in Fig. 13.5 belongs to the set of decision trees, which correspond to this bundle. All such decision trees belong to the considered set, and this set does not contain any other decision trees. Then the set of decision trees corresponding to the node Θ coincides with the union of sets of decision trees corresponding to the bundles starting in Θ and the set containing one decision tree depicted in Fig. 13.4, where b is the most common decision for Θ. We denote by $D(\Theta)$ the set of decision trees corresponding to the node Θ.

The following proposition shows that the graph $\Delta(T)$ can represent all decision trees for table T.

Proposition 1 ([31]) *Let T be a decision table and Θ a node in the graph $\Delta(T)$. Then the set $D(\Theta)$ coincides with the set of all decision trees for table Θ.*

13.2.3 Cost Functions

We will consider cost functions which are given in the following way: values of considered cost function ψ, which are nonnegative numbers, are defined by induction on pairs (T, Γ), where T is a decision table and Γ is an α-decision tree for T. Let Γ be an α-decision tree represented in Fig. 13.4. Then $\psi(T, \Gamma) = \psi^0(T, b)$,

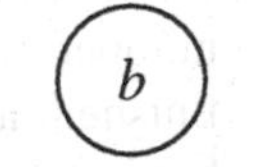

Fig. 13.4 Trivial decision tree

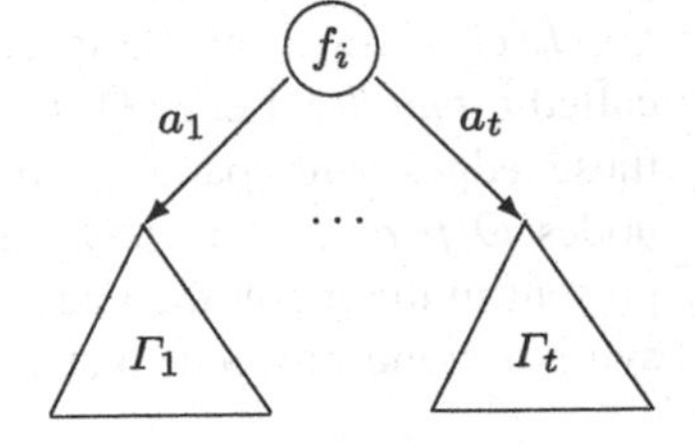

Fig. 13.5 Aggregated decision tree

where ψ^0 is a nonnegative function depending on T and b. Let Γ be an α-decision tree depicted in Fig. 13.5. Then $\psi(T, \Gamma) = F(N(T), \psi(T(f_i, a_1), \Gamma_1), \ldots, \psi(T(f_i, a_t), \Gamma_t))$. Here, $N(T)$ is the number of rows in table T, and $F(n, \psi_1, \psi_2, \ldots)$ is an operator which transforms the considered tuple of nonnegative numbers into a nonnegative number. Note that the number of variables $\psi_1, \psi_2, \ldots$ is not bounded from above.

The considered cost function will be called *monotone* if for any natural t, and any nonnegative numbers $a, c_1, \ldots, c_t, d_1, \ldots, d_t$, satisfying the inequalities $c_1 \leq d_1, \ldots, c_t \leq d_t$, the inequality $F(a, c_1, \ldots, c_t) \leq F(a, d_1, \ldots, d_t)$ holds.

We will say that ψ is *bounded from below* if, ψ^0 is constant and $\psi(T, \Gamma) \geq \psi^0$ for any decision table T and any α-decision tree Γ for T.

Now we take a closer view of some monotone cost functions, which are bounded from below with exception of the number of misclassifications.

Number of nodes: $\psi(T, \Gamma)$ is the number of nodes in α-decision tree Γ. For this cost function $\psi^0 \equiv 1$ and

$$F(n, \psi_1, \psi_2, \ldots, \psi_t) = 1 + \sum_{i=1}^{t} \psi_i.$$

Depth: $\psi(T, \Gamma)$ is the maximum length of a path from the root to a terminal node of Γ. For this cost function $\psi^0 \equiv 0$ and

$$F(n, \psi_1, \psi_2, \ldots, \psi_t) = 1 + \max\{\psi_1, \ldots, \psi_t\}.$$

Total Path length: For an arbitrary row t of table T we denote by $l(r)$ the length of the path from the root to a terminal node v of Γ such that r is in $T(v)$. Then $\psi(T, \Gamma) = \sum_r l(r)$, where we take the sum on all rows r of table T. For this cost function $\psi^0 \equiv 0$ and

$$F(n, \psi_1, \psi_2, \ldots, \psi_t) = n + \sum_{i=1}^{t} \psi_i.$$

Note that the *average depth* of Γ is equal to the total path length divided by $N(T)$.

Number of Misclassifications: $\psi(T, \Gamma)$ is the number of rows r in T for which the terminal node v of Γ such that v is in $T(v)$ is labeled with a decision different from the decision attached to r. For this cost function, $\psi^0(T, b)$ is equal to number of rows in T, which are labeled with decisions different from b, and

$$F(n, \psi_1, \psi_2, \ldots, \psi_t) = \sum_{i=1}^{t} \psi_i.$$

13.2.4 Uncertainty Measures

An uncertainty measure is a function $\mathcal{U}$ defined on the set of decision tables. For any decision table T, $\mathcal{U}(T) \geq 0$ and $\mathcal{U}(T) = 0$ if and only if T is degenerate.

Let T be a decision table with $N = N(T)$ rows, m attributes, and k different decisions $d_1, \ldots, d_k$. For each such decision d_i for $i \in \{1, \ldots, k\}$ we say $N_i = N_i(T)$ is number of rows in T labeled with the decision d_i. Furthermore we say p_i is the ratio N_i/N for table T. We define the following five uncertainty measures decision table T :

Entropy	$\mathrm{Ent}(T) = -\sum_{i=1}^{k} p_i \log_2 p_i$ (with the assumption that $0 \log_2 0 = 0$)
Gini Index	$\mathrm{Gini}(T) = 1 - \sum_{i=1}^{k} p_i^2$
ME	$\mathrm{ME}(T) = N(T) - \max_{1 \leq j \leq k} N_j(T)$
RME	$\mathrm{RME}(T) = \mathrm{ME}(T)/N(T)$
R	$R(T)$ is the number of ordered pairs of rows in T with different decisions. $R(T)$ and Gini index are related i.e., $R(T) = N^2(T) \cdot \mathrm{Gini}(T)/2$

13.2.5 Transformation of Functions

Let f and g be two functions from a set A onto C_f and C_g respectively, where C_f and C_g are finite sets of nonnegative integers. Let $B_f = \{m_f, m_f + 1, \ldots, M_f\}$ and $B_g = \{n_g, n_g + 1, \ldots, N_g\}$ where $m_f = \min\{m : m \in C_f\}$ and $n_g = \min\{n : n \in C_g\}$. Furthermore, M_f and N_g are natural numbers such that $m \leq M_f$ and $n \leq N_g$ for any $m \in C_f$ and $n \in C_g$, respectively.

We define two functions $\mathcal{F} : B_g \to B_f$ and $\mathcal{G} : B_f \to B_g$ as following:

$$\mathcal{F}(n) = \min\{f(a) : a \in A,\ g(a) \leq n\},\ \forall n \in B_g, \tag{13.1}$$

$$\mathcal{G}(m) = \min\{g(a) : a \in A, f(a) \leq m\},\ \forall m \in B_f. \tag{13.2}$$

It is clear that both $\mathcal{F}$ and $\mathcal{G}$ are nonincreasing functions.

The following proposition states that the functions $\mathcal{F}$ and $\mathcal{G}$ can be used interchangeably and we can evaluate $\mathcal{F}$ using $\mathcal{G}$ and vice versa, i.e., it is enough to know only one function to evaluate the other.

Proposition 2 ([31]) *For any $n \in B_g$ and for any $m \in B_f$,*

$$\mathcal{F}(n) = \min\{m \in B_f : \mathcal{G}(m) \leq n\},$$

$$\mathcal{G}(m) = \min\{n \in B_g : \mathcal{F}(n) \leq m\}.$$

Proposition 2 allows us to transform the function $\mathcal{G}$ given by a tuple $(\mathcal{G}(m_f), \mathcal{G}(m_f + 1), \ldots, \mathcal{G}(M_f))$ into the function $\mathcal{F}$ and vice versa. We know that $\mathcal{G}(m_f) \geq \mathcal{G}(m_f + 1) \geq \cdots \geq \mathcal{G}(M_f)$, to find the minimum $m \in B_f$ such that $\mathcal{G}(m) \leq m$ we can use binary search which requires $O(\log |B_f|)$ comparisons of numbers. So to find the value $\mathcal{F}(n)$ for $n \in B_g$, it is enough to make $O(\log |B_f|)$ operations of comparison.

13.3 Relationships

This section presents algorithms for studying the relationships between various parameters for decision tree optimization. We consider two kinds of relationships; relationships between an uncertainty measure (determining tree accuracy) and a cost function and relationships between two different cost functions for decision tree. Figure 13.6 summarizes the kinds of relationships, the numbers in brackets as labels on edges in this figure represent the publication references. It is important to note that all relationships, previously considered, between a cost function and an uncertainty measure $\mathcal{U} = \text{R}$. In this chapter, we generalize the algorithm for computing such relationships and use $\mathcal{U} = \text{ME}$ as uncertainty measure and depth as cost function.

We also consider the algorithm for computing the relationships between two cost functions, the number of nodes and the number of misclassifications (previously considered in a conference paper [32]).

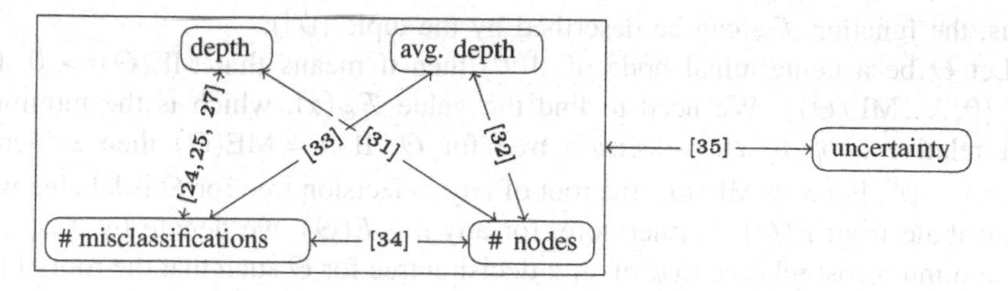

Fig. 13.6 Relationship among various parameters for decision tree construction

13.3.1 Relationship Between Cost and Uncertainty of Decision Trees

We consider ME as uncertainty measure, defining the accuracy of decision trees. Let T be a decision table with m columns labeled with conditional attributes $f_1, \ldots, f_m$ and ψ be a monotone and bounded from below cost function. In the following, we provide an algorithm to compute relationship between the cost and accuracy of decision trees for T.

We represent the relationship using the function $\mathcal{F}_{\psi,T}$, which is defined on the set $\{0, \ldots, \mathrm{ME}(T)\}$. For any $\alpha \in \{0, \ldots, \mathrm{ME}(T)\}$, the value of $\mathcal{F}_{\psi,T}(\alpha)$ is equal to the minimum cost of an α-decision tree for T, relative to the cost function ψ. This function can be represented by the tuple

$$(\mathcal{F}_{\psi,T}(0), \ldots, \mathcal{F}_{\psi,T}(\mathrm{ME}(T))).$$

13.3.1.1 Computing the Function $\mathcal{F}_{\psi,T}$

For decision table T, we construct the graph $\Delta(T)$ and for each node Θ of the graph $\Delta(T)$, we compute the function $\mathcal{F}_\Theta = \mathcal{F}_{\psi,\Theta}$ (we compute the $(\mathrm{ME}(\Theta) + 1)$-tuple describing this function).

A node of $\Delta(T)$ is called *terminal* if there are no edges leaving this node. We will move from the terminal nodes, which are labeled with numbers, to the node T.

Let Θ be a terminal node of $\Delta(T)$ which is labeled with a number b that is the most common decision for Θ. The decision tree depicted in Fig. 13.4 is the only 0-decision tree for Θ. Since $\mathrm{ME}(\Theta) = 0$, we should consider only one value of α—the value 0. It's clear that the minimum cost of 0-decision tree for Θ is equal to ψ^0. Thus, the function $\mathcal{F}_\Theta$ can be described by the tuple (ψ^0).

Let Θ be a nonterminal node of $\Delta(T)$ then it means that $\mathrm{ME}(\Theta) > 0$. Let $\alpha \in \{0, \ldots, \mathrm{ME}(\Theta)\}$. We need to find the value $\mathcal{F}_\Theta(\alpha)$, which is the minimum cost relative to ψ of an α-decision tree for Θ. If $\alpha = \mathrm{ME}(\Theta)$ then evidently $\mathcal{F}_\Theta(\alpha) = \psi^0$. For $\alpha < \mathrm{ME}(\Theta)$, the root of any α-decision tree for Θ is labeled with an attribute from $E(\Theta)$. Furthermore, for any $f_i \in E(\Theta)$, we denote by $\mathcal{F}_\Theta(\alpha, f_i)$ the minimum cost relative to ψ of an α-decision tree for Θ such that the root of this tree is labeled with f_i. It is clear that

$$\mathcal{F}_\Theta(\alpha) = \min\{\mathcal{F}_\Theta(\alpha, f_i) : f_i \in E(\Theta)\}. \tag{13.3}$$

Let $f_i \in E(\Theta)$ and $E(\Theta, f_i) = \{a_1, \ldots, a_t\}$. Then any α-decision tree Γ for Θ with the attribute f_i attached to the root can be represented in the form depicted in Fig. 13.5, where $\Gamma_1, \ldots, \Gamma_t$ are α-decision trees for $\Theta(f_i, a_1), \ldots, \Theta(f_i, a_t)$. Since ψ

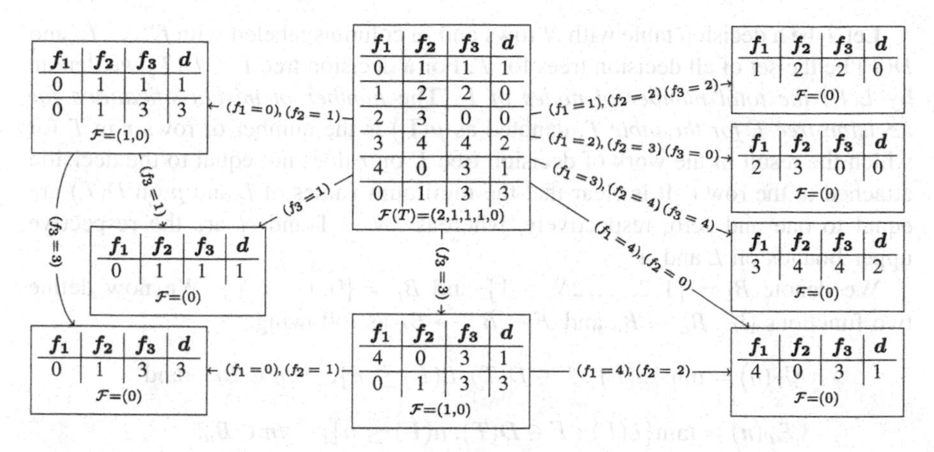

Fig. 13.7 DAG with relationships between uncertainty (ME) and depth

is a monotone cost function, tree Γ will have the minimum cost if the costs of trees $\Gamma_1, \ldots, \Gamma_t$ are minimum. Therefore,

$$\mathcal{F}_{\Theta}(\alpha, f_i) = F(N(\Theta), \mathcal{F}_{\Theta(f_i,a_1)}(\alpha), \ldots, \mathcal{F}_{\Theta(f_i,a_t)}(\alpha)). \tag{13.4}$$

If for some $j, 1 \leq j \leq t$, we have $\alpha > \mathrm{ME}(\Theta(f_i, a_j))$ then we set $\mathcal{F}_{\Theta(f_i,a_j)}(\alpha) = \psi^0$, since the decision tree depicted in Fig. 13.4, where b is the most common decision for $\Theta(f_i, a_j)$, is an α-decision tree for $\Theta(f_i, a_j)$. The cost of this tree is ψ^0. Since ψ is a cost function, which is bounded from below, the cost of any α-decision tree for $\Theta(f_i, a_j)$ is at least ψ^0.

The formulas (13.3) and (13.4) allow us to find the value of $\mathcal{F}_{\Theta}(\alpha)$ if we know the values of $\mathcal{F}_{\Theta(f_i,a_j)}(\alpha)$, where $f_i \in E(\Theta)$ and $a_j \in E(\Theta, f_i)$. When we reach the node T we will obtain the function $\mathcal{F}_T = \mathcal{F}_{\psi,T}$.

Figure 13.7 is the directed acyclic graph (DAG) as an illustration of running the considered algorithm where we use ME as the uncertainty measure and depth of decision tree as the cost functions.

13.3.2 Relationships Between a Pair of Cost Functions

In the following, we present the algorithm for computing relationships between two cost functions for decision trees. Let us consider, for example, the situation for computing relationships for *number of nodes* and *number of misclassifications*.

Let T be a decision table with N rows and m columns labeled with $f_1, \ldots, f_m$ and $D(T)$ be the set of all decision trees for T. For a decision tree $\Gamma \in D(T)$ we denote by $L(\Gamma)$ *the total number of nodes of* Γ. The *number of misclassifications for decision tree* Γ *for the table* T, denoted as $\mu(\Gamma)$ is the number of rows r in T for which the result of the work of decision tree Γ on r does not equal to the decision attached to the row r. It is clear that the minimum values of L and μ on $D(T)$ are equal to one and zero, respectively, whereas $2N - 1$ and N are the respective upper bounds on L and μ.

We denote $B_L = \{1, 2, \ldots, 2N - 1\}$ and $B_\mu = \{0, 1, \ldots, N\}$. We now define two functions $\mathcal{G}_T : B_L \rightarrow B_\mu$ and $\mathcal{F}_T : B_\mu \rightarrow B_L$ as following:

$$\mathcal{G}_T(n) = \min\{\mu(\Gamma) : \Gamma \in D(T), L(\Gamma) \leq n\}, \quad \forall n \in B_L, \text{ and}$$

$$\mathcal{F}_T(n) = \min\{L(\Gamma) : \Gamma \in D(T), \mu(\Gamma) \leq n\}, \quad \forall n \in B_\mu.$$

We now describe the algorithm to construct the function $\mathcal{G}_\Theta$ for every node (subtable) Θ from the graph $\Delta(T)$. For simplicity, we assume that $\mathcal{G}_\Theta$ is defined on the set B_L. We begin from the terminal nodes of $\Delta(T)$ and move upward to the root node T.

Let Θ be a terminal node. It means that all nodes of Θ are labeled with the same decision b and the decision tree Γ_b as depicted in Fig. 13.4 belongs of $D(\Theta)$. It is clear that $L(\Gamma_b) = 1$ and $\mu(\Gamma_b) = 0$ for the table Θ. Therefore, $\mathcal{G}_\Theta(n) = 0$ for any $n \in B_L$.

Let us consider a nonterminal node Θ and a bundle of edges labeled with pairs $(f_i, a_1), \ldots, (f_i, a_t)$, which start from this node. Let these edges enter into nodes $\Theta(f_i, a_1), \ldots, \Theta(f_i, a_t)$, respectively, to which the functions $\mathcal{G}_{\Theta(f_i, a_1)}, \ldots, \mathcal{G}_{\Theta(f_i, a_t)}$ are already attached.

We correspond to this bundle (f_i-bundle) the function $\mathcal{G}_\Theta^{f_i}$, for any $n \in B_L$, $n > t$, $\mathcal{G}_\Theta^{f_i}(n) = \min \sum_{j=1}^{t} \mathcal{G}_{\Theta(f_i, aj)}(n_j)$ where the minimum is taken over all $n_1, \ldots, n_t$ such that $1 \leq n_j \leq 2N - 1$ for $j = 1, \ldots, t$ and $n_1 + \cdots + n_t + 1 \leq n$. (Computing $\mathcal{G}_\Theta^{f_i}$ is a nontrivial task. We describe the method in detail in the following Sect. 13.3.3). The minimum number of nodes for decision tree such that f_i is attached to the root is $t + 1$ therefore for $n \in B_L$, such that $1 \leq n \leq t$, $\mathcal{G}_\Theta^{f_i}(n) = N(\Theta) - N_{\text{mcd}}(\Theta)$, where $N_{\text{mcd}}(\Theta)$ is the number of nodes in Θ labeled with *the most common decision* for Θ.

It is not difficult to prove that for all $n \in B_L$, $n > t$,

$$\mathcal{G}_\Theta(n) = \min\left\{\mathcal{G}_\Theta^{f_i}(n) : f_i \in E(\Theta)\right\}.$$

We can use the following proposition to construct the function $\mathcal{F}_T$ (can be proved using the method of transformation of functions as described in Sect. 13.2.5).

Proposition 3 *For any* $n \in B_\mu$, $\mathcal{F}_T(n) = \min\{p \in B_L : \mathcal{G}_T(p) \leq n\}$.

Note that to find the value $\mathcal{F}_T(n)$ for some $n \in B_\mu$ it is enough to make $O(\log |B_L|) = O(\log(2N - 1))$ operations of comparisons.

13.3.3 Computing $\mathcal{G}_{\Theta}^{f_i}$

Let Θ be a nonterminal node in $\Delta(T)$, $f_i \in E(\Theta)$, and $E(\Theta, f_i) = \{a_1, \ldots, a_t\}$. Furthermore, we assume that the functions $\mathcal{G}_{\Theta(f_i, a_j)}$, $j = 1, \ldots, t$, have already been computed. Let the values for $\mathcal{G}_{\Theta(f_i, a_j)}$ be given by the tuple of pairs $((1, \mu_1^j), (2, \mu_2^j), \ldots, (2N - 1, \mu_{2N-1}^j))$. We need to compute values of $\mathcal{G}_{\Theta}^{f_i}$ for all $n \in B_L$:

$$\mathcal{G}_{\Theta}^{f_i}(n) = \min \sum_{j=1}^{t} \mathcal{G}_{\Theta(f_i, a_j)}(n_j) \quad \text{for} \quad 1 \leq n_j \leq 2N - 1, \quad \text{s.t.,} \quad \sum_{i=1}^{t} n_i + 1 \leq n.$$

We construct a layered directed acyclic graph (DAG) $\delta(\Theta, f_i)$ to compute values of $\mathcal{G}_{\Theta}^{f_i}$ as following.

The DAG $\delta(\Theta, f_i)$ contains nodes arranged in $t + 1$ layers $(l_0, l_1, \ldots, l_t)$. Each node has a pair of labels and each layer $l_j (1 \leq j \leq t)$ contains at most $j(2N - 1)$ nodes. The first entry of labels for nodes in a layer l_j is an integer from $\{1, 2, \ldots, j(2N - 1)\}$. The layer l_0 contains only one node labeled with (0, 0).

Each node in a layer $l_j (0 \leq j < t)$ has exactly $2N - 1$ outgoing edges to nodes in layer l_{j+1}. These edges are labeled with the corresponding pairs in $\mathcal{G}_{\Theta(f_i, a_{j+1})}$. A node with label x as a first entry in its label-pair in a layer l_j connects to nodes with labels $x + 1$ to $x + 2N - 1$ (as a first entry in their label-pairs) in layer l_{j+1}, with edges labeled as $(1, \mu_1^{j+1})$, $(2, \mu_2^{j+1})$, $\ldots$, $(2N - 1, \mu_{2N-1}^{j+1})$, respectively.

The values of function $\mathcal{G}_{\Theta}^{f_i}(n)$ for $n \in B_L$ can be easily computed using the DAG $\delta(\Theta, f_i)$ for $\Theta \in \Delta(T)$ and for the considered bundle of edges for the attribute $f_i \in E(\Theta)$ as following:

Each node in layer l_1 gets its second value copied from the corresponding second value in incoming edge label to the node (since there is only one incoming edge for each in layer l_1). Let (k, μ) be a node in layer l_j, $2 \leq j \leq t$. Let $E = \{(v_1, \mu_1), (v_2, \mu_2), \ldots, (v_r, \mu_r)\}$ be the set of incoming nodes to (k, μ) such that (α_1, β_1), (α_2, β_2), $\ldots$, (α_r, β_r) are the labels of these edges between the nodes in E and $(k, \cdot)$, respectively. It is clear that $k = v_i + \alpha_i$, $1 \leq i \leq r$. Then $\mu = \min_{1 \leq i \leq r}\{\mu_i + \beta_i\}$. We do this for every node layer-by-layer till all nodes in $\delta(\Theta, f_i)$ have received their second label.

Once we finish computing the second value of label pairs of layer l_t, we can use these labels to compute $\mathcal{G}_{\Theta}^{f_i}(n)$. It is clear that the nodes in layer l_t have labels as

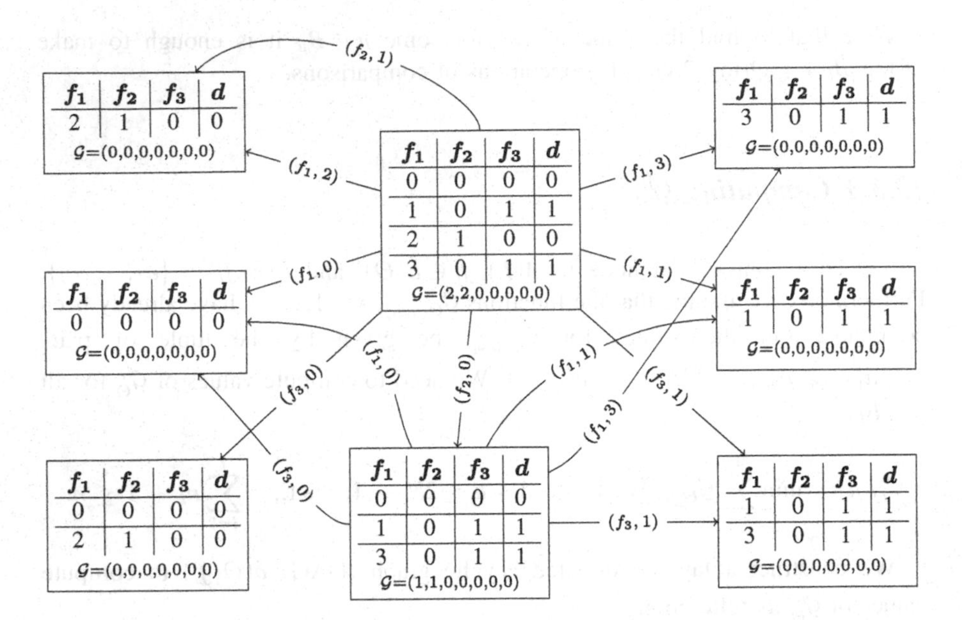

Fig. 13.8 DAG with relationship between the no. of nodes and no. of misclassifications

$(t, \mu(t)), \ldots, (t(2N-1), \mu(t(2N-1)))$, respectively. For $n \in B_L$ such that $1 \leq n \leq t$, $\mathcal{G}^{f_i}_{\Theta}(n) = N(\Theta) - N_{\text{mcd}}(\Theta)$. For $n \in B_L$, such that $t < n \leq 2N-1$, $\mathcal{G}^{f_i}_{\Theta}(n) = \min_{t<k\leq n} \mu(k-1)$.

Let us consider the time complexity of the considered algorithm. The DAG $\delta = \delta(\Theta, f_i)$ has $t+1$ layers and each layer l_j has at most $j(2N-1)$ nodes. Therefore, total number of nodes in δ is $O(t^2N)$. Since every node has $2N-1$ outgoing edges (except the nodes in layer l_t), the number of edges in δ is $O(t^2N^2)$. So, to build the graph δ, we need $O(t^2N^2)$ time (proportional to the number of nodes and edges in δ). To find the second labels we need a number of additions and comparisons bounded from above by the number of edges – $O(t^2N^2)$. Similarly, to find values of $\mathcal{G}^{f_i}_{\Theta}$ we need $O(N^2)$ comparisons. Therefore, the total time is $O(t^2N^2)$.

Table 13.1 UCI datasets used for experiments

Dataset	# rows	# attributes
BREAST-CANCER	266	10
CARS	1,728	6
FLAGS	194	27
NURSERY	12,960	8
SPECT-TEST	169	22
TIC-TAC-TOE	169	22

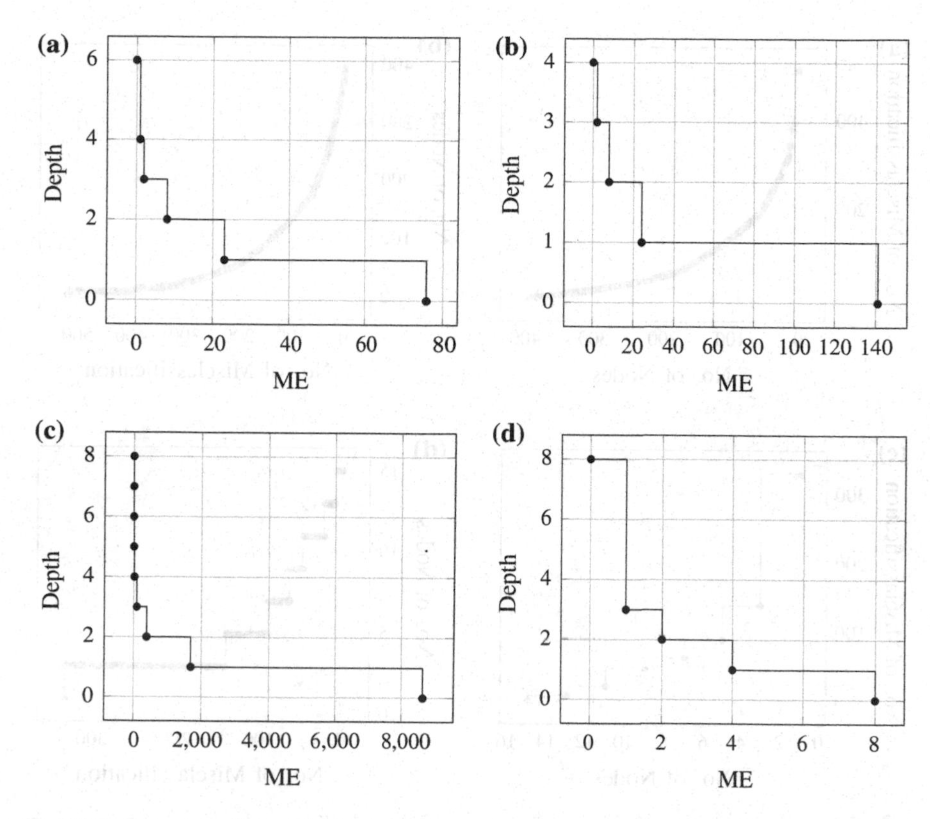

Fig. 13.9 Plots for relationships between uncertainty and cost function. **a** BREAST-CANCER dataset. **b** FLAGS dataset. **c** NURSERY dataset. **d** SPECT-TEST dataset

Figure 13.8 shows the directed acyclic graph (DAG) as an illustration of running the algorithm for computing relationship between number of nodes and number of misclassifications for the given decision table.

13.4 Experimental Results

In what follows, we present some experimental results for demonstration of relationships on datasets acquired from UCI ML Repository [26] (see Table 13.1 for information about datasets used in the experiments). We have made sure that datasets do not have missing values or duplicate rows by removing all such rows. First, we show results for algorithms computing relationships between an uncertainty measure and a cost function (we have fixed here uncertainty measure as ME and depth as cost function) (see Fig. 13.9). The second set of figures depicts

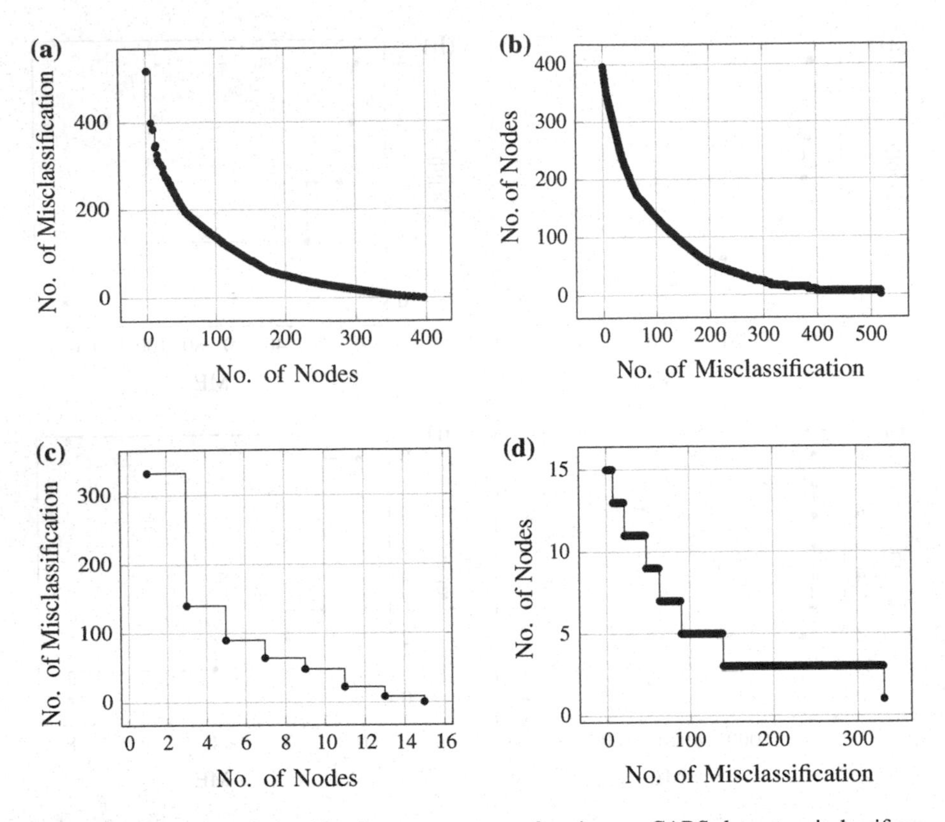

Fig. 13.10 Plots for relationships between two cost functions. **a** CARS dataset: misclassifications versus number of nodes. **b** CARS dataset: misclassifications versus number of nodes. **c** TIC-TAC-TOE dataset: misclassifications versus number of nodes. **d** TIC-TAC-TOE dataset: misclassifications versus number of nodes

experimental results for relationships between two different cost functions (see Fig. 13.10).

13.5 Conclusions and Future Work

This chapter is devoted to the consideration of algorithms for computing relationships between a pair of cost functions and between an uncertainty measure and a cost function. We present, in details, these algorithms together with simple examples to demonstrate how these algorithms work and some experimental results using standard datasets. These algorithms have been implemented in the software system called DAGGER. Further studies in this direction will be devoted to consideration of relationships between space and time complexity of decision trees corresponding Boolean functions.

Acknowledgments Author is greatly indebted to Prof. Colette Faucher for the valuable comments and suggestions which immensely improve the quality and readability of the paper. Author also wishes to acknowledge Prof. Mikhail Moshkov and Dr. Igor Chikalov for continuous help and support in theoretical and computational aspects of the ongoing research at KAUST.

References

1. Hastie, T., Tibshirani, R., Friedman, J.H.: The Elements of Statistical Learning: Data Mining, Inference, and Prediction. Springer, New York (2001)
2. Breiman, L., Friedman, J., Olshen, R., Stone, C.: Classification and Regression Trees. Wadsworth and Brooks, Monterey (1984)
3. Moshkov, M.J.: Time complexity of decision trees. In Peters, J.F., Skowron, A. (eds.) T. Rough Sets. Volume 3400 of Lecture Notes in Computer Science, pp. 244–459. Springer, Heidelberg (2005)
4. Hyafil, L., Rivest, R.L.: Constructing optimal binary decision trees is NP-complete. Inf. Process. Lett. **5**(1), 15–17 (1976)
5. Riddle, P., Segal, R., Etzioni, O.: Representation design and brute-force induction in a Boeing manufacturing domain. Appl. Artif. Intell. **8**, 125–147 (1994)
6. Garey, M.R.: Optimal binary identification procedures. SIAM J. Appl. Math. **23**, 173–186 (1972)
7. Martelli, A., Montanari, U.: Optimizing decision trees through heuristically guided search. Commun. ACM **21**(12), 1025–1039 (1978)
8. Schumacher, H., Sevcik, K.C.: The synthetic approach to decision table conversion. Commun. ACM **19**(6), 343–351 (1976)
9. Breitbart, Y., Reiter, A.: A branch-and-bound algorithm to obtain an optimal evaluation tree for monotonic boolean functions. Acta Inf. **4**, 311–319 (1975)
10. Chai, B.B.: Binary linear decision tree with genetic algorithm. In: Proceedings of the International Conference on Pattern Recognition (ICPR '96). Volume 7472 of ICPR '96., pp. 530–534. IEEE Computer Society, Washington, DC (1996)
11. Heath, D.G., Kasif, S., Salzberg, S.: Induction of oblique decision trees. In Bajcsy, R. (ed.) IJCAI, pp. 1002–1007. Morgan Kaufmann, Massachusetts (1993)
12. Boryczka, U., Kozak, J.: New algorithms for generation decision trees—ant-miner and its modifications. In Abraham, A., Hassanien, A.E., de Leon Ferreira de Carvalho, A.C.P., Snásel, V. (eds.) Foundations of Computational Intelligence, pp. 229–262. Springer, Berlin (2009)
13. Quinlan, J.R.: Induction of decision trees. Mach. Learn. **1**(1), 81–106 (1986)
14. Moret, B.M.E., Thomason, M.G., Gonzalez, R.C.: The activity of a variable and its relation to decision trees. ACM Trans. Program. Lang. Syst. **2**(4), 580–595 (1980)
15. Alkhalid, A., Chikalov, I., Moshkov, M.: Comparison of greedy algorithms for α-decision tree construction. In Yao, J., Ramanna, S., Wang, G., Suraj, Z. (eds.) RSKT, pp. 178–186. Springer, Berlin (2011)
16. Alkhalid, A., Chikalov, I., Moshkov, M.: Decision tree construction using greedy algorithms and dynamic programming—comparative study. In: Proceedings of 20th International Workshop on Concurrency, Specification and Programming, Pultusk. Białystok University of Technology, Poland (2011)
17. Alkhalid, A., Chikalov, I., Moshkov, M.: Comparison of greedy algorithms for decision tree construction. In Filipe, J., Fred, A.L.N. (eds.) KDIR, pp. 438–443. SciTePress (2011)
18. Fayyad, U.M., Irani, K.B.: The attribute selection problem in decision tree generation. In: AAAI, pp. 104–110. MIT Press, Cambridge (1992)
19. Kononenko, I.: On biases in estimating multi-valued attributes. In: IJCAI, pp. 1034–1040. Morgan Kaufmann Publishers Inc. San Francisco, CA, USA (1995)

20. Martin, J.K.: An exact probability metric for decision tree splitting and stopping. Mach. Learn. **28**(2–3), 257–291 (1997)
21. Mingers, J.: Expert systems—rule induction with statistical data. J. Oper. Res. Soc. **38**, 39–47 (1987)
22. Alkhalid, A., Chikalov, I., Moshkov, M.: On algorithm for building of optimal α-decision trees. In: Szczuka, M.S., Kryszkiewicz, M., Ramanna, S., Jensen, R., Hu, Q. (eds.) RSCTC, pp. 438–445. Springer, Heidelberg (2010)
23. Alkhalid, A., Chikalov, I., Moshkov, M.: A tool for study of optimal decision trees. In Yu, J., Greco, S., Lingras, P., Wang, G., Skowron, A. (eds.) RSKT, vol. LNCS 6401., 353–360. Springer, Berlin (2010)
24. Alkhalid, A., Chikalov, I., Hussain, S., Moshkov: Extensions of dynamic programming as a new tool for decision tree optimization. In Ramanna, S., Jain, L.C., Howlett, R.J. (eds.) Emerging Paradigms in Machine Learning. Volume 13 of Smart Innovation, Systems and Technologies, pp. 11–29. Springer, Berlin (2013)
25. Alkhalid, A., Amin, T., Chikalov, I., Hussain, S., Moshkov, M., Zielosko, B.: Dagger: a tool for analysis and optimization of decision trees and rules. In Ficarra, F.V.C., Kratky, A., Veltman, K.H., Ficarra, M.C., Nicol, E., Brie, M. (eds.) Computational Informatics, Social Factors and New Information Technologies: Hypermedia Perspectives and Avant-Garde Experiencies in the Era of Communicability Expansion, pp.29–39. Blue Herons (2011)
26. Frank, A., Asuncion, A.: UCI Machine Learning Repository (2010)
27. Chikalov, I., Hussain, S., Moshkov, M.: Relationships between depth and number of misclassifications for decision trees. In Kuznetsov, S.O., Ślęzak, D., Hepting, D.H., Mirkin, B. (eds.) Rough Sets, Fuzzy Sets, Data Mining and Granular Computing. Volume 6743 of Lecture Notes in Computer Science, pp. 286–292. Springer, Berlin (2011)
28. Pawlak, Z.: Theoretical Aspects of Reasoning about Data. Kluwer Academic Publishers, Dordrecht (1991)
29. Skowron, A., Rauszer, C.: The discernibility matrices and functions in information systems. In: Slowinski, R. (ed.) Intelligent Decision Support. Handbook of Applications and Advances of the Rough Set Theory, pp. 331–362. Kluwer Academic Publishers, Dordrecht (1992)
30. Nguyen, H.S.: From optimal hyperplanes to optimal decision trees. Fundamenta Informaticae **34**(1–2), 145–174 (1998)
31. Chikalov, I., Hussain, S., Moshkov, M.: Relationships for cost and uncertainty of decision trees. In Skowron, A., Suraj, Z. (eds.) Rough Sets and Intelligent Systems—Professor ZdzisÅ‘aw Pawlak in Memoriam. Volume 43 of Intelligent Systems Reference Library, pp. 203–222. Springer, Berlin (2013)
32. Chikalov, I., Hussain, S., Moshkov, M.: Relationships between number of nodes and number of misclassifications for decision trees. In: Proceedings of 8th International Conference of Rough Sets and Current Trends in Computing, RSCTC, pp. 212–218. Springer, Berlin (2012)
33. Chikalov, I., Hussain, S., Moshkov, M.: Relationships between average depth and number of nodes for decision trees. In: Proceedings of 7th International Conference on Intelligent Systems and Knowledge Engineering (ISKE2012). Springer, Berlin (2012)
34. Chikalov, I., Hussain, S., Moshkov, M.: Average depth and number of misclassifications for decision trees. In Popova-Zeugmann, L. (ed.) Proceedings of 21st International Workshop on Concurrency, Specification and Programming, pp. 160–169 (2012)
35. Chikalov, I., Hussain, S., Moshkov, M.: On cost and uncertainty of decision trees. In Yao, J., Yang, Y., Slowinski, R., Greco, S., Li, H., Mitra, S., Polkowski, L. (eds.) Proceedings of 8th International Conference of Rough Sets and Current Trends in Computing, RSCTC, pp. 190–197. Springer, Berlin (2012)

About the Editors

Professor Colette Faucher is a professor of Computer Science and Artificial Intelligence at Polytech'Marseille, a School of Industrial Engineering and Computer Science in Marseilles, France. She received her Ph.D., in Artificial Intelligence in 1991 and her Habilitation to Direct Research in 2002. Her research interests focus on Knowledge Representation and Reasoning, as well as Cognitive Modelling. She has extensively worked on Classification Reasoning and Concept Formation, inspired by psychological works. Her research is indeed strongly based on works stemming from Psychology, Cognitive and Developmental, Sociology and Anthropology. She is currently working on the modeling of the notion of Culture in order to take into account that notion in the framework of human tasks, both individual and collective, Culture being so important in nowadays world of globalization.

She has been the Chair of many conferences and a member of program committees of various conferences, as well as a member of several editorial boards of international journals.

Dr. Lakhmi C. Jain serves as Adjunct Professor in the Division of Information Technology, Engineering and the Environment at the University of South Australia, Australia and University of Canberra, Australia.

Dr. Jain founded the KES International for providing a professional community the opportunities for publications, knowledge exchange, cooperation and teaming. Involving around 5,000 researchers drawn from universities and companies world-wide, KES facilitates international cooperation and generate synergy in teaching and research. KES regularly

C. Faucher and L. C. Jain (eds.), *Innovations in Intelligent Machines-4*,
Studies in Computational Intelligence 514, DOI: 10.1007/978-3-319-01866-9,

provides networking opportunities for professional community through one of the largest conferences of its kind in the area of KES. www.kesinternational.org

His interests focus on the artificial intelligence paradigms and their applications in complex systems, security, e-education, e-healthcare, unmanned air vehicles and intelligent agents.

Printed in the United States
By Bookmasters

Printed in the United States
By Bookmasters